Lecture Notes in Computer Science 9018

Commenced Publication in 1973
Founding and Former Series Editors:
Gerhard Goos, Juris Hartmanis, and Jan van Leeuwen

More information about this series at http://www.springer.com/series/7407

António Gaspar-Cunha · Carlos Henggeler Antunes
Carlos Coello Coello (Eds.)

Evolutionary Multi-Criterion Optimization

8th International Conference, EMO 2015
Guimarães, Portugal, March 29 – April 1, 2015
Proceedings, Part I

Springer

Editors
António Gaspar-Cunha
Institute for Polymers and Composites/I3N
University of Minho
Guimarães
Portugal

Carlos Coello Coello
CINVESTAV-IPN
Depto. de Computacíon
Col. San Pedro Zacatenco
Mexico

Carlos Henggeler Antunes
Dept. of Electrical and Computer Engg.
University of Coimbra
Coimbra
Portugal

ISSN 0302-9743
Lecture Notes in Computer Science
ISBN 978-3-319-15933-1
DOI 10.1007/978-3-319-15934-8

ISSN 1611-3349 (electronic)

ISBN 978-3-319-15934-8 (eBook)

Library of Congress Control Number: 2015932656

LNCS Sublibrary: SL1 – Theoretical Computer Science and General Issues

Springer Cham Heidelberg New York Dordrecht London

Printed on acid-free paper

Springer International Publishing AG Switzerland is part of Springer Science+Business Media (www.springer.com)

Preface

EMO is a biennial international conference series devoted to the theory and practice of evolutionary multi-criterion optimization.

The first EMO took place in 2001 in Zürich (Switzerland), with later conferences taking place in Faro (Portugal) in 2003, Guanajuato (Mexico) in 2005, Matsushima-Sendai (Japan) in 2007, Nantes (France) in 2009, Ouro Preto (Brazil) in 2011, and Sheffield (UK) in 2013. The proceedings of this series of conferences have been published as a volume in Lecture Notes in Computer Science (LNCS), respectively, in volumes 1993, 2632, 3410, 4403, 5467, 6576, and 7811.

The 8th International Conference on Evolutionary Multi-Criterion Optimization (EMO 2015) took place in Guimarães, Portugal, from March 29 to April 1, 2015. The event was organized by the University of Minho. Following the success of the two previous EMO conferences, a special track was offered aiming to foster further cooperation between the EMO and the multiple criteria decision making (MCDM). Also, a special track on real-world applications (RWA) was endorsed.

EMO 2015 received 90 full-length papers, which were submitted to a rigorous single-blind peer-review process, with a minimum of three referees per paper. Following this process, a total of 68 papers were accepted for presentation and publication in this volume, from which 40 were chosen for oral and 24 for poster presentation. The selected papers were distributed through the different tracks as follows: 46 main track, 6 MCDM track, and 16 RWA track.

The conference benefitted from the presentations of plenary speakers on research subjects fundamental to the EMO field: Thomas Stützle, from the IRIDIA laboratory of Université libre de Bruxelles (ULB), Belgium; Murat Köksalan, from the Industrial Engineering Department of Middle East Technical University, Ankara, Turkey; Luís Santos, from the University of São Paulo and Embraer, Brazil; Carlos Fonseca, from the University of Coimbra, Portugal.

From the beginning, this conference provided significant advances in relevant subjects of evolutionary multi-criteria optimization. This event aimed to continue these type of developments, being the papers presented focused on: theoretical aspects, algorithms development, many-objectives optimization, robustness and optimization under uncertainty, performance indicators, multiple criteria decision making, and real-world applications.

Finally, we would express our gratitude to the plenary speakers for accepting our invitation, to all the authors who submitted their work, to the members of the International Program Committee for their hard work, to the members of the Organizing Committee, particularly Lino Costa, and to the Track Chairs Kaisa Miettinen, Salvatore Greco, and

Robin Purshouse. We would like to acknowledge the support of the School of Engineering of the University of Minho. We would also like to thank Alfred Hofmann and Anna Kramer at Springer for their support in publishing these proceedings.

March 2015   António Gaspar-Cunha
                                                                    Carlos Henggeler Antunes
                                                                    Carlos Coello Coello

Organization

Committees

General Chairs

António Gaspar-Cunha   University of Minho, Portugal
Carlos Henggeler Antunes   University of Coimbra, Portugal
Carlos Coello Coello   CINVESTAV-IPN, Mexico

MCDM Track Chairs

Kaisa Miettinen   University of Jyväskylä, Finland
Salvatore Greco   University of Catania, Italy

Real-world Applications Track Chairs

Robin Purshouse   University of Sheffield, UK
Carlos Henggeler Antunes   University of Coimbra, Portugal

EMO Steering Committee

Carlos Coello Coello   CINVESTAV-IPN, Mexico
David Corne   Heriot-Watt University, UK
Kalyanmoy Deb   Michigan State University, USA
Peter Fleming   University of Sheffield, UK
Carlos M. Fonseca   University of Coimbra, Portugal
Hisao Ishibuchi   Osaka Prefecture University, Japan
Joshua Knowles   University of Manchester, UK
Kaisa Miettinen   University of Jyväskylä, Finland
J. David Schaffer   Binghamton University, USA
Lothar Thiele   ETH Zürich, Switzerland
Eckart Zitzler   PH Bern, Switzerland

Local Organization Committee

António Gaspar-Cunha   University of Minho, Portugal
Pedro Oliveira   University of Porto, Portugal
Lino Costa   University of Minho, Portugal
M. Fernanda P. Costa   University of Minho, Portugal
Isabel Espírito Santo   University of Minho, Portugal
A. Ismael F. Vaz   University of Minho, Portugal
Ana Maria A.C. Rocha   University of Minho, Portugal

Program Committee

Ajith Abraham	MIR Labs, USA
Adiel Almeida	Federal University of Pernambuco, Brazil
Maria João Alves	University of Coimbra, Portugal
Helio Barbosa	Laboratório Nacional de Computação Científica, Brazil
Matthieu Basseur	LERIA Lab., France
Juergen Branke	University of Warwick, UK
Dimo Brockhoff	INRIA Lille - Nord Europe, France
Marco Chiarandini	University of Southern Denmark, Denmark
Sung-Bae Cho	Yonsei University, South Korea
Leandro Coelho	Pontifical Catholic University of Parana, Brazil
Salvatore Corrente	University of Catania, Italy
M. Fernanda P. Costa	University of Minho, Portugal
Clarisse Dhaenens	University of Lille 1, France
Yves De Smet	Université libre de Bruxelles, Belgium
Alexandre Delbem	Institute of Mathematical and Computer Sciences, Brazil
Michael Doumpos	Technical University of Crete, Greece
Michael Emmerich	Leiden University, Netherlands
Andries Engelbrecht	University of Pretoria, South Africa
Isabel Espírito Santo	University of Minho, Portugal
Jonathan Fieldsend	University of Exeter, UK
José Figueira	CEG-IST, Portugal
Peter Fleming	University of Sheffield, UK
António Gaspar-Cunha	University of Minho, Portugal
Martin Josef Geiger	Helmut-Schmidt-Universität, Germany
Ioannis Giagkiozis	University of Sheffield, UK
Christian Grimme	University of Münster, Germany
Walter Gutjahr	University of Vienna, Austria
Francisco Herrera	University of Granada, Spain
Jin-Kao Hao	University of Angers, France
Evan Hughes	White Horse Radar Limited, UK
Masahiro Inuiguchi	Osaka University, Japan
Hisao Ishibuchi	Osaka Prefecture University, Japan
Yaochu Jin	University of Surrey, UK
Laetitia Jourdan	University of Lille 1, France

Daniel Vanderpooten	Paris Dauphine University, France
Fernando J. Von Zuben	University of Campinas, Brazil
Tobias Wagner	Technische Universität Dortmund, Germany
Elizabeth Wanner	CEFET-MG, Brazil
Farouk Yalaoui	University of Technology of Troyes, France
Gary Yen	Oklahoma State University, USA
Qingfu Zhang	University of Essex, UK
Andre de Carvalho	University of São Paulo, Brazil
Alexis Tsoukias	CNRS, France

Support Institution

School of Engineering, University of Minho

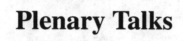

Plenary Talks

Interactive Approaches in Multiple Criteria Decision Making and Evolutionary Multi-objective Optimization

Murat Köksalan

Industrial Engineering, Middle East Technical University
06800 Ankara, Turkey
koksalan@metu.edu.tr

Abstract. The developments related to Multiple Criteria Decision Making (MCDM) go back a long time. It has been over 50 years since MCDM became an active research area. There have been major developments during this time both in theory and applications.

Evolutionary Multi-objective Optimization (EMO) is a relatively new field that has enjoyed a fast growth. EMO, although heuristic in nature, can successfully handle many complex problems. EMO has started independently from MCDM and there was very little interaction between the researchers in early developments of EMO. Earlier approaches mostly addressed two objectives and attempted to generate the whole PO set. Preference-based approaches that attempt to converge preferred regions are more recent. Efforts to combine forces from the two areas are also more recent.

Although approaches have been developed to characterize the whole Pareto Optimal (PO) frontier/set in both MCDM and EMO, this task is neither useful nor feasible for many complex practical problems. Many of the PO solutions may be less attractive than many dominated solutions for the decision maker (DM). In complex problems from practice, the available computational budget would be more wisely spent if the search is concentrated in the regions of interest to the DM. Therefore, obtaining preference information from the DM and using the obtained information to converge the preferred solutions is important. Incorporating preference information in the solution process has been well-developed in MCDM and is being addressed in EMO in recent years.

In this talk, I will briefly review the historical developments in MCDM. Then I will concentrate on preference-based approaches in general and interactive approaches in particular. I will cover some interactive approaches developed for different types of MCDM problems. I will also cover some of the interactive approaches developed in the EMO field. Multi-objective combinatorial optimization (MOCO) problems are computationally complex and there is a growing interest in this field. I will briefly cover some preference-based approaches in MOCO and emphasize the potential of EMO to address these problems.

Towards Automatically Configured Multi-objective Optimizers

Thomas Stützle

IRIDIA-CoDE, Université Libre de Bruxelles (ULB), Brussels, Belgium
stuetzle@ulb.ac.be

Abstract. The design of algorithms for computationally hard problems is a time-consuming and difficult task. This is in large part due to the large number of degrees of freedom in defining and selecting algorithm components and settings of numerical parameters but also due to a number of aggravating circumstances such as the NP-hardness of most of the problems to be solved and the difficulty of algorithm analysis due to stochasticity and heuristic biases. Even when tackling specific problems or problem classes by off-the-shelf optimizers such as high performing integer programming solvers, their performance can often be improved substantially by using appropriate settings of the parameters that influence search behavior.

While traditionally algorithm design and the choice of specific parameter settings has usually been done manually, over the recent years various automatic algorithm configuration methods have been developed to effectively search large parameter spaces and to support algorithm designers as well as practitioners. These automatic algorithm configuration methods have shown to be able to identify new algorithm designs and performance improving parameter settings in a number of applications and proved in this way to be instrumental for developing high-performance algorithms.

In this talk, we will first introduce the scope and potential impact automatic algorithm configuration methods have and give an overview of the main existing techniques. We will then illustrate the successful application of automatic algorithm configuration methods by a number of case studies where they have been crucial to obtain improved algorithm designs and reach or surpass state-of-the-art performance. In particular, we show how these methods can be applied to automatically configure algorithms for multi-objective optimization and we demonstrate the performance gains that can be achieved for various types of multi-objective optimizers ranging from the two-phase and Pareto local search framework, over multi-objective ant colony optimization algorithms to multi-objective evolutionary algorithms. Next, we show how the same methodology that appeared to be successful for configuring multi-objective optimizers can be used to improve the anytime behavior of algorithms. Finally, we argue that automatic algorithm configuration will transform the way optimization algorithms are developed in the future and give an outlook on future research challenges.

A Review of Evolutionary Multiobjective Optimization Applications in Aerospace Engineering

Luis Santos

Universidada de São Paulo, São Paulo, Brazil
lccs13@yahoo.com

Abstract. Evolutionary Multiobjective Optimization (EMO) has been applied to several relevant problems in Aerospace Engineering for several years. To estabilish a basis of comparison a 10-year span of publications of the aerospace field is analyzed. This basis of publications is comprised by the publications of the professional societies AIAA, ICAS, SAE and their related journals. From the papers selected several aspects will be compared such as:
- The choice of the evolutionary, or bio-inspired methods, such as genetic algorithms or particle swarm.
- The number of objective functions and their type (continuous or discrete)
- The number of design variables and their type (continuous or discrete)
- The number of constraints, the type of constraints, and how the constraints are implemented.
- Convergence criteria and computational cost
- The use of surrogate methods
- and any other relevant aspects regarding applications.

The analysis of these parameters will provide an idea of current level of use and application of EMO methods in Aerospace, providing the EMO research community with a reference to the current industrial practice in the field. The analysis of the data presented aims to encourage potentially novel applications incorporating the advances of the latest EMO research. That may serve as a guide of cooperation between researchers of both fields.

Performance Evaluation of Multiobjective Optimization Algorithms: Quality Indicators and the Attainment Function

Carlos M. Fonseca

CISUC, Department of Informatics Engineering, University of Coimbra
Pólo II, 3030-290 Coimbra, Portugal
cmfonsec@dei.uc.pt

Abstract. The development of improved optimization algorithms and their adoption by end users are intrinsically dependent on the ability to evaluate how well they perform on the problem classes of interest. In the absence of theoretical guarantees, performance must be evaluated experimentally. Beyond the selection of suitable, representative problem instances, which is a crucial step in the design of such experiments, analysis of the results must take both the experimental conditions and the nature of the data collected into account.

A posteriori approaches to multiobjective optimization typically lead to discrete approximations of the true Pareto-optimal front of the given problem in the form of sets of mutually non-dominated points in objective space. When the algorithm is stochastic, such non-dominated point sets are *random*, and vary according to some probability distribution.

In the literature, two main approaches have been proposed to deal with non-dominated point set distributions: *quality indicators* and the *attainment function*. Quality indicators map non-dominated point sets to real values, and make subsequent data analysis simpler by side-stepping the set nature of the data. In contrast, the attainment-function approach addresses the non-dominated point set distribution directly. Distributional aspects such as location, variability, and dependence, are captured by the moments of the set distribution, which can be estimated from the raw non-dominated point set data.

In this presentation, quality indicators and the attainment function are reviewed as tools for the performance evaluation of stochastic multiobjective optimization algorithms. Complexity issues concerning the computation, visualization, and size of the moment estimates, as the number of objectives, number of runs, and size of the Pareto-front approximations grow are highlighted. Recent results relating the statistical distributions of some unary quality indicators to the attainment function are presented, establishing a link between the two approaches. A discussion of opportunities for further work concludes the presentation.

Contents – Part I

Contents – Part II

Mulit-Criterion Decision Making (MCDM)

Real World Applications

Theory and Hyper-Heuristics

A Multimodal Approach for Evolutionary Multi-objective Optimization (MEMO): Proof-of-Principle Results

Cem C. Tutum and Kalyanmoy Deb(✉)

Department of Electrical and Computer Engineering, Michigan State University,
428 S. Shaw Lane, 2120 EB, East Lansing, MI 48824, USA
tutum@msu.edu, kdeb@egr.msu.edu
http://www.egr.msu.edu/~kdeb

Abstract. Most evolutionary multi-objective optimization (EMO) methods use domination and niche-preserving principles in their selection operation to find a set of Pareto-optimal solutions in a single simulation run. However, classical generative multi-criterion optimization methods repeatedly solve a parameterized single-objective problem to achieve the same. Due to lack of parallelism in the classical generative methods, they have been reported to be slow compared to efficient EMO methods. In this paper, we use a specific scalarization method, but instead of repetitive independent applications, we formulate a multimodal scalarization of multiple objectives and develop a niche-based evolutionary algorithm to find multiple Pareto-optimal solutions in a single simulation run. Proof-of-principle results on two to 10-objective problems from our proposed multimodal approach are compared with standard evolutionary multi/many-objective optimization methods.

Keywords: Multimodal optimization · Achievement scalarization function · Genetic algorithms · Multi-objective optimization

1 Introduction

The success in solving multi-objective optimization problems using evolutionary algorithms comes from a balance of three aspects in their selection operation: (i) emphasis of non-dominated solutions in a population, (ii) emphasis of less crowded solutions in a population and (iii) emphasis of elites in a population [3,6]. The parallel search ability introduced by the population approach and recombination operator of an EMO algorithm aided by the above properties of its selection operator causes it to move towards the speed. On the other hand, classical generative multi-objective optimization methods base their search by solving a series of parameterized single-objective problem serially [2,19]. One criticism of the generative approaches is their lack of parallelism which makes them computationally expensive compared to EMO methods [21].

© Springer International Publishing Switzerland 2015
A. Gaspar-Cunha et al. (Eds.): EMO 2015, Part I, LNCS 9018, pp. 3–18, 2015.
DOI: 10.1007/978-3-319-15934-8_1

When a finite set of Pareto-optimal solutions is the target, the multi-objective optimization problem can be considered as a single-objective multimodal problem in which each Pareto-optimal solution is targeted as an independent optimal solution. Although the idea is intriguing, what we need is a suitable scalarization method that will allow us to consider such a mapping. In this paper, we use achievement scalarization function (ASF) method [24] and construct a multimodal problem for a multi-objective optimization problem. We then suggest an efficient niching-based evolutionary algorithm [9,16] for finding multiple optimal solutions in a single simulation run. The resulting multimodal EA (we call MEMO) is shown to solve two to 10-objective optimization problems for finding hundreds of Pareto-optimal solution efficiently. These proof-of-principle results are encouraging and suggest immediate extension of the approach to solve more complex multi/many-objective optimization problems.

In the remainder of the paper, we give a brief summary of niching-based evolutionary multimodal optimization methods in Section 2. Section 3 discusses classical generative multi-criterion optimization methods. In Section 4, we present the proposed MEMO algorithm. Results on multi and many-objective problems are shown and compared with NSGA-II [8] and recently proposed NSGA-III [10] in Section 5. Finally, conclusions and extensions to this study are discussed in Section 6.

2   Multimodal Optimization

Multimodal function optimization problems have multiple global and/or local optima and the task in a multimodal optimization algorithm is to find as many such optima as possible in a single simulation. The advantage of finding multiple optimal for a single-objective optimization problem is that multiple solutions provides the user with a flexibility of implementing a different solution as and when the situation demands and also their knowledge provides additional insights for a better understanding the problem at hand.

Since evolutionary algorithms processes a population at every generation, they are ideally suitable for finding and maintaining multiple optimal solutions. Despite some early studies [1], Goldberg and Richardson [16] suggested a niching-based genetic algorithm with the help of a sharing function concept to solve multi-modal problems. Later, many other principles, such as crowding [13], clearing [20], and restricted tournament selection [17], clustering [25] have been suggested. In these methods, the usual selection operator in an EA is restricted to compare neighboring solutions alone, thereby allowing to form multiple niches hopefully around each optimal solution. Recently, Deb and Saha suggested a multiobjectivization strategy for solving multimodal problems [11]. A good review of niching-based EAs can be found elsewhere [5].

3 Classical Generative Methods for Multi-objective Problem Solving

Multi and many-objective optimization problems give rise to a set of Pareto-optimal solutions, thereby making them more difficult to be solved than single-objective optimization problems. While efficient evolutionary multi-objective optimization (EMO) methods [15,18,22] were developed since early nineties to find multiple Pareto-optimal solutions in a single simulation run, classical generative methods were suggested since early seventies [2]. In the generative principle, a multi-objective optimization problem is scalarized to a single-objective function by using one or more parameters. Weighted-sum approach use relative weights for objective functions; epsilon-constraint approach uses a vector of ϵ-values for converting objective functions into constraints; Tchebyshev method use a weight vector for forming the resulting objective function. The idea is then to solve the parameterized single-objective optimization problem with associated constraints repeatedly for different parameter values one at a time. The above scalarization methods, if solved to their optimality, are guaranteed to converge to a Pareto-optimal solution every time [19]. However, some scalarization methods are not capable of finding certain Pareto-optimal solutions in a problem no matter what parameter values are chosen [14]. These methods (such as weighted-sum or Lp-norm (for $p \neq \infty$) methods) are relatively unpopular and methods capable of finding each and every Pareto-optimal solution for certain combination of parameter values, such as epsilon-constraint and Tchebyshev method are popular. Here, we use one such popular scalarization method that also guarantees to find any Pareto-optimal solution.

3.1 Achievement Scalarizing Function (ASF) Method

The achievement scalarizing (ASF) method [24] requires one reference point $\mathbf{Z} = (z_1, z_2, \ldots, z_M)^T$ and one weight vector $\mathbf{w} = (w_1, w_2, \ldots, w_M)^T$ as parameters. Both these vectors are associated with the objective space and are illustrated in Figure 1.

For any variable vector $\mathbf{x}$, the corresponding objective vector $\mathbf{f}$ is shown in the figure. For the supplied $\mathbf{w}$-vector (representing inverse of the relative preference of objectives), the following ASF function is minimized to find a Pareto-optimal solution:

$$\text{Minimize ASF}(\mathbf{x}) = \max_{j=1}^{M} \left(\frac{f_i(\mathbf{X}) - z_i}{w_i} \right), \qquad (1)$$
$$\text{subject to } \mathbf{x} \in \mathbf{X}.$$

For the illustrated objective vector $\mathbf{F} = (f_1, f_2)^T$, two components within brackets are $(f_1(\mathbf{x}) - z_1)/w_1$ and $(f_2(\mathbf{x}) - z_2)/w_2$. If the point $\mathbf{F}$ lies on the weight vector ($\mathbf{ZW}$), that is, the point $\mathbf{G}$, the above two components will be identical and the ASF function is equal to any of the two terms (say, it is equal to the first term, $\text{ASF}(\mathbf{G}) = (f_1(\mathbf{x}) - z_1)/w_1$). However, since the objective vector $\mathbf{F}$ lies below the weight vector, the maximum of the two components will be $\text{ASF}(\mathbf{F}) = (f_1(\mathbf{x}) - z_1)/w_1$. This is identical to $\text{ASF}(\mathbf{G})$. In fact, for all objective vectors on

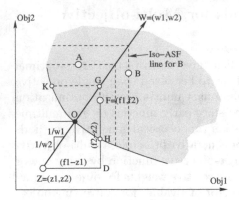 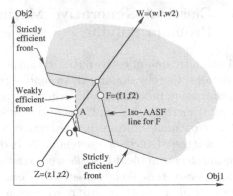

Fig. 1. The achievement scalarizing function (ASF) approach is illustrated

Fig. 2. The augmented achievement scalarizing function (AASF) approach is illustrated

the line **GH**, the ASF value will be identical to the that at **F**. The same argument can be made with any point on the line **GK**. The figure also shows two other objective vectors (**A** and **B**) and their respective iso-ASF lines. Clearly, the ASF value for **F** is smaller than that of **A**, which is smaller than ASF value of **B**. Thus, as the iso-ASF line comes closer to the reference point **Z**, the ASF value gets smaller. Since the problem in Equation 1 is a minimization problem, the ASF optimization will result in point **O** as the optimal solution.

It turns out that the above ASF optimization problem can result in a *weakly* efficient point as well. To ensure finding a *strict* efficient point, the following augmented scalarizing function (AASF) was suggested [19]:

$$\text{Minimize AASF}(\mathbf{x}) = \max_{j=1}^{M} \left(\frac{f_i(\mathbf{X}) - z_i}{w_i} \right) + \rho \sum_{i=1}^{M} \left(\frac{f_i(\mathbf{X}) - z_i}{w_i} \right), \quad (2)$$
$$\text{subject to } \mathbf{x} \in \mathbf{X}.$$

A small value of ρ ($\approx 10^{-4}$) is suggested. The augmented term makes the iso-AASF lines inclined as shown in Figure 2 and avoids finding weakly efficient points. For the example shown, the weight vector intersects the weakly efficient front at **A**, but the AASF value at this point is not the minimum possible value. The point **O** has a smaller AASF value as the corresponding iso-AASF lines intersect the weight vector closer to the reference point **Z**. Interestingly, the use of AASF for a scenario depicted in Figure 1 still finds the identical efficient solution **O**.

4   Multimodal Approach for Solving Multi-objective Optimization (MEMO)

In our proposed multimodal approach, we plan to scalarize the multi-objective problem with an augmented achievement scalarizing function (AASF), requiring a reference point **Z** and a weight vector **w**. We have noticed that for a single reference point **Z**, we can find a single Pareto-optimal solution by minimizing the corresponding AASF problem. Thus, if we specify multiple reference points ($\mathbf{Z}^{(k)}$, $k = 1, 2, \ldots, K$) systematically distributed in the objective space, we can obtain multiple Pareto-optimal solutions, each corresponding to one reference point. This principle is the main crux of our proposed multimodal approach, which we formulate next.

4.1 Formulating a Multimodal Problem

In the recent past, decomposition-based methods have been proposed to solve multi- and many-objective optimization problems [10,26]. In both these methods, Das and Dennis's [4] strategy for specifying a set of reference directions or reference points were suggested. On an M-dimensional linear hyperplane making equal angle to all objective axes and intersecting each axis at one, we specify $K = \binom{(M+p-1)}{p}$ structured points, where p is one less than the number of points along each edge of the hyperplane. Each of the reference point $\mathbf{Z}^{(k)} = \left(z_1^{(k)}, z_2^{(k)}, \ldots, z_M^{(M)}\right)^T$ would then satisfy the following condition: $\sum_{i=1}^{M} z_i^{(k)} = 1$. Figure 3 shows distribution of 15 such reference points on a $M = 3$-objective problem with $p = 5$. For each population member $\mathbf{x}$, we can calculate the AASF value (AASF$(\mathbf{x}, \mathbf{Z}^{(k)})$ with respect to each reference point $\mathbf{Z}^{(k)}$. Thereafter, we can assign the minimum AASF value over all reference points as the fitness to the population member $\mathbf{x}$:

$$\text{Fitness}(\mathbf{x}) = \min_{k=1}^{K} \text{AASF}(\mathbf{x}, \mathbf{Z}^{(k)}). \tag{3}$$

This way, a separate minimum will be created at the efficient point corresponding to each reference point.

The formulation of the AASF problem require an automatically defined weight vector $\mathbf{w}$ for each reference point. The resulting multimodal AASF would then depend on the chosen weight vector. One idea would be to use a uniform weight vector, meaning a constant $w_i = 1/\sqrt{M}$ for all i. However, such a weight vector

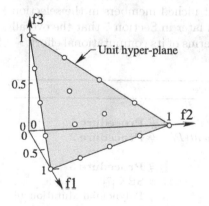

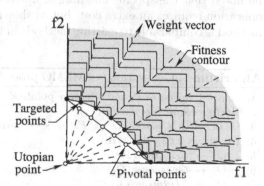

Fig. 3. A diverse set of reference points is created using Das and Dennis's strategy for $M = 3$ and $p = 5$

Fig. 4. The multimodal achievement scalarizing function having eight minima is illustrated

would emphasize the extreme efficient vectors more than the intermediate ones. We suggest the following combination of reference points and weight vectors for our purpose. Instead of using different reference points and a single weight vector, we suggest using a single reference point and multiple weight vectors. First, we rename K diverse reference points created using Das and Dennis's approach as *pivotal* points, $p_i^{(k)} = z_i^{(k)}$ for all $k = 1, 2, \ldots, K$ in each objective i. Then, an utopian point is set as the only reference point: $z_i = f_i^* - \epsilon_i$, where f_i^* is the minimum value of the i-th objective and ϵ_i is a small positive number (0.01 is used here). Thereafter, each weight vector corresponding to k-th pivotal point is calculated as follows:

$$w_i^{(k)} = p_i^{(k)} - z_i, \quad \text{for all } i = 1, 2, \ldots, M. \tag{4}$$

The weight vector thus created is normalized to convert it into a unit vector. Due to the use of an utopian vector as $\mathbf{Z}$, the above weight values are always strictly positive, thereby satisfying the non-negativity requirement of weights for AASF scalarization process of solving multi-objective optimization problems. Figure 4 shows the contour plot of the above multimodal ASF for a two-objective ZDT2 problem for $p = 7$. The figure indicates that the ASF has eight different minimum solutions on the efficient front. It is interesting to note how the whole objective space is classified into eight non-overlapping niches. A similar multimodal function also results using AASF. A suitable niching-based EA should now find these eight efficient solutions in a single simulation run.

4.2   Proposed Multimodal Evolutionary Algorithm: MEMO

The basic framework of the proposed multi- or many-objective MEMO algorithm is given in Algorithm 1. The main idea is to preserve individuals corresponding to different pivotal points. One advantage of MEMO over NSGA-II or NSGA-III is that it does not require non-dominated sorting of solutions into different fronts. Only the first front needs to be identified for the sake of normalization. It should be noted that the specific handling of different niched members in the selection operation requires an extra cost, but as shown later in Section 5 that the overall method is comparable to existing methods in terms of its computational effort.

Algorithm 1. Generation t of MEMO procedure

1: $gen = 1$, initialize P_t and reference points $Z^{1,2,\ldots,H}$
2: $Archieve\_and\_Normalize(P_t)$ # **Procedure 1**
3: $[MinAASF, ClusterID] = FitnessAssignment(P_t)$ # **Procedure 2**
4: **for** $gen = 2$ **to** gen_{max} **do**
5:     $Q_t \leftarrow Selection(P_t)$ # **Procedure 3**
6:     $Q_t \leftarrow Crossover(Q_t)$ # SBX [7]
7:     $Q_t \leftarrow Mutation(Q_t)$ # Polynomial Mutation [6]
8:     $[MinAASF, ClusterID] = FitnessAssignment(Q_t)$
9:     $P_{t+1} \leftarrow Merge\_and\_Reduce(P_t, Q_t)$ # **Procedure 4**
10: **end for**

Procedure 1: *Archieve_and_Normalize( )* The objective values need to be normalized in order to handle scaled multi-objective optimization problems where each objective function produces different magnitude of values. Normalization is also needed for comparing MEMO population members with the chosen reference points, which are usually introduced on the normalized hyper-plane. This is usually the case in real-world problems. Archiving the extreme points and the normalization procedure is identical to normalization process of NSGA-III procedure [10].

Procedure 2: *FitnessAssignment( )* The computation of minimum AASF as fitness (Eq. 3) is also used for the associating a point to a pivotal point to incorporate the new niching concept in the selection and *Merge_and_Reduce* operators (Procedures 2 and 4, respectively).

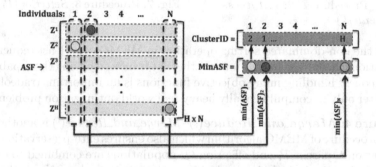

Fig. 5. Fitness and cluster identification (reference point association) assignment scheme used in Procedure 2 (*FitnessAssignment()*)

As the first step, for each individual **x**, an AASF vector is first calculated for all pivotal points, as shown in Figure 5. Then, the minimum value of each column (Eq. 3) is assigned as the fitness value of that individual in $MinAASF$ array and the corresponding row index of this minimum value indicates the associated pivotal point of **x**. This is recorded in $ClusterID$ array. In the initial generations, depending on the complexity of the problem, some individuals of the population might be associated to the same cluster and therefore some of the clusters may be empty as illustrated in Figure 9 as having third and seventh clusters being empty while the first, second, fifth and sixth clusters having more than one individual associated.

Procedure 3: *Selection()* Binary tournament selection is used for the reproduction process. As mentioned before, the selection operator preserves different cluster members in a population. At each tournament, individuals from the same cluster are shuffled first and then compared pairwise. Two solutions are compared with respect to their $MinAASF$ value and the one with the smaller value is selected. To make our procedure to find strict Pareto-optimal solutions only, we use AASF, instead of ASF in solving all problems of this paper. Limiting these pairwise comparisons to only the same cluster individuals brings an extra cost of having slightly a bigger offspring population as compared to fixed-size EMO algorithms. However

```
1: tour_max = 2
2: for k = 1 to k = tour_max do
1: initialize AASF_{HxN}          3:    for cluster = 1 to cluster = H do
2: for i = 1 to i = H do          4:       tour_k = [ ]
3:    for j = 1 to j = N do       5:       tour_k ←
4:       AASF(i, j) ←             6:          suffle(cluster_individual_indices)
5:          compute_AASF(Z^i, P_t) 7:       N_cluster ← size(tour_k)
6:    end for                     8:       for j = 1 to j = N_cluster do
7: end for                        9:          Q_t ←
8: for i = 1 to i = N do         10:             min(Fitness(tour_k[j], tour_k[j+1]))
9:    [Fitness_i, ClusterID_i] ← 11:       end for
10:       min(AASF(row_{(1:H)}, col_i)) 12:    end for
11: end for                      13: end for
12: return [MinAASF, ClusterID]  14: return Q_t
```

Fig. 6. Procedure 2: *Fitness*   **Fig. 7.** Procedure 3: *Selection*(P_t)
Assignment(P_t)

skipping the non-dominated sorting operation in MEMO or at least reducing the computational load by searching for only the first front and having the capability of simultaneously handling many-objective functions is an important tradeoff. This may matter only in computationally heavy real world optimization problems.

Procedure 4: *Merge_and_Reduce*( ) *Merge_and_Reduce*( ) is another vital niching procedure of MEMO algorithm which also ensures elite-preservation. First, at generation t, parent P_t and offspring Q_t populations are combined to create a new population R_t of which size can be bigger than $2 x N_{fit}$ where N_{fit} is the size of P_t (population size). However, the size of the next parent population P_{t+1} obtained at the end of this procedure is again reduced to $N_{fit} = N$.

After normalizing and evaluating the combined population R_t, niching procedure which is composed of two steps is applied. First, all clusters are filled with the individuals having the right *ClusterID* value which was determined at *FitnessAssignment* procedure (Section 4.2). In the same step, members of the same cluster are sorted with respect to their fitness, i.e. *MinAASF* value. This is represented in Figure 9 with circles of the same color (members of the same cluster) being sorted from smaller radius to larger radius. In the last step, best solutions of each cluster (i.e., those encircled with dashed line and indicated as *counter* = 1) are put into P_{t+1} and this procedure is repeated for the next best solutions of all clusters (*counter* = 2, 3, ...) until $N_{fit} = H$ solutions are stored in P_{t+1}. In the early generations, some of the clusters might be empty but all of them will eventually be filled with at least one solution as the algorithm converges and produces better spread solutions. At the end, each individual of the final population will be an optimum solution with respect to a reference point and all together form the efficient optimal set for the multi-objective optimization problem. As discussed in the original study [23], the overall complexity of the proposed MEMO algorithm is $O(N^2)$.

1: $R_t \leftarrow P_t \cup Q_t$
2: initialize P_{t+1}
3: $N_{fit} \leftarrow size(P_t)$
4: $Archieve\_and\_Normalize(R_t)$
5: $[MinAASF, ClusterID] =$
6:   $FitnessAssignment(R_t)$
7: $clusters \leftarrow \{\ \}$ # Niching (Fig.9) starts ...
8: **for** $i = 1$ **to** $i = H$ **do**
9:   $j \leftarrow 1, 2, ..., size(clusters_i)$
10:   $clusters_i \leftarrow sort(clusters_i\{Fitness_j\})$
      # Local sorting
11: **end for**
12: $counter = 1$
13: $N_{non-empty} \leftarrow size(clusters_i \neq 0)$
14: **while** $size(P_{t+1}) \leq N_{fit}$ **do**
15:   **for** $k = 1$ **to** $k = N_{non-empty}$ **do**
16:     $P_{t+1} \leftarrow clusters_k\{individual_{counter}\}$
17:   **end for**
18:   $counter \leftarrow counter + 1$
19: **end while**   # ... Niching ends
20: **return** P_{t+1}

Fig. 8. Procedure 4: $Merge\_and\_Reduce$ (P_t, Q_t)

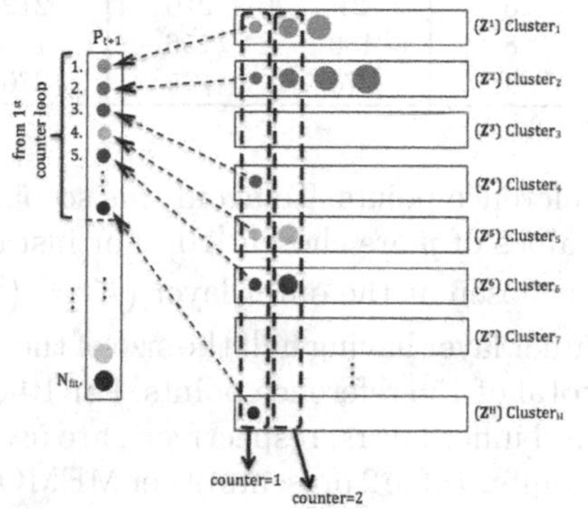

Fig. 9. Niching scheme used in Procedure 4 ($Merge\_and\_Reduce()$)

5   Results and Discussions

In this section, we present the simulation results of MEMO algorithm on two to 10-objective optimization problems. More results can be found in the original study [23]. The results of two objective problems are compared with NSGA-II [8], whereas the rest of the results are compared with NSGA-III [10]. As a performance measure of convergence as well as diversity of obtained solutions, we use the hypervolume (HV) metric [28] for two-objective problems and the inverse generational distance (IGD) metric [27] for higher objective problems. The smaller the IGD value indicated better performance whereas it is opposite for the HV predictor. For each problem, the algorithms are executed 20 times with different initial populations. Best, median and worst HV and IGD performance values evaluated using the population members from the final generation are reported. Table 1 shows the number of chosen reference points (H) for different number of objectives (M) for different test problems. The population size (N) of MEMO is kept the same as H, whereas the population size (N') of NSGA-III is the smallest multiple of four but equal or higher than H.

For two-objective problems, reference points are created using $p = 31$ in Das and Dennis's method resulting in $R = \binom{(2\,|\,31-1)}{31}$ or 32 points. We have used $p = 12$ for three-objective problems ($H = 91$) and $p = 6$ for five-objective problems ($H = 210$). For problems having more than five objectives, to have at least one intermediate reference point, $p \geq M$ is required and it results in a large number

Table 1. The number reference points and the population size used for different objectives in MEMO and NSGA-III algorithms

Number of objectives (M)	Number of reference points (H)	MEMO population size (N)	NSGA-III population size (N')
2	32	32	32
3	91	91	92
5	21	210	212
8	156	156	156
10	275	275	276

Table 2. Parameter values used for MEMO and NSGA-III where n is the number of variables

Parameters	MEMO	NSGA-III
SBX [7], p_c	1	1
Poly. mutation [6], p_m	$1/n$	$1/n$
SBX η_c	30	30
Mut. η_m	20	20

reference points. To avoid this scenario, two layers of reference points with small values of p are chosen [10]. For instance, for the eight-objective problem, $p = 3$ is chosen at the outer layer ($K = \binom{8+3-1}{3} = 120$) and $p = 2$ is chosen for the inner layer having half the size of the outer layer ($K = \binom{8+2-1}{2} = 36$), resulting in total of 156 reference points. For 10-objective problems, $p = 3$ and $p = 2$ in outer and inner layers, respectively, are used, resulting in $H = 220 + 55 = 275$ reference points. Table 2 presents other MEMO and NSGA-III parameters used in this study.

5.1   Two-Objective Problems

To start with, we show results of our MEMO procedure on two-objective ZDT1, ZDT2 and ZDT3 problems in Figure 10.

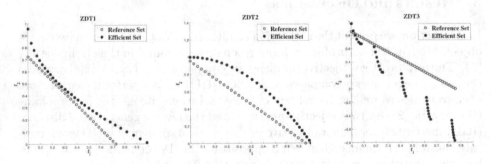

Fig. 10. ZDT1, ZDT2, and ZDT3 results using proposed MEMO algorithm

Performance of ZDT4 is similar. A comparison of its performance with NSGA-II on all four ZDT problems is presented in Table 3. ZDT1 to ZDT3 are run for 200 generations and ZDT4 is run for 500 generations. Figures and the table indicate similar performance of MEMO to the NSGA-II procedure.

5.2   Three-Objective Problems

Next, we choose original DTLZ1 and DTLZ2 and their scaled versions [10,12]. Figure 11 shows the MEMO-obtained efficient fronts for DTLZ1 and DTLZ2 (black

Table 3. Best, median and worst hypervolume values for two-objective ZDT1 to ZDT4 problems using MEMO and NSGA-II algorithms

		MEMO	NSGA-II
ZDT1		8.555×10^{-1}	8.482×10^{-1}
		8.481×10^{-1}	8.365×10^{-1}
		8.534×10^{-1}	8.298×10^{-1}
ZDT2		5.218×10^{-1}	$\mathbf{5.250 \times 10^{-1}}$
		5.172×10^{-1}	$\mathbf{5.229 \times 10^{-1}}$
		5.096×10^{-1}	$\mathbf{5.198 \times 10^{-1}}$
ZDT3		$\mathbf{1.037 \times 10^{0}}$	1.034×10^{0}
		$\mathbf{1.035 \times 10^{0}}$	1.029×10^{0}
		$\mathbf{1.034 \times 10^{0}}$	1.007×10^{0}
ZDT4		$\mathbf{8.601 \times 10^{-1}}$	8.551×10^{-1}
		$\mathbf{8.568 \times 10^{-1}}$	8.492×10^{-1}
		8.382×10^{-1}	$\mathbf{8.397 \times 10^{-1}}$

Table 4. Best, median and worst IGD values obtained for MEMO and NSGA-III on three-objective DTLZ1 and DTLZ2 problems

	MaxGen	MEMO	NSGA-III
DTLZ1	400	$\mathbf{1.413 \times 10^{-4}}$	4.880×10^{-4}
		$\mathbf{2.360 \times 10^{-4}}$	1.308×10^{-3}
		$\mathbf{2.008 \times 10^{-3}}$	4.880×10^{-3}
DTLZ2	250	$\mathbf{1.045 \times 10^{-3}}$	1.262×10^{-3}
		1.538×10^{-3}	$\mathbf{1.357 \times 10^{-3}}$
		2.311×10^{-3}	$\mathbf{2.114 \times 10^{-3}}$

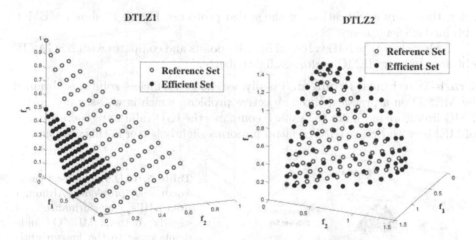

Fig. 11. DTLZ1 and DTLZ2 results using proposed MEMO algorithm

circles), together with the set of reference points (gray circles). As it was explained in Table 1, a nicely distributed set of 91 efficient points can be observed from both figures. Table 4 summarizes the performance of MEMO and recently proposed NSGA-III [10] methods using the IGD metric. MEMO works slightly better in DTLZ1, whereas NSGA-III works slightly better in DTLZ2.

Real world optimization problems usually involve objective values having different order of magnitudes. To simulate such a problem, we scale i-th objective by 10^i. MEMO and NSGA-III still use the same methodology [4] to create the reference set. The adaptive normalization procedure enables the use of this reference

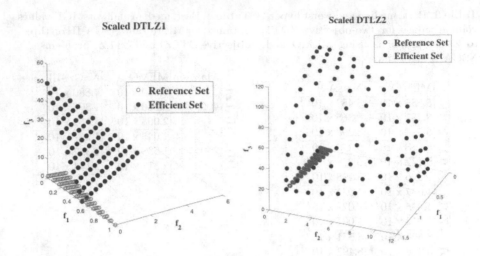

Fig. 12. Scaled DTLZ1 and Scaled DTLZ2 results using MEMO algorithm

set without any modification for the scaled problems. Figure 12 shows MEMO-obtained efficient points.

Table 6 shows the IGD values of MEMO points and compares with NSGA-III-obtained points. MEMO performs slightly better.

Crash-Worthiness Problem. Finally, we show the efficient solutions obtained by MEMO on a practical three-objective problem, which is well-studied in the EMO literature in Figure 13. Table 5 compares the IGD values at generation 200 of MEMO with NSGA-III. The latter performs slightly better in this problem.

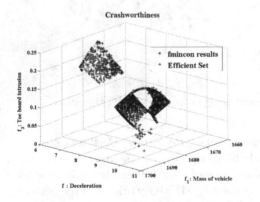

Table 5. IGD values for Crash-worthiness problem. Although NSGA-III's performance is slightly better, MEMO finds points close to the known optimized solutions.

	MEMO	NSGA-III
Best	7.220×10^{-2}	$\mathbf{1.9 \times 10^{-3}}$
Median	7.735×10^{-2}	$\mathbf{2.2 \times 10^{-3}}$
Worst	7.838×10^{-2}	$\mathbf{2.6 \times 10^{-3}}$

Fig. 13. Crash-worthiness results using MEMO

5.3 Many-objective Problems

Next, we apply MEMO algorithm to many-objective version of DTLZ1 and DTLZ2 problems. Table 7 presents IGD values for 5, 8 and 10-objective problems. Although

Table 6. Best, median and worst IGD values for three-objective Scaled DTLZ1 and DTLZ2 problems using MEMO and NSGA-III algorithms. A scaling factor of 10^i, $i=1$, 2 and 3 is used.

Problem	M	Scaling factor	MaxGen	MEMO	NSGA-III
Scaled DTLZ1				**3.441 x 10^{-4}**	3.853 x 10^{-4}
	3	10^i	400	**6.696 x 10^{-4}**	1.214 x 10^{-3}
				2.071 x 10^{-3}	1.103 x 10^{-2}
Scaled DTLZ2				**6.983 x 10^{-4}**	1.347 x 10^{-3}
	3	10^i	250	**1.334 x 10^{-3}**	2.069 x 10^{-3}
				5.346 x 10^{-3}	**5.284 x 10^{-3}**

Table 7. Best, median and worst IGD values obtained for MEMO and NSGA-III on M-objective DTLZ1 and DTLZ2 problems

	M	Max Gen	MEMO	NSGA-III
DTLZ1	5	600	**3.703 x 10^{-4}**	5.116 x 10^{-4}
			1.735 x 10^{-3}	**9.799 x 10^{-4}**
			3.707 x 10^{-3}	**1.979 x 10^{-3}**
	8	750	5.509 x 10^{-3}	**2.044 x 10^{-3}**
			7.414 x 10^{-3}	**3.979 x 10^{-3}**
			8.683 x 10^{-3}	8.721 x 10^{-3}
	10	1000	7.206 x 10^{-3}	**2.215 x 10^{-3}**
			9.244 x 10^{-3}	**3.462 x 10^{-3}**
			1.201 x 10^{-2}	**6.869 x 10^{-3}**
DTLZ2	5	350	3.584 x 10^{-3}	**4.254 x 10^{-3}**
			4.499 x 10^{-3}	4.982 x 10^{-3}
			5.561 x 10^{-3}	5.862 x 10^{-3}
	8	500	4.379 x 10^{-2}	**1.371 x 10^{-2}**
			5.872 x 10^{-2}	**1.571 x 10^{-2}**
			6.915 x 10^{-2}	**1.811 x 10^{-2}**
	10	750	5.401 x 10^{-2}	**1.350 x 10^{-2}**
			6.093 x 10^{-2}	**1.528 x 10^{-2}**
			6.701 x 10^{-2}	**1.697 x 10^{-2}**

Table 8. Best, median and worst IGD values for M-objective Scaled DTLZ1 and DTLZ2 problems using MEMO and NSGA-III algorithms

	M	Scal. fact.	Max Gen	MEMO	NSGA-III
Scaled DTLZ1	5	10^i	600	**6.405 x 10^{-4}**	1.099 x 10^{-3}
				1.155 x 10^{-3}	2.500 x 10^{-3}
				3.647 x 10^{-3}	3.921 x 10^{-2}
	8	3^i	750	1.183 x 10^{-2}	**4.659 x 10^{-3}**
				5.449 x 10^{-2}	**1.051 x 10^{-2}**
				1.097 x 10^{-1}	**1.167 x 10^{-1}**
	10	2^i	1000	3.912 x 10^{-2}	**3.403 x 10^{-3}**
				7.133 x 10^{-2}	**5.577 x 10^{-3}**
				1.328 x 10^{-1}	**3.617 x 10^{-2}**
Scaled DTLZ2	5	10^i	350	**9.603 x 10^{-3}**	1.005 x 10^{-2}
				3.188 x 10^{-2}	**2.564 x 10^{-2}**
				6.417 x 10^{-2}	**8.430 x 10^{-2}**
	8	3^i	500	5.482 x 10^{-2}	**1.582 x 10^{-2}**
				8.901 x 10^{-2}	**1.788 x 10^{-2}**
				1.083 x 10^{-1}	**2.089 x 10^{-2}**
	10	3^i	750	7.794 x 10^{-2}	**2.113 x 10^{-2}**
				2.320 x 10^{-1}	**3.334 x 10^{-2}**
				5.224 x 10^{-1}	**2.095 x 10^{-1}**

NSGA-III's performance is slightly better, MEMO also manages to find small IGD metric values.

Figure 14 shows the parallel coordinate plots 10-objective for DTLZ1 and DTLZ2 problems solved using MEMO.

Table 8 presents MEMO's performance on scaled version of many-objective DTLZ1 and DTLZ2 problems. The performance is comparable to NSGA-III's performance on these problems.

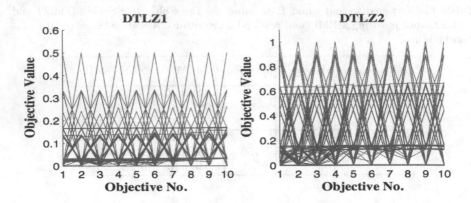

Fig. 14. Parallel coordinate plots for 10-objective DTLZ1 and DTLZ2 problems using MEMO

6   Conclusions and Extensions

In this paper, we have suggested a procedure for converting a multi/many-objective optimization problem into a suitable multi-modal scalarized single-objective problem in which every optimum corresponds to a different Pareto-optimal solution of the original multi-objective problem. Thereafter, we have suggested a niching-based multimodal evolutionary algorithm to find multiple solutions simultaneously. Results on two to 10-objective test problems and on a practical design problem have been compared with other state-of-the-art EMO methodologies and comparable results have been reported.

The conversion is interesting and should open further viable avenues for using other scalarization methods for finding multiple Pareto-optimal solutions. On a similar spirit, other evolutionary multimodal optimization methods can also be tried and this may boost the research on multimodal optimization. Moreover, an immediate extension of the idea for handling constrained multi-objective problems will be important.

Acknowledgments. The authors acknowledge the support provided by the Department of Electrical and Computer Engineering, Michigan State University for executing this study.

References

1. Cavicchio, D.J.: Adaptive Search Using Simulated Evolution. PhD thesis: University of Michigan, Ann Arbor (1970)
2. Chankong, V., Haimes, Y.Y.: Multiobjective Decision Making Theory and Methodology. North-Holland, New York (1983)
3. Coello, C.A.C., VanVeldhuizen, D.A., Lamont, G.: Evolutionary Algorithms for Solving Multi-Objective Problems. Kluwer, Boston (2002)

4. Das, I., Dennis, J.E.: Normal-boundary intersection: A new method for generating the Pareto surface in nonlinear multicriteria optimization problems. SIAM Journal of Optimization 8(3), 631–657 (1998)
5. Das, S., Maity, S., Qu, B.-Y., Suganthan, P.N.: Real-parameter evolutionary multimodal optimization - A survey of the state-of-the-art. Swarm and Evolutionary Computation 1(2), 71–88 (2011)
6. Deb, K.: Multi-objective optimization using evolutionary algorithms. Wiley, Chichester (2001)
7. Deb, K., Agrawal, R.B.: Simulated binary crossover for continuous search space. Complex Systems 9(2), 115–148 (1995)
8. Deb, K., Agrawal, S., Pratap, A., Meyarivan, T.: A fast and elitist multi-objective genetic algorithm: NSGA-II. IEEE Transactions on Evolutionary Computation 6(2), 182–197 (2002)
9. Deb, K., Goldberg, D.E.: An investigation of niche and species formation in genetic function optimization. In: Proceedings of the Third International Conference on Genetic Algorithms, pp. 42–50 (1989)
10. Deb, K., Jain, H.: An evolutionary many-objective optimization algorithm using reference-point based non-dominated sorting approach, Part I: Solving problems with box constraints. IEEE Transactions on Evolutionary Computation 18(4), 577–601 (2014)
11. Deb, K., Saha, A.: Multimodal optimization using a bi-objective evolutionary algorithms. Evolutionary Computation Journal 20(1), 27–62 (2012)
12. Deb, K., Thiele, L., Laumanns, M., Zitzler, E.: Scalable test problems for evolutionary multi-objective optimization. In: Abraham, A., Jain, L., Goldberg, R. (eds.) Evolutionary Multiobjective Optimization, pp. 105–145. Springer, London (2005)
13. DeJong, K.A.: An Analysis of the Behavior of a Class of Genetic Adaptive Systems. PhD thesis. University of Michigan, Ann Arbor (1975). Dissertation Abstracts International 36(10), 5140B (University Microfilms No. 76–9381)
14. Ehrgott, M.: Multicriteria Optimization. Springer, Berlin (2000)
15. Fonseca, C.M., Fleming, P.J.: Genetic algorithms for multiobjective optimization: Formulation, discussion, and generalization. In: Proceedings of the Fifth International Conference on Genetic Algorithms, pp. 416–423, Morgan Kaufmann, San Mateo (1993)
16. Goldberg, D.E., Richardson, J.: Genetic algorithms with sharing for multimodal function optimization. In: Proceedings of the First International Conference on Genetic Algorithms and Their Applications, pp. 41–49 (1987)
17. Harik, G.: Finding multi-modal solutions using restricted tournament selection. In: Proceedings of the Sixth International Conference on Genetic Algorithms (ICGA 1995), pp. 24–31 (1997)
18. Horn, J., Nafploitis, N., Goldberg, D.E.: A niched Pareto genetic algorithm for multiobjective optimization. In: Proceedings of the First IEEE Conference on Evolutionary Computation, pp. 82–87 (1994)
19. Miettinen, K.: Nonlinear Multiobjective Optimization. Kluwer, Boston (1999)
20. Petrowski, A.: A clearing procedure as a niching method for genetic algorithms. In: Proceedings of Third IEEE International Conference on Evolutionary Computation ICEC 1996, pp. 708–803. IEEE Press, Piscataway (1996)
21. Shukla, P., Deb, K.: Comparing classical generating methods with an evolutionary multi-objective optimization method. In: Coello Coello, C.A., Hernández Aguirre, A., Zitzler, E. (eds.) EMO 2005. LNCS, vol. 3410, pp. 311–325. Springer, Heidelberg (2005)

22. Srinivas, N., Deb, K.: Multi-objective function optimization using non-dominated sorting genetic algorithms. Evolutionary Computation Journal **2**(3), 221–248 (1994)
23. Tutum, C.C., Deb, K.: A multimodal approach for evolutionary multi-objective optimization: MEMO. COIN Report Number 2014018, Computational Optimization and Innovation Laboratory (COIN), Electrical and Computer Engineering, Michigan State University, East Lansing (2014)
24. Wierzbicki, A.P.: The use of reference objectives in multiobjective optimization. In: Fandel, G., Gal, T. (eds.) Multiple Criteria Decision Making Theory and Applications, pp. 468–486. Springer, Berlin (1980)
25. Yin, X., Germay, N.: A fast genetic algorithm with sharing scheme using clustering analysis methods in multimodal function optimization. In: Proceedings of International Conference on Artificial Neural Networks and Genetic Algorithms, pp. 450–457 (1993)
26. Zhang, Q., Li, H.: MOEA/D: A multiobjective evolutionary algorithm based on decomposition. IEEE Transactions on Evolutionary Computation **11**(6), 712–731 (2007)
27. Zhang, Q., Zhou, A., Zhao, S.Z., Suganthan, P.N., Liu, W., Tiwari, S.: Multiobjective optimization test instances for the cec-2009 special session and competition. Technical report, Nanyang Technological University, Singapore (2008)
28. Zitzler, E., Thiele, L.: Multiobjective optimization using evolutionary algorithms - A comparative case study. In: Eiben, A.E., Bäck, T., Schoenauer, M., Schwefel, H.-P. (eds.) PPSN 1998. LNCS, vol. 1498, pp. 292–301. Springer, Heidelberg (1998)

Unwanted Feature Interactions Between the Problem and Search Operators in Evolutionary Multi-objective Optimization

Chad Byers[✉], Betty H.C. Cheng, and Kalyanmoy Deb

Michigan State University, East Lansing, MI 48824, USA
{byerscha,chengb}@msu.edu, kdeb@egr.msu.edu

Abstract. Providing self-reconfiguration at run-time amidst adverse environmental conditions is a key challenge in the design of dynamically adaptive systems (DASs). Prescriptive approaches to manually preload these systems with a limited set of strategies/solutions before deployment often result in brittle, rigid designs that are unable to scale and cope with environmental uncertainty. Alternatively, a more scalable and adaptable approach is to embed a search process within the DAS capable of exploring and *generating* optimal reconfigurations at run time. The presence of multiple competing objectives, such as cost and performance, means there is no single optimal solution but rather a *set* of valid solutions with a range of trade-offs that must be considered. In order to help manage competing objectives, we used an evolutionary multi-objective optimization technique, NSGA-II, for generating new network configurations for an industrial remote data mirroring application. During this process, we observed the presence of a hidden search factor that restricted NSGA-II's search from expanding into regions where valid optimal solutions were known to exist. In follow-on empirical studies, we discovered that a variable-length genome design causes unintended interactions with crowding distance mechanisms when using discrete objective functions.

Keywords: NSGA-II · Diversity maintenance · Crowding distance · Discrete objectives · Granularity · Variable-length genome

1 Introduction

Dynamically adaptive system (DAS) are intended to address the challenges posed by adverse environmental conditions [12,14] and varying user requirements. Unlike traditional software systems that can be taken offline and modified by hand, DASs must self-reconfigure at *run time* to avoid staggering financial penalties and/or critical data loss [10]. Smart energy grids, telecommunication systems, smart traffic systems, and similar emerging applications necessitate the deployment of DASs to cope with the various forms of environmental and system uncertainty commonly faced by these applications. These real-world applications

© Springer International Publishing Switzerland 2015
A. Gaspar-Cunha et al. (Eds.): EMO 2015, Part I, LNCS 9018, pp. 19–33, 2015.
DOI: 10.1007/978-3-319-15934-8_2

often contain multiple competing concerns (e.g., cost vs. performance vs. reliability) where trade-offs exist among solutions for dynamic reconfiguration. Evolutionary search techniques, such as genetic algorithms, that rely on biological principles of parallel search provide one approach for the generation of candidate solutions. This paper provides insight into how the underlying solution's encoding may have unintended interactions with specific operators of evolutionary search that produce artificial barriers within the solution space.

In order to automate the generation of DAS configurations, a search-based technique can be embedded within the DAS that is capable of discovering optimal reconfiguration strategies at *run time*. One such evolutionary search-based technique, genetic algorithms (GAs) [5,6], explores a large number of solutions in parallel and uses stochasticity to avoid becoming trapped in suboptimal regions of the solution space. In previous work, we developed Plato [15], a GA-based reconfiguration tool used to evolve overlay networks [1] for a remote data mirroring (RDM) application. In the original Plato tool, user-specified weighting coefficients established a prioritization among the problem's dimensions. Upon further inspection, we found that with this approach, large regions of the solution space were unreachable by evolutionary search, and solutions were often misaligned with the user's weights and therefore ill-fit for their intended environment. To mitigate these issues, we developed Targeting Plato [2] that successfully returned solutions from regions previously unreachable by Plato by targeting desired, user-specified solution qualities. Without *a priori* knowledge of the solution space, two critical disadvantages for using weighting coefficients and target values to guide evolutionary search exist: (1) a trial-and-error approach may be necessary to identify the correct combination of search parameters yielding good solutions, and (2) multiple iterations of search may be necessary to obtain a diverse solution set.

This paper proposes an approach to incorporate multi-objective evolutionary algorithms (MOEAs) into the decision-making process of DASs to generate target reconfigurations at run-time in response to changing environments and requirements. Unlike the GA-based approaches previously mentioned, MOEAs do not require users to specify desired solution characteristics or a prioritization among the problem's dimensions (i.e., objectives) in order to guide evolutionary search. Instead, MOEAs are able to return a *diverse* suite of *Pareto-optimal* solutions whose quality along one particular dimension cannot be improved upon without sacrificing quality along another dimension. As a result, MOEAs can be harnessed for DAS applications to explore the solution space landscape and inform the end user where solution tradeoffs occur.

While leveraging a commonly-used MOEA named NSGA-II [1] for our industrial RDM application, we observed that this algorithm's search coverage was significantly limited when compared to the overall Pareto surface where additional novel solutions existed. In an empirical study, we investigated the potential causes that might prevent search from expanding into these novel regions by performing a set of experiments that targeted specific search operators. Our

---

[1] Non-dominated Sorting Genetic Algorithm [4].

results revealed that in a discrete optimization problem where the number of solution elements can freely evolve (i.e., variable-length genome), an artificial selective pressure is created that selects against valid regions of the objective space. Specifically, solutions with fewer elements are able to mutate farther distances in the solution space than solutions with a greater number of elements. As a result, solutions with fewer elements are prioritized by NSGA-II's crowding distance operator and search is biased into regions containing these solutions. Despite nearly 20 years of algorithm developmental studies in EMO, this aspect of limited search ability of an EMO due to interactions between varying genome size and its diversity preserving operator has not been studied in depth.

The remainder of this paper is organized as follows. In Section 2, we provide background regarding RDM systems as well as Plato and Targeting Plato. Section 3 discusses our initial observation of the underlying problem when NSGA-II was applied to the RDM problem, where its search coverage was compared to the entire Pareto surface. The series of experiments investigating the root cause of NSGA-II's restricted search performance is provided in Section 4, and its impact on similar application domains and evolutionary approaches is discussed in Section 5. Lastly, Section 6 overviews related work, and Section 7 summarizes our findings and outlines future directions for this work.

2 Background

This section overviews topics fundamental to the approach described in this paper. First, we describe the RDM application, necessitate the use of DASs within this domain, and describe the challenges in their design. Next, we provide an overview of genetic algorithms and discuss their utility within the original Plato dynamic reconfiguration tool for navigating vast, complex solution spaces.

2.1 Remote Data Mirroring

In the RDM application [9] provided by an industrial collaborator, the objective is to *copy* critical data residing at primary sites and *remotely store* (mirror) this data on one or more secondary sites across a network in order to mitigate the presence of site/link failure and ensure file synchronization [10,11]. Two critical design decisions for an RDM solution, referred to as an **overlay network** [1], are (1) the subset of network links to include from the underlying base network and (2) which of two RDM networking protocol types should be used on each active link. A *synchronous* protocol requires that each critical data item is received and applied at all secondary site(s) before proceeding at the primary site. In contrast, *asynchronous* protocols coalesce data items at the primary site that are later sent to secondary sites in batch form and applied atomically after a specified length of time. Table 1 presents the elapsed time between each batch, the amount of data at risk should a failure occur (in GB), and the proportion of bandwidth consumed for synchronous (P1) and asynchronous (P2-P7) protocols.

In our experiments, we evolve overlay network solutions for a fully-connected underlying network containing 26 remote data mirrors. The order of complexity for this problem encompasses $2^{n(n-1)/2}$ network constructions. For 26 remote data mirrors and 7 different RDM protocols, there are $7 * 2^{325}$ possible overlay network configurations. The RDM application contains multiple competing objectives where trade-offs must be made among solutions' operational cost, performance in bandwidth consumption, and reliability in the face of failure.

Table 1. Properties of synchronous and asynchronous RDM protocols [10]

Protocol Type	Communication Protocol	Interval	Data at Risk (GB)	Bandwidth
Synchronous	P1	0 minutes	0.0	1.0
Asynchronous	P2	1 minute	0.35	0.9098
	P3	5 minutes	0.6989	0.8623
	P4	1 hour	1.7782	0.7271
	P5	4 hours	2.3802	0.5732
	P6	12 hours	2.8573	0.4380
	P7	24 hours	3.1584	0.3967

2.2 Plato

Plato [15] is a genetic algorithm-based tool to support RDM reconfiguration at run time according to high-level, user-specified objectives. Within Plato, an evolved overlay network solution is encoded as a vector where each element maps to a specific connection (link) in the base network and stores (1) a boolean flag for whether the connection is used in the solution, and (2) the specific RDM protocol used by the active connection.

Three competing objectives are targeted during optimization: Cost (f_{cost}), Performance (f_{perf}), and Reliability (f_{reliab}). The aggregate formulas for determining these objective values are given in Equations (1)-(7) and were derived from studies for optimizing data recovery systems [10]. In these equations, a candidate solution vector ($\mathbf{x}$) contains N total links from the underlying base network. For each link ($\mathbf{x}_i$), the $\mathbf{x}_i^{flag}$ indicates that the link is active (1) or inactive (0) and $\mathbf{x}_i^{risk}$ and $\mathbf{x}_i^{bandwidth}$ correspond to the data at risk and bandwidth consumed by the link's RDM protocol, respectively. The operational expense of an underlying network link is denoted as C_i, while properties of a particular RDM protocol, such as the bandwidth consumed (ex. $P1^{bandwidth}$), refer to the values in Table 1. To avoid biasing search along objectives with larger ranges of values, each objective is normalized between 0.0 and 1.0.

In the original Plato tool, a user's high-level goals were incorporated into a linear-weighted sum (e.g., $\alpha_{cost}f_{cost} + \alpha_{perf}f_{perf} + \alpha_{reliab}f_{reliab}$) to guide evolutionary search towards regions of desired solutions. As environmental conditions and/or requirements change at run time, the system responds by automatically updating these coefficients to evolve new network reconfigurations. Upon closer inspection, we found that the "surface" containing valid solutions is non-convex and cannot be detected [3] by the linear-weighted sum approach used by Plato. As a result, Plato's evolved solutions were often misaligned with the user's weights and therefore ill-fit for their intended environment.

$$\text{Minimize} \quad (f_{cost}, f_{perf}, f_{reliab}) \tag{1}$$

$$f_{cost}(\mathbf{x}) = \frac{\sum_{i=0}^{N} C_i x_i^{flag}}{\sum_{i=0}^{N} C_i} \tag{2}$$

$$f_{perf}(\mathbf{x}) = \frac{f_{efficiency}(\mathbf{x}) - P1^{bandwidth}}{P1^{bandwidth} - P7^{bandwidth}} \tag{3}$$

$$f_{efficiency}(\mathbf{x}) = \frac{\sum_{i=0}^{N} x_i^{bandwidth} x_i^{flag}}{\sum_{i=0}^{N} P1^{bandwidth} x_i^{flag}} \tag{4}$$

$$f_{reliab}(\mathbf{x}) = 0.5 \times f_{reliab1}(\mathbf{x}) + 0.5 \times f_{reliab2}(\mathbf{x}) \tag{5}$$

$$f_{reliab1}(\mathbf{x}) = 1.0 - \frac{\sum_{i=0}^{N} x_i^{flag}}{N} \tag{6}$$

$$f_{reliab2}(\mathbf{x}) = \frac{\sum_{i=0}^{N} x_i^{risk} x_i^{flag}}{\sum_{i=0}^{N} P7^{risk} x_i^{flag}} \tag{7}$$

To mitigate the aforementioned issues, we developed Targeting Plato [2] where a user specified the desired, *target values* of each objective to be optimized instead of specifying a relative prioritization via weighting coefficients. In this approach, candidate solutions were rewarded for their proximity to the ideal solution's target values. While Targeting Plato provided a more intuitive method for domain experts and expanded search coverage into regions previously unreachable using Plato, these techniques (1) required *a priori* knowledge of the solution space to ensure suboptimal solutions were not returned, (2) were highly dependent on the correct specification of user inputs (e.g., weights and target values), and (3) rewarded solutions for maximizing a combined objective function. As a result, these approaches often sacrificed search exploration for exploitation and returned solution sets often lacking in diversity and trade-offs made among the objectives.

3   Problem Definition

Domain experts often seek to optimize multiple competing (orthogonal) objectives simultaneously in order to assess where trade-offs exist among the problem dimensions. Multi-objective evolutionary algorithms (MOEAs) differ from traditional genetic algorithms in that competing objectives/dimensions are not collapsed into a single objective function but instead are treated individually in order to find a *diverse* set of *Pareto-optimal* solutions.

3.1   Original NSGA-II

For this work, we use the Non-Dominated Sorting Genetic Algorithm (NSGA-II) [4] whose design is particularly well-suited to mitigate the drawbacks of both

Plato and Targeting Plato through the incorporation of two main operators: (1) *non-dominated sorting* and (2) a *crowding-distance* operator. Non-dominated sorting mitigates the concern of suboptimal solutions and ensures that solutions approach true Pareto-optimality by giving priority to solutions whose objective measures dominate (i.e., improve upon) the objective measures of other solutions in the population. NSGA-II's use of a crowding distance operator mitigates the lack of diversity problem by giving priority to non-dominated solutions located in less crowded (i.e., novel) regions of the objective space. In addition, NSGA-II does not rely upon user-specified weighting coefficients or target values.

Using the original implementation of NSGA-II [4], we performed a series of runs to assess its ability to evolve overlay network solutions comparable to the experimental solutions found in previous work [15]. Each run contained a population of 500 candidate solutions employing tournament selection ($k = 5$), two-point crossover, and a 5% mutation rate for a total of 1,000 generations, equating to roughly 3 minutes of wall-clock time. To provide adequate statistical significance, 30 replicate runs were evaluated with each run using a unique random seed.

Using the three objective measures (Cost, Performance, and Reliability), we plotted a three-dimensional point for each solution returned by NSGA-II (colored red in Figure 1). We observed that solutions were clustered around three distinct extrema in the objective space and that, despite the use of NSGA-II's crowding-distance operator, solutions were predominantly located near one another. Moreover, a high Cost measure correlates to more active network links and therefore, a broader range of evolvable Performance levels should exist since each link can support one of seven different RDM protocols. As shown in Figure 1, however, NSGA-II was unable to return solutions with a diverse set of Performance measures as the majority of networks with high Cost only have a Performance level near 0.70. These observations suggested that NSGA-II was not returning a solution set representative of the entire Pareto-optimal surface.

3.2   Epsilon-Constrained NSGA-II (Pareto Surface)

To assess the true shape of the underlying Pareto-optimal surface, we performed a series of additional runs using the epsilon-constraint method for NSGA-II. Using this method, search continues to seek Pareto-optimal solutions that minimize all three objective values described in Equation (1), however, subject to the constraint that they are located within a set of user-specified boundaries. By performing an exhaustive sweep of the objective space using interval sizes of 0.01 along both the Cost and Performance dimensions, we obtained a fine-grained sampling of 10,000 regions across the Pareto-optimal surface.

In Figure 1, we plotted solutions (colored grey) returned by epsilon-constrained NSGA-II and confirmed that the original NSGA-II returned only a limited subset of the solutions on the true Pareto-optimal surface. This observation was troubling for several reasons. First, NSGA-II was unable to return solutions from a large region of the search space where Pareto-optimal solutions were demonstrated to exist, suggesting that hidden factors may be hindering search. Second, despite NSGA-II's use of a crowding-distance operator designed to coerce solutions into unoccupied, novel regions of the objective space, the returned solutions are clustered closely together. Third, the overall shape of the returned solution set tapers towards three distinct regions within the objective space, indicating the potential presence of an artificial selective pressure biasing solutions toward the extremes of each objective.

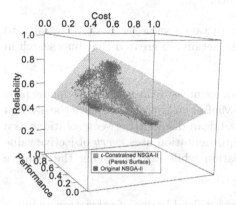 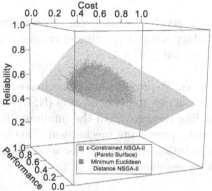

Fig. 1. Original NSGA-II solutions (red) compared against solutions on Pareto surface (grey)

Fig. 2. NSGA-II-MinEuc solutions (green) compared against solutions on Pareto surface (grey)

4   Experimental Design and Results

Next, we describe a series of experiments that were performed to determine the leading causes of (i) artificial basins-of-attraction that solutions evolve towards, (ii) high solution crowding/clustering, and (iii) large regions of undiscovered non-dominated solutions. Each of the experimental treatments comprise 30 replicate runs using the same experimental parameters discussed in Section 3, unless stated otherwise.

4.1   NSGA-II (Minimum Euclidean Crowding-Distance)

Problem/Motivation: Upon analyzing the number of unique Pareto fronts maintained for each generation, we observed that the original NSGA-II rapidly converged to a *single* Pareto front. Consequently, the distinguishing selection factor among solutions becomes the crowding distance operator. In the original NSGA-II [4], crowding distance is assigned by sorting each Pareto front by an objective measure and either (1) awarding positive infinity to "boundary solutions" possessing the minimum/maximum objective values or (2) summing the distance between adjacent solutions' objective values for non-boundary solutions.

While priority is given to solutions maximizing their crowding distance, this implementation may become noisy and overestimate how crowded a solution truly is in the objective space. Adjacent solutions within a *single* objective may be quite distant when their *additional* objective values are taken into consideration. As a result, the original crowding distance operator does not store the distance to the nearest individual solution but rather stores the shortest distances to *any* solution within each objective. In addition, boundary solutions receive the maximum achievable crowding distance thereby producing artificial advantageous regions of the objective space.

Hypothesis 1: By assigning a crowding distance value of positive infinity to boundary solutions, artificial basins of attraction are created that bias search in the original NSGA-II.

Methods: To provide a more accurate distance measure and avoid producing false optima, a new implementation (NSGA-II-MinEuc) was used to replace the original crowding distance with the minimum Euclidean distance between solutions as a diversity-preserving mechanism. This implementation uses *every* objective value of a solution during its distance calculation while also removing the positive infinity assignment bias.

Results: In Figure 2, we observed that the artificial basins-of-attraction anomaly is no longer present with the removal of the positive infinity assignment, and the returned solution set is more evenly distributed in the objective space. Despite similar high levels of solution clustering witnessed previously, NSGA-II-MinEuc is able to *dynamically* respond to boundary solutions in novel areas without an explicit reward/bias. Therefore, in our remaining experiments, we leverage the minimum Euclidean crowding distance operator as we address NSGA-II's restricted search coverage.

4.2   Epsilon-Constrained NSGA-II (Offspring Distance)

Problem/Motivation: From our previous observations, we were able to conclude that the hidden factor restricting search coverage (1) affected the crowding distance values of solutions since NSGA-II quickly converged to a single Pareto front and (2) placed a negative selective pressure on large networks since search coverage tapered off as Cost increased. Taken together, these results suggested that an evolved overlay network's size (i.e. number of active links) might affect the crowding distance its offspring are able to attain.

Hypothesis 2: The number of active networks links of an evolved solution produces a difference in the parent-to-offspring crowding distance values across the objective space.

Methods: To determine whether this differential existed, we used epsilon-constrained NSGA-II to iterate across the objective space in increments of 0.01 in order to measure the average crowding distance between parent and offspring solutions. For each parent solution, we generated 50 offspring solutions and recorded the average Euclidean crowding distance to the parent. We required 100 parent solutions to be discovered within each increment to allow search to potentially discover different solutions with similar Cost and Performance values. In this experiment, we removed the crossover operator to measure two important aspects of search: (1) the average parent-to-offspring mutation distance and (2) the average distance solutions mutate away from where the crossover operator initially places them in the objective space. Therefore, these results indirectly measure how evolved solutions would be affected had crossover been included.

Results: After plotting the average parent-to-offspring distance across the Pareto surface in Figure 3 and applying a color scheme to visualize its topography, we observed that smaller networks with fewer active links (i.e. low Cost) are, on average, able to mutate *farther* from their parents than networks with more active links (i.e. high Cost). In addition, we plotted where solutions were returned by NSGA-II-MinEuc in Figure 4. These results confirmed our initial hypothesis and were a result of the objective functions listed in Equations (1)-(7) being formulated as ratios and normalized, a common approach for many optimization problems. For example, a mutation altering 3 links in a *small* network containing 10 links has a much greater impact (e.g., 30% change in the network's objective values) when compared to altering 3 links in a *large* network containing 100 or 500 links. The ratios of large networks experience greater "inertia," or *resistance to change*, in their objective measures and therefore, receive worse crowding distance values than networks with fewer active links that are capable of moving around the objective space more easily.

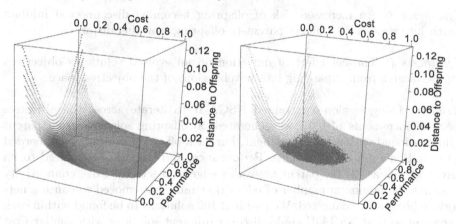

Fig. 3. Average parent-to-offspring Euclidean crowding distance

Fig. 4. Locations where NSGA-II-MinEuc solutions were returned

4.3   Epsilon-Constrained NSGA-II (Additional Factors)

Problem/Motivation: One remaining issue left to address is to determine whether NSGA-II is selecting for regions associated with large crowding distances between parent and offspring solutions; what additional factors are preventing NSGA-II's search from expanding into surrounding regions where even larger crowding distance values were shown to exist?

One factor that may prevent search from expanding into low Cost regions associated with larger parent-to-offspring crowding distance values is an increased risk of offspring solutions becoming disconnected. In our RDM application, it is critical that evolved networks remain *connected* meaning that, a path exists from any data mirror to all other data mirrors to ensure copies of critical data items can be distributed should a site failure occur. In our NSGA-II implementation, a disconnected overlay network is considered dominated by any connected solution, regardless of its objective measures.

A second factor that may prevent search from expanding along the Performance dimension is a decreased probability of evolved networks adopting the same RDM protocol across an increasing number of their active links. With seven RDM protocols, the probability of mutation alone producing networks with optimal Performance is $(1/7)^N$. In our experiments, a 26-mirror base network requires *minimally* 25 active links to ensure a connected network resulting in a probability of $(1/7)^{25}$. Although selection and the crowding distance operator work to maintain these solutions, the disruptive effect of crossover and mutation counteract solutions gaining additional identical links. Moreover, each time a novel solution is selected, there is an increased likelihood that its offspring will mutate a shorter distance than its current nearest neighbor causing the region to become more crowded and disadvantageous in subsequent generations.

Hypothesis 3: An increased risk of offspring becoming disconnected inhibits search into regions with higher parent-to-offspring crowding distances.

Hypothesis 4: The net effect of mutation on an evolved solution's objectives opposes search from expanding into novel regions of the objective space.

Methods: Using epsilon-constrained NSGA-II to iterate across the objective space in increments of 0.01, we generated 50 offspring solutions from parent solutions found within each increment. For each offspring solution, we recorded (1) mutation's net effect on Cost, Performance, and Reliability compared to its parent as well as (2) the parent network's edge connectivity. Edge connectivity measures the minimum number of edges that must be removed to cause a network to become disconnected. We required 100 solutions to be found within each increment so that NSGA-II could discover different solutions with similar Cost and Performance measures.

Results: In Figure 5, we observed that the percentage of active links required to disconnect a network decreases with respect to network Cost, matching our expectations since fewer active links increase the likelihood that a critical link is removed in mutation. In addition, we observed a sudden drop in edge connectivity where minimum spanning tree networks have achieved the lowest possible

Cost since the removal of *any* network link disconnects the network. More importantly, Figure 5 demonstrates that NSGA-II-MinEuc returned solutions up until this boundary, thus providing a key insight for why search did not expand further along the Cost dimension.

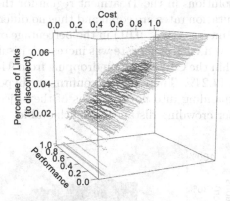

Fig. 5. Percentage of active links to remove and disconnect networks

In addition, we confirmed that mutation opposes the expansion of search into novel regions along the Performance and Reliability dimension as solutions with low objective values experience a net *increase* after mutation is applied and similarly, solutions with high objective values experience a net *decrease*. As a result, these evolutionary operators produce a "funneling" effect as they work to return solutions to a zero net effect point where there is equal probability of activating/deactivating a network link and substituting among RDM protocols.

4.4 Increased Mutation Probability

Problem/Motivation: In previous experiments, we determined that network size affected the parent-to-offspring mutation distance as well as the distance similar networks are able to mutate from one another. While the original NSGA-II appears to have responded to this differential, we have not formally tested whether this factor affected search.

Hypothesis 5: NSGA-II's search forgoes expanding into novel regions of the objective space in favor of regions associated with greater mutation distances

Methods: To test our hypothesis, we divided the objective space into two equal regions: (1) a *Control* region (Cost $\leq$ 0.50) where we maintained the original mutation rate of 5% and (2) a *Treatment* region (Cost > 0.50) where we increased the mutation rate solutions were exposed to. If our hypothesis is correct, we expect a majority of the final NSGA-II solutions to reside within the Treatment region where an evolved network is more *likely* (not guaranteed) to experience a greater number of mutations and therefore a greater change in their objectives and crowding distance, than networks found in the Control region.

Results: In Figure 6, the 5% mutation rate treatment provides a baseline measure (5.94% ± 1.33%) of solutions in the Treatment region for the original NSGA-II when both regions' mutation rates are equal and thus no difference is present. We observed a significant increase ($p \ll 0.01$) in the percentage of solutions found in the Treatment region as its mutation rate was increased, resulting in the number of solutions found within the Control region dropping from 94.06% in the original NSGA-II to as low as 17.24%. These results confirm our hypothesis that NSGA-II's search forgoes expanding into novel regions of the objective space in favor of regions where higher crowding distances are achievable.

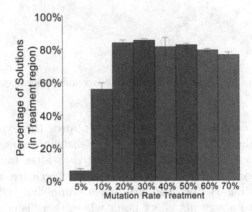

Fig. 6. Percentage of solutions found in the Treatment region

5 Discussion

The experiments in this paper have demonstrated the presence of a hidden interaction between the underlying genome representation and the crowding distance operator found in MOEA approaches such as NSGA-II. Two factors of our application caused this interaction to occur, namely, (1) a *variable-length genome* and (2) objective functions formulated as *ratios*. By allowing the number of solution elements to evolve freely, offspring generated from solutions with fewer elements are more likely to experience a larger change to their objective measures. As a result, offspring having a similar number of elements to their parents but that produce larger crowding distances are likely to be created and accepted by the NSGA-II algorithm. Therefore, NSGA-II's search is unable to locate novel, undiscovered regions of the Pareto front where differently structured solutions reside since these solutions cannot simultaneously achieve similar crowding distance values compared to the current parent members of the population.

The implications of this hidden interaction are significant since, depending on the impact of this interaction, the set of Pareto-optimal solutions returned by NSGA-II may only encompass an extremely limited region of the overall Pareto surface for a given application. More importantly, without knowledge of the underlying Pareto surface, this hidden interaction would be difficult to detect, and the original returned solution set might easily have been accepted as sufficient. In various application domains, the number of solution elements is often a free variable evolved in order to explore different designs and their associated trade-offs. Similarly, a common and often necessary task when normalizing an objective is to formulate the objective as a ratio. As such, this interaction becomes more difficult to detect and also more likely to occur as the dimensionality of the problem increases.

6   Related Work

To the best of the authors' knowledge, the interaction of variable-length genomes and the crowding distance operator as well as its effect on search coverage has not been documented in the literature. Ishibuchi et al. [7, 8] examined how discrete objective functions with two different *granularities* (width of intervals within an objective) affected search when using popular multi-objective optimization algorithms including, NSGA-II, SPEA2, MOEA/D, and SMS-EMOA. A two-objective 0/1 knapsack problem [16] with integer profit values for each knapsack item was used in their experiments. By applying rounding factors of different sizes (e.g., round by 10, 100, etc.) to each objective, they explored the effect of different granularity combinations on search. Their results demonstrated that when the granularities of the problem's dimensions vary, search is biased towards particular regions of the objective space.

While these results strongly corroborate our findings and explore a similar problem, several key differences distinguish our work. First, the granularities of each objective were established by applying *ad hoc* rounding factors and were not attributes of the original application. As a result, both of their objectives had *uniform granularity* whereby the interval width was constant within each objective. Second, the granularity of each objective in [7, 8] was *static* during search meaning that the interval width did not change from one generation to the next. In contrast, the granularities of the objectives in our work were not established but rather they *emerged* from our application thus concealing the hidden interaction and making it more difficult to detect. Third, the granularity of objectives was *non-uniform* since the interval width was dependent on the number of links within an evolved solution. Smaller networks experienced larger granularities during mutation and larger networks experienced smaller granularities. Lastly, the granularity of objectives was *dynamic* and evolved with respect to network size over time. For example, two networks (network A − 10 expensive links, network B = 30 inexpensive links) with the *same* Cost value of 0.30 will experience different granularities in their Cost objective.

These key differences between our two problems also lead to orthogonal results and conclusions. Ishibuchi et al. [7, 8] observed that search was biased towards objectives with finer granularity whereas in this paper, solutions are biased towards the coarse granularity region of each objective. Also, this study found that discrete objectives with coarse granularities improve the search ability of NSGA-II with many dimensions (objectives), whereas the coarse granularity regions of the objective space in our work hinders NSGA-II's search coverage. The combined results of these two independent studies should enable the community to make more informed decisions about which MOEAs to use for problems with similar characteristics or at least should make the researcher more inquisitive of returned solutions.

7   Conclusions and Future Work

In this paper, we first explored the use of NSGA-II for an industrial remote data mirroring application. In this process, we observed the presence of a hidden interaction preventing search from reaching regions on the Pareto surface where optimal solutions were known to exist. Through a series of experiments, we determined that the root cause was the restricted search power of NSGA-II due to an unfavorable interaction between a variable-length genome representation and the crowding distance operator. Solutions with fewer elements experience greater changes to their objective values due to a more coarse-grained granularity and are able to achieve greater crowding distances. As a result, evolutionary search is hindered from exploring regions of the objective space where larger, Pareto-optimal solutions are known to exist. Recently [13], objective function granularity has been highlighted as an important future research area for determining its effects on search performance/dynamics as well as problem analysis. As a newly discovered interaction, we believe this phenomenon is limited to NSGA-II, but is potentially applicable to any EMO procedure, although further studies are needed to confirm this point. These results provide a key insight in this area and raise the level of awareness for researchers exploring multi-objective optimization for domains where the solution's size and/or number of features may evolve freely during search.

Future directions for this work include (1) finding methods, such as epsilon-dominance, that can be incorporated to handle this unintended interaction, (2) surveying other EMO approaches that utilize a different diversity-preserving mechanism and comparing their performance on this problem, and (3) designing a formal test problem to evaluate existing and new EMO algorithms on this rather unexplored interaction between search space properties and diversity preserving operators. Several variables considered for further study include the granularity level within each objective, variation in granularity across an objective, and dynamic changes made to granularity during search.

Acknowledgments. This work has been supported in part by NSF grants IIP-0700329, CCF-0820220, DBI-0939454, CNS-0854931, CNS-0915855. Any opinions, findings, and conclusions or recommendations expressed in this material are those of the author(s) and do not necessarily reflect the views of the National Science Foundation or other research sponsors.

References

1. Andersen, D., Balakrishnan, H., Kaashoek, F., Morris, R.: Resilient overlay networks. SIGOPS Oper. Syst. Rev. **35**(5), 131–145 (2001)
2. Byers, C.M., Cheng, B.H.: Mitigating uncertainty within the dimensions of a remote data mirroring problem. Tech. Rep. MSU-CSE-14-10, Computer Science and Engineering, Michigan State University, East Lansing, Michigan, September 2014
3. Das, I., Dennis, J.: A closer look at drawbacks of minimizing weighted sums of objectives for Pareto set generation in multicriteria optimization problems. Structural Optimization **14**(1), pp. 63–69 (1997)
4. Deb, K., Pratap, A., Agarwal, S., Meyarivan, T.: A fast and elitist multiobjective genetic algorithm: NSGA-II. Trans. Evol. Comp. **6**(2), 182–197 (2002)
5. Goldberg, D.E.: Genetic Algorithms in Search. Optimization and Machine Learning, 1st edn. Addison-Wesley Longman Publishing Co. Inc., Boston (1989)
6. Holland, J.H.: Genetic algorithms. Scientific American, July 1992
7. Ishibuchi, H., Yamane, M., Nojima, Y.: Effects of discrete objective functions with different granularities on the search behavior of emo algorithms. In: Soule, T., Moore, J.H. (eds.) GECCO, pp. 481–488. ACM (2012)
8. Ishibuchi, H., Yamane, M., Nojima, Y.: Difficulty in evolutionary multiobjective optimization of discrete objective functions with different granularities. In: Purshouse, R.C., Fleming, P.J., Fonseca, C.M., Greco, S., Shaw, J. (eds.) EMO 2013. LNCS, vol. 7811, pp. 230–245. Springer, Heidelberg (2013)
9. Ji, M., Veitch, A.C., Wilkes, J.: Seneca: remote mirroring done write. In: USENIX Annual Technical Conf., General Track, pp. 253–268. USENIX (2003)
10. Keeton, K., Santos, C., Beyer, D., Chase, J., Wilkes, J.: Designing for disasters. In: Proceedings of the 3rd USENIX Conf. on File and Storage Technologies, Berkeley, CA, USA, pp. 59–62 (2004)
11. Keeton, K., Wilkes, J.: Automatic design of dependable data storage systems (2003)
12. Kephart, J.O., Chess, D.M.: The vision of autonomic computing. Computer **36**(1), 41–50 (2003)
13. McClymont, K.: Recent advances in problem understanding: Changes in the landscape a year on. In: Proceedings of the 15th Annual Conference Companion on Genetic and Evolutionary Computation, GECCO 2013 Companion, pp. 1071–1078, ACM, New York (2013)
14. McKinley, P.K., Sadjadi, S.M., Kasten, E.P., Cheng, B.H.C.: Composing adaptive software. Computer **37**(7), 56–64 (2004)
15. Ramirez, A.J., Knoester, D.B., Cheng, B.H., McKinley, P.K.: Applying genetic algorithms to decision making in autonomic computing systems. In: Proceedings of the 6th International Conference on Autonomic Computing, ICAC 2009, pp. 97–106. ACM, New York (2009)
16. Zitzler, E., Thiele, L.: Multiobjective evolutionary algorithms: A comparative case study and the strength pareto approach. Trans. Evol. Comp. **3**(4), 257–271 (1999)

Neutral but a Winner! How Neutrality Helps Multiobjective Local Search Algorithms

Aymeric Blot[1,2], Hernán Aguirre[4], Clarisse Dhaenens[2,3], Laetitia Jourdan[2,3], Marie-Eléonore Marmion[2,3(✉)], and Kiyoshi Tanaka[4]

[1] ENS Rennes, Ker Lann, Université Rennes 1, Rennes, France
aymeric.blot@ens-rennes.fr
[2] Inria Lille - Nord Europe, DOLPHIN Project-team, Lille, France
{clarisse.dhaenens,laetitia.jourdan,marie-eleonore.marmion}@lifl.fr
[3] Université Lille 1, LIFL, UMR CNRS 8022, Villeneuve-d'Ascq, France
[4] Faculty of Engineering, Shinshu University, Nagano, Japan
{ahernan,ktanaka}@shinshu-u.ac.jp

Abstract. This work extends the concept of neutrality used in single-objective optimization to the multi-objective context and investigates its effects on the performance of multi-objective dominance-based local search methods. We discuss neutrality in single-objective optimization and fitness assignment in multi-objective algorithms to provide a general definition for neutrality applicable to multi-objective landscapes. We also put forward a definition of neutrality when Pareto dominance is used to compute fitness of solutions. Then, we focus on dedicated local search approaches that have shown good results in multi-objective combinatorial optimization. In such methods, particular attention is paid to the set of solutions selected for exploration, the way the neighborhood is explored, and how the candidate set to update the archive is defined. We investigate the last two of these three important steps from the perspective of neutrality in multi-objective landscapes, propose new strategies that take into account neutrality, and show that exploiting neutrality allows to improve the performance of dominance-based local search methods on bi-objective permutation flowshop scheduling problems.

Keywords: Neutrality · Multi-objective optimization · Local search · Permutation flowshop scheduling

1 Introduction

In the single-objective context, solving large optimization problems with local search approaches allows to obtain good solutions in a reasonable time [6]. These local search methods are based on a neighborhood relation that enables to perform local improvements. It has been shown that such methods are sensitive to the properties of the landscape of the problem studied, and that it is crucial to analyze and understand such properties in order to improve the performance of the algorithms.

© Springer International Publishing Switzerland 2015
A. Gaspar-Cunha et al. (Eds.): EMO 2015, Part I, LNCS 9018, pp. 34–47, 2015.
DOI: 10.1007/978-3-319-15934-8_3

This work focuses on neutrality, a property that characterizes neighboring solutions having the same fitness. In single objective-optimization it is known that the degree of neutrality of a landscape impacts the behavior of local search methods. There are also several studies showing that exploiting neutrality in a local search method can improve performance of the method [7].

In the multi-objective context, there are also efficient local search methods that have been proposed to approximate the Pareto optimal set, such as the Dominance based Multi-objective Local Search (DMLS) [4]. However, not much is known about neutrality, its effects, and how to take advantage of it in order to improve the performance of multi-objective algorithms. Indeed, the performance of a DMLS algorithm is closely related to the geometry of the landscape of the problem to solve. Moreover, the Pareto dominance relation induces landscapes where many solutions cannot be compared with many others (solutions equivalent in term of quality), and one major difficulty of DMLS algorithms is, at each iteration, to choose a selected neighbor which may be *equivalent*, i.e. incomparable with the current explored solution.

This work extends the concept of neutrality to multi-objective optimization with the aim to analyze whether exploiting neutrality is also beneficial in a multi-objective context. We discuss neutrality in single-objective optimization and fitness assignment in multi-objective algorithms to provide a general definition for neutrality applicable to multi-objective landscapes. We also put forward a definition of neutrality when Pareto dominance is used to compute fitness of solutions. Then, we analyze existing DMLS from the point of view of neutrality, in order to propose new efficient schemes. We focus on strategies that take into account neutrality, particularly during the neighborhood exploration and the creation of the candidate set of solutions to update the archive, showing that exploiting neutrality allows to improve the performance of dominance-based local search methods on bi-objective permutation flowshop scheduling problems.

The paper is organized as follows. Section 2 states background definitions of multi-objective combinatorial optimization. It presents the problem that will be used as an illustration, the Permutation Flowshop Scheduling Problem (PFSP) and Dominance based Multi-objective Local Search approaches (DMLS). In Section 3, we propose a multi-objective concept of neutrality, and analyze its integration in existing DMLS algorithms. This leads us to propose several improvements to DMLS algorithms to efficiently incorporate this notion. In Section 4, experiments are conducted in order to emphasize the importance of taking care of neutrality in DMLS algorithms and to measure the impact of our propositions on the Permutation Flowshop Scheduling Problem (PFSP). At last, Section 5 gives the conclusions of the presented work and future research interests.

2   Background

This work investigates the effects of neutrality focusing on Dominance-based Local Search Methods using the Bi-objective Permutation Flowshop Scheduling Problem as an illustrative example. This section describes the optimization problem and the local search methods used, together with necessary notation to better understand the paper.

2.1   The Bi-objective Permutation Flowshop Scheduling Problem

The Permutation Flowshop Scheduling Problem (PFSP) is a multi-objective combinatorial optimization (MoCO) problem widely investigated in the literature. The PFSP consists in scheduling a set of N jobs $\{J_1, \ldots, J_N\}$, on a set of M machines $\{M_1, \ldots, M_M\}$. Machines are critical resources that can only process one job at a time. A job J_i is composed of M tasks $\{t_{i,1}, \ldots, t_{i,M}\}$ for the M machines respectively. A processing time $p_{i,j}$ is associated to each task $t_{i,j}$, and a due date d_i is associated with each job J_i. The operating sequence is the same on every machine. Therefore, a schedule may be represented as a permutation of jobs $\pi = \{\pi_1, \ldots, \pi_N\}$. Ω is the set of the feasible solutions.

The two objectives, f_1 and f_2, considered in this paper are the makespan $C_{\max}$ (eq. 1), i.e. the total completion time, and the total tardiness T (eq. 2). Both objectives have to be minimized.

$$f_1 = C_{\max} = \max_{i \in \{1, \ldots, N\}} \{C_{\pi_i}\} \tag{1}$$

$$f_2 = T = \sum_{i=1}^{N} \{\max \{0, C_{\pi_i} - d_{\pi_i}\}\} \tag{2}$$

The feasible outcome vectors of the *objective space* are compared using *Pareto dominance* $\succ$. In this minimization context, a solution $x \in \Omega$ is said to dominate a solution $y \in \Omega$, denoted by $x \succ y$, if they satisfy relation (3).

$$\forall i \in \{1, 2\}, \ f_i(x) \leq f_i(y) \bigwedge \exists i \in \{1, 2\}, \ f_i(x) < f_i(y) \tag{3}$$

If solution y is non-dominated by solution x we denote $y \not\succ x$.

This paper focuses on multi-objective local search methods based on a neighborhood definition. The neighborhood considered in this paper uses the insertion operator, where a job located at position i is inserted at position $j \neq i$ and the jobs located between positions i and j are shifted. The number of neighbors per solution is then $(N-1)^2$.

2.2   Dominance-Based Multi-objective Local Search

In the literature, numerous methods have been proposed to solve MoCO problems. The Dominance-based Multi-objective Local Search algorithms represent a class of local search approaches designed to approximate the Pareto front of a MoCO [4] problem. DMLS algorithms keep an archive of mutually non-dominated solutions and uses a neighborhood structure to improve the solutions of the archive. The main steps of a DMLS algorithm are as follows.

Step 1. Initialize the archive with a randomly created solution x, $\mathcal{A} \leftarrow \{x\}$.

Step 2. Select from the archive a set of solutions for exploration, $\mathcal{X} \subseteq \mathcal{A}$.

Step 3. For each solution $x \in \mathcal{X}$, explore the neighborhood of x until a neighbor z fulfilling a required criterion is found. During exploration of x, in addition to z, neighbor solutions x' non-dominated by x are collected as candidate solutions to update the archive, $\mathcal{C} = \{x' \in \mathcal{N}(x) \mid x' \not\prec x\} \cup \{z\}$.

Step 4. Update the archive $\mathcal{A}$ with the collected candidate solutions $\mathcal{C}$ making sure that only non-dominated solutions remain in the archive.

Step 5. If termination criterion is not met, repeat from Step 2.

Step 6. Return the archive $\mathcal{A}$.

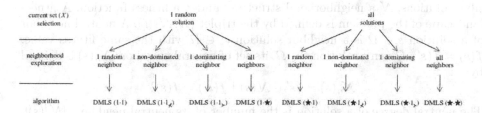

Fig. 1. Nomenclature of DMLS algorithms

Several strategies are defined for Step 2 and Step 3, which lead to different configurations of DMLS algorithms. A specific nomenclature and classification of DMLS algorithms was proposed by Liefooghe et al. [4], as shown in Figure 1. In Step 2, the candidate set for exploration $\mathcal{X}$ can be obtained by selecting from the archive either *one* solution randomly (DMLS $(1 \cdot \;)$) or *all* solutions (DMLS $(\bigstar \cdot \;)$). In Step 3, the neighborhood exploration of each solution $x \in \mathcal{X}$ can be either *exhaustive* or *partial*. If it is exhaustive (DMLS $(\; \cdot \bigstar)$), all the neighbors are visited and all non-dominated neighbors $x' \not\prec x$ are collected in the candidate set $\mathcal{C}$ to update the archive. If the exploration is partial, different strategies can be used. A possible partial exploration strategy is to accept a random neighbor whatever its quality (DMLS $(\; \cdot 1)$). This strategy corresponds to a random search. Other strategy is to explore the neighborhood of a solution until a dominating neighbor $z \succ x$ is found (DMLS $(\; \cdot 1_{\succ})$). A third partial exploration strategy is to explore the neighborhood of a solution until a non-dominated solution is found $z \not\prec x$ (DMLS $(\; \cdot 1_{\not\prec})$), in which case z could be a dominating solution $z \succ x$ or mutually non-dominated with respect to x, $z \not\prec x$ and $x \not\prec z$.

Liefooghe et al. [4] experimented on the bi-objective PFSP showing that some DMLS configurations perform better than others. In the rest of the paper, only the following configurations DMLS $(1 \cdot 1_{\not\prec})$, DMLS $(1 \cdot 1_{\succ})$ and DMLS $(\bigstar \cdot 1_{\succ})$ are considered.

3   Neutrality Extended to Multi-objective Optimization

In this section, we discuss the concept of neutrality in single objective optimization, propose a definition of neutrality in the multi-objective context, particularly for Pareto dominance based approaches, clarify how neutrality has been used so far in the DMLS algorithm, and propose new strategies for DMLS aiming to further exploit neutrality in multi-objective optimization.

3.1   Neutrality in Single-objective Optimization

In single-objective optimization, neutrality arises when neighboring solutions have the same quality. More formally, let us denote Ω the space of the admissible solutions, $\mathcal{N}$ a neighborhood structure, and f a fitness function. A fitness landscape of the problem is defined by the triplet $(\Omega, \mathcal{N}, f)$. A neutral neighbor of a solution $s \in \Omega$ is a neighbor solution $s' \in \Omega$ with the same fitness value, $f(s) = f(s')$. Given a solution $s \in \Omega$, its set of neutral neighbors $N_n(s)$ is defined by:

$$\mathcal{N}_n(s) = \{s' \in \mathcal{N}(s) \mid f(s') = f(s)\}$$

The neutral degree of a solution is the number of its neutral neighbors $|\mathcal{N}_n(s)|$. A fitness landscape is said to be neutral if there are many solutions with a high neutral degree $|\mathcal{N}_n(s)|$.

Neutral solutions can be considered in the design of local search algorithms [1, 7,10] either to escape from a local optimum or to explore more widely the search space. Since equivalent solutions have proved to be useful in single-objective optimization, we propose to study the effects of exploiting equivalent solutions in multi-objective optimization.

3.2   Neutrality in Multi-Objective Optimization

The definition of neutrality in single-objective optimization is based on neighbor solutions that have same fitness values. In order to give a definition of neutrality in multi-objective optimization, we need to characterize neutral neighbors in this context. Particularly, what means equal fitness of two solutions.

In multi-objective optimization there are various approaches to compute fitness of solutions. These include Pareto dominance, Pareto dominance and density estimation, scalarization functions, and indicator functions such as the hypervolume I_{HV} or the epsilon indicator $I_{\epsilon+}$. In general, we can say that fitness f is a function of the n single-objective fitness values $f_1, f_2, \cdots, f_n$ computed for a solution, i.e. $f = g(f_1, f_2, \cdots, f_n)$. Thus, a similar definition used for neutrality in single-objective optimization can be used for neutrality in multi-objective optimization. Namely, a multi-objective neutral neighbor of a solution $s \in \Omega$ is a neighbor solution $s' \in \Omega$ with the same fitness value $f(s) = f(s')$, where $g(f_1, f_2, \cdots, f_n)$ is a function of the single-objective fitness values and $f = g(f_1, f_2, \cdots, f_n)$.

It should be noted that each approach to compute fitness in multi-objective optimization implies a different fitness function and therefore a different landscape. This also means that the set of neutral neighbors of a solution might vary depending on the approach used to compute fitness. However, all approaches aim to find the Pareto optimal set of the problem or a good approximation of it. It will be very interesting to study the effects of neutrality in the different approaches to multi-objective optimization. In this work, we restrict our attention to approaches that use Pareto dominance to determine fitness of solutions.

Given a a solution x to explore based on a neighborhood structure $\mathcal{N}$, Pareto dominance implies three types of neighbors x' respect to x: dominating neighbors $x' \succ x$, dominated neighbors $x \succ x'$, or mutually non-dominated neighbors $x \nprec x'$ and $x' \nprec x$, as shown in Figure 2. These latter neighbors are non-comparable solutions. Therefore, they can be viewed as *equivalent* neighbors, or same fitness neighbors, that define the neutral neighbors in the multi-objective context when fitness of solutions is computed using Pareto dominance. More precisely, given a solution $s \in \Omega$, its set of neutral neighbors is defined by:

$$\mathcal{N}_e(s) = \{s' \in \mathcal{N}(s) \mid s \nprec s' \text{ and } s' \nprec s\}$$

Note that this definition includes the case where two neighbors have the same objective vector $(s' \in \mathcal{N}(s), \forall i \in [1, n], f_i(s) = f_i(s'))$.

The motivation to extend neutrality from single- to multi-objective optimization comes from the fact that single-objective local search algorithms can benefit from equivalent/neutral solutions. These solutions allow to continue the search when it is trapped in a local optimum without degrading. Similarly, in multi-objective optimization that uses Pareto dominance to establish fitness of the individuals, a local search algorithm can be trapped in a Pareto local optimum (PLO) and some equivalent/neutral neighbors can help to escape from it. In the following, we use the term *neutral* to qualify these equivalent neighbors.

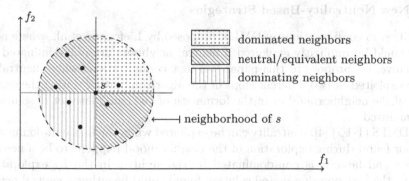

Fig. 2. Neighborhood in bi-objective optimization

3.3 Neutrality in DMLS Algorithms

Section 2.2 briefly described the DMLS algorithms for multi-objective optimization. Analyzing these algorithms, we can see that some of them can exploit neutral neighbors to approach the Pareto front, but require that the neighbors survive several steps of the algorithm. For example, during Step 3 DMLS $(1 \cdot 1_{\nparallel})$ and DMLS $(\bigstar \cdot 1_{\nparallel})$ algorithms can generate at most one neutral neighbor solutions per $x \in \mathcal{X}$ if and only if during exploration a dominating solution is not found first. On the other hand, DMLS $(1 \cdot 1_{\succ})$ and DMLS $(\bigstar \cdot 1_{\succ})$ can generate more than one neutral neighbor solution per $x \in \mathcal{X}$ until the first dominating solution is found or the whole neighborhood has been explored if there is no dominating solution. The neutral neighbors found in Step 3 become part of the candidate solution set $\mathcal{C}$ to update the archive. In Step 4 these neutral neighbors could be included in the archive only if they are non dominated by all members of the current archive. Then in Step 2 of the next iteration the newly found neutral neighbors can be selected for exploration. Thus, DMLS algorithms in order to exploit a neutral neighbor of x also requires that it is non-dominated by the archive.

Neutrality seems to be exploited to increase the performance of DMLS as equivalent neighbors may be candidates to be archived. However, it is not clear the contribution of neutral neighbors to the effectiveness of DMLS algorithms. In this paper, we want to clarify and show the impact of using neutral neighbors in multi-objective local search. To do so, we will analyze two configurations of DMLS denoted DMLS $(1 \cdot \overline{1_{\succ}})$ and DMLS $(\bigstar \cdot \overline{1_{\succ}})$ where neutrality is never exploited and compare them with already existing strategies for DMLS algorithms that explore to some degree neutrality. In DMLS $(1 \cdot \overline{1_{\succ}})$ and DMLS $(\bigstar \cdot \overline{1_{\succ}})$ the neighborhood of each selected solution is explored until a dominating solution is found and only this neighbor represents a candidate to be archived, thus never exploiting the neutral neighbors of a solution.

3.4 New Neutrality-Based Strategies

In addition to configurations of DMLS, proposed by Liefooghe et al., where neutrality could be implicitly exploited if neutral neighbors are non-dominated by the archive, we propose in this paper two new configurations where neutrality can be exploited in two different steps of the algorithm: either during the exploration of the neighborhood or in the formation of the candidate set of solutions to be archived.

In DMLS $(1 \cdot 1_{\nparallel})$ [4], neutrality can be exploited when the first non-dominated neighbor found during exploration of the neighborhood happens to be a neutral neighbor and later it is non-dominated by the archive. In the $1_{\nparallel}$ exploration strategy, the first non-dominated solution found could be either a neutral neighbor or a dominating neighbor. It is arguable whether the first neutral neighbor found would be the best to improve later the Pareto front in the archive, so that neutrality could be exploited. Similarly, it is also arguable whether the first dominating neighbor could be the best dominating neighbor. Therefore, we propose a $k_{\nprec}$ exploration strategy, where the neighborhood of a solution is explored until k different non-dominated neighbors have been found. This strategy gives the opportunity to explore more widely the neighborhood of a solution. Indeed more chance to find one or more dominating neighbors is given. In addition, since all non-dominated neighbors are collected in the candidate set $\mathcal{C}$ to update the archive, this strategy increases the likelihood of finding neutral neighbors that can become part of the archive, diversifies the Pareto front, and emphasizes the exploitation of neutrality. The new DMLS with the $k_{\nprec}$ exploration strategy is denoted DMLS $(1 \cdot k_{\nprec})$, where the number k is an integer defined from 1 to the neighborhood size.

The candidate set of solutions $\mathcal{C}$ considered to update the archive is a key element when dealing with neutrality. Indeed, Liefooghe et al. take into account all neutral neighbors visited during the neighborhood exploration when a $1_{\succ}$ strategy is used. In this paper, in addition to collect neutral neighbors in the candidate set $\mathcal{C}$ of solutions to update the archive, we propose to use some of them for further exploration, before they are used to update the archive.

DMLS $(\bigstar \cdot 1_{\succ})$ [4] is a configuration where all solutions of the archive are selected to be explored until a dominating neighbor is found for each one. This algorithm may integrate a large number of solutions in $\mathcal{C}$ during a single step of archiving. We modify Step 3 of the DMLS $(\bigstar \cdot 1_{\succ})$ algorithm. When a non-dominated neighbor x' is found, we check Pareto dominance between x' and those already in the set $\mathcal{X}$ selected for exploration. If no solution in $\mathcal{X}$ is dominated by x', then x' is added to the exploration set $\mathcal{X}$. Thus, the exploration set $\mathcal{X}$ grows dynamically as neutral neighbors are found. This strategy allows to explore neutral neighbors that could not be archived in Step 4 after finding the dominating neighbors of solutions in $\mathcal{X}$. This proposed DMLS, denoted DMLS $(\bigstar + \mathcal{X}_{\nprec} \cdot 1_{\succ})$, takes advantage of neutrality more intensively than the known configurations of DMLS. Figure 3 gives the complete nomenclature of DMLS algorithms with the proposed configurations (in the red boxes) and the most used DMLS configurations. Note that in this figure the definition of the candidate set of solutions $\mathcal{C}$ used to update the archive is explicitly described.

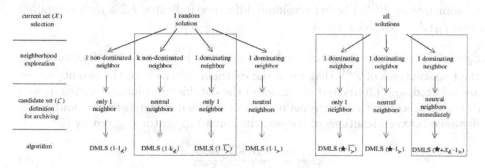

Fig. 3. Nomenclature of DMLS algorithms with the proposed configurations

4 Experiments and Discussion

The aim of this section is to compare performance of the different DMLS configurations studied in this work and clarify the effects of neutrality on the Permutation Flowshop Scheduling Problem (PFSP).

4.1 Experimental Protocol

Instances. Experiments are realized on classical Muti-objective PFSP instances. These instances have been proposed by Minella et al. [8] as an extension of the well-known random generated instances of Taillard [9], in which due dates have been added. In the following, these instances are denoted by a triplet $(JJJ \times MM \times NN)$, where JJJ is the number of jobs, MM is the number of machines, and NN is the identifier of the $(JJJ \times MM)$ instance.

Performance Assessment. In order to rank the different algorithms and observe the behavior and influence of the neutral neighbors on performance, three complementary indicators are used as recommended in [3]. Namely, unary ε-indicator $I^1_{\varepsilon+}$, hypervolume difference indicator I^-_H, and Spread. These indicators are based on set Z^{all} that is the union of the final sets of solutions obtained by all algorithms and on the reference set R that contains the Pareto set of Z^{all}. The three performance indicators are explained below, where A stands for the set of solutions found by an algorithm.

ε-indicator $I^1_{\varepsilon+}$ The unary ε-indicator is computed using the binary version given by (4) and the reference set R, with $I^1_{\varepsilon+}(A) = I_{\varepsilon+}(A, R)$.

$$I_{\varepsilon+}(A, R) = \inf_{\varepsilon \in \mathbb{R}} \{\forall z^1 \in R, \exists z^2 \in A, \forall i \in 1 \ldots n, z^1_i \le \varepsilon + z^2_i\} \qquad (4)$$

Hypervolume difference indicator I^-_H The hypervolume indicator I_H is computed by the measure of the hypervolume between a set of solutions and the point $z = (z_1, \ldots, z_n)$ where z_k is the upper bound of the kth objective considering all solutions of Z^{all}. The hypervolume difference indicator I^-_H is then computed with $I^-_H(A) = I_H(R) - I_H(A)$.

Spread. The spread indicator used in this paper is computed as follows. First, the two solutions of Z^{all} that reach the extrema relatively to the two objectives are selected, and filtered out of the set of the considered solutions. Given d_f and d_l the distances to those extreme points, $\bar{d}$ the mean of the distances, and d_i the distance between solutions of the set, the spread indicator is given by (5).

$$\Delta = \frac{d_f + d_l + \sum |d_i - \bar{d}|}{d_f + d_l + \sum \bar{d}} \qquad (5)$$

Experimental Design. All DMLS implementations are realized under the ParadisEO 2.0 software framework [5]. Most of the performance assessment indices are computed using the PISA platform [2] and its performance assessment module. The spread indicator has been developed to be automatically computed into PISA. The results are then verified with the *Friedman* statistical test, and a global ranking is computed using the *Wilcoxon* statistical test.

Seven instances have been selected from Minella et al., spanning over seven problem sizes. The seven algorithms of Figure 3 are compared. For the parameter k in DMLS $(1 \cdot k_{\nearrow})$, two different values *low* and *high* have been tested, leading to two versions of this algorithm: DMLS $(1 \cdot k_{\nearrow}^{low})$ and DMLS $(1 \cdot k_{\nearrow}^{high})$. The *low* and *high* values of k depend on the number of jobs of the instance as the size of the neighborhood depends on it. Parameter k has been set arbitrarily according to the number of jobs: $k = 5$ and 10 for 20 jobs, $k = 15$ and 25 for 50 jobs, and $k = 20$ and 50 for 100 jobs.

For each instance, 20 executions have been recorded for each algorithm. A maximal runtime has been fixed for each size of instance corresponding to $N \times M$ seconds. Those runtimes were sufficient for all algorithms to converge, even if they did not reach a natural termination.

4.2 Experimental Results

Table 1 shows the rankings computed with the indicator $I_{\varepsilon+}^1$ for each instance with respect to the final Pareto local set R. Similarly, Table 2 and Table 3 show the rankings computed with I_H^- indicator and *spread* indicator, respectively.

The Friedman statistical tests give a p-value of $2.449e^{-6}$ for the $I_{\varepsilon+}^1$ indicator, $4.796e^{-6}$ for I_H^- and $1.758e^{-5}$ for *spread*. Thus, the behavior of the all algorithms is statistically different on the three indicators, and ranks give valuable information about performance. These tables allow several observations.

Table 1. Rankings according to $I_{\varepsilon+}^1$

Instance	$(1 \cdot \overline{1_{\succ}})$	$(\bigstar \cdot \overline{1_{\succ}})$	$(1 \cdot 1_{\nearrow})$	$(1 \cdot k_{\nearrow}^{low})$	$(1 \cdot k_{\nearrow}^{high})$	$(1 \cdot 1_{\succ})$	$(\bigstar \cdot 1_{\succ})$	$(\bigstar + \varkappa_{\nearrow} \cdot 1_{\succ})$
$(020 \times 05 \times 01)$	7	8	1	5	2	6	4	3
$(020 \times 10 \times 01)$	8	7	2	6	5	4	3	1
$(020 \times 20 \times 01)$	8	7	4	5	2	6	3	1
$(050 \times 10 \times 01)$	7	8	2	4	5	6	3	1
$(050 \times 20 \times 01)$	8	7	5	2	4	6	3	1
$(100 \times 10 \times 01)$	7	8	3	5	4	6	2	1
$(100 \times 20 \times 01)$	7	8	3	6	5	4	2	1
mean	7.42	7.57	2.85	4.71	3.85	5.42	2.85	1.28

No neutrality exploitation. Tables 1, 2 and 3 show that algorithms DMLS $(1 \cdot \overline{1_{\succ}})$ and DMLS $(\bigstar \cdot \overline{1_{\succ}})$ share the worse results on the three indicators. These two methods select during the neighborhood exploration one single dominating neighbor for each explored solution, without any neutrality consideration,

Table 2. Rankings according to I_H^-

Instance	$(1 \cdot \overline{1_\succ})$	$(\star \cdot \overline{1_\succ})$	$(1 \cdot 1_{\not\star})$	$(1 \cdot k_{\not\star}^{low})$	$(1 \cdot k_{\not\star}^{high})$	$(1 \cdot 1_\succ)$	$(\star \cdot 1_\succ)$	$(\star + \mathcal{X}_{\not\star} \cdot 1_\succ)$
(020 × 05 × 01)	8	7	3	4	1	6	5	2
(020 × 10 × 01)	8	7	2	6	5	4	3	1
(020 × 20 × 01)	8	7	4	5	2	6	3	1
(050 × 10 × 01)	7	8	2	4	5	6	3	1
(050 × 20 × 01)	8	7	5	2	4	6	3	1
(100 × 10 × 01)	7	8	3	6	4	5	2	1
(100 × 20 × 01)	7	8	3	6	5	4	2	1
mean	7.57	7.42	3.14	4.71	3.71	5.28	3.00	1.14

Table 3. Rankings according to the spread indicator

Instance	$(1 \cdot \overline{1_\succ})$	$(\star \cdot \overline{1_\succ})$	$(1 \cdot 1_{\not\star})$	$(1 \cdot k_{\not\star}^{low})$	$(1 \cdot k_{\not\star}^{high})$	$(1 \cdot 1_\succ)$	$(\star \cdot 1_\succ)$	$(\star + \mathcal{X}_{\not\star} \cdot 1_\succ)$
(020 × 05 × 01)	8	7	3	5	6	4	1	2
(020 × 10 × 01)	7	8	3	6	2	4	5	1
(020 × 20 × 01)	8	7	4	6	3	5	2	1
(050 × 10 × 01)	7.5	7.5	6	3	5	4	1	2
(050 × 20 × 01)	8	7	6	5	3	4	2	1
(100 × 10 × 01)	8	7	3	6	5	4	1	2
(100 × 20 × 01)	8	7	6	5	4	3	2	1
mean	7.78	7.21	4.42	5.14	4.00	4.00	2.00	1.42

i.e. not even one neutral neighbor is considered as candidate solution to update the archive. If this strategy allows to quickly optimize at the beginning of the search, it does not allow to obtain a good approximation of the whole Pareto front.

Considering neutrality to update the archive. One way to take profit from neutrality during the search is, as exposed before, to collect neutral neighbors during the neighborhood exploration and use them as candidates to update the archive. A strategy may consist in exploring the neighborhood until a dominating neighbor is reached (as in the worst versions), but keeping all equivalent neighbors encountered. This leads to DMLS $(1 \cdot 1_\succ)$ and DMLS $(\star \cdot 1_\succ)$. Note that DMLS $(1 \cdot 1_\succ)$ is ranked better than DMLS $(1 \cdot \overline{1_\succ})$ and DMLS $(\star \cdot 1_\succ)$ is ranked better than DMLS $(\star \cdot \overline{1_\succ})$. This shows that considering neutrality in the candidate set to update the archive seems to be effective. Methods DMLS $(1 \cdot 1_{\not\star})$, DMLS $(1 \cdot k_{\not\star}^{low})$ and DMLS $(1 \cdot k_{\not\star}^{high})$ also consider neutral neighbors as interesting neighbors to update the archive and explore the neighborhood until one or several (k) non-dominated solutions (i.e. either neutral or dominating solutions) are found. Results show that these strategies perform better than methods that only consider the first dominating neighbor encountered. Moreover, when k neighbors are required, several potential interesting solutions may become part of the candidate set to update the archive, which improves the quality of the obtained approximation of the Pareto optimal set. Note that a high value of k leads to better results.

Considering neutrality before archiving. Another way to exploit neutral neighbors is to add them dynamically to the set of solutions $\mathcal{X}$ to be explored in addition to the set of candidate solutions to update the archive, as explained in section 3.3. This leads to the method DMLS ($\bigstar + \mathcal{X}_{\not{\ell}} \cdot 1_{\succ}$) which obtained the best results over all the tested methods. This method outperforms the method DMLS ($\bigstar \cdot 1_{\succ}$) that also takes into account neutral neighbors to update the archive. This is because the dynamical insertion of neutral neighbors into the set of solutions to be explored allows the method to go deeper in the search. In addition, it also saves some computational effort.

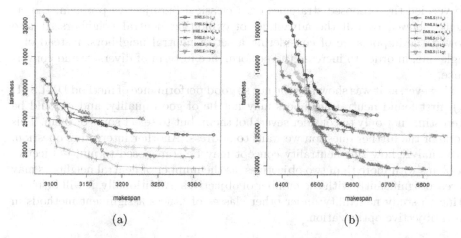

Fig. 4. Pareto Fronts for the instances $050 \times 10 \times 01$ (a) and $100 \times 20 \times 01$ (b)

These above observations are complemented by Figure 4 (a) and (b) that show the Pareto fronts obtained by each method on two instances. These figures illustrate the good average performance of DMLS ($\bigstar + \mathcal{X}_{\not{\ell}} \cdot 1_{\succ}$) and show that DMLS ($\bigstar \cdot 1_{\succ}$) is able to produce very good results on large instances. These two figures allow to confirm also that DMLS ($1 \cdot k_{\not{\ell}}^{high}$) outperforms DMLS ($1 \cdot k_{\not{\ell}}^{low}$) as indicated on the previous tables.

In summary, these experiments show that non considering neutral neighbors (method DMLS ($1 \cdot \overline{1_{\succ}}$) and DMLS ($\bigstar \cdot \overline{1_{\succ}}$)) is less efficient than considering them. In particular, the diversity of the Pareto front produced is greatly impacted. Also, as shown by the not so good performance of method DMLS ($1 \cdot 1_{\not{\ell}}$), in particular in terms of spread, the first found neutral neighbors are not always of good quality and it may be important to consider several of them in order to improve results. Moreover, the exploitation of neutral neighbors that may be dominated by the archive could lead to improve the performance of local search Pareto dominance based approaches.

5 Conclusion

Neutrality has obviously a critical role in multi-objective combinatorial optimization, and furthermore in local search algorithms. Small changes in the way neutral neighbors are handled greatly modify the general behavior of algorithms. This is why the understanding of the relation between local search and neutrality is very important in multi-objective as well as in single-objective optimization. This paper extended the concept of neutrality to multi-objective optimization, focused the discussions about the neutrality in the context of dominance-based multi-objective local search algorithms, and proposed new strategies to improve the behavior of those algorithms towards the exploitation of neutral neighbors. We verified the proposed strategies on a classical bi-objective problem. Experiments showed overall the advantage of exploiting neutral neighbors. It also showed the importance of considering a set of neutral neighbors, instead of a single one, in order to increase the performance in term of diversity and convergence.

However, as it was shown by the not so good performance of method DMLS $(1 \cdot 1_{\not\prec})$, first found neutral neighbors may not be of good quality, and it could be interesting, not only to consider several of them, but to select some of them. This is one of the further question we want to address. Another interesting question, is to analyze how this neutrality concept may be transposed to multi-objective problems with more than two objectives, as the number of neutral neighbors may increase significantly with the number of objectives. Additionally, it will be interesting to study neutrality under other classes of fitness assignment methods in multi-objective optimization.

References

1. Barnett, L.: Netcrawling - optimal evolutionary search with neutral networks. In: Proceedings of the 2001 Congress on Evolutionary Computation, CEC 2001, pp. 30–37. IEEE Press (2001)
2. Bleuler, S., Laumanns, M., Thiele, L., Zitzler, E.: PISA – a platform and programming language independent interface for search algorithms. In: Fonseca, C.M., Fleming, P.J., Zitzler, E., Deb, K., Thiele, L. (eds.) EMO 2003. LNCS, vol. 2632, pp. 494–508. Springer, Heidelberg (2003)
3. Knowles, J., Thiele, L., Zitzler, E.: A Tutorial on the Performance Assessment of Stochastic Multiobjective Optimizers. TIK Report 214, Computer Engineering and Networks Laboratory (TIK), ETH Zurich (February 2006)
4. Liefooghe, A., Humeau, J., Mesmoudi, S., Jourdan, L., Talbi, E.G.: On dominance-based multiobjective local search: design, implementation and experimental analysis on scheduling and traveling salesman problems. J. Heuristics 18(2), 317–352 (2012)
5. Liefooghe, A., Jourdan, L., Talbi, E.G.: A software framework based on a conceptual unified model for evolutionary multiobjective optimization: Paradiseo-moeo. European Journal of Operational Research 209(2), 104–112 (2011)

6. Lourenco, H., Martin, O., Stutzle, T.: Iterated local search. In: Glover, F., Kochenberger, G. (eds.) Handbook of Metaheuristics, International Series in Operations Research & Management Science, vol. 57, pp. 321–353. Kluwer Academic Publishers, Norwell (2002)
7. Marmion, M.-E., Dhaenens, C., Jourdan, L., Liefooghe, A., Verel, S.: NILS: a neutrality-based iterated local search and its application to flowshop scheduling. In: Merz, P., Hao, J.-K. (eds.) EvoCOP 2011. LNCS, vol. 6622, pp. 191–202. Springer, Heidelberg (2011)
8. Minella, G., Ruiz, R., Ciavotta, M.: A review and evaluation of multiobjective algorithms for the flowshop scheduling problem. INFORMS Journal on Computing 20(3), 451–471 (2008)
9. Taillard, E.: Benchmarks for basic scheduling problems. European Journal of Operational Research 64(2), 278–285 (1993)
10. Verel, S., Collard, P., Clergue, M.: Scuba search : when selection meets innovation. In: Evolutionary Computation, 2004. CEC2004 Evolutionary Computation, 2004. CEC2004., pp. 924–931. IEEE Press, Portland (Oregon) United States (2004)

To DE or Not to DE? Multi-objective Differential Evolution Revisited from a Component-Wise Perspective

Leonardo C.T. Bezerra[✉], Manuel López-Ibáñez, and Thomas Stützle

IRIDIA, Université Libre de Bruxelles (ULB), Brussels, Belgium
{lteonaci,manuel.lopez-ibanez,stuetzle}@ulb.ac.be

Abstract. Differential evolution (DE) research for multi-objective optimization can be divided into proposals that either consider DE as a stand-alone algorithm, or see DE as an algorithmic component that can be coupled with other algorithm components from the general evolutionary multiobjective optimization (EMO) literature. Contributions of the latter type have shown that DE components can greatly improve the performance of existing algorithms such as NSGA-II, SPEA2, and IBEA. However, several experimental factors have been left aside from that type of algorithm design, compromising its generality. In this work, we revisit the research on the effectiveness of DE for multi-objective optimization, improving it in several ways. In particular, we conduct an iterative analysis on the algorithmic design space, considering DE and environmental selection components as factors. Results show a great level of interaction between algorithm components, indicating that their effectiveness depends on how they are combined. Some designs present state-of-the-art performance, confirming the effectiveness of DE for multi-objective optimization.

Keywords: Multi-objective optimization · Evolutionary algorithms · Differential evolution · Component-wise design

1 Introduction

Differential evolution (DE) [15] plays an important role in single-objective optimization and has led to the development of a number of effective optimization algorithms for both constrained and unconstrained continuous problems [6]. In particular, one of the most attractive features of DE is its simplicity and its ability to outperform classical genetic algorithms (GAs) [13]. As a result, a number of research proposals have extended DE algorithms to tackle multi-objective optimization problems (MOPs) in the Pareto sense [6,10,14]. In general, extensions follow different paths on how to adapt DE to deal with Pareto optimality, and these stand-alone algorithms have been compared to well-known GA-based algorithms such as NSGA-II [7] or SPEA2 [20] to test their effectiveness. Interestingly, two research groups independently proposed the same DE algorithm at

© Springer International Publishing Switzerland 2015
A. Gaspar-Cunha et al. (Eds.): EMO 2015, Part I, LNCS 9018, pp. 48–63, 2015.
DOI: 10.1007/978-3-319-15934-8_4

about the same time: DEMO [14] and GDE3 [10]. To highlight the effectiveness of this algorithm, we remark that it ranked among the top five best-performing algorithms at the 2009 CEC competition on multi-objective optimization [18].

In the most comprehensive study conducted so far on DE for multi-objective optimization, Tušar and Filipič [17] have considered DEMO as a template for instantiating DE algorithms. Concretely, DEMO uses DE for exploring the decision space, but uses the environmental selection strategy of NSGA-II. The authors then considered the possibility of using other environmental selection approaches, and compared three top-performing GA-based algorithms, NSGA-II, SPEA2, and IBEA [19] with DE versions of these algorithms, aliased DEMO^{NS-II}, DEMO^{SP2} and DEMO^{IB}. By performing pairwise comparisons between algorithms that differ only in the underlying search mechanism (GA or DE), the DE operators were shown to obtain more accurate approximations of the Pareto front and DEMO^{SP2} was found to best balance convergence and diversity [16].

We extend here this excellent earlier work by carrying out a more profound component-wise analysis [3,4] of the design of DE algorithms for MOPs. Our analysis shows that a more fine-grained view of DE components can lead to new insights. In the original analysis only the environmental selection strategy was a component to be set in the DEMO template. However, the DE-part of DEMO differs from traditional GAs in more than one component. In addition to the *DE variation operator*, there is an *online replacement strategy*, i.e., newly generated solutions are compared to existing solutions as soon as they are created, enforcing a higher convergence pressure. In fact, the latter component was found to be the key improvement of DEMO over earlier DE adaptations to MOPs [14]. However, when we consider the DEMO versions that use environmental selection strategies from IBEA and SPEA2 instead of the original DEMO algorithm that uses the environmental selection from NSGA-II, we show that the online replacement strategy is not always beneficial to the effectiveness of the DEMO versions. In other words, while DEMO was an improvement over existing NSGA-II based DE algorithms because of its online replacement strategy, the other DEMO versions present the same (or, sometimes, worst) performance than versions of IBEA and SPEA2 that simply use the DE variation operator.

Furthermore, we consider several factors that affect the conclusions in the original analysis. First, in the original paper, the quality indicator used by IBEA and DEMO^{IB} was the binary hypervolume difference, whereas strong evidence points to a better performance of IBEA when using the binary epsilon indicator [2,19]. Second, the analysis conducted in the original paper was done using the default parameter settings traditionally adopted by the EMO community for the benchmarks considered. However, we have recently shown that tuning the numerical parameters of EMO algorithms can significantly improve their performance [2], altering their relative performance. Finally, although the original paper considered a representative number of benchmark functions, they all used the same number of variables. In this work, we consider several different problem sizes to ensure scalability issues do not compromise the generality of our results.

Algorithm 1. componentWiseDE template
___
1: Initialize(pop)
2: **repeat**
3:    Variate(pop)
4:    Reduce(pop)
5: **until** termination criteria met
Output: pop

Algorithm 2. DE variation	**Algorithm 3.** GA variation
Input: pop	**Input:** pop
1: **repeat**	1: pool ← Select(pop)
2: $trial$ ← DE_operator($target$)	2: pop$_{new}$ ← GA_operators(pool)
3: OnlineReplace(pop, $target$, $trial$)	3: pop ← pop ∪ pop$_{new}$
4: **until** #offspring produced	

The remainder of this paper is organized as follows. Section 2 presents our component-wise approach to differential evolution, and how we instantiate both DE-based and GA-based algorithms using a flexible template. Section 3 presents the intermediate algorithmic designs we use in this work to understand the contribution of the individual DE components we consider. The experimental setup used for this assessment is given in Section 4. We split the discussion of the results in two parts. In Section 5, we compare algorithms grouped by environmental selection strategy. In Section 6, we compare all algorithms among themselves and to a well-known efficient EMO algorithm, SMS-EMOA [1]. We do so to put the results in perspective, since we have recently shown that SMS-EMOA performs consistently well for the experimental setup considered here [2]. Finally, we conclude and discuss future work in Section 7.

2   Differential Evolution from a Component-Wise View

Several articles in the literature propose how to adapt DE algorithms to multi-objective optimization. However, the differences among most of these algorithms are quite small. From a very high-level perspective, multi-objective DE algorithms can be represented using the template defined by Algorithms 1 and 2. The general template displayed in Algorithm 1 could actually represent any of the most used evolutionary computation approaches (GA, DE or evolution strategies). Starting from an initial population (line 1), variation operators and environmental selection are applied to a population to promote evolution, until a given stopping criterion is reached.

In DE algorithms, the variation procedure is carried out as displayed in Algorithm 2. The DE operator produces a trial vector from an existing target vector of the population. Although the single-objective optimization literature presents many different strategies for this operation, the multi-objective DE algorithms proposed so far use the *DE/rand/1/bin* approach [15]. The most significant difference between the existing DE proposals is encapsulated in procedure OnlineReplace (line 3). In earlier algorithms, the trial vector x_{trial} only replaced the target vector x_{target} if x_{trial} dominated x_{target}. In this case, no environmental replacement is necessary, since the population size is always constant. Later, algorithms considered the option of adding the trial vector to the population in case both trial and target vectors were nondominated. In this case, the population size might double at each iteration, and hence environmental replacement strategies are employed after the variation is concluded, to reduce the population to its original size. While this prevents algorithms from early stagnation, it may as well slow down their convergence. We refer to these two replacement versions as *online replacement strategies*, since trial solutions may replace target solutions during the variation stage, before the actual population management represented by procedure Reduce happens. However, some multi-objective DE algorithms do not consider online replacement at all. In this case, solutions are created by the DE operator, but are only compared to the population altogether, when procedure Reduce is executed. These three different options for online solution replacement are listed in the bottom part of Table 1.

The three different DEMO versions considered by Tušar and Filipič [17] can be easily instantiated using this template as follows (all three versions use DE variation and (non)dominance online solution replacement):

DEMO^{NS-II} uses environmental selection strategy proposed for NSGA-II, i.e., nondominated sorting with tie-breaking according to crowdedness.

DEMO^{SP2} uses the environmental selection strategy proposed for SPEA2, i.e., sorting according to dominance strength and tie-breaking according to nearest neighbor density estimation.

DEMO^{IB} uses the environmental selection strategy proposed for IBEA, i.e., sorting according to the binary ϵ-indicator (I_ϵ).

In an analogous fashion, the original GA-based algorithms NSGA-II, SPEA2 and IBEA can be instantiated using the same template. To do so, instead of a DE-based variation, we use a traditional GA variation approach, outlined by Algorithm 3. The mating selection (line 1) is done according to the fitness of the individuals, which is computed using the same strategies adopted for the environmental replacement in the respective GA-based algorithms. Besides the previously discussed algorithms, the component-wise template presented here could also be used to instantiate other algorithms. We will discuss this in more detail in the next section.

3 Investigating Intermediate Designs

As explained in the previous section, the three original DEMO versions [17] comprise more than a single atomic DE-related algorithmic component. Concretely, it is a combination of the DE variation operator and an online replacement strategy. Although the DEMO versions of NSGA-II, SPEA2, and IBEA have indeed shown performance improvements over the original algorithms, it

Table 1. Algorithmic options of a component-wise multi-objective DE template

Component	Domain	Description
Variate	DE variation, GA variation	Underlying variation options
Reduce	NSGA-II, SPEA2, IBEA	Environmental selection approaches
OnlineReplace	dominance, (non)dominance none	Online solution replacement criterion (this component only takes effect when DE variation is used)

remains unclear how each of these individual components contribute to these performance gains. To properly assess the effectiveness of these components, we propose a set of intermediate algorithmic designs: $\mathbf{DE^{NS\text{-}II}}$, $\mathbf{DE^{SP2}}$, and $\mathbf{DE^{IB}}$ which are identical to the DEMO variants except that they do not use online solution replacement. Moreover, the only difference between these DE versions and the original versions of NSGA-II, SPEA2 and IBEA is the use of the DE variation operator. For instance, considering the case of NSGA-II, $\mathrm{DE^{NS\text{-}II}}$, and $\mathrm{DEMO^{NS\text{-}II}}$, the first uses traditional GA selection and variation, while the latter two use DE variation. However, while $\mathrm{DEMO^{NS\text{-}II}}$ may replace solutions as soon as they are created, $\mathrm{DE^{NS\text{-}II}}$ replaces solutions only at the environmental selection stage (procedure Reduce of Algorithm 1).

In the next section, we present the experimental setup in which we use these intermediate designs to properly investigate the effectiveness of the DE operators used by the different DEMO versions.

4   Experimental Setup

The benchmark sets we consider here include all unconstrained DTLZ [8] and WFG [9] functions (DTLZ1-7 and WFG1-9). Since both benchmark sets offer scalability as to the number of variables and objectives, we explore this feature to increase the representativeness of our investigation. We consider versions of these problems with three and five objectives. Concerning the number of variables n, we consider problems with $n \in \{20, 21, \ldots, 60\}$. Furthermore, to ensure that numerical parameters do not affect our performance assessment of the DE components, we initially tune all algorithms, but we use disjoint sets for tuning and testing to prevent overfitting. More precisely, we use problems with sizes $n_{testing} = \{30, 40, 50\}$ for testing, and problems with sizes $n \in \{20, 21, \ldots, 60\} \setminus n_{testing}$ for tuning. For both testing and tuning, experiments are run on a single core of Intel Xeon E5410 CPUs, running at 2.33GHz with 6MB of cache size under Cluster Rocks Linux version 6.0/CentOS 6.3. The remaining details about tuning and testing are given below.

Table 2. Parameter space for tuning all MOEAs for continuous optimization

Parameter	$\mu = \|\text{pop}\|$	$\lambda = \|\text{pop}_{\text{new}}\|$	t_{size}	p_c, p_m	η_c, η_m	CR	F
			GA variation			DE variation	
Domain	$\{10, 20, \ldots, 100\}$	1 or $\lambda_r \cdot \mu$ $\lambda_r \in [0.1, 2]$	$\{2, 4, 8\}$	$[0, 1]$	$\{1, 2, \ldots, 50\}$	$[0, 1]$	$[0.1, 2]$

4.1 Tuning Setup

The automatic parameter configuration tool we use in this work is irace [11]. Although it was originally proposed for configuring single-objective optimization algorithms, it can be adapted for multi-objective optimization by using the hypervolume indicator [12]. Concretely, for each problem considered by irace, candidate configurations are run for a maximum number of function evaluations (10 000, following [2]). The approximation fronts they produce are then normalized to the range $[1, 2]$ to prevent issues due to dissimilar domains. Finally, we compute the hypervolume for each front using $r_i = 2.1$, $i = 1, \ldots, M$ as reference point, where M is the number of objectives considered.

The parameter space we consider for tuning all algorithms is given in Table 2. Parameter μ applies to both DE-based and GA-based algorithms. The following six parameters (λ, t_{size}, p_c, p_m, η_c, η_m) only apply to GA-based algorithms. In particular, we highlight that all GA-based algorithms use SBX crossover and polynomial mutation, as commonly done in the literature [1, 8, 9]. Parameter t_{size} controls the size of the deterministic tournament used for mating selection. The probability of applying the crossover operator to a given pair of individuals is controlled by parameter p_c. Analogously, the probability of applying the mutation operator to a given individual is controlled by parameter p_m. In addition, we consider two different mutation schemes: (i) *bitwise*, which sets the mutation probability per variable $p_v = 1/n$; and (ii) *fixed*, where p_v becomes a parameter $\in [0.01, 1]$. Finally, η_c and η_m are the distribution indices for the SBX crossover and polynomial mutation, respectively. The remaining two parameters (CR and F) in Table 2 concern DE variation. They control the number of variables affected by the operator (parameter CR) and the strength of the changes (parameter F).

There are two additional parameters that concern only SPEA2 and IBEA. The original version of SPEA2 contains an additional parameter k for its k-th nearest neighborhood density estimation strategy in the mating selection. Here, besides the default value, which is computed according to the population size and we denote with $k_{\text{method}} = default$, we also give irace the possibility of configuring k directly, with $k \in \{1, 2, \ldots, 9\}$. For IBEA, as previously discussed, several different binary quality indicators can be used. Here we allow irace to select between the two most commonly adopted [19], the binary hypervolume indicator (I_H^-) and the binary ϵ-indicator (I_ϵ). Additionally, irace is given the flexibility to set different quality indicators for mating and for environmental selection if that leads the algorithm to better performance. Algorithms are tuned for each benchmark set (DTLZ or WFG) and for each number of objectives

(3 or 5); that is, for each algorithm X, we obtain four tuned variants: X_{D3}, X_{D5}, X_{W3} and X_{W5}. For brevity, the tuned settings for all algorithms considered in this work are provided as supplementary material [5].

4.2   Testing Setup

For comparing the tuned algorithms, we run each algorithm 25 times and evaluate them based on the relative hypervolume of the approximation fronts they produce w.r.t. the Pareto optimal fronts. Since the latter are typically infinite, we generate, for each problem instance, a Pareto front with 10 000 Pareto-optimal solutions following the methodology described in the papers where the benchmarks were proposed [8,9]. Given an approximation front A generated by an algorithm when applied to a problem instance and the Pareto front P of the same problem instance, the relative hypervolume of A equals $I_H(A)/I_H(P)$. A relative hypervolume of 1.0 means the algorithm was able to perfectly approximate the Pareto front for the problem considered.

The comparison is done visually by means of boxplots, and analytically through rank sums. Since we generate a large set of results, we only discuss the most representative ones here. In particular, many of the DTLZ problems have been found to be easy for EMO algorithms, creating a ceiling effect in the results. For this reason, we focus the discussion on the WFG benchmark and provide the analysis on the DTLZ benchmark as supplementary material [5]. Additionaly, due to the large amount of results we produce, we present here the results for $n = 40$. Similar results were found for $n \in \{30, 50\}$, and are also provided as supplementary material.

5   Experimental Analysis Grouped by Environmental Selection Strategy

To investigate how each algorithm component individually affects the performance of the different DEMO versions, we first conduct an analysis where algorithms are grouped by the environmental selection strategy they employ.

5.1   NSGA-II, DE^{NS-II}, and DEMO^{NS-II}

The boxplots of the relative hypervolume achieved by the algorithms that use the environmental selection strategy proposed for NSGA-II are given in Figures 1 and 2. For the 3-objective problems (Figure 1), we observe very heterogeneous results. For some problems such as WFG7 and WFG8 there is almost no difference between the algorithms, indicating that the DE components are unable to improve the performance of the original NSGA-II. However, for problems such as WFG1, WFG2, WFG4, and WFG6, the performance of NSGA-II can be improved, sometimes by a large margin, such as for WFG1 and WFG2. When we consider the effectiveness of the DE components, we see that sometimes using both components (as in DEMO^{NS-II}) is beneficial (e.g., WFG1, WFG5,

Table 3. Sum of ranks depicting the overall performance of algorithms grouped by environmental selection strategy. Algorithms in boldface present rank sums not significantly higher than the lowest ranked for a significance level of 95%.

3 objectives			5 objectives		
DEMO$^{\text{NS-II}}$w3 (**1259.5**)	**DE$^{\text{NS-II}}$w3** (**1321**)	NSGA-IIw3 (1469.5)	**DE$^{\text{NS-II}}$w5** (**1257**)	DEMO$^{\text{NS-II}}$w5 (1393)	NSGA-IIw5 (1400)
DEMO$^{\text{SP2}}$w3 (**1281**)	**SPEA2w3** (**1299.5**)	DE$^{\text{SP2}}$w3 (1469.5)	**DEMO$^{\text{SP2}}$w5** (**1259**)	**DE$^{\text{SP2}}$w5** (**1346.5**)	SPEA2w5 (1444.5)
DE^{IB}w3 (**1212**)	**DEMO^{IB}w3** (**1246.5**)	IBEAw3 (1591.5)	**DEMO^{IB}w5** (**1215.5**)	**DE^{IB}w5** (**1225.5**)	IBEAw5 (1609)

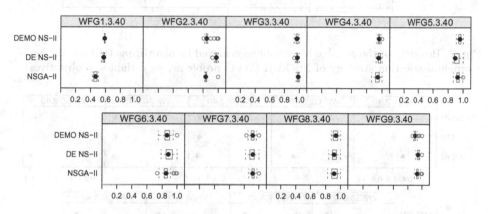

Fig. 1. Boxplots of the relative hypervolume achieved by algorithms that use the environmental selection strategy of NSGA-II (WFG problems, 40 variables, 3 objectives)

and WF8), but for other problems it is better to use the DE variation without the online replacement strategy as in DE$^{\text{NS-II}}$ (e.g., WFG2, WFG6, and WFG9). Particularly for WFG9, using both components simultaneously worsens the performance of NSGA-II. When we aggregate results for all runs and sizes of 3-objective WFG problems in a rank sum analysis (Table 3), we see that both DE-based algorithms improve over NSGA-II, but no significant difference can be found among DEMO$^{\text{NS-II}}$ and DE$^{\text{NS-II}}$ using Friedman's test at 95% confidence level.

The performance shown by NSGA-II, DE$^{\text{NS-II}}$, and DEMO$^{\text{NS-II}}$ on the 5-objective WFG problems (see Fig. 2) is quite different. This time, using both DE components (DEMO$^{\text{NS-II}}$) is only beneficial for problems WFG1, WFG4, WFG5, and WFG8. In the other problems, the online replacement leads to results worse even than the ones achieved by the original NSGA-II. However, when we consider only the DE variation (DE$^{\text{NS-II}}$), we see that the performance of NSGA-II is improved for most functions, except for WFG2 and WFG5. When we aggregate results for all 5-objective problems, we see that DE$^{\text{NS-II}}$ indeed ranks first, with significantly lower rank sums than the remaining algorithms (Table 3).

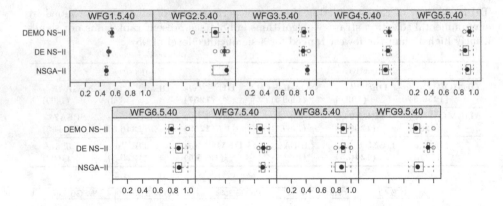

Fig. 2. Boxplots of the relative hypervolume achieved by algorithms that use the environmental selection strategy of NSGA-II (WFG problems, 40 variables, 5 objectives)

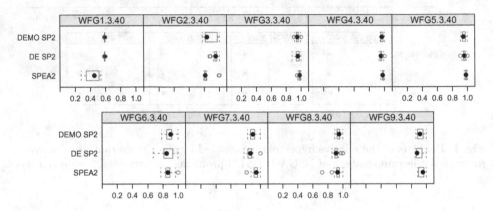

Fig. 3. Boxplots of the relative hypervolume achieved by algorithms that use the environmental selection strategy of SPEA2 (WFG problems, 40 variables, 3 objectives)

5.2   SPEA2, DE^{SP2}, and DEMO^{SP2}

The boxplots of the relative hypervolume achieved by the algorithms that use the environmental selection strategy proposed for SPEA2 are given in Figures 3 and 4. This time the 3-objective problems (Figure 3) show a more clear separation between problems for which DE components lead to improvements and problems for which they worsen the performance of the original SPEA2. For the first group (WFG1, WFG2, and WFG6), we see that there is no pattern as to whether the online replacement is a suitable component for improving SPEA2. However, for the problems where DE components do not lead to performance improvements, typically the version that uses online replacement (that is, DEMO^{SP2}) shows better results than the version that does not use it (that

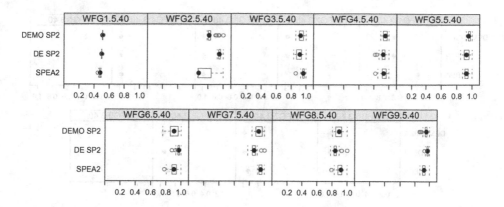

Fig. 4. Boxplots of the relative hypervolume achieved by algorithms that use the environmental selection strategy of SPEA2 (WFG problems, 40 variables, 5 objectives)

is, DE^{SP2}). When we aggregate results for all 3-objective problems, we see that SPEA2 and $DEMO^{SP2}$ show equivalent results, while DE^{SP2} shows significantly higher rank sums than both.

For the 5-objective WFG problems (see Figure 4), the online replacement component plays a more important role than in the 3-objective problems. For most problems, the performance of DE^{SP2} and $DEMO^{SP2}$ is quite different: while $DEMO^{SP2}$ outperforms SPEA2 for most problems, DE^{SP2} worsens the performance of SPEA2 for nearly half of the problems considered. The main exception is WFG2, where DE^{SP2} has the best performance among all algorithms. When all 5-objective problems are considered (Table 3), $DEMO^{SP2}$ ranks first with rank sums significantly lower than DE^{SP2} and SPEA2, which respectively rank second and third. Despite its erratic behavior, DE^{SP2} also presents significantly lower rank sums than SPEA2.

5.3 IBEA, DE^{IB}, and $DEMO^{IB}$

The boxplots of the relative hypervolume achieved by the algorithms that use the environmental selection strategy proposed for IBEA are given in Figures 5 and 6. The results for the 3-objective problems achieved by these indicator-based versions are far more homogeneous than the results shown before for NSGA-II and SPEA2 environmental selection strategies. In almost all situations, DE^{IB} and $DEMO^{IB}$ perform nearly identically. Moreover, the DE-based variants always outperform the GA-based version, except for problems WFG3–WFG5, where the original IBEA was already very effective. These results indicate that, for 3-objective problems, the online replacement component is not an effective component when combined with the indicator-based environmental selection strategy proposed by IBEA.

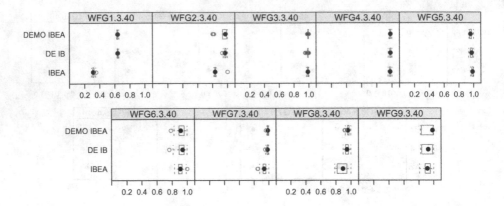

Fig. 5. Boxplots of the relative hypervolume achieved by algorithms that use the environmental selection strategy of IBEA (WFG problems, 40 variables, 3 objectives)

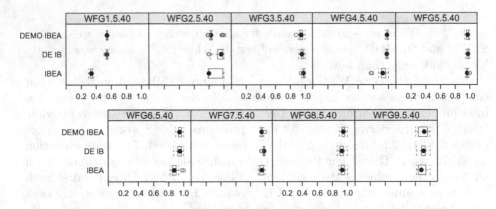

Fig. 6. Boxplots of the relative hypervolume achieved by algorithms that use the environmental selection strategy of IBEA (WFG problems, 40 variables, 5 objectives)

The results for the 5-objective problems (see Figure 6) are somehow consistent with the results on the 3-objective problems. However, on the 5-objective problems, online replacement leads to performance changes. For some problems, such as WFG2 and WFG7, DE^{IB} finds better results than $DEMO^{IB}$. The opposite happens for problems WFG8 and WFG9. When we aggregate across all problems (Table 3), we see that these two algorithms get nearly the same rank sum, and that IBEA gets significantly worse rank sums.

5.4   Overall Remarks

Overall, the DE operator leads algorithms to better results on problems WFG1, WFG2, WFG6, and WFG9. As common characteristics, WFG1 and WFG2 present convex geometry, WFG1 and WFG9 present some form of bias, and WFG6 and WFG9 present a complex non-separable reduction [9]. As for the online replacement component, the only problem for which we can say that it is beneficial is the WFG8 problem. However, since the DE operator typically worsens the performance of the original algorithms for this problem, we see that the online replacement is only weakening the effects of the DE operator. Although these results might seem to contradict the results presented by the authors of DEMO, we see that the environmental selection strategy from NSGA-II represents a special case here. DEMO$^{\text{NS-II}}$ in fact improves over DE$^{\text{NS-II}}$ and NSGA-II, particularly for functions where NSGA-II faces difficulties [9]. However, this is most likely explained by the poor performance of NSGA-II rather than by the effectiveness of the online replacement strategy.

6   Comparison to SMS-EMOA

In this section we compare all algorithms with SMS-EMOA. In a recent comparison using the same experimental setup, SMS-EMOA was found to be very effective for the benchmarks considered in this work [2].

For the 3-objective problems (Figure 7) we see that, in general, the DE-based algorithms are never clearly worse than SMS-EMOA, except for the WFG6 problem. Particularly for WFG1 and WFG2, the differential evolution operator leads to a significant performance improvement. However, the online replacement is not effective for these two problems regardless of the environmental selection strategy employed, and often worsens the performance of the algorithms. When we aggregate across all 3-objective problems considered (Table 4), we see that DE^{IB} and DEMO^{IB} achieve significantly lower rank sums than all other algorithms. DEMO$^{\text{SP2}}$ and SPEA2 rank second, along with SMS-EMOA. These results confirm that DE algorithmic components can indeed lead to significant performance improvements, but that the interactions between them and the environmental selection are also significant.

The comparison between all algorithms for 5-objective problems is given in Figure 8. This time the environmental selection strategy becomes very important for the effectiveness of the algorithms. As expected, dominance-based approaches (NSGA-II and SPEA2) are not as effective for many-objective scenarios, and hence even the DE versions of these algorithms are not able to perform as well as the indicator-based algorithms. However, the performance improvements provided by the DE variation to IBEA is such that both DE^{IB} and DEMO^{IB} become the top-performing algorithms, even though IBEA itself did not perform as competitively as SMS-EMOA. These results indicate that, if coupled with proper many-objective search mechanisms, DE algorithmic components can possibly improve state-of-the-art algorithms, such as SMS-EMOA.

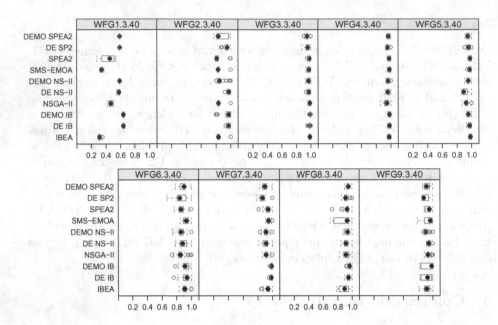

Fig. 7. Relative hypervolume boxplots: 3-objective WFG problems with 40 variables

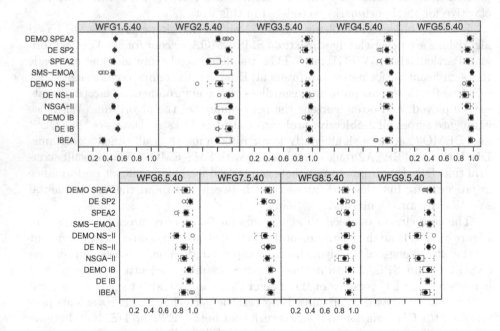

Fig. 8. Relative hypervolume boxplots: 5-objective WFG problems with 40 variables

Table 4. Sum of ranks depicting the overall performance of all algorithms. ΔR is the critical rank sum difference for Friedman's test with 95% confidence. Algorithms in boldface present rank sums not significantly higher than the lowest ranked.

3 objectives ($\Delta R = 271$)	5 objectives ($\Delta R = 265$)
DE^{IB}w3 (2532)	**DE^{IB}w5 (2493)**
DEMO^{IB}w3 (2535)	**DEMO^{IB}w5 (2506)**
DEMO^{SP2}w3 (3738.5)	SMS-EMOAw5 (2891.5)
SPEA2w3 (3764.5)	IBEAw5 (3930)
SMS-EMOAw3 (3798)	DEMO^{SP2}w3 (3932.5)
IBEAw3 (3924)	DE^{NS-II}w5 (4089)
DEMO^{NS-II}w3 (3972.5)	DE^{SP2}w5 (4123.5)
DE^{NS-II}w3 (4094.5)	SPEA2w5 (4271.5)
DE^{SP2}w3 (4325.5)	DEMO^{NS-II}w5 (4426.5)
NSGA-IIw3 (4440.5)	NSGA-IIw5 (4461)

7   Conclusions

This paper has examined how the individual components of DE interact with the components of various EMO algorithms. In particular, we studied the underlying variation operator (GA or DE), the environmental selection strategy (NSGA-II, SPEA2, or IBEA), and the use of an online replacement strategy. For the DTLZ benchmark, results presented a ceiling effect, and hence we focused our analysis on the WFG benchmark. For both three or five objectives, results showed that the DE-operator improves the algorithms in most problems and that there is a strong interaction between this component and environmental selection. However, for the online replacement component, results almost always indicated that this component is not effective, except when combined with NSGA-II environmental selection.

These results represent a significant contribution of our investigation. Before our work, it was believed that the online replacement component was critical to the effectiveness of multi-objective DE algorithms [14]. Furthermore, this result reinforces the value of the component-wise design approach [2], which advocates that components should be jointly investigated to account for interactions. In fact, the component-wise design of effective DE-based algorithms is an important next step for this research. Here, we have shown that, when coupled with the environmental selection strategy from IBEA and used with numerical parameters properly tuned, a very effective algorithm can be devised. Concretely, this DEMO^{IB} algorithm has consistently outperformed SMS-EMOA, an algorithm that was recently shown to be very effective on the benchmarks considered here. It is then natural to envision the possibility of designing even more effective algorithms if a large set of components is considered, as in the automatic component-wise design methodology [2,12].

Acknowledgments. The research presented in this paper has received funding from the COMEX project within the Interuniversity Attraction Poles Programme of the Belgian Science Policy Office. L.C.T. Bezerra, M. López-Ibáñez and T. Stützle acknowledge support from the Belgian F.R.S.-FNRS, of which they are a FRIA doctoral fellow, a postdoctoral researcher and a senior research associate, respectively.

References

1. Beume, N., Naujoks, B., Emmerich, M.: SMS-EMOA: Multiobjective selection based on dominated hypervolume. Eur. J. Oper. Res. **181**(3), 1653–1669 (2007)
2. Bezerra, L.C.T., López-Ibáñez, M., Stützle, T.: Automatic component-wise design of multi-objective evolutionary algorithms. Tech. Rep. TR/IRIDIA/2014-012, IRIDIA, Université Libre de Bruxelles, Belgium, Brussels (2014)
3. Bezerra, L.C.T., López-Ibáñez, M., Stützle, T.: Automatic design of evolutionary algorithms for multi-objective combinatorial optimization. In: Bartz-Beielstein, T., Branke, J., Filipič, B., Smith, J. (eds.) PPSN 2014. LNCS, vol. 8672, pp. 508–517. Springer, Heidelberg (2014)
4. Bezerra, L.C.T., López-Ibáñez, M., Stützle, T.: Deconstructing multi-objective evolutionary algorithms: An iterative analysis on the permutation flowshop. In: Pardalos, P.M., Resende, M.G.C., Vogiatzis, C., Walteros, J.L. (eds.) LION 2014. LNCS, vol. 8426, pp. 57–172. Springer, Heidelberg (2014)
5. Bezerra, L.C.T., López-Ibáñez, M., Stützle, T.: To DE or not to DE? Multi-objective differential evolution revisited from a component-wise perspective, (2015). http://iridia.ulb.ac.be/supp/IridiaSupp2015-001/
6. Das, S., Suganthan, P.N.: Differential evolution: a survey of the state-of-the-art. IEEE Trans. Evol. Comput. **15**(1), (2011)
7. Deb, K., Pratap, A., Agarwal, S., Meyarivan, T.: A fast and elitist multi-objective genetic algorithm: NSGA-II. IEEE Trans. Evol. Comput. **6**(2), 182–197 (2002)
8. Deb, K., Thiele, L., Laumanns, M., Zitzler, E.: Scalable test problems for evolutionary multiobjective optimization. In: Abraham, A., et al. (eds.) Evolutionary Multiobjective Optimization, pp. 105–145. Advanced Information and Knowledge Processing, Springer, London (2005)
9. Huband, S., Hingston, P., Barone, L., While, L.: A review of multiobjective test problems and a scalable test problem toolkit. IEEE Trans. Evol. Comput. **10**(5), 477–506 (2006)
10. Kukkonen, S., Lampinen, J.: GDE3: the third evolution step of generalized differential evolution. In: IEEE CEC, pp. 443–450. IEEE Press (2005)
11. López-Ibáñez, M., Dubois-Lacoste, J., Stützle, T., Birattari, M.: The irace package, iterated race for automatic algorithm configuration. Tech. Rep. TR/IRIDIA/2011-004, IRIDIA, Université Libre de Bruxelles, Belgium (2011)
12. López-Ibáñez, M., Stützle, T.: The automatic design of multi-objective ant colony optimization algorithms. IEEE Trans. Evol. Comput. **16**(6), 861–875 (2012)
13. Price, K., Storn, R.M., Lampinen, J.A.: Differential Evolution: A Practical Approach to Global Optimization. Springer, New York (2005)
14. Robič, T., Filipič, B.: DEMO: Differential evolution for multiobjective optimization. In: Coello Coello, C.A., Hernández Aguirre, A., Zitzler, E. (eds.) EMO 2005. LNCS, vol. 3410, pp. 520–533. Springer, Heidelberg (2005)
15. Storn, R., Price, K.: Differential evolution - a simple and efficient heuristic for global optimization over continuous spaces. J. Glob. Optim. **11**(4), 341–359 (1997)
16. Tušar, T.: Design of an Algorithm for Multiobjective Optimization with Differential Evolution. M.sc. thesis, Faculty of Computer and Information Science, University of Ljubljana (2007)

17. Tušar, T., Filipič, B.: Differential evolution versus genetic algorithms in multiobjective optimization. In: Obayashi, S., Deb, K., Poloni, C., Hiroyasu, T., Murata, T. (eds.) EMO 2007. LNCS, vol. 4403, pp. 257–271. Springer, Heidelberg (2007)
18. Zhang, Q., Suganthan, P.N.: Special session on performance assessment of multiobjective optimization algorithms/CEC 2009 MOEA competition, (2009). http://dces.essex.ac.uk/staff/qzhang/moeacompetition09.htm
19. Zitzler, E., Künzli, S.: Indicator-based selection in multiobjective search. In: Yao, X., Burke, E.K., Lozano, J.A., Smith, J., Merelo-Guervós, J.J., Bullinaria, J.A., Rowe, J.E., Tiňo, P., Kabán, A., Schwefel, H.-P. (eds.) PPSN 2004. LNCS, vol. 3242, pp. 832–842. Springer, Heidelberg (2004)
20. Zitzler, E., Laumanns, M., Thiele, L.: SPEA2: Improving the strength pareto evolutionary algorithm for multiobjective optimization. In: Giannakoglou, K.C., et al. (eds.) EUROGEN, pp. 95–100. CIMNE, Barcelona (2002)

Model-Based Multi-objective Optimization: Taxonomy, Multi-Point Proposal, Toolbox and Benchmark

Daniel Horn[1]([✉]), Tobias Wagner[2], Dirk Biermann[2],
Claus Weihs[1], and Bernd Bischl[1]

[1] Chair of Computational Statistics, Technische Universität Dortmund,
Vogelpothsweg 87, 44227 Dortmund, Germany
{dhorn,weihs,bischl}@statistik.uni-dortmund.de
[2] Institute of Machining Technology (ISF), Technische Universität Dortmund,
Baroper Str. 303, 44227 Dortmund, Germany
{wagner,biermann}@isf.de

Abstract. Within the last 10 years, many model-based multi-objective
optimization algorithms have been proposed. In this paper, a taxonomy
of these algorithms is derived. It is shown which contributions were made
to which phase of the MBMO process. A special attention is given to the
proposal of a set of points for parallel evaluation within a batch. Proposals for four different MBMO algorithms are presented and compared
to their sequential variants within a comprehensive benchmark. In particular for the classic ParEGO algorithm, significant improvements are
obtained. The implementations of all algorithm variants are organized
according to the taxonomy and are shared in the open-source R package
mlrMBO.

Keywords: Expected improvement · Hypervolume · Kriging · Performance indicator · Surrogate model

1 Introduction

In recent years, the use of surrogate models for partly replacing the actual objective function allowed multi-objective optimization techniques to be applied to real-
world problems in an efficient way [16]. The resulting combinations of surrogate
models and optimization algorithms are denoted as model-based multi-objective

We acknowledge partial support by the Mercator Research Center Ruhr under grant
Pr-2013-0015 *Support-Vektor-Maschinen für extrem große Datenmengen*, by the Ger-
man Research Foundation (DFG) within the Collaborative Research Center SFB
823 *Statistical modelling of nonlinear dynamic processes*, project C2, and within
the Collaborative Research Center SFB 708 *3D-Surface Engineering*, project C4. In
addition, the authors acknowledge support by the French national research agency
(ANR) within the Modèles Numérique project NumBBO – Analysis, Improvement
and Evaluation of Numerical Blackbox Optimizers.

A. Gaspar-Cunha et al. (Eds.): EMO 2015, Part I, LNCS 9018, pp. 64–78, 2015.
DOI: 10.1007/978-3-319-15934-8_5

optimization (MBMO) algorithms in the following. In the early algorithms, surrogate models have been fitted, and have then been used for the optimization in replacement of the actual objective functions. No sequential update has been performed. If a validation is performed at all, only the finally selected solution has been evaluated on the actual problem.

Since 2005, sequential approaches – using the surrogate to decide on new points to evaluate and update the model in an iterative fashion – have been proposed. Most of these approaches are based on ideas of the popular Efficient Global Optimization (EGO) procedure [13]. Early work in the multi-objective scenario has either scalarized the objectives [15] to allow EGO to be directly used or has optimized EGO's figure of merit for different models in parallel using MOEA [11,19]. Later, also set-based improvement criteria, specifically designed for multi-objective optimization, have been defined [1,9,14,18,23]. Until now, the algorithms as a whole were considered as a contribution to the field of MBMO. In order to better distinguish the actual contributions, a first taxonomy of existing MBMO approaches is introduced in this paper.

Due to the enormous growth of parallel computing power and the advantages of performing real experiments in batches, allowing more than one point to be proposed per iteration (batch processing) is of great interest. Right now, only one multi-objective approach exists [23] (see [3] for a comparison of methods and a new approach in the single-objective case). As a consequence, possibilities to integrate batch proposals into existing MBMO algorithms are proposed in the paper. In particular for set-based improvement criteria in MBMO, this is done for the first time, to the best of our knowledge.

The taxonomy is introduced in section 2. In section 3, it is shown how the existing algorithms can be classified using the concepts of the taxonomy. The ideas for allowing a batch proposal within specific algorithm classes are proposed in section 4. All covered algorithms are integrated into the R toolbox mlrMBO for model-based optimization (MBO), whose software design closely reflects the presented taxonomy. The toolbox is briefly presented in section 5. The MBMO algorithms are compared on a comprehensive benchmark, which is described and evaluated in section 6. The paper is concluded by a summary of the results and an outlook on possible further improvements.

2   Taxonomy

The taxonomy of the MBMO approaches is based on the standard procedure of a sequential MBO algorithm, whose phases are shown on the left of Fig. 1. First, an initial design is evaluated on the actual, expensive objective function in order to train the surrogate model. In principle, all available design-of-experiment (DOE) techniques can be used. Due to its connection to the established Kriging models [13], however, Latin Hypercube Sampling (LHS) is applied in almost all existent MBMO approaches, and hence explicitly mentioned as an option.

For model fitting, two approaches are established. In the straightforward variant, an individual surrogate model is built for each objective function. In order to

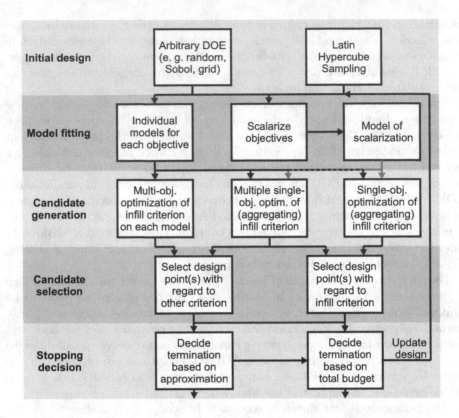

Fig. 1. Phases and tasks within a generalized MBMO algorithm

allow established single-objective model-based optimization criteria [12], such as the expected improvement (EI), the probability of improvement (PI) or the lower confidence bound (LCB), to be directly used, the objectives can be scalarized before the surrogate model is fitted. In this case, the multi-objective problem is effectively reduced to a single-objective one. To still obtain an approximation of the complete Pareto frontier, the parameterization or type of the scalarization function can be varied over the iterations of the MBMO algorithm.

The candidate generation represents the step where most of the contributions have been made. In case of a single model, which predicts the value of a scalarization of the objectives, established criteria for generating candidates can be used [3,10,12]. If individual models for each objective are available, three different strategies can be pursued. Due to the current focus on single-point proposals and set-based multi-objective optimization, mainly criteria for an internal single-objective optimization of an aggregating infill criterion on the model [21] have been proposed within recent work. Also specific algorithms for performing the internal optimization have been designed [20]. A single optimum solution is found, which is then evaluated in order to update the training set for the model.

If a batch of solutions is desired, two alternative options can be used. In the first one, the internal optimization is performed by an MOEA, which operates on single-objective infill criteria for each model. The final Pareto front approximation then provides the candidates for the batch processing. In the second variant, different single-objective subproblems (e.g. scalarizations) are optimized in parallel based on an established infill criterion. By collecting the respective optimum solutions, a batch of candidates is compiled. This approach has only been performed within a single algorithm [23] until now.

In the last step, the candidate set is reduced to the desired size. It is thus only required in case of a multi-point proposal in the candidate generation step. As the outcome of the MOEA can be filtered to obtain a mutually nondominated set, another, aggregating infill criterion has to be chosen for the selection if the number of solutions exceeds the desired one. Compared to the direct optimization of this criterion, the MOEA allows multiple points to be found, improves exploration of the search space and prevents effects of oversearching. On the other hand, it may result in suboptimal solutions with regard to the final selection criterion. If more subproblems than the desired batch size have been internally optimized, potentially two selection approaches can be used. The former uses a similar, global variant of the internal criteria within in the subproblems, whereas the latter decides based on a completely different, e.g., space-filling, explorative criterion. Both approaches improve exploration while also retaining the optimality, at least with regard to the defined subproblems.

After each iteration, it is checked whether the optimization process can be terminated. This decision is usually based on a budget of evaluations fixed beforehand. Recently, however, a new method to estimate the uncertainty of the current Pareto front approximation has been proposed [2]. In case the desired approximation quality is obtained before the total budget is spent, the remaining evaluations can be skipped to save expensive resources.

3   Considered MBMO Algorithms

In this subsection, the algorithms considered in the following benchmark study are described as instantiations of the taxonomy. We only omitted algorithms applying complex and tedious indicator-based improvement criteria (cf. paragraph on Direct Indicator-Based MBMO), as recent studies have shown that conceptually similar (with regard to the taxonomy), but computationally cheaper variants, provide a comparable or even better performance [20]. By these means, the generality of the taxonomy is demonstrated. A summary of all approaches and their classification is provided in Table 1. All algorithms optimize a function $f : \mathbf{R}^d \to \mathbf{R}^m$, where d is the decision space dimension and m the number of objective functions. For specific evaluation, $f(x) = y$ holds.

Scalarization-Based MBMO. Two scalarization-based MBMO algorithms using the augmented Tchebycheff norm

$$u(x) = -\max\left[w(f(x) - i)\right] + \rho w^T (f(x) - i) \tag{1}$$

Table 1. Summary of the approaches from literature considered in the benchmark

Algorithm	Initial design	Model fitting	Candidate Generation	Candidate selection	Stopping decision
ParEGO [15]	LHS	Model of scalarization	single-objective optimization of EI	One point, same criterion	Total budget
MOEA/D-EGO [23]	LHS	Models for each objective	Multiple single-objective optimizations of scalarizations	Multi point, same criterion (on clusters of subproblems)	Total budget
Multi-EGO [11]	LHS	Models for each objective	Multi-objective optimization of individual EI	Multi point, space-filling selection	Total budget
MOEA using Surrogates [19]	Other DOE (Sobol)	Models for each objective	Multi-objective optimization of model prediction	Multi point, space-filling selection	Total budget
MSPOT [22]	LHS	Models for each objective	Multi-objective optimization of model prediction	One point, hypervolume contribution	Total budget
SMS-EGO [18]	LHS	Models for each objective	Single-objective optimization of the hypervolume contribution	One point, same criterion	Total budget
ε-EGO [20]	LHS	Models for each objective	Single-objective optimization of the additive ε-indicator	One point, same criterion	Total budget

with ideal point i and weight vector w ($\sum_{j=1}^m w_j = 1$) do exist, which differ in model fitting, candidate generation, and candidate selection. ParEGO [15] randomly chooses w from a uniformly distributed set in each iteration. The surrogate model is fitted to the respective scalarization, and the EI is optimized on this model. Only the optimum solution is evaluated on the actual problem. In contrast, MOEA/D-EGO [23] fits models for each objective. In the internal optimization, the EI for all weight vectors is maximized in parallel, and finally the solutions obtaining the highest EI within N predefined weight vector clusters are evaluated. As a consequence, the distribution of chosen solutions with respect to the corresponding w can suffer a bias towards balanced components [7].

Pareto-Based MBMO. The algorithms summarized under the term Pareto-based MBMO are using a multi-objective optimization of infill criteria on each objective in order to obtain a candidate set for evaluation. In Multi-EGO [11], the EI is used, whereas MSPOT [22] or Voutchkov's and Keane's surrogate-based MOEA [19] directly optimize the model predictions $\hat{y}(x)$. The final selection from the Pareto front approximation is either distance- (Multi-EGO, Surrogate-based MOEA) or indicator-based (MSPOT).

Direct Indicator-Based MBMO. In indicator-based MBMO algorithms, the contribution of an additional point to the indicator value of the current Pareto front approximation $\mathcal{Y}^*_{\text{approx}}$ is formulated as a single-objective criterion for the internal optimization. In the literature, two approaches can be distinguished. The first ones directly evaluate $\hat{y}(x)$ or simple combinations with the associated uncertainty $\hat{s}(x)$, such as the LCB $l(x) = \hat{y}(x) + \lambda\hat{s}(x)$. The algorithms are denoted as direct indicator-based (DIB) approaches. Examples are the SMS-EGO [18] for the hypervolume and the ε-EGO [20] for the additive ε-indicator. Whereas the former is enhanced by a check for nondominance based on additive ε-dominance ($\preceq_\varepsilon$), i.e., nondominance by an additional gap of ε, and a respective penalty $\Psi(x) = \max_{\{y^{(i)} \in \mathcal{Y}^*_{\text{approx}} | y^{(i)} \preceq_\varepsilon l(x)\}} -1 + \prod_{j=1}^m \left(1 + \max(l_j(x) - y_j^{(i)}, 0)\right)$, the latter only uses the respective indicator.

In addition, also more complex indicator-based infill criteria have been proposed. Their criteria analytically compute the expected improvement of the respective indicator by tediously integrating over the objective space [1,9]. Despite improvements regarding their complexity [8], these indicators are hard to implement. They are thus excluded from this benchmark study.

4   Batch Proposal for Parallel Evaluation

The structure implied by the taxonomy allows the realization of single phases to be easily replaced. This was used to propose N points for a batch evaluation within different MBMO algorithms originally designed for a single-point setup. To accomplish this, the candidate generation and selection steps of these algorithms were modified.

4.1   ParEGO

ParEGO was enhanced to a multi-point proposal by increasing the number of weight vectors randomly drawn in each iteration. If N points are desired, cN ($c > 1$) weight vectors are selected. Then, the pairwise distance between all weight vectors is calculated, and one vector of the pair resulting in the minimum distance is eliminated. This procedure is repeated until the set is reduced to the desired size. This greedy reduction of the larger set ensures that the selected weights cover the weight space in an almost uniform way.

In the following, the scalarizations implied by each weight vector are computed and individual models for each scalarization are fitted and optimized with respect to a single-objective infill criterion. The respective optima of each model build the batch to be evaluated. As the fitting and optimization of each model are mutually independent, they can be computed in parallel.

4.2   Pareto-Based MBMO

For the Pareto-based MBMO algorithms, the candidate generation by means of an internal multi-objective optimization already produces enough candidates for the batch evaluation. In the algorithms relying on a distance-based selection [11,19], a multi-point proposal is already realized. Hence, particularly the indicator-based candidate selection of MSPOT was enhanced to a multi-point proposal. To accomplish this, a greedy selection was used. Until the number of desired candidates for the batch evaluation is reached, the point of the candidate set having the highest contribution to the indicator is selected, added to the Pareto front approximation, and the contributions of the remaining points is updated. Consequently, the advantage of the multi-objective candidate generation to produce a set instead of single points is not only used for improving the exploration of the decision space, but also for obtaining a well-spread batch of solutions.

4.3   Direct Indicator-Based MBMO

For integrating a multi-point proposal within SMS- and ε-EGO, the concept of simulated evaluations was used. The optimization of the respective infill criterion is performed in its standard way, but the optimum solution is not directly evaluated on the actual, expensive problem. Instead, the LCB $l(x^*)$ of the optimum solution x^* is added to the current Pareto front approximation without refitting the model. Based on the updated approximation, the criterion is optimized again, and the procedure is repeated until N points for a batch evaluation have been found. As the contribution to the indicator in the vicinity of the simulated point vanishes, particularly due to the optimistic bias implied by the LCB, it is likely that the following optimization will focus on different areas of the objective space. Hence, a batch of solutions distributed over the Pareto front is expected.

5   The mlrMBO R Software Package

The mlrMBO package [4] is based on the mlr package for machine learning in R [6]. It is designed as an encompassing toolbox for general MBO techniques, including single- and multi-objective, as well as single- and multi-point methods. Not only Kriging can be used as a surrogate model, but every regression method integrated into mlr. In the single-objective case, the package allows the optimization of mixed decision spaces, including integer, categorical and dependent parameters[1]. Extensive logging into a well structured archive enables the post-hoc inspection of runs. The archive contains the Pareto front and set, as well as all evaluations made in optimization process. By these means, visualizations of the runs are possible (at least for bi-objective problems). This is useful for a deeper understanding of algorithmic aspects in order to derive potential improvements. As real-world runs on, e.g., complicated simulators, often introduce technical problems, the package contains various error-handling mechanisms.

The setup of an MBMO algorithm by means of the toolbox is done by special control objects which closely follow the structure of the taxonomy. The supplementary material to this paper [5] includes a simple and documented example.

---

[1] We plan to soon provide this feature also in the multi-objective case.

6 Experiments

The improvements obtained by the proposed contributions are evaluated by means of a comprehensive benchmark. To focus on specific results, our expectations are first formulated as research hypotheses. Then, the design of the experimental study is described. In the main part of this section, the hypotheses are checked using statistical testing and the respective observations are discussed.

Page limitations restrict the evaluation to the main hypotheses. The complete source code of the experimental study, tables including all indicator values, convergence plots, as well as empirical attainment surfaces on the bi-objective problems can be found in the supplementary material [5]. In order to exploit the full information provided by the benchmark, we strongly recommend to take this material into account.

6.1 Research Hypotheses

Within this paper, the benchmark results are analyzed with regard to three research hypotheses:

1. MBMO can significantly improve the approximation quality compared to model-free approaches in case of a strictly restricted budget of evaluations.
2. Compared to a single-point proposal, a multi-point proposal can significantly reduce clock time and preparation effort while not significantly deteriorating the results with regard to the budget of evaluations.
3. The structure of mlrMBO implied by the taxonomy allows the realizations of specific steps of the algorithm to be exchanged, benchmarked, and finally improved in a simple and efficient way.

In addition to the new candidate selection methods for the multi-point proposal, the last hypothesis is tested by exchanging the infill criteria for the candidate generation and selection in ParEGO and MSPOT.

6.2 Experimental Setup

Algorithms. ParEGO, SMS-EGO, ε-EGO, and MSPOT were implemented using the mlrMBO toolbox. Hence, all different classes of MBMO algorithms (Pareto-, scalarization-, and indicator-based) are covered. As MOEA/D-EGO applies more complex candidate generation and selection phases, and hence would result in additional implementation and space requirements, it is omitted within these experiments. Multi-EGO and MOEA using surrogates are adressed by considering their alternative infill criterion within MSPOT.

The initial design size of the algorithms was set to $n_{\text{init}} = 4d$. Kriging models were fitted with a Matern5/2 kernel. A total budget of $n_{\text{total}} = 40d$ was allowed, resulting in $36d$ points proposed over the iterations. The small n_{init} was chosen intentionally in order to have a high number of sequential evaluations while still operating under a severely restricted total number of evaluations.

In ParEGO, ρ of equation 1 was set to $\rho = 0.05$. The number of uniform steps used for generating the weight vectors was adjusted in a way that approximately $100,000$ weight vectors result in total. The ideal point i was estimated using the minimum objective values of the currently seen observations.

In SMS-EGO, the gap of the additive ε-dominance was estimated using the adaptive formula

$$\varepsilon = \frac{\Delta \mathcal{Y}^*_{\text{approx}}}{|\mathcal{Y}^*_{\text{approx}}| + c \cdot (n_{\text{total}} - n)}, \quad \Delta \mathcal{Y}^*_{\text{approx}} = \max(\mathcal{Y}^*_{\text{approx}}) - \min(\mathcal{Y}^*_{\text{approx}}),$$

where n is the current number of evaluations and $c = 1 - 1/(2^m)$ corresponds to the idealized probability of a random solution being non-dominated. min and max are vectorized operations, i. e, the minimum (maximum) for each dimension is returned. $|\mathcal{Y}^*_{\text{approx}}|$ denotes the number of observation in $|\mathcal{Y}^*_{\text{approx}}|$. As reference point for the hypervolume computations, $r = \max(\mathcal{Y}^*_{\text{approx}}) + \mathbf{1}$ was used.

For the evaluation of the first hypothesis, all considered MBMO algorithms are tested against NSGA-II and random search. NSGA-II was taken from the R package MCO and was run with a population size $P = n_{\text{init}}$ for 10 generations. This allows a direct comparison to the MBMO algorithms. As variation operators, simulated binary crossover (SBX) and polynomial mutation (PM) are applied with their standard parameters $p_c = 1$, $\eta_c = 15$, $p_m = \frac{1}{d}$, and $\eta_m = 20$. Random search acts as a baseline. It starts with the same initial design as the MBMO algorithms and randomly proposes the remaining points.

The second hypothesis is analyzed by implementing the candidate generation and selection concepts of section 4 into mlrMBO. The number of points in a batch was set to $N = 4$. To achieve a balanced set of weight vectors in parallel ParEGO, $cN = 20$ ($c = 5$) weight vectors were randomly drawn and reduced using the distance-based filter. As a consequence of the batch evaluation, only $9d$ iterations of the sequential procedure were performed.

As examples for investigating the third hypothesis, also the LCB was considered as infill criterion for optimizing the model of the scalarization within ParEGO. In addition, the multi-objective optimization for generating the candidates in MSPOT was also performed based on the EI and the LCB. As in the direct indicator-based (DIB) MBMO, the factor λ of the LCB was computed based on a given probability level p ($p = 0.5$ in this study) by $\lambda = -\Phi^{-1}(0.5 \sqrt[m]{p})$.

Due to a full factorial combination of infill criteria and the single- and multi-point candidate selection, in total 4 variants of ParEGO and 6 variants of MSPOT were considered. For SMS-EGO and ε-EGO, one single- and one multi-point variant were assessed, respectively. Hence, 14 MBMO instantiations were benchmarked. For all algorithms, including NSGA-II and random search, 20 runs were performed. All runs with the same index were based on the same initial design, except for NSGA-II which used a random initial population for technical reasons.

The internal single-objective optimization tasks were solved using a focusing random search. It performs large random searches on the decision space, which can be evaluated in parallel to reduce technical overhead when querying the machine learning model, and iteratively shrinks the boundaries of the

Table 2. Test functions designed by combining global optimization problems

Name	d	m	Internal test functions
gomop-22	2	2	Branin, 3-Hump-Camel ($x \in [-2, 2]^2$)
gomop-25	2	5	Branin, 3-Hump-Camel ($x \in [-2, 2]^2$), Hartman,
			Goldstein-Price, 6-Hump-Camel ($x_1 \in [-2, 2]$, $x_2 \in [-1, 1]$)
gomop-52	5	2	Hartman, Rastrigin ($x \in [-0.5, 0.5]^5]$)
gomop-55	5	5	Hartman, Rastrigin ($x \in [-0.5, 0.5]^5$), Rosenbrock,
			Zahkharov ($x \in [-1, 1]^5$), Powell ($x \in [-1, 1]^5$)

sample space around the best obtained point by a factor of 0.5, enforcing local convergence. Additionally, restarts of the whole approach were performed, for a further global optimization effect. In the experiments, a random set of 1,000 points is evaluated within each of the three focusing steps and three restarts are performed, resulting in total in 9,000 evaluations for each internal optimization.

For the multi-objective optimization in MSPOT, again the NSGA-II was applied. For the internal optimization, the population size 100 and 90 generations were specified in order to also allow 9000 evaluations of the surrogate models.

Test Functions. All algorithms were evaluated on 9 test functions. Two settings, ($d = 2$, $d = 5$) and ($m = 2$, $m = 5$), of both, decision and objective space, were considered, respectively. As established test functions, zdt1, zdt2, and zdt3 with $d = 5$ decision and $m = 2$ objective space dimensions, as well as dtlz1 with $d = 5$ and both $m = 2$ and $m = 5$, were used. In addition, the concept of combined multi-objective problems from single-objective problems [17,20] was utilized in order to design 4 additional test functions. These test functions are based on established global optimization functions and are summarized in Table 2[2]. In order to unify the box constraints of the decision spaces, the respective bounds of each single-objective test function were mapped to $[0, 1]^d$.

Performance Assessment. The final Pareto front approximations of the algorithms were compared using three performance indicators: R2, hypervolume, and additive ε [24]. The R2 and the hypervolume indicator were used in their unary variant. Hence, the ε and R2 indicators have to be minimized, whereas the hypervolume has to be maximized.

For each test function, the reference sets for the binary ε-indicator were built from the Pareto-optimal solutions of the union of all available Pareto front approximations. All approximations and reference sets are normalized to the interval $[1, 2]^m$ with respect to the ideal and nadir points given in table 3 before computing the indicators.

All indicators are recommended for performance assessment based on their favorable theoretical properties [24]. As we mainly compare algorithm variants

---

[2] For further information: http://www.sfu.ca/~ssurjano/optimization.html

Table 3. Nadir and ideal points for each test function

	gomop-22	gomop-25	gomop-52	gomop-55
Ideal	(0, 0)	(0, -5, 1, 0, -1.1)	(-3.5, 35)	(-3.5, 8.5, 35, 0, 0)
Nadir	(40, 2.5)	(125, 0, 15, 6, 3.1)	(0, 125)	(0, $3 \cdot 10^6$, 150, 2000, 350)
	dtlz2-52	dtlz2-55	zdt1-52	zdt2-52	zdt3-52
Ideal	(0, 0)	(0, 0, 0, 0, 0)	(0, 0)	(0, 0)	(0, -1)
Nadir	(2, 2)	(1.25, 1.25, 1.25, 1.25, 1.25)	(1, 10)	(1, 10)	(1, 10)

within their respective MBMO class to check our hypotheses, only the metric corresponding to the internal selection mechanism of the respective MBMO class is shown in the result tables.

6.3 Observations

The results of the experiments are summarized in Tables 4, 5, and 6. Significant improvements ($p = 0.05$) to the baseline algorithms with respect to independent pairwise Wilcoxon tests are indicated by subscripts (r random search, n NSGA-II). In addition, superscripts are added in order to provide information regarding the comparison of the multi-point variants with their original counterpart shown in the left column of each table. $^+$ means no significant deterioration, whereas $^{++}$ corresponds to a significant improvement.

Hypothesis 1. Random search and NSGA-II were outperformed by almost all MBMO algorithms on almost all test functions. The use of kriging models can thus drastically reduce the number of evaluations required to solve multi-objective optimization problems. Surprisingly, the original ParEGO (1-ei) was not able to outperform these baselines on 4 test functions.

Hypothesis 2. The second hypothesis has to be considered separately for the different algorithms. For ε-EGO (cf. Table 4, left), a significant deterioration of the multi-point compared to the single-point variant was observed on only one test function. Hence, the simulated evaluation strategy can be applied to reduce clock time and preparation effort without a significant loss of approximation

Table 4. Results of the indicator-based EGO variants with regard to their indicator

	dib-1-eps	dib-4-eps	dib-1-sms	dib-4-sms
gomop-22	0.035_{rn}	$\mathbf{0.029}^+_{rn}$	$\mathbf{1.152}_{rn}$	1.136_{rn}
gomop-25	$\mathbf{0.074}_{rn}$	0.075^+_{rn}	$\mathbf{1.252}_{rn}$	1.235_{rn}
gomop-52	$\mathbf{0.098}_{rn}$	0.121^+_{rn}	0.982_{rn}	0.959_{rn}
gomop-55	$\mathbf{0.230}$	0.246^+	1.169_{rn}	$\mathbf{1.221}^{++}_{rn}$
dtlz2-52	$\mathbf{0.003}_{rn}$	0.004_{rn}	$\mathbf{1.011}_{rn}$	1.007_{rn}
dtlz2-55	$\mathbf{0.135}_{rn}$	0.137^+_{rn}	1.476_{rn}	$\mathbf{1.492}^{++}_{rn}$
zdt1-52	0.024_{rn}	$\mathbf{0.023}^+_{rn}$	$\mathbf{1.171}_{rn}$	1.169_{rn}
zdt2-52	0.043_{rn}	$\mathbf{0.038}^{++}_{rn}$	$\mathbf{1.133}_{rn}$	1.132_{rn}
zdt3-52	0.048_{rn}	$\mathbf{0.046}^+_{rn}$	$\mathbf{1.105}_{rn}$	1.103^+_{rn}

Table 5. Results of the ParEGO variants with regard to the R2 indicator

	1-ei	4-ei		1-lcb	4-lcb
gomop-22	**0.051**$_{rn}$	0.051$^+_{rn}$		0.051$_{rn}$	**0.049**$^+_{rn}$
gomop-25	0.061	**0.058**$^+$		0.043$_{rn}$	**0.043**$^{++}_{rn}$
gomop-52	**0.176**	0.177$^+$		**0.103**$_{rn}$	0.108$^+_{rn}$
gomop-55	**0.066**	0.068$^+$		**0.042**$_{rn}$	0.042$^+_{rn}$
dtlz2-52	0.123	**0.123**$^+$		**0.110**$_{rn}$	0.110$_{rn}$
dtlz2-55	**0.023**$_{rn}$	0.023$^+_{rn}$		**0.024**$_{rn}$	0.024$^+_{rn}$
zdt1-52	**0.039**$_{rn}$	0.040$_{rn}$		0.032$_{rn}$	**0.032**$^+_{rn}$
zdt2-52	0.052$_{rn}$	**0.051**$^+_{rn}$		0.045$_{rn}$	**0.045**$^+_{rn}$
zdt3-52	0.070$_{rn}$	**0.070**$^+_{rn}$		0.059$_{rn}$	**0.059**$^+_{rn}$

quality. The same holds for the use of multiple weight vectors for generating batch evaluations in ParEGO (cf. Table 5) which did not result in significant deteriorations, except on ZDT1 (ei) and DTLZ2 with $m = 2$ (lcb).

For SMS-EGO (cf. Table 4, right) and MSPOT (cf. Table 6), however, this result could not be confirmed. Only on 2 to 3 of the 9 test functions considered in this study, the multi-point variants were not significantly worse. On two test functions, however, a batch evaluation led to improved results for SMS-EGO.

Hypothesis 3. Also for the third hypothesis, the different MBMO algorithms have to be considered separately. For MSPOT, the exchange of the infill criterion does generally not result in significant performance differences. Only on two of the GOMOP functions, the LCB deteriorates the results compared to mean prediction and EI. Hence, it is possible to exchange specific steps of the algorithm without detoriating the algorithm's performance.

The exchange of the EI and the LCB in ParEGO obtained excellent improvements. On almost all test functions, the results using the LCB are better, sometimes by far margins. The same held for the multi-point variants. Here the taxonomy allowed us to construct a new algorithm variant, that outperforms its original counterpart.

Table 6. Results of the MSPOT variants with regard to the hypervolume indicator

	1-mean	4-mean		1-ei	4-ei		1-lcb	4-lcb
gomop-22	**1.148**$_{rn}$	1.142$_{rn}$		**1.146**$_{rn}$	1.141$^+_{rn}$		1.136$_{rn}$	**1.142**$^+_{rn}$
gomop-25	**1.246**$_{rn}$	1.225$_{rn}$		**1.245**$_{rn}$	1.216$_{rn}$		**1.248**$_{rn}$	1.226$_{rn}$
gomop-52	**0.907**$_{rn}$	0.874$_{rn}$		**0.908**$_{rn}$	0.862$_{rn}$		**0.904**$_{rn}$	0.880$_{rn}$
gomop-55	**1.145**$_{rn}$	1.127$^+_{rn}$		**1.143**$_{rn}$	1.124$^+_{rn}$		1.126$_{rn}$	1.126$^+_{rn}$
dtlz2-52	**1.003**$_{rn}$	0.997$_{rn}$		**1.002**$_{rn}$	0.006$_{rn}$		**1.002**$_{rn}$	0.997$_{rn}$
dtlz2-55	1.414$_{rn}$	1.416$^+_{rn}$		1.409$_{rn}$	1.409$^+_{rn}$		1.411$_{rn}$	**1.414**$^+_{rn}$
zdt1-52	**1.116**$_{rn}$	1.091$_r$		**1.116**$_{rn}$	1.094$_r$		**1.115**$_{rn}$	1.099$_r$
zdt2-52	**1.057**$_{rn}$	1.029$_r$		**1.056**$_{rn}$	1.034$_r$		**1.055**$_{rn}$	1.029$_r$
zdt3-52	**1.051**$_{rn}$	1.022$_r$		**1.054**$_{rn}$	1.034$_r$		**1.051**$_{rn}$	1.022$_r$

General Recommendations. The original one-point ParEGO using the EI performs worse compared to all considered MBMO algorithms. By exchanging the EI with the LCB, however, the approach becomes competetive. The new variant can thus be recommended as a standard choice for the future.

In case of a one-point proposal, SMS-EGO (dib-1-sms) performs better or comparable on almost all test cases. It can be proposed as a general recommendation. If a multi-point proposal is desired, the respective variants of the SMS-EGO (dib-4-sms) and ParEGO (parego-4-lcb) show a comparable performance. As ParEGO only requires a single model, can be parallized without simulated evaluations, and is much faster to compute, in particular on many-objective problems, it can recommended for this case.

6.4 Discussion

The experiments showed two main results: (1) For ParEGO and ε-EGO, no significant deterioration of the results can be observed due to the multi-point proposal; (2) The change from the EI to the LCB significantly improved ParEGO.

Regarding the first result, the infill criteria of ParEGO and ε-EGO still have minor conceptual issues which inhibit the exploitation of the additional information obtained by more frequent updates. ParEGO draws the weight vectors for the scalarization at random. Hence, the implied search directions can point to regions already crowded with observations. By choosing more weight vectors per iteration in a space-filling way, the coverage of the Pareto front is improved. In comparison to MOEA/D-EGO, which evaluates all weight vectors in each generation and chooses based on the maximum EI values of a predefined, fixed clustering, the proposed procedure does not suffer from a systematic bias towards certain regions [7,20]. In ε-EGO, the optimization of an indicator based on two sets is reduced to one based on a set and a single solution. This may hinder the finetuning of $\mathcal{Y}^*_{approx}$ with regard to the global indicator.

Main result (2) can be caused by the properties of the fitness landscape implied by the EI. It has plateaus whose size increases with decreasing uncertainty of the model. The maxima of the EI lie within small basins surrounded by these plateaus. They are hard to find for both local optimization algorithms and global sampling strategies, such as the focusing random search. In particular, if a weight vector pointing to crowded region is selected, the EI can show values far below 10^{-6}, even after only 2-3 iterations. By switching to the LCB, a global trend is available which can be exploited during the internal optimization.

7   Conclusions and Outlook

In this paper, a taxonomy for MBMO algorithms was presented for the first time. Based on this taxonomy, an R toolbox was designed and some established MBMO algorithms were implemented. In order to allow batch processing, the candidate generation step of all considered algorithms was enhanced to a multi-point proposal. In addition, the internal infill and optimization criteria were exchanged and different variants of the MBMO algorithms were compared within a comprehensive benchmark.

For ParEGO and ε-EGO, the multi-point variants did not significantly deteriorate the results. They even improved the approximation quality in some cases. Moreover, the change from the EI to the LCB could improve the results of the internal optimization within ParEGO.

In future work, the scalability of the multi-point proposal with the batch size N has to be further evaluated. Moreover, systematic problems, such as the random choice of the weight vector in ParEGO, should be tackled. A simple strategy would be to redraw a weight vector in case of too low EI values. In addition, different shifts of the ideal (ParEGO) or reference point (SMS-EGO) can be used for constructing different subproblems for multi-point proposals. The simulated evaluation strategy used in the DIBs can be combined with fake observations and a refit of the model in order to improve exploration or exploitation of certain regions. The framework created by the taxonomy and the R toolbox make this possible in a structured and convenient way.

References

1. Bautista, D.C.: A Sequential Design For Approximating The Pareto Front Using the Expected Pareto Improvement Function. Ph.D. thesis, Ohio State University (2009)
2. Binois, M., Ginsbourger, D., Roustant, O.: Quantifying uncertainty on pareto fronts with gaussian process conditional simulations. European Journal of Operational Research, 1–9 (2014) (accepted, available online)
3. Bischl, B., Wessing, S., Bauer, N., Friedrichs, K., Weihs, C.: MOI-MBO: multi-objective infill for parallel model-based optimization. In: Pardalos, P.M., Resende, M.G.C., Vogiatzis, C., Walteros, J.L. (eds.) LION 2014. LNCS, vol. 8426, pp. 173–186. Springer, Heidelberg (2014)
4. Bischl, B., Horn, D., Bossek, J., Richter, J., Lang, M.:Package mlrMBO (2014). https://github.com/berndbischl/mlrMBO
5. Bischl, B., Horn, D., Wagner, T.: Model-Based Multi-Objective Optimization: Taxonomy, Multi-Point Proposal, Toolbox and Benchmark: Supplementary Material (2014). http://dx.doi.org/10.6084/m9.figshare.1275926
6. Bischl, B., Lang, M., Bossek, J., Judt, L., Richter, J., Kuehn, T., Studerus, E.: Package mlr: Machine Learning in R (2014). http://cran.r-project.org/web/packages/mlr/index.html
7. Brockhoff, D., Wagner, T., Trautmann, H.: On the properties of the R2 indicator. In: Soule, T., et al. (eds.) Proc. 2012 Genetic and Evolutionary Computation Conference (GECCO 2012), Philadelphia, US, pp. 465–472 (2012)
8. Couckuyt, I., Deschrijver, D., Dhaene, T.: Fast calculation of multiobjective probability of improvement and expected improvement criteria for pareto optimization. Journal of Global Optimization, 1–21 (2014) (accepted, available online)
9. Emmerich, M.T.M., Deutz, A., Klinkenberg, J.W.: Hypervolume-based expected improvement: Monotonicity properties and exact computation. In: Corne, D., et al. (eds.) Proc. IEEE Congress on Evolutionary Computation (CEC), pp. 2147–2154. IEEE (2011)

10. Janusevskis, J., Le Riche, R., Ginsbourger, D., Girdziusas, R.: Expected improvements for the asynchronous parallel global optimization of expensive functions: Potentials and challenges. In: Hamadi, Y., Schoenauer, M. (eds.) LION 2012. LNCS, vol. 7219, pp. 413–418. Springer, Heidelberg (2012)
11. Jeong, S., Obayashi, S.: Efficient global optimization (EGO) for multi-objective problem and data mining. In: Corne, D., et al. (eds.) Proc. IEEE Congress on Evolutionary Computation (CEC 2005), pp. 2138–2145. IEEE (2005)
12. Jones, D.R.: A taxonomy of global optimization methods based on response surfaces. Journal of Global Optimization 21(4), 345–383 (2001)
13. Jones, D.R., Schonlau, M., Welch, W.J.: Efficient global optimization of expensive black-box functions. Journal of Global Optimization 13(4), 455–492 (1998)
14. Keane, A.J.: Statistical improvement criteria for use in multiobjective design optimization. AIAA Journal 44(4), 879–891 (2006)
15. Knowles, J.: ParEGO: A hybrid algorithm with on-line landscape approximation for expensive multiobjective optimization problems. IEEE Transactions on Evolutionary Computation 10(1), 50–66 (2006)
16. Knowles, J.D., Nakayama, H.: Meta-modeling in multiobjective optimization. In: Branke, J., Deb, K., Miettinen, K., Słowiński, R. (eds.) Multiobjective Optimization. LNCS, vol. 5252, pp. 245–284. Springer, Heidelberg (2008)
17. Okabe, T., Jin, Y., Olhofer, M., Sendhoff, B.: On test functions for evolutionary multi-objective optimization. In: Yao, X., et al. (eds.) PPSN 2004. LNCS, vol. 3242, pp. 792–802. Springer, Heidelberg (2004)
18. Ponweiser, W., Wagner, T., Biermann, D., Vincze, M.: Multiobjective optimization on a limited amount of evaluations using model-assisted S-metric selection. In: Rudolph, G., Jansen, T., Lucas, S., Poloni, C., Beume, N. (eds.) PPSN X. LNCS, vol. 5199, pp. 784–794. Springer, Heidelberg (2008)
19. Voutchkov, I., Keane, A.J.: Multiobjective optimization using surrogates. In: Parmee, I.C. (ed.) Proc. 7th Intl. Conf. Adaptive Computing in Design and Manufacture (ACDM), Bristol, UK, pp. 167–175 (2006)
20. Wagner, T.: Planning and Multi-Objective Optimization of Manufacturing Processes by Means of Empirical Surrogate Models. Vulkan Verlag, Essen (2013)
21. Wagner, T., Emmerich, M., Deutz, A., Ponweiser, W.: On expected-improvement criteria for model-based multi-objective optimization. In: Schaefer, R., Cotta, C., Kołodziej, J., Rudolph, G. (eds.) PPSN XI. LNCS, vol. 6238, pp. 718–727. Springer, Heidelberg (2010)
22. Zaefferer, M., Bartz-Beielstein, T., Naujoks, B., Wagner, T., Emmerich, M.: A case study on multi-criteria optimization of an event detection software under limited budgets. In: Purshouse, R.C., Fleming, P.J., Fonseca, C.M., Greco, S., Shaw, J. (eds.) EMO 2013. LNCS, vol. 7811, pp. 756–770. Springer, Heidelberg (2013)
23. Zhang, Q., Liu, W., Tsang, E., Virginas, B.: Expensive multiobjective optimization by MOEA/D with gaussian process model. IEEE Transactions on Evolutionary Computation 4(3), 456–474 (2010)
24. Zitzler, E., Thiele, L., Laumanns, M., Fonseca, C.M., da Fonseca, V.G.: Performance assessment of multiobjective optimizers: An analysis and review. Transactions on Evolutionary Computation 8(2), 117–132 (2003)

Temporal Innovization: Evolution of Design Principles Using Multi-objective Optimization

Sunith Bandaru[1]($\boxtimes$) and Kalyanmoy Deb[2]

[1] Virtual Systems Research Centre, University of Skövde, 541 28 Skövde, Sweden
sunith.bandaru@his.se
[2] Department of Electrical and Computer Engineering, Michigan State University,
428 S. Shaw Lane, 2120 EB, East Lansing, MI 48824, USA
kdeb@egr.msu.edu

Abstract. Multi-objective optimization yields multiple solutions each of which is no better or worse than the others when the objectives are conflicting. These solutions lie on the Pareto-optimal front which is a lower-dimensional slice of the objective space. Together, the solutions may possess special properties that make them optimal over other feasible solutions. Innovization is the process of extracting such special properties (or design principles) from a trade-off dataset in the form of mathematical relationships between the variables and objective functions. In this paper, we deal with a closely related concept called temporal innovization. While innovization concerns the design principles obtained from the trade-off front, temporal innovization refers to the evolution of these design principles during the optimization process. Our study indicates that not only do different design principles evolve at different rates, but that they start evolving at different times. We illustrate temporal innovization using several examples.

1 Introduction

Evolutionary algorithms (EAs) are ideal for multi-objective optimization problems since they evolve a population of randomly initialized solutions by iteratively applying operators that mimic the natural evolution process and converge to a set of near Pareto-optimal solutions (or trade-off solutions) all of which are high-performing with respect to the conflicting objectives. These Pareto-optimal solutions are special in some sense because they lie on a lower-dimensional manifold of the objective space. It is therefore natural to assume that they may possess exclusive properties which make them Pareto-optimal. While analytically deriving such properties from the optimization problem may not always be possible, an alternate approach is to first obtain a representative trade-off dataset of (near) Pareto-optimal solutions using an MOEA and then apply machine learning techniques to extract mathematical relationships that are valid on either a part or whole of the dataset. Since the trade-off dataset usually contains columns corresponding to the variable values and the corresponding function values, the extracted relationships depict correlations between these entities.

© Springer International Publishing Switzerland 2015
A. Gaspar-Cunha et al. (Eds.): EMO 2015, Part I, LNCS 9018, pp. 79–93, 2015.
DOI: 10.1007/978-3-319-15934-8_6

In our past works we have shown that such mathematical properties do exist and can be obtained from the trade-off dataset through a process called *innovization* - innovation through optimization. The original innovization methodology [7] involved manually plotting various combinations of the columns in the trade-off datasets, visually identifying correlations and using mathematical functions to perform regression on the correlated parts of the dataset. The method was tedious and prone to errors. However, our recent works [1,8] have dealt with automating the innovization process by using clustering methods to automatically identify the correlations. This new *automated innovization* approach is capable of generating multiple significant relationships and has been successfully applied to practical engineering design problems. The obtained relationships are referred to as design principles because they are extremely useful for the designer in understanding how different variables should vary for maintaining Pareto-optimal operation/performance of the system or design.

1.1 Temporal Innovization and Human Evolution

Trade-off solutions are the end result of an MOEA and hence the design principles obtained through automated innovization pertain to the final generation of the MOEA. However, all MOEAs start with a randomly initialized population of solutions which are evolved using operators that, to some extent, mimic the natural process of evolution over several generations. Ignoring complex phenomena such as dynamic environments, cooperative individuals, sexual reproduction and interspecies interactions, the evolutionary optimization process can be viewed similar to the natural process of human evolution.

Homo sapiens acquired various anthropological features during the process of human evolution. There is sufficient documented evidence showing that these features evolved *gradually* over millions of years [11], rather than appearing out as a single event, driven by the natural mechanisms of reproduction, genetic mutation and natural selection. Despite the relative simplicity of MOEAs, the design principles that Pareto-optimal solutions possess can be thought of as somewhat analogous to the anthropological features. Just as the anthropological features distinguish present day humans and make them high-performing when compared to their ancestors, the design principles make Pareto-optimal solutions high-performing among all other feasible solutions.

Temporal innovization refers to the study of evolution of design principles over generations of an MOEA. The detailed procedure was laid out in [5]. The goal of this paper is to perform temporal innovization on various engineering design problems in order to support and extend the results provided in [5]. Specifically, we investigate if there exists a gradual evolution of design principles over generations of an MOEA, in the same way that anthropological features of humans developed gradually over millions of years. Archaeological evidence also shows that human evolution involved a hierarchy of keys developments. In other words, different anthropological features appeared at different times in the evolutionary time-line. The vertebrae emerged first, followed by the appearance of the first limbs, and so on. Adequate development of certain features was essential for some other feature to appear. If design principles indeed resemble the anthropological features, there is reason to believe that they too may exhibit hierarchical evolution in addition to gradual evolution over the generations of an MOEA. In this paper, we also investigate whether such hierarchy occurs in the evolution of design principles.

The paper is organized as follows. In Section 2 we describe the methodology for performing a temporal innovization study and generating the evolution plots. We use this approach on three engineering design problems in Section 3 and interpret the evolution plots. In Section 4, we use a visual method for the extraction of design principles on two topology optimization problems. Temporal innovization is also performed visually.

2   Methodology

MOEAs being stochastic by nature, the route that evolving solutions take while converging towards the Pareto-optimal front may differ between runs, even if the trade-off front obtained at the end is approximately the same in all the runs. In order to account for this statistical variance, the first task is to obtain generation-wise population datasets from multiple runs of the same evolutionary algorithm. In this paper we use NSGA-II [4] to solve the multi-objective problem at hand. Multiple runs are executed with uniformly distributed seed values for the random number generator.

One of the trade-off datasets is randomly chosen and the design principles are extracted using the automated innovization algorithm developed in [8]. The details of this algorithm are irrelevant to discussion in this paper. However, it suffices to mention that the design principles obtained have the following mathematical form, where the quantity c on the right-hand side is allowed different values in different clusters of solutions but remains *approximately* constant within each cluster [1,8].

$$\prod_{j=1}^{N} \phi_j(\mathbf{x})^{a_j b_j} = c, \tag{1}$$

where ϕ_j's are N symbolic entities (variables $\mathbf{x}$, objective functions $f(\mathbf{x})$, etc.) called basis functions which can have a Boolean exponent a_j and a real valued exponent b_j. Each design principle is associated with a *significance value* which indicates the percentage of trade-off solutions for which that design principle remains invariant, i.e. takes a (approximately) constant value c.

For recording the evolutionary time-line of design principles, the non-dominated solutions from each of the runs at each generation t are stored. Next, each design principle (DPi) is checked for its presence in the combined data at each generation. The significance of DPi at generation t, denoted by $S_t^{\text{DP}i}$, is calculated as the proportion of points satisfying the design principle to the total non-dominated points in the final generation ($NGEN$). The stepwise procedure for calculating $S_t^{\text{DP}i}$ for a given design principle DPi is presented below:

Step 0: Set $t \leftarrow 0$.

Step 1: Collect solutions at generation t from all runs into the set $\mathbf{P}_t$. Thereafter, remove the dominated points from $\mathbf{P}_t$.

Step 2: Evaluate DPi at all solutions in $\mathbf{P}_t$ to obtain c-values and collect them in set $\mathbf{C}_t$.

Step 3: Every element $c \in \mathbf{C}_t$ is checked for its association with any of the $\mathcal{C}$ clusters of DPi using the criterion,

$$c \in \text{cluster } k \Leftrightarrow \mu_c^{(k)} - s\,\sigma_c^{(k)} \leq c \leq \mu_c^{(k)} + s\,\sigma_c^{(k)},$$

where $\mu_c^{(k)}$ and $\sigma_c^{(k)}$ are respectively the mean and standard deviation for the k-th cluster. The number of elements E_t in $\mathbf{C}_t$ that belong to any of the $\mathcal{C}$ clusters is recorded. A variation of s standard deviations is allowed in the c-values. In this paper $s = 4$ is used and recommended.

Step 4: Calculate the significance of DPi in the current generation t as,

$$S_t^{\text{DP}i} = \frac{E_t}{|\mathbf{P}_{NGEN}|} \times 100\%,$$

where $|.|$ represents the set size.

Step 5: If $t = NGEN$ **Stop** else $t \leftarrow t + 1$ and **Goto Step 1**.

Thereafter, a plot of the significance value of each design principle with generation is used to reveal the relative order in which design principles appear during the optimization process. The plot also shows which of the design principles evolve faster and which ones evolve at a slow rate.

3 Results I: Design Principles Through Automated Innovization

The procedure described above is now illustrated on three engineering design problems, namely, car side impact problem, metal cutting problem and MEMS resonator design problem. While the first two problems are relatively simple mathematical models of complex design problems, the latter is directly a real-world problem which takes all practical design considerations into account. The difference between the three becomes clear in the following sections.

3.1 Car Side Impact Problem

A car is subjected to a regulatory side impact test. Various impact loads, rib deflections and a quantity called viscous criterion (V*C) are measured for the crash test dummy. The velocities of B-pillar midpoint and front door are measured on the vehicle structure. The following decision variables are to be optimized.

$$
\begin{aligned}
0.5 \ &\leq x_1 : \text{Thickness of B-Pillar inner} & \leq 1.5 \text{ mm} \\
0.45 \ &\leq x_2 : \text{Thickness of B-Pillar reinforcement} & \leq 1.35 \text{ mm} \\
0.5 \ &\leq x_3 : \text{Thickness of floor side inner} & \leq 1.5 \text{ mm} \\
0.5 \ &\leq x_4 : \text{Thickness of cross members} & \leq 1.5 \text{ mm} \\
0.875 \ &\leq x_5 : \text{Thickness of door beam} & \leq 2.625 \text{ mm} \\
0.4 \ &\leq x_6 : \text{Thickness of door beltline reinforcement} & \leq 1.2 \text{ mm} \\
0.4 \ &\leq x_7 : \text{Thickness of roof rail} & \leq 1.1 \text{ mm}
\end{aligned}
$$

Table 1. Design principles for the car side impact problem

Notation	Design principles	Significance
DP1	$x_6^{1.0000} = \text{constant}$	77.44 %
DP2	$x_7^{1.0000} = \text{constant}$	80.28 %
DP3	$f_1^{0.3652} f_2^{1.0000} x_5^{-0.7699} = \text{constant}$	70.33 %
DP4	$x_2^{1.0000} x_4^{-0.1012} x_5^{-0.9360} = \text{constant}$	70.33 %
DP5	$f_2^{1.0000} x_3^{0.1293} x_5^{-0.8856} = \text{constant}$	70.12 %
DP6	$x_2^{1.0000} x_5^{-0.8748} = \text{constant}$	70.93 %
DP7	$f_1^{1.0000} x_3^{-0.2952} x_5^{-0.9675} = \text{constant}$	71.75 %
DP8	$f_2^{-0.6684} x_2^{1.0000} x_5^{-0.9887} = \text{constant}$	70.53 %
DP9	$x_1^{0.1166} x_2^{1.0000} x_5^{-0.9331} = \text{constant}$	72.15 %
DP10	$f_2^{1.0000} x_1^{0.2113} x_5^{-0.7592} = \text{constant}$	72.15 %
DP11	$f_1^{0.1161} x_2^{1.0000} x_5^{-0.8927} = \text{constant}$	71.14 %

The objectives are to minimize the weight of the vehicle and the average rib deflection on the dummy. The complete formulation is provided in [6].

$$
\begin{aligned}
\text{Minimize} \quad & f_1(\mathbf{x}) = \text{Weight} \\
\text{Minimize} \quad & f_2(\mathbf{x}) = (D_{ur} + D_{mr} + D_{lr})/3 \\
\text{Subject to} \quad & \text{Abdomen Load} \leq 1 \text{ kN}, \\
& \{VC_{upper}, VC_{middle}, VC_{lower}\} \leq 0.32 \text{ m/s}, \\
& \{D_{upper}, D_{middle}, D_{lower}\} \text{ (Rib deflections)} \leq 32 \text{ mm}, \\
& F \text{ (Pubic force)} \leq 4 \text{ kN}, \\
& V_{MBP} \text{ (Velocity of B-pillar midpoint)} \leq 9.9 \text{ mm/ms}, \\
& V_{FD} \text{ (Velocity of front door)} \leq 15.7 \text{ mm/ms}.
\end{aligned}
\tag{2}
$$

The above problem is solved using NSGA-II and the obtained trade-off dataset is provided as input to the automated innovization algorithm. The obtained design principles and their significance values are shown in Table 1. Ten runs of the NSGA-II algorithm are performed and the generation-wise datasets for $NGEN = 100$ generations are obtained. Using the procedure described in Section 2, the significance values are calculated for all 11 design principles at all generations and are plotted together as shown in Figure 1.

3.2 Metal Cutting Problem

In this problem [14], a steel bar is to be machined using a carbide tool of nose radius $r_n = 0.8$ mm on a lathe with $P^{max} = 10$ kW rated motor to remove 219912 mm^3 of material. A maximum cutting force of $F_c^{max} = 5000$ N is allowed. The motor has a transmission efficiency η. The total operation time (T_p) and the used tool life (ξ) are to be minimized by optimizing the cutting speed (v), the feed rate (f) and the depth of cut (a) while maintaining a surface roughness of $R^{max} = 50\mu m$.

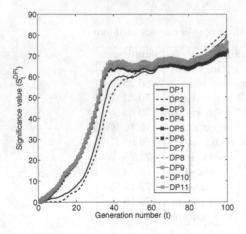
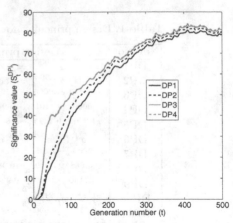

Fig. 1. Evolution of the 11 design principles shown in Table 1 for the car side impact problem.

Fig. 2. Evolution of the four design principles shown in Table 2 for the metal cutting problem

The problem is formulated as,

$$
\begin{aligned}
\text{Minimize} \quad & f_1(\mathbf{x}) = T_p(\mathbf{x}) \\
\text{Minimize} \quad & f_2(\mathbf{x}) = \xi(\mathbf{x}) \\
\text{Subject to} \quad & P(\mathbf{x}) \le \eta P^{max}, \\
& F_c(\mathbf{x}) \le F_c^{max}, \\
& R(\mathbf{x}) \le R^{max}, \\
& 250 \le v \le 400 \text{ m/min}, \\
& 0.15 \le f \le 0.55 \text{ mm/rev}, \\
& 0.5 \le a \le 6 \text{ mm},
\end{aligned}
\tag{3}
$$

where

$$
T_p(\mathbf{x}) = 0.15 + 219912 \left(\frac{1 + \frac{0.20}{T(\mathbf{x})}}{MRR(\mathbf{x})} \right) + 0.05, \quad \xi(\mathbf{x}) = \frac{219912}{MRR(\mathbf{x})T(\mathbf{x})} \times 100,
$$

$$
T(\mathbf{x}) = \frac{5.48 \times 10^9}{v^{3.46} f^{0.696} a^{0.460}}, \qquad\qquad F_c(\mathbf{x}) = \frac{6.56 \times 10^3 f^{0.917} a^{1.10}}{v^{0.286}},
$$

$$
P(\mathbf{x}) = \frac{vF_c(\mathbf{x})}{60000}, \qquad MRR(\mathbf{x}) = 1000vfa, \ R(\mathbf{x}) = \frac{125 f^2}{r_n}.
$$

The trade-off dataset obtained by solving Equation (3) using NSGA-II is used to generate the four design principles shown in Table 2 through automated innovization.

Again, ten runs of the NSGA-II algorithm are performed to obtain generation-wise datasets for $NGEN = 500$ generations. Figure 2 shows the plot of significance values versus generation number for all four design principles.

Both Figures 1 and 2 clearly show that design principles evolve in a gradual manner, as hypothesized in Section 1.1, slowly increasing in significance as the

Table 2. Design principles for the metal cutting problem

Notation	Design principles	Significance
DP1	$f^{1.0000} = $ constant	80.63 %
DP2	$\xi^{-0.3558}v^{1.0000} = $ constant	80.85 %
DP3	$T_p^{0.9950}\xi^{-0.2712}v^{1.0000} = $ constant	81.42 %
DP4	$T_p^{1.0000}v^{0.2391} = $ constant	81.38 %

population converges close to the true Pareto-optimal front. However, the secondary hypothesis that there may exist a hierarchy in the evolution of design principles cannot be verified in these examples. In fact, all design principles shown in the two figures evolve in exactly the same way for each problem. The design principles also appear very early during the optimization process. The reason for this is that these problems are only simplified versions of practical problems that are much more complex. In order to exhibit hierarchy during evolution, the design problem should be closer to the real-world, so that the multi-objective optimizer being used 'struggles' to build design principles during optimization thus bringing their hierarchy into the picture. In the next section, we reproduce the results obtained in [5] where the problem was shown to possess a hierarchy in the evolution of design principles.

3.3 MEMS Resonator Design Problem

The MEMS (MicroElectroMechanical System) component design problem involves the minimization of the power consumption f_1 (same as applied voltage, V) and the minimization of the total area f_2 of the device. Figure 3 is a schematic of the MEMS model showing the 14 design variables. The complete problem formulation can be found in [9,10]. The problem is known to be highly non-linear in terms of the two objectives and involves 24 constraints (10 linear and 14 non-linear), making it rather difficult for NSGA-II to optimize and therefore to build design principles. Ten trade-off datasets are generated using NSGA-II. Their progress towards trade-off front at some specific generations is shown in Figure 4.

The design principles obtained through automated innovization are discussed in detail in [5]. Figure 5 shows the significance values for each of the 13 design principles at various generations. The evolution history shown in the figure reveals the time at which each principle started to evolve during the optimization process. The evolution is shown around 10% significance value. Clearly, a gradual evolution pattern of DPs can be seen along with the hierarchy (DP2, DP13, DP11, DP12, DP1, DP5, DP4, DP3, DP7, DP10, DP8, DP9, DP6). This information of some design principles evolving earlier than others may provide valuable knowledge about building a design from scratch in an optimal manner.

Discrete Variables in Evolution. The MEMS resonator design involves a discrete variable N_c representing the number of teeth on the rotor comb. Though no design principles involving this variable were obtained through automated innovization, a simple manual innovization process reveals two design principles

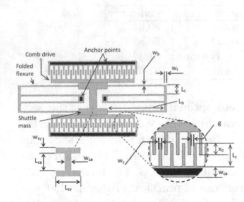

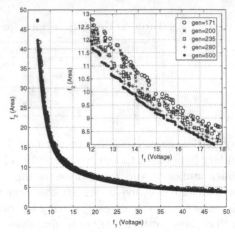

Fig. 3. MEMS resonator model showing the design variables

Fig. 4. Progress of solutions with generations towards trade-off front

that do not fit the form in Equation (1) that relate N_c to the objectives f_1 and f_2. These are,

$$B1 \equiv N_c - \frac{1.265 \times 10^5}{f_1^{3.829}} \quad \text{and}$$
$$B2 \equiv 2.012\, f_2 - N_c. \tag{4}$$

The significance values of B1 and B2 with generations are calculated as was done for the other design principles. Their evolution curves are also shown in Figure 5. B2 starts to evolve at around 130 generations and its evolution curve stays well before those of others. B1 appears in the population after B2 around the 160 generation mark and it too keeps evolving before the other design principles.

Coming back to human evolution, the anthropological description of the present day human too involves many discrete variables, like the number of eyes, limbs, fingers, backbones, etc. Taking the example of the human eye, it is now widely believed [13] that eyes initially appeared in the form of photo-receptive proteins that sense light and can only distinguish between bright and dark. Evidence for their existence can still be seen in certain green algae and unicellular organisms like euglenids [12]. The earliest eyes formed over millions of years as groups of photo-receptor cells came together and gradually depressed into what later became eye sockets. The number of such 'eyespots' was fixed very early on in the evolution of different species. In case of all ancestors of present day humans and related species, its value has always remained two. Though the current form of the human eye evolved only recently (relative to when life forms first appeared), the number of eyes was decided very early on in the evolution of man. Same is the case with number of limbs, fingers, etc.

Extending the analogy with human evolution to engineering design, it appears that whenever there are discrete variables involved, the evolution of optimal designs follows a path where design principles involving those discrete variables appear

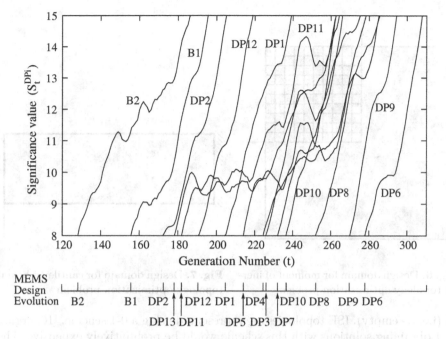

Fig. 5. Evolution of design principles for the MEMS design problem. The design principles evolve gradually but also maintain a hierarchy.

before any of the other design principles. This is at least the case in the MEMS resonator design problem as seen above. More problems need to be studied for evolution as described in this section to gather empirical evidence in support of this theory.

4 Results II: Design Features Through Superposition

In certain engineering design problems where the trade-off designs can be represented schematically to show sufficient detail, the visual identification of optimal design features may prove to be more insightful than mathematical design principles obtained using automated innovization. In this section, we perform the temporal innovization task visually, since the nature of the problems used does not allow the generation of evolution plots.

Both problems used in this section concern structural topology optimization where the interest is in finding the topology (i.e. the distribution or layout of material inside) of a design domain subjected to loads and boundary conditions, so as to extremize one or more objectives. Recent research in topology optimization has been in the development of finite element based numerical methods as they can deal with complex topologies involving different types of materials [15]. Among numerical methods, especially popular are the so-called ISE (Isotropic Solid or Empty) topologies, in which blocks of finite elements (called ground elements) can either contain the given isotropic material (i.e. be solid) or contain no material at

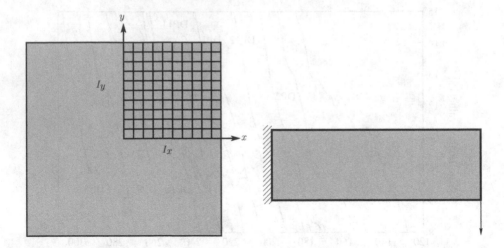

Fig. 6. Design domain for moment of iner-   **Fig. 7.** Design domain for cantilever beam
tia topology optimization problem   topology optimization problem

all (i.e. be empty). ISE topologies are represented using a 0-1 scheme. [16] argues
that obtaining solutions with this scheme would be prohibitively expensive. The
problem can be avoided by allowing ground elements to have intermediate densities and penalizing them using a power law [2].

4.1   Moment of Inertia Problem

The problem involves topology optimization of a square domain as shown in
Figure 6 in order to, (i) maximize moment of inertia about the x-axis (I_x) and,
(ii) minimize moment of inertia about the y-axis (I_y). The quarter problem is
solved using a grid of size 10×10 as shown in Figure 6 and a 0-1 representation,
as described above. In the NSGA-II framework the 100 variables are represented
using a binary string. For population members containing disconnected regions
(identified after mirroring the 10×10 grid to full the whole domain), the clustering
approach suggested in [3] is used to identify the largest cluster of cells with material and connected to the nearest axis. 100 random initial population members
are evolved over 500 generations using the two-dimensional crossover operator [3]
with $p_c = 0.8$ and bit-wise mutation with $p_m = 0.01$. The obtained trade-off front
and a few designs are shown in Figure 8.

   This problem is well-suited for extracting design features by superposition
of solutions and studying their evolution. The procedure is very similar to that
described in Section 2. The only difference is that after obtaining the set $\mathbf{P}_t$ of
non-dominated solutions at the generation t, corresponding cell values from all
solutions $\mathbf{P}_t$ are added to obtain a single matrix $\mathbf{M}_t$ of size 20×20. Thus, if a
particular cell contains material (i.e. has a value of 1) in all non-dominated solu-
tions of generation t, then its value in the resultant matrix becomes $|\mathbf{P}_t|$. The non-
dominated solutions are therefore said to be superposed.

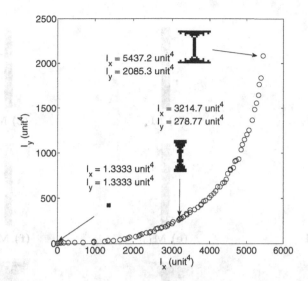

Fig. 8. Trade-off front for the moment of inertia topology optimization problem

The status of design feature evolution at generation t can be observed visually by the plotting the matrix $\mathbf{M}_t$, such that cells with the largest value are shown in black and those with the lowest value are shown in white. For the present problem, the evolution of design features is shown in Figure 9. It can be seen clearly that during the initial generations, solutions in which material is concentrated along the vertical centerline are non-dominated. Up until $t = 30$ generations the material only spreads outwards from the vertical. Thereafter, evolution requires that the material should also be pushed outwards from the horizontal centerline (seen at $t = 40$). The features at the corners of the domain, do not form until the later generations. Looking at $\mathbf{M}_{500}$ it can be said that, for maximizing I_x and minimizing I_y, anything resembling an I-section would be close to Pareto-optimality. Though, in this case this property is intuitive, the study here revealed how exactly the I-section is formed during evolution.

4.2   Cantilever Beam Problem

In this problem, the topology of a cantilever beam carrying an end load of one unit is to be optimized within a rectangular domain as shown in Figure 7 for, (i) minimizing the compliance (or maximizing stiffness) and, (ii) minimizing the weight of the structure. Since the calculation of these objectives involves finite element simulations and the use 0-1 scheme may lead to indeterminate solutions in NSGA-II, the power-law approach of allowing intermediate densities suggested in [2] is employed. The intermediate densities are however not penalized as the present problem has two objectives and retaining such solutions may lead to better diversity. A grid of size 30×10 is imposed on the domain. The 300 variables are initialized for a population size of 400 and evolved using SBX ($p_c = 0.9$) and polynomial mutation ($p_m = 0.05$) for 2000 generations. Finite element analysis is performed

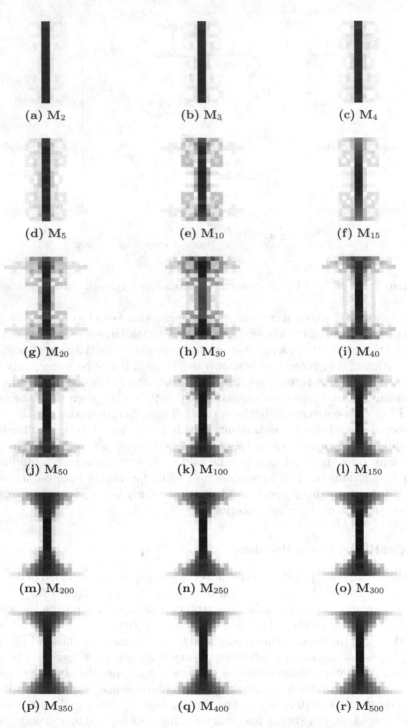

Fig. 9. Evolution of design features for moment of inertia topology optimization problem. M_t is obtained by superposing all non-dominated solutions from generation t.

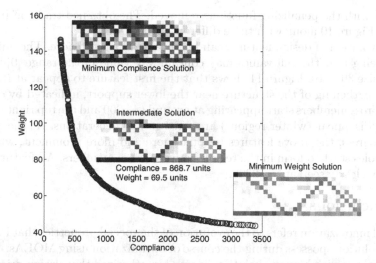

Fig. 10. Trade-off front for the cantilever beam topology optimization problem

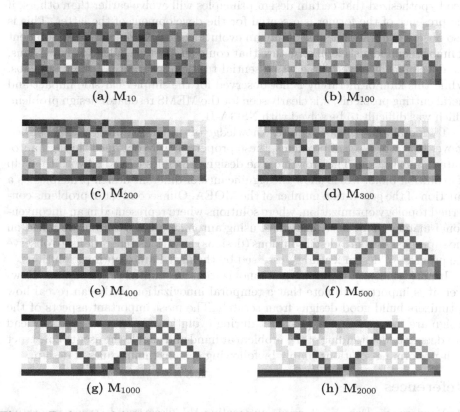

(a) M_{10}

(b) M_{100}

(c) M_{200}

(d) M_{300}

(e) M_{400}

(f) M_{500}

(g) M_{1000}

(h) M_{2000}

Fig. 11. Evolution of design features for cantilever beam topology optimization problem. M_t is obtained by superposing all non-dominated solutions from generation t.

using [17] with the penalizing functions removed. The obtained trade-off front is shown in Figure 10 along with three different structures.

The evolution of design features can be obtained just as before. The only difference being that the cell values may contain any value in the range $[0, 1]$ and $\mathbf{M}_t$ is of size 30×10. Figure 11 shows that the first feature to appear at $t = 100$ is the strengthening of the structure near the lower support, indicated by darker regions. Cross members start appearing at around $t = 200$ and the top right corner begins to disappear (whiter regions) around $t = 300$ generations. As the generations progress, the above features become more and more prominent, while at $t = 500$ holes start to form in the region between cross members. All features are eventually clearly seen at $t = 2000$.

5   Conclusions

Temporal innovization refers to the evolution of the special properties that Pareto-optimal solutions possess during the course of optimization using MOEAs. Based on the analogy with human evolution, it was hypothesized that design principles (like anthropological features of humans) evolve gradually over generations. It was also hypothesized that certain design principles will evolve earlier than others, if the presence of the former is essential for the development of the latter. This is also observed in the time-line of human evolution. For example, the development of limbs required a skeletal structure that could support the added weight. Thus, the development of vertebrae was essential to the appearance of the first limbs. While this kind of hierarchy is not observed for the simpler car side impact and metal cutting problems, it is clearly seen for the MEMS resonator design problem, which was difficult to be solved with NSGA-II.

Temporal innovization requires knowledge of the special properties described above. In our first set of problems, these properties were obtained through automated innovization. The growth of the design principles is studied with the help of evolution plots, which show the significance of different design principles as a function of the generation number of the MOEA. Our second set of problems concerned topology optimization, where solutions where represented in an unconventional manner. Therefore, instead of using automated innovization, we relied on the superposition of trade-off solutions (designs) at various generations, to observe the progress of design features possessed by the near Pareto-optimal solutions.

It may be argued that temporal innovization has limited practical uses. However, it is important to note that a temporal innovization study can reveal how optimizers build good designs from scratch. The most important aspects of the design are the ones that appear first during evolution. Such knowledge can lead to a deeper understanding of the problem at hand and enable the user to construct high-performing solutions simply by following certain thumb-rules of design.

References

1. Bandaru, S., Deb, K.: Towards automating the discovery of certain innovative design principles through a clustering-based optimization technique. Engineering Optimization **43**(9), 911–941 (2011)

2. Bendsøe, M.: Optimal shape design as a material distribution problem. Structural and Multidisciplinary Optimization **1**(4), 193–202 (1989)
3. Datta, D., Deb, K.: Design of optimum cross-sections for load-carrying members using multi-objective evolutionary algorithms. In: Proceedings of International Conference on Systemics, Cybernetics and Informatics, pp. 571–577 (2005)
4. Deb, K., Agarwal, S., Pratap, A., Meyarivan, T.: A fast and elitist multi-objective genetic algorithm: NSGA-II. IEEE Transactions on Evolutionary Computation **6**(2), 182–197 (2002)
5. Deb, K., Bandaru, S., Tutum, C.C.: Temporal evolution of design principles in engineering systems: Analogies with human evolution. In: Coello, C.A.C., Cutello, V., Deb, K., Forrest, S., Nicosia, G., Pavone, M. (eds.) PPSN 2012, Part II. LNCS, vol. 7492, pp. 1–10. Springer, Heidelberg (2012)
6. Deb, K., Gupta, S., Daum, D., Branke, J., Mall, A., Padmanabhan, D.: Reliability-based optimization using evolutionary algorithms. IEEE Trans. on Evolutionary Computation **13**(5), 1054–1074 (2009)
7. Deb, K., Srinivasan, A.: Innovization: Innovating design principles through optimization. In: Proceedings of the 8th Annual Conference on Genetic and Evolutionary Computation, GECCO 2006, pp. 1629–1634. ACM, New York (2006)
8. Deb, K., Bandaru, S., Greiner, D., Gaspar-Cunha, A., Tutum, C.C.: An integrated approach to automated innovization for discovering useful design principles: Case studies from engineering. Applied Soft Computing **15**, 42–56 (2014)
9. Fedder, G., Iyer, S., Mukherjee, T.: Automated optimal synthesis of microresonators. In: Proceedings of the Ninth Int. Conf. Solid State Sens. Actuators, Chicago, IL, pp. 1109–1112, April 1997
10. Fedder, G., Mukherjee, T.: Physical design for surface-micromachined MEMS. In: Proceedings of the Fifth ACM SIGDA Physical Design Workshop, Virginia, USA, April 1996
11. Haeckel, E.: The evolution of man, vol. 1. Kessinger Publishing (1879)
12. Kreimer, G.: The green algal eyespot apparatus: A primordial visual system and more? Current Genetics **55**(1), 19–43 (2009)
13. Land, M., Fernald, R.: The evolution of eyes. Annual Review of Neuroscience **15**(1), 1–29 (1992)
14. Quiza Sardiñas, R., Rivas Santana, M., Alfonso Brindis, E.: Genetic algorithm-based multi-objective optimization of cutting parameters in turning processes. Engineering Applications of Artificial Intelligence **19**(2), 127–133 (2006)
15. Rozvany, G.: Aims, scope, methods, history and unified terminology of computer-aided topology optimization in structural mechanics. Structural and Multidisciplinary Optimization **21**(2), 90–108 (2001)
16. Rozvany, G.: A critical review of established methods of structural topology optimization. Structural and Multidisciplinary Optimization **37**(3), 217–237 (2009)
17. Sigmund, O.: A 99 line topology optimization code written in matlab. Structural and Multidisciplinary Optimization **21**(2), 120–127 (2001)

MOEA/D-HH: A Hyper-Heuristic
for Multi-objective Problems

Richard A. Gonçalves[✉], Josiel N. Kuk,
Carolina P. Almeida, and Sandra M. Venske

Department of Computer Science, UNICENTRO, Guarapuava, Brazil
{richard,josiel,carol,ssvenske}@unicentro.br

Abstract. Hyper-Heuristics is a high-level methodology for selection or
automatic generation of heuristics for solving complex problems. Despite
the hyper-heuristics success, there is still only a few multi-objective
hyper-heuristics. Our approach, MOEA/D-HH, is a multi-objective selection hyper-heuristic that expands the MOEA/D framework. It uses an
innovative adaptive choice function proposed in this work to determine
the low level heuristic (Differential Evolution mutation strategy) that
should be applied to each individual during a MOEA/D execution. We
tested MOEA/D-HH in a well established set of 10 instances from the
CEC 2009 MOEA Competition. MOEA/D-HH is compared with some
important multi-objective optimization algorithms and the results
obtained are promising.

Keywords: Hyper-heuristic · Choice function · MOEA/D

1 Introduction

The use of heuristics and meta-heuristics have been quite effective in solving
complex optimization problems. Currently there are several heuristics and metaheuristics available for adoption by users, but the choice of the best heuristic to
be applied to solve each problem is difficult and can vary for each instance. In
order to alleviate this problem, Hyper-Heuristics emerged. A Hyper-heuristic is
a high-level methodology for automatic selection or generation of heuristics for
solving complex problems [2].

Although hyper-heuristics have been successfuly applied to solve various optimization problem, its use to solve multi-objective optimization problems (MOPs)
is incipient. So, the main objective of this work is to developed a multi-objective
hyper-heuristic, called MOEA/D-HH (MOEA/D Hyper-Heuristic). The proposed
approach is a selection hyper-heuristic that uses an innovative adaptive choice
function to determine the Differential Evolution (DE) [21] mutation strategy
(low level heuristic) applied to each individual during a MOEA/D-DRA execution.

MOEA/D, and its variants as the MOEA/D-DRA, decomposes a MOP into
a number of single objective optimization subproblems. The objective of each

© Springer International Publishing Switzerland 2015
A. Gaspar-Cunha et al. (Eds.): EMO 2015, Part I, LNCS 9018, pp. 94–108, 2015.
DOI: 10.1007/978-3-319-15934-8_7

subproblem is a (linear or nonlinear) weighted aggregation of all the individual objectives in the MOP. MOEA/D-HH was tested in a set of 10 unconstrained MOPs from the CEC 2009 MOEA Competition.

The main contributions of this paper are: (i) the proposal of a new multi-objective hyper-heuristic and (ii) an innovative adaptive choice function.

The remainder of this paper is organized as follows. An overview of Hyper-Heuristics is presented in 2 while the fundamental concepts related to Differential Evolution algorithm are shown in Section 3. Section 4 gives a brif overview of multi-objective Optimization and the MOEA/D-DRA framework. MOEA/D-HH and its new adaptive choice function are detailed in Section 5. Experiments and results are presented and discussed in Section 6. Section 7 concludes the paper and presents the directions for the future work.

2   Hyper-Heuristics

Hyper-Heuristics is an increasingly important topic of research in the heuristic optimisation community, as it permits the design of flexible solvers by means of the automatic selection and/or generation of heuristic components (low level heuristics) [22]. One important distinction between heuristic and hyper-heuristics is that the first searches the space of solutions while the latter search the space of heuristics [2].

The hyper-heuristics are classified into generation hyper-heuristics or selection hyper-heuristics. Generation hyper-heuristics automatically produce new heuristics from a set of heuristic components. On the other hand, selection hyper-heuristics use automated methods for choosing heuristics [2]. The generated or selected heuristics are called low level heuristics. In this work a choice function based selection hyper-heuristic is investigated for multi-objective optimization.

Decisions taken by the hyper-heuristics (selection or generation) are based on feedback from the search process. This learning process can be classified as on-line or off-line. In on-line hyper-heuristics (used in this work) information is obtained during the execution of the algorithm. While in off-line hyper-heuristics the information is obtained before the search process starts (*a priori*) [17].

Our proposed algorithm is an on-line selection hyper-heuristic. A general on-line selection hyper-heuristic is briefly described as follows [2]. An initial set of solutions is generated and iteratively improved using low level heuristics until a termination criterion is satisfied. During each iteration, the heuristic selection decides which low level heuristic will be used next based on feedback obtained in previous iterations. After the selected heuristic is applied to the current solution, a decision is made whether to accept the new solution or not using an acceptance method.

In the literature on Hyper-Heuristics there are several techniques for selection of heuristics [22], among them the Choice Function [5] [16]. The Choice Function is one of the most successful selection heuristics described in the hyper-heuristic literature. The Choice Function chooses a heuristic by the combination of three terms: performance of each heuristic, the heuristic performance when it is applied following the last heuristic applied (performance of heuristic pairs) and information about the last time it was applied. The first two terms favor intensification while the latter favors diversification.

The heuristic selection mechanism used in our work is a new adaptive choice function (described in Subsection 5.1) and the low level heuristics used correspond to Differential Evolution mutation and crossover strategies (see Section 3). The acceptance method used corresponds to the MOEA/D population update step (described in Section 5).

Multi-objective hyper-heuristics can be found in [1,3,9,10,16,22].

3   Differential Evolution

DE is a stochastic, population-based search strategy developed by Storn and Price [21]. DE has three control parameters: N, F and CR. N is the population size. The scaling factor parameter (F) is used to scale the difference between vectors, which is further added to a target vector $\hat{x}$. This process is called *mutation* in DE. The vector resulting from mutation, named *trial vector*, is combined with a parent vector $\mathbf{x}^p$ in the crossover operation, according to the Crossover Rate parameter (CR). Finally, the offspring is compared with its parent vector to decide (based on their fitness) who will "survive" to the next generation.

There are some variations to the basic DE, they differ specially in the way the target vector is selected (x), the number of difference vectors used (y), and the way that the crossover point is determined (z). The notation adopted to characterize the variations is DE/x/y/z. In this paper we used DE/*rand*/1/*bin*, DE/*rand*/2/*bin*, DE/*current − to − rand*/1/*bin*, DE/*current − to − rand*/2/*bin*, and DE/*nonlinear* variations.

For DE/*rand*/1/*bin* and DE/*rand*/2/*bin*, a random individual (*rand*) is selected for target vector $\hat{\mathbf{x}}$, there is one, $y = 1$, (or there are two, $y = 2$) pair(s) of solutions which is(are) randomly chosen to calculate the differential mutation and the binomial crossover ($z = bin$) is used (the probability of choosing a component from the parent vector or the trial vector is given by a binomial distribution). For DE/*current−to−rand*/1/*bin* and DE/*current−to−rand*/2/*bin*, the current individual is selected for the target vector $\hat{\mathbf{x}}$ and there is one, $y = 1$, (or there are two, $y = 2$) pair(s) of solutions randomly chosen to calculate the differential mutation and the binomial crossover ($z = bin$) [21]. The last mutation strategy, called in this work DE/*nonlinear*, was proposed in [19] for MOEA/D framework, and disregards the values of CR and F. It is a hybrid operator based on polynomials, where each polynomial represents the offspring which takes the form $\boldsymbol{\rho}(w) = w^2\mathbf{c}_a + w\mathbf{c}_b + \mathbf{c}_c$, where w is generated based on an interpolation probability, P_{inter} [19]. Assuming that $rand \in U[0,1]$ (i.e., *rand* is a value between 0 and 1 randomly generated by a uniform distribution) we have $w \in U[0,2]$, if $rand \leq P_{inter}$ and $w \in U[2,3]$ otherwise. Individuals $\mathbf{c}_a$, $\mathbf{c}_b$ e $\mathbf{c}_c$ are defined as $\mathbf{c}_a = (\mathbf{x}_c - 2\mathbf{x}_b + \mathbf{x}_a)/2$, $\mathbf{c}_b = (4\mathbf{x}_b - 3\mathbf{x}_c - \mathbf{x}_a)/2$, $\mathbf{c}_c = \mathbf{x}_c$, where $\mathbf{x}_c$, $\mathbf{x}_b$ and $\mathbf{x}_a$ are individuals randomly chosen from the current population, in accordance with the scope defined by MOEA/D-DRA algorithm.

DE/$rand$/1/bin was chosen because this strategy usually presents slow convergence speed and good exploration capability. Therefore, it seems more suitable for solving multimodal problems than strategies relying on the best solution found so far [18]. DE/$current - to - rand$/1/bin enables the algorithm to solve rotated problems more effectively [18]. Some works in DE, such as [18] and [20], point that two-difference-vectors-based strategies may provide better perturbations than one-difference-vector-based strategies. Thus, DE/$rand$/2/bin and DE/$current-to-rand$/2/bin were also included. DE/$nonlinear$ is a new hybrid mutation operator that includes a non-linear part for the DE mutation operator. This operator was tested with MOEA/D in [19] providing a robust performance with better results for many test problems.

In MOEA/D-HH (see Section 5), the mutation strategies are used as low level heuristics.

4   Multi-objective Optimization and MOEA/D-DRA

General multi-objective Optimization Problem is defined as Min (or Max) $\mathbf{f}(\mathbf{x}) = (f_1(\mathbf{x}), ..., f_M(\mathbf{x}))$ subject to $g_i(\mathbf{x}) \leq 0$, i = $\{1, ..., G\}$, and $h_j(\mathbf{x}) = 0$, j = $\{1, ..., H\}$ $\mathbf{x} \in \Omega$, and where the integer M ≥ 2 is the number of objectives. A solution minimizes (or maximizes) the components of the objective vector $\mathbf{f}(\mathbf{x})$ where $\mathbf{x}$ is a n-dimensional decision variable vector $\mathbf{x} = (x_1, ..., x_n) \in \Omega$.

In this work we consider the CEC 2009 multi-objective benchmark (10 instances) [24] to be minimized in the evolutionary process. To solve the multi-objective benchmark we propose a new algorithm called MOEA/D-HH, which is a multi-objective selection hyper-heuristic based on the MOEA/D-DRA, a MOEA/D variant. MOEA/D is based on conventional aggregation approaches [8], it decomposes a MOP into a number of single objective optimization subproblems. The objective of each subproblem is a linear (or nonlinear) weighted aggregation of all individual objectives in the MOP. Neighborhood relations among these subproblems depend on distances among their aggregation weight vectors. Each subproblem is simultaneously optimized using mainly information from its neighboring subproblems. Some versions of MOEA/D, such as [23] and [14], treat equally all the subproblems and each of them receives about the same amount of computational effort. These subproblems, however, may have different computational difficulties, therefore, it is very reasonable to assign different amounts of computational effort to different problems. In MOEA/D-DRA (MOEA/D with Dynamical Resource Allocation) [25] is defined and computed a utility value for each subproblem. Computational efforts are distributed to these subproblems based on their utilities.

5   Proposed Approach

MOEA/D-HH is a selection hyper-heuristic that uses an adaptive choice function to determine the DE mutation strategy applied to each individual during a MOEA/D-DRA [25] execution.

MOEA/D-HH includes five strategies (DE/*rand*/1/*bin* , DE/*rand*/2/*bin*, DE/*current−to−rand*/1/*bin*, DE/*current−to−rand*/2/*bin* and DE/*nonlinear*) into the pool of low level heuristics. Algorithm 1 presents the pseudocode of the proposed approach.

Algorithm 1. Pseudocode of MOEA/D-HH

1: Generate N weight vectors $\boldsymbol{\lambda}^i = (\lambda_1^i, \lambda_2^i, ...\lambda_M^i), i = 1, ..., N$
2: For $i = 1, \cdots, N$, define the set of indices $B^i = \{i_1, \cdots, i_C\}$ where $\{\boldsymbol{\lambda}^{i_1}, .., \boldsymbol{\lambda}^{i_C}\}$ are the C closest weight vectors to $\boldsymbol{\lambda}^i$ (by the Euclidean distance)
3: Generate an initial population $P^0 = \{\mathbf{x}^1, \cdots, \mathbf{x}^N\}$, $\mathbf{x}^i = (x_1^i, x_2^i, ...x_n^i)$
4: Evaluate each individual in the initial population P^0 and associate $\mathbf{x}^i$ with $\boldsymbol{\lambda}^i$
5: Initialize $\mathbf{z}^* = (z_1^*, \cdots, z_M^*)$ by setting $z_j^* = min_{1 \leq i \leq N} f_j(\mathbf{x}^i)$
6: $g = 1$
7: **repeat**
8:    Let all the indices of the subproblems whose objectives are MOP individual objectives f_i compose the initial I. By using 10-tournament selection based on π^i, select other $N/5M$ indices and add them to I.
9:    **for** each individual $\mathbf{x}^i$ in I **do**
10:       s = argmax$_{i=1...LLH}$ CF(i) where LLH is the number of low level heuristics
11:       **if** $rand < \delta$ **then** //*Determining the scope* (rand in U[0,1])
12:          $scope = B^i$
13:       **else**
14:          $scope = \{1, \cdots, N\}$
15:       **end if**
16:       Generate a new solution $\mathbf{y}$ by strategy s (repair it if necessary)
17:       Apply polynomial mutation to produce $\mathbf{y}'$ (repair it if necessary)
18:       Update $\mathbf{z}^*$, $z_j^* = min(z_j^*, f_j(\mathbf{y}'))$
19:       **for** each subproblem k (k randomly selected from *scope*) **do**
20:          **if** $g^{te}(\mathbf{y}' \mid \boldsymbol{\lambda}^k, \mathbf{z}^*) < g^{te}(\mathbf{x}^k \mid \boldsymbol{\lambda}^k, \mathbf{z}^*)$ **then**
21:             **if** a new replacement may occur **then**
22:                Replace $\mathbf{x}^k$ by $\mathbf{y}'$ and increment n_r
23:             **end if**
24:          **end if**
25:       **end for**
26:       reward = $g^{te}(\mathbf{x}^i \mid \boldsymbol{\lambda}^k, \mathbf{z}^*) - g^{te}(\mathbf{y}' \mid \boldsymbol{\lambda}^k, \mathbf{z}^*)$
27:       Update CF(s) accordingly to Algorithm 2
28:    **end for**
29:    **if** g modulo 50 == 0 **then**
30:       Update the utility π^i of each subproblem i
31:    **end if**
32:    $g = g + 1$;
33: **until** $g >$MAX-EV

The first steps of MOEA/D-HH initialize various data structures (steps 1 to 6). The weight vectors $\boldsymbol{\lambda}^i$, $i = 1, ..., N$, representing coefficients associated with each objective, are generated using a uniform distribution. The neighborhood

$(B^i = \{i_1, \cdots, i_C\})$ of weight vector $\boldsymbol{\lambda}^i$ is comprised by the indices of the C weight vectors closest to $\boldsymbol{\lambda}^i$. The initial population is randomly generated and evaluated. Each individual $(\mathbf{x}^i)$ is associated with the ith weight vector. The empirical ideal point $(\mathbf{z}^*)$ is initialized as the minimum value of each objective found in the initial population and the generation (g) is set to 1.

After the initialization steps, the algorithm enters its main loop (steps 7 to 33). The first step of the main loop is to determine which individuals from the population will be processed. A 10-tournament selection based on the utility value of each subproblem (π^i, calculated accordingly to Equation 1) is used to determine these individuals.

$$\pi^i = \begin{cases} 1, & \text{if } \Delta^i > 0.001 \\ (0.95 + 0.05 * \Delta^i/0.001) * \pi^i, & \text{otherwise} \end{cases} \tag{1}$$

Where Δ^i is the relative decrease of the objective function value of subproblem i.

Step 10 selects the low level heuristic s used to generate a new individual. The selection is performed based on the Choice Function value of each strategy i ($CF(i)$), which is calculated and updated in step 27. The heuristic chosen is the one with higher $CF(i)$ value. In this work, the low level heuristics used correspond to DE mutation and crossover strategies.

The low level heuristics are applied considering individuals randomly selected from $scope$. In this work, $scope$ can swap from the neighborhood to the entire population (and vice-versa) along the evolutionary process of MOEA/D-HH. It is composed by the indices of chromosomes from either the neighborhood B^i (with probability δ) or from the entire population (with probability $1 - \delta$). Applying the chosen low level heuristic, a modified chromosome $\mathbf{y}$ is generated in step 16.

The polynomial mutation in step 17 generates $\mathbf{y}' = (y'_1, \cdots, y'_n)$ from $\mathbf{y}$ in the following way [6]:

$$y'_d = \begin{cases} y_d + \sigma_d \cdot (y_d^{(U_p)} - y_d^{(L_w)}), & \text{with probability } p_m \\ y_d, & \text{with probability } 1 - p_m \end{cases} \tag{2}$$

with

$$\sigma_d = \begin{cases} (2 \cdot rand)^{\frac{1}{\tau+1}} - 1, & \text{if } rand < 0.5 \\ 1 - (2 - 2 \cdot rand)^{\frac{1}{\tau+1}}, & \text{otherwise} \end{cases} \tag{3}$$

where $rand \in U[0,1]$. The distribution index τ and the mutation rate p_m are two DE parameters. Remembering $y_d^{(L_w)}$ and $y_d^{(U_p)}$ are the lower and upper bounds of the dth decision variable, respectively.

In step 18, if the new chromosome $\mathbf{y}'$ has an objective value better than the value stored in the empirical ideal point, $\mathbf{z}^*$ is updated with this value.

The next steps involve the population update process (steps 19 to 25) which is based on the comparison of the fitness of individuals. In the MOEA/D framework, the fitness of an individual is measured accordingly to a decomposition function. In this work the *Tchebycheff function* is used. Using this decomposition method, each subproblems have the form:

$$\text{Min } g^{te}(\mathbf{x} \mid \boldsymbol{\lambda}, \mathbf{z}^*) = \max_{1 \leq j \leq M} \{\lambda_j \mid f_j(\mathbf{x}) - z_j^* \mid\} \tag{4}$$

$$\text{subject to } \mathbf{x} \in \Omega$$

where g^{te} is the Tchebycheff function, $\mathbf{f}(\mathbf{x}) = (f_1(\mathbf{x}), ..., f_M(\mathbf{x}))$ is set of functions to be minimized, and $\boldsymbol{\lambda} = (\lambda_1, ..., \lambda_M)$ is the weight vector considered.

Accordingly to what is selected for the *scope* (steps 12 or 14), the neighborhood or the entire population is updated. The population update is as follows: if a new replacement may occur, (i.e. while $n_r < NR$ and there are unselected indices in *scope*), a random index (k) from *scope* is chosen. If $\mathbf{y}'$ has a better Tchebycheff value than $\mathbf{x}^k$ (both using the kth weight vector - $\boldsymbol{\lambda}^k$) then $\mathbf{y}'$ replaces $\mathbf{x}^k$ and the number of updated chromosomes (n_r) is incremented. To avoid the proliferation of $\mathbf{y}'$ to a great part of the population, a maximum number of updates (NR) is used.

Then, the reward obtained by the application of the selected low level heuristic is calculated as the difference between the Tchebycheff value of the parent and child. This reward is used in step 27 to update the Choice Function values. Two kinds of Choice Functions are investigated in this work: the Modified Choice Function proposed in [7] and a New Choice Function proposed in this work (see Subsection 5.1).

If the current generation is a multiple of 50, then the utility value of each subproblem is updated using Equation 1.

The evolutionary process stops when the maximum number of evaluations (MAX-EV) is reached and MOEA/D-HH outputs the Pareto set and Pareto front approximations.

5.1   Proposed Choice Function

One of the most important parts of a hyper-heuristic is the selection method, which is responsible for choosing the best heuristic for each stage of the optimization process. In this paper we use two adaptive versions of the Choice Function [11] as the selection method: the Modified Choice Function [7] and a new version proposed in this paper.

The Choice Function scores heuristics based on a combination of different measures and the heuristic to apply at each stage is chosen based on these scores. In this work, three different measures (functions) are used. The first measure (f_1) records the previous performance of each individual heuristic. The second measure (f_2) considers the pair-wise relationship between heuristics. The last measure (f_3) corresponds to the time elapsed since each heuristic was last selected by the Choice Function. The classical Choice Function is presented in Equation 5 [11].

$$CF(i) = \alpha * f_1(i) + \beta * f_2(i) + \gamma * f_3(i) \tag{5}$$

One difficult that emerges when using the classical version of the Choice Function is how to set the parameters α, β and γ appropriately. In order to avoid this difficulty we used the Modified Choice Function proposed in [7]. This version substitutes parameters α and β for the parameter ϕ and the parameter γ is relabelled as δ. In this way, ϕ is associated with the intensification of the selection of the best heuristics while the parameter δ is associated with the diversification of the selection. Parameter ϕ is set to 0.99 each time an heuristic successfully improves a solution and decreased by 0.01 otherwise. Parameter δ is set to 1 - ϕ. The Modified Choice Function is shown in Equation 6 [7].

$$CF(i) = \phi * f_1(i) + \phi * f_2(i) + \delta * f_3(i) \qquad (6)$$

But the Modified Choice Function still suffers from two problems: (i) the measures used by f_1 and f_2 can be in different scales from the measures used in f_3 and (ii) the raw accumulation of previous rewards in f_1 and f_2 is detrimental to the on-line adaptation of the selection: a heuristic that becomes the best heuristic has to offset the accumulated rewards of the previous best heuristic before it is systematically selected. So in this paper a new Choice Function is proposed. In order to deal with (i) we propose a scale factor (SF) while to deal with (ii) we propose the use of mean values for f_1 and f_2 instead of the accumulated values.

The update procedure of the new Choice Function is described in Algorithm 2. *mean* is a boolean variable that determines if the mean value or the accumulated value is used in f_1 and f_2. The Modified Choice Function happens SF is set to 1.0 and *mean* is false.

6 Experiments and Results

In this section we present the experiments conducted to evaluate our proposed approach, considering all the unconstrained (bound constrained) instances from the CEC 2009 multi-objective benchmark (10 instances) [24]. The search space dimension n is defined as 30 for all the instances. Table 1 shows the characteristics of each instance.

The experiments' set is composed by three parts: (i) empirical setting of the choice function parameters, Subsection 6.1, (ii) investigation of the benefits of the proposed Hyper-Heuristic when compared with single low level heuristic versions, Subsection 6.2, and (iii) comparison with recent literature algorithms, Subsection 6.3. Inverted Generational Distance (IGD) and Hypervolume (HV) are used to assess convergence, uniformity and spread performance [13]. All tables of this section use the following color conventions: dark gray cells indicate the best results while light gray cells mark statistical equivalent result accordingly to the Wilcoxon [4] rank sum test with 95% confiability.

6.1 Choice Function Parameter Setting

This section presents the effects of the choice function parameters *mean* and SF. Six versions of the algorithm are tested: CF1, CF0.5, CF5, CFMean1,

---

Algorithm 2. Pseudocode of the New Choice Function

1: $f_1(s) = f_1(s)$ + reward
2: timesUsed(s)++
3: f_2(lastOperator, s) = f_2(lastOperator, s) + lastReward + reward
4: timesUsedPair(lastOperator, s)++
5: **for** i = 1 ... LLH **do**
6:     $f_3(i) = f_3(i)$ + timeSpentOn(s)
7: **end for**
8: $f_3(s) = 0$
9: **if** reward > 0 **then**
10:     $\phi = 0.99$
11:     $\delta = 0.01$
12: **else**
13:     $\phi = \phi$ - 0.01
14:     $\delta = 1.0 - \phi$
15: **end if**
16: lastReward = reward
17: lastOperator = s
18: **for** i = 1 ... LLH **do**
19:     **if** mean is true **then**
20:         CF(i) = SF * (ϕ * $f_1(i)$/timesUsed(i) + ϕ * f_2(s, i)/timesUsedPair(s, i))
                + δ * $f_3(i)$
21:     **else**
22:         CF(i) = SF * (ϕ * $f_1(i)$ + ϕ * f_2(s, i)) + δ * $f_3(i)$
23:     **end if**
24: **end for**

CFMean0.5 and CFMean5. The *mean* parameter is false for the first three versions and true otherwise. The value at the end of the versions' names represents the different values for the SF parameter of the function choice (1, 0.5 and 5). It is important to note that version CF1 corresponds to the use of the Modified Choice Function while all other versions uses the proposed Choice Function.

Table 2 shows the mean values for the IGD metric. The version that obtained best results set *mean* to true and *SF* to 0.5 (CFMean0.5). This version obtained best results for UF1, UF2, UF3, UF6 and UF7. Table 3 shows the mean values for the Hypervolume (HV) metric. Again the version that obtained best results was CFMean0.5. This version obtained best results for UF1, UF2, UF6 and UF7. So, for the remainder of this paper only the CFMean0.5 version will be considered and it will be renamed as MOEA/D-HH. Table 4 presents the parameter values used in MOEA/D-HH.

6.2   MOEA/D-HH × Single Low Level Heuristics

This section presents the comparison of the proposed algorithm (MOEA/D-HH) with the differential evolution strategies (low level heuristics) applied individually. Tables 5 and 6 demonstrate that MOEA/D-HH is better than or equal to individual low level heuristics for all CEC 2009 MOEA competition instances,

Table 1. Characteristics of unconstrained functions considered

Function	Objectives	Search space range	Properties of Pareto Front
UF1	2	$[0, 1] \times [-1, 1]^{n-1}$	Concave
UF2	2	$[0, 1] \times [-1, 1]^{n-1}$	Concave
UF3	2	$[0, 1]^n$	Concave
UF4	2	$[0, 1] \times [-2, 2]^{n-1}$	Convex
UF5	2	$[0, 1] \times [-1, 1]^{n-1}$	21 points front
UF6	2	$[0, 1] \times [-1, 1]^{n-1}$	One isolated point and 2 disconnected parts
UF7	2	$[0, 1] \times [-1, 1]^{n-1}$	Continuous straight line
UF8	3	$[0, 1]^2 \times [-2, 2]^{n-2}$	Parabolic
UF9	3	$[0, 1]^2 \times [-2, 2]^{n-2}$	Planar
UF10	3	$[0, 1]^2 \times [-2, 2]^{n-2}$	Parabolic

with the exception of UF8 and UF9 in the IGD metric and UF6 in the HV metric. Therefore, the results of the MOEA/D-HH surpass those obtained by any of the individual strategies.

6.3   MOEA/D-HH: Comparison with Literature

In the case of comparison with literature, we use the IGD-metric and empirical attainment function (EAF) plots [15]. An empirical attainment function plot graphically describes the probabilistic distribution of the outcomes obtained by a stochastic algorithm in the objective space [15]. We evaluate the IGD-metric

Table 2. Mean IGD values for CF1, CF0.5, CF5, CFMean1, CFMean0.5 and CFMean5

	CF1	CF0.5	CF5	CFMean1	CFMean0.5	CFMean5
UF1	$2.22e - 03$	$1.97e - 03$	$2.12e - 03$	$1.81e - 03$	$1.74e - 03$	$2.01e - 03$
UF2	$3.21e - 03$	$3.14e - 03$	$3.13e - 03$	$3.17e - 03$	$3.10e - 03$	$3.17e - 03$
UF3	$2.63e - 02$	$2.06e - 02$	$2.01e - 02$	$5.17e - 03$	$5.15e - 03$	$6.06e - 03$
UF4	$5.36e - 02$	$5.42e - 02$	$5.45e - 02$	$5.50e - 02$	$5.46e - 02$	$5.51e - 02$
UF5	$2.62e - 01$	$2.91e - 01$	$3.04e - 01$	$2.85e - 01$	$2.90e - 01$	$2.96e - 01$
UF6	$1.19e - 01$	$1.04e - 01$	$1.25e - 01$	$1.27e - 01$	$8.70e - 02$	$1.36e - 01$
UF7	$1.77e - 03$	$1.81e - 03$	$1.75e - 03$	$1.69e - 03$	$1.69e - 03$	$1.83e - 03$
UF8	$4.68e - 02$	$4.63e - 02$	$4.63e - 02$	$4.43e - 02$	$5.03e - 02$	$4.46e - 02$
UF9	$3.96e - 02$	$4.67e - 02$	$5.65e - 02$	$4.58e - 02$	$4.25e - 02$	$5.11e - 02$
UF10	$4.45e - 01$	$4.37e - 01$	$4.54e - 01$	$4.26e - 01$	$4.35e - 01$	$4.15e - 01$

Table 3. Mean HV values for CF1, CF0.5, CF5, CFMean1, CFMean0.5 and CFMean5

	CF1	CF0.5	CF5	CFMean1	CFMean0.5	CFMean5
UF1	$1.98e - 03$	$1.73e - 03$	$1.92e - 03$	$1.52e - 03$	$1.56e - 03$	$2.01e - 03$
UF2	$3.10e - 03$	$2.83e - 03$	$2.96e - 03$	$2.97e - 03$	$3.04e - 03$	$3.06e - 03$
UF3	$9.53e - 03$	$8.02e - 03$	$7.79e - 03$	$4.32e - 03$	$4.28e - 03$	$3.71e - 03$
UF4	$5.42e - 02$	$5.44e - 02$	$5.41e - 02$	$5.50e - 02$	$5.38e - 02$	$5.47e - 02$
UF5	$2.61e - 01$	$2.76e - 01$	$3.09e - 01$	$2.75e - 01$	$2.90e - 01$	$2.91e - 01$
UF6	$7.53e - 02$	$7.35e - 02$	$7.31e - 02$	$7.79e - 02$	$6.84e - 02$	$7.73e - 02$
UF7	$1.68e - 03$	$1.63e - 03$	$1.53e - 03$	$1.61e - 03$	$1.58e - 03$	$1.59e - 03$
UF8	$4.46e - 02$	$4.46e - 02$	$4.44e - 02$	$4.38e - 02$	$4.55e - 02$	$4.38e - 02$
UF9	$3.01e - 02$	$3.01e - 02$	$2.99e - 02$	$2.97e - 02$	$2.96e - 02$	$3.06e - 02$
UF10	$4.32e - 01$	$4.31e - 01$	$4.61e - 01$	$4.24e - 01$	$4.37e - 01$	$4.19e - 01$

Table 4. Parameters used in different versions of the proposed approach

	Values	Description
DE Parameters		
N	600	Population size (for instances with 2-objectives).
	1000	Population size (for instances with 3-objectives).
CR	1.0	Crossover rate.
F	0.5	Scaling factor.
P_{inter}	0.75	Interpolation probability.
p_m	1/30	Polynomial mutation probability.
τ	20	Distribution index of polynomial mutation.
MAX-EV	300,000	Maximum number of evaluations.
MOEA/D Parameters		
C	20	Number of weight vectors in the neighborhood.
NR	2	Maximal number of solutions replaced by each offspring.
δ	0.9	Probability that parent solutions are selected from the neighborhood.
Choice Function Parameters		
Mean	true	Choice Function with f_1 and f_2 calculated using the average values of the rewards (proposed in this paper).
SF	0.5	Scale parameter to balance f_1 and f_2 versus f_3.

of the final approximation over 30 independent executions of our approach,
MOEA/D-HH, MOEA/D-DRA, MOEA/D-DRA-CMX+SPX and NSGA-II for

Table 5. Mean IGD values for MOEA/D-HH and single low level heuristics

	MOEA/D-HH	rand/1/bin	rand/2/bin	current-to-rand/1	current-to-rand/2	nonlinear
UF1	$1.74e-03$	$2.56e-03$	$2.51e-03$	$4.14e-03$	$1.91e-03$	$2.97e-03$
UF2	$3.10e-03$	$5.05e-03$	$4.84e-03$	$8.70e-03$	$4.59e-03$	$2.50e-03$
UF3	$5.15e-03$	$1.51e-02$	$7.71e-03$	$6.74e-02$	$4.67e-03$	$1.14e-02$
UF4	$5.46e-02$	$5.72e-02$	$5.15e-02$	$6.04e-02$	$5.55e-02$	$5.66e-02$
UF5	$2.90e-01$	$3.44e-01$	$2.90e-01$	$3.84e-01$	$2.96e-01$	$3.79e-01$
UF6	$8.70e-02$	$1.43e-01$	$7.15e-02$	$1.92e-01$	$9.79e-02$	$1.78e-01$
UF7	$1.69e-03$	$1.73e-03$	$2.64e-03$	$4.45e-03$	$2.33e-03$	$4.18e-03$
UF8	$5.03e-02$	$4.66e-02$	$5.24e-02$	$4.80e-02$	$4.69e-02$	$7.29e-02$
UF9	$4.64e-02$	$4.64e-02$	$4.06e-02$	$5.65e-02$	$4.01e-02$	$1.32e-01$
UF10	$4.35e-01$	$4.51e-01$	$1.01e+00$	$4.13e-01$	$7.60e-01$	$4.58e-01$

Table 6. Mean HV values for MOEA/D-HH and single low level heuristics

	MOEA/D-HH	rand/1/bin	rand/2/bin	current-to-rand/1	current-to-rand/2	nonlinear
UF1	$1.56e-03$	$2.54e-03$	$2.35e-03$	$1.44e-03$	$1.65e-03$	$2.36e-03$
UF2	$3.04e-03$	$4.34e-03$	$4.53e-03$	$8.42e-03$	$4.48e-03$	$2.44e-03$
UF3	$4.28e-03$	$4.56e-03$	$5.53e-03$	$6.95e-02$	$3.06e-03$	$1.00e-02$
UF4	$5.38e-02$	$5.65e-02$	$5.07e-02$	$5.98e-02$	$5.46e-02$	$5.59e-02$
UF5	$2.90e-01$	$2.98e-01$	$2.84e-01$	$3.68e-01$	$2.93e-01$	$3.62e-01$
UF6	$6.84e-02$	$7.20e-02$	$6.58e-02$	$1.96e-01$	$6.42e-02$	$1.75e-01$
UF7	$1.58e-03$	$1.48e-03$	$2.23e-03$	$3.14e-03$	$2.11e-03$	$4.02e-03$
UF8	$4.55e-02$	$4.47e-02$	$5.11e-02$	$4.62e-02$	$4.56e-02$	$6.44e-02$
UF9	$2.96e-02$	$3.18e-02$	$3.31e-02$	$3.44e-02$	$2.88e-02$	$1.54e-01$
UF10	$4.37e-01$	$4.41e-01$	$1.04e+00$	$4.07e-01$	$7.52e-01$	$4.56e-01$

each CEC09 test instance. Table 7 presents median, standard deviation (std), minimum (min) and maximum (max) of IGD-metric values. The IGD-metric values of the MOEA/D-DRA-CMX+SPX are those described in [12,26]. Table 7 shows that MOEA/D-HH is the best method in UF2, UF4, UF6 and UF9. Moreover, MOEA/D-HH is the second best method in UF3, UF7 and UF8.

Table 7. The IGD statistics based on 30 independent runs for our approach, MOEA/D-HH, MOEA/D-DRA, MOEA/D-DRA-CMX+SPX and NSGA-II. Dark gray cells emphasize the best results while light gray cells emphasize the second best results.

CEC09	Algorithm	Median	Std	Min	Max
UF1	MOEA/D-HH	0.001558	0.000532	0.001114	0.002817
	MOEA/D-DRA	0.001503	0.000090	0.001417	0.001757
	MOEAD-CMX-SPX	0.004171	0.000263	0.003985	0.005129
	NSGA-II	0.095186	0.003249	0.088511	0.103222
UF2	MOEA/D-HH	0.003037	0.000767	0.001898	0.005170
	MOEA/D-DRA	0.003375	0.001039	0.002326	0.006638
	MOEAD-CMX-SPX	0.005472	0.000412	0.005149	0.006778
	NSGA-II	0.035151	0.001479	0.032968	0.039164
UF3	MOEA/D-HH	0.004280	0.003064	0.001706	0.012040
	MOEA/D-DRA	0.001488	0.004131	0.001086	0.014019
	MOEAD-CMX-SPX	0.005313	0.013093	0.004155	0.068412
	NSGA-II	0.089894	0.016815	0.062901	0.126556
UF4	MOEA/D-HH	0.053836	0.003800	0.049133	0.064506
	MOEA/D-DRA	0.060765	0.004757	0.051492	0.070912
	MOEAD-CMX-SPX	0.063524	0.004241	0.055457	0.075361
	NSGA-II	0.080935	0.002809	0.074034	0.084683
UF5	MOEA/D-HH	0.289896	0.068554	0.164365	0.510875
	MOEA/D-DRA	0.220083	0.089484	0.146933	0.511464
	MOEAD-CMX-SPX	0.379241	0.135554	0.211058	0.707093
	NSGA-II	0.214958	0.051622	0.154673	0.331853
UF6	MOEA/D-HH	0.068445	0.046913	0.036113	0.215999
	MOEA/D-DRA	0.207831	0.287571	0.053371	0.823381
	MOEAD-CMX-SPX	0.248898	0.185717	0.056972	0.792910
	NSGA-II	0.080177	0.006460	0.067996	0.090331
UF7	MOEA/D-HH	0.001576	0.000415	0.001229	0.002907
	MOEA/D-DRA	0.001569	0.001364	0.001336	0.008796
	MOEAD-CMX-SPX	0.004745	0.003307	0.003971	0.014662
	NSGA-II	0.048873	0.001959	0.044634	0.051968
UF8	MOEA/D-HH	0.045544	0.012282	0.037872	0.080418
	MOEA/D-DRA	0.040352	0.003788	0.033777	0.050412
	MOEAD-CMX-SPX	0.056872	0.003366	0.051800	0.065620
	NSGA-II	0.112219	0.002742	0.109836	0.121190
UF9	MOEA/D-HH	0.029626	0.034824	0.027201	0.146364
	MOEA/D-DRA	0.137856	0.039018	0.025008	0.139986
	MOEAD-CMX-SPX	0.144673	0.054285	0.033314	0.151719
	NSGA-II	0.106841	0.000681	0.105806	0.108729
UF10	MOEA/D-HH	0.436591	0.045099	0.348434	0.525347
	MOEA/D-DRA	0.406094	0.066770	0.210500	0.553572
	MOEAD-CMX-SPX	0.467715	0.038698	0.391496	0.533234
	NSGA-II	0.257846	0.012541	0.234047	0.288036

In oder to graphically compare the algorithmic behaviors of MOEA/D-HH and MOEA/D-DRA, empirical attainment functions were plotted. Figures 1 and 2 present EAFs for MOEA/D-HH and MOEA/D-DRA for instances UF2 and UF4. For UF2, MOEA/D-HH is better around the extreme value for the first objective. MOEA/D-HH is better than MOEA/D-DRA for UF4 in both extremes and in some areas of the middle of the approximation frontiers.

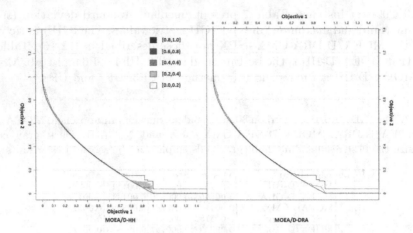

Fig. 1. Attainment Function Comparison of MOEA/D-HH and MOEA/D-DRA for UF2

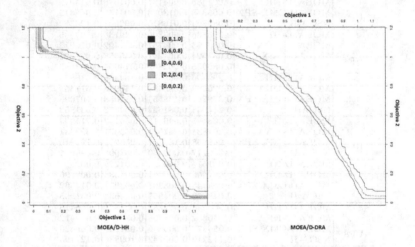

Fig. 2. Attainment Function Comparison of MOEA/D-HH and MOEA/D-DRA for UF4

7   Conclusions

In this paper, we proposed a new multi-objective hyper-heuristic based on choice function to adaptively select appropriate low level heuristics (operators) in the MOEA/D framework. Our pool of low level heuristics was constituted by five commonly used DE operators.

We also proposed an innovative adaptive choice function that tackles two important problems that may happen: (i) difference in the scale of the functions that compose the choice function and (ii) if a heuristic is chosen frequently, its accumulated reward may dominate the rewards from other heuristics.

We conducted experimental studies on test instances from the CEC 2009 MOEA Competition. The results demonstrate that the use of the proposed hyper-heuristic is beneficial, i.e. it improves the performance of the MOEA/D framework using single low level heuristics.

The experiments conducted also showed that proposed adaptive choice function performs better than a modified choice function recently proposed in [7]. The proposed approach (MOEA/D-HH) was favourably compared with state-of-the-art methods.

Future work includes the investigation of the contribution of each low level heuristic and the effectiveness of the adopted performance measures. It also includes the proposal of other selection heuristics and analysis of the performance of MOEA/D-HH in many objective problems.

Acknowledgments. The authors acknowledge CAPES and Fundação Araucária for the partial financial support with project number 23.116/2012.

References

1. Baños, R., Ortega, J., Gil, C., Márquez, A.L., De Toro, F.: A hybrid meta-heuristic for multi-objective vehicle routing problems with time windows. Comput. Ind. Eng. **65**(2), 286–296 (2013)
2. Burke, E.K., Gendreau, M., Hyde, M., Kendall, G., Ochoa, G., Ozcan, E., Qu, R.: Hyper-heuristics. J. Oper. Res. Soc. **64**(12), 1695–1724 (2013)
3. Burke, E.K., Silva, J.L., Silva, A., Soubeiga, E.: Multi-objective hyper-heuristic approaches for space allocation and timetabling. In: Meta-heuristics: Progress as Real Problem Solvers, p. 129. Springer (2003)
4. Conover, W.J.: Practical Nonparametric Statistics, 3rd edn. Wiley (1999)
5. Cowling, P.I., Kendall, G., Soubeiga, E.: A hyperheuristic approach to acheduling a sales summit. In: Burke, E., Erben, W. (eds.) PATAT 2000. LNCS, vol. 2079, pp. 176–190. Springer, Heidelberg (2001)
6. Deb, K.: Multi-Objective Optimization using Evolutionary Algorithms. Wiley-Interscience Series in Systems and Optimization. John Wiley & Sons, Chichester (2001)
7. Drake, J.H., Özcan, E., Burke, E.K.: An improved choice function heuristic selection for cross domain heuristic search. In: Coello, C.A.C., Cutello, V., Deb, K., Forrest, S., Nicosia, G., Pavone, M. (eds.) PPSN 2012, Part II. LNCS, vol. 7492, pp. 307–316. Springer, Heidelberg (2012)
8. Ehrgott, M.: A discussion of scalarization techniques for multiple objective integer programming. Ann. Oper. Res. **147**, 343–360 (2006)
9. Gomez, J.C., Terashima-Marín, H.: Approximating multi-objective hyper-heuristics for solving 2D irregular cutting stock problems. In: Sidorov, G., Hernández Aguirre, A., Reyes Garcia, C.A. (eds.) MICAI 2010, Part II. LNCS, vol. 6438, pp. 349–360. Springer, Heidelberg (2010)
10. Kateb, D.E., Fouquet, F., Bourcier, J., Traon, Y.L.: Artificial mutation inspired hyper-heuristic for runtime usage of multi-objective algorithms. CoRR abs/1402.4442 (2014). http://arxiv.org/abs/1402.4442

11. Kendall, G., Soubeiga, E., Cowling, P.: Choice function and random hyperheuristics. In: Proceedings of the Fourth Asia-Pacific Conference on Simulated Evolution and Learning, SEAL, pp. 667–671. Springer (2002)
12. Khan Mashwani, W., Salhi, A.: A decomposition-based hybrid multiobjective evolutionary algorithm with dynamic resource allocation. Appl. Soft. Comput. **12**(9), 2765–2780 (2012)
13. Knowles, J., Thiele, L., Zitzler, E.: A Tutorial on the Performance Assessment of Stochastic Multiobjective Optimizers. TIK Report 214, Computer Engineering and Networks Laboratory (TIK), ETH Zurich, February 2006
14. Li, H., Zhang, Q.: Multiobjective optimization problems with complicated Pareto sets, MOEA/D and NSGA-II. IEEE Trans. Evol. Comput. **13**(2), 284–302 (2009)
15. Lpez-Ibez, M., Paquete, L., Sttzle, T.: Exploratory analysis of stochastic local search algorithms in biobjective optimization. In: Bartz-Beielstein, T., Chiarandini, M., Paquete, L., Preuss, M. (eds.) Experimental Methods for the Analysis of Optimization Algorithms, pp. 209–222. Springe, Heidelberg (2010)
16. Maashi, M., Özcan, E., Kendall, G.: A multi-objective hyper-heuristic based on choice function. Expert Systems with Applications **41**(9), 4475–4493 (2014)
17. Pappa, G., Ochoa, G., Hyde, M., Freitas, A., Woodward, J., Swan, J.: Contrasting meta-learning and hyper-heuristic research: the role of evolutionary algorithms. Genetic Programming and Evolvable Machines **15**(1), 3–35 (2014)
18. Qin, A.K., Huang, V.L., Suganthan, P.N.: Differential evolution algorithm with strategy adaptation for global numerical optimization. IEEE Trans. Evol. Comput. **13**, 398–417 (2009)
19. Sindhya, K., Ruuska, S., Haanp, T., Miettinen, K.: A new hybrid mutation operator for multiobjective optimization with diferential evolution. Soft Comput. **15**(10), 2041–2055 (2011)
20. Storn, R.: On the usage of differential evolution for function optimization. In: NAFIPS 1996, pp. 519–523. IEEE (1996)
21. Storn, R., Price, K.: Differential evolution – a simple and efficient heuristic for global optimization over continuous spaces. J. Global Optim. **11**(4), 341–359 (1997)
22. Vazquez-Rodriguez, J.A., Petrovic, S.: A mixture experiments multi-objective hyper-heuristic. J. Oper. Res. Soc. **64**(11), 1664–1675 (2013)
23. Zhang, Q., Li, H.: MOEA/D: A multi-objective evolutionary algorithm based on decomposition. IEEE Trans. Evol. Comput. **11**(6), 712–731 (2007)
24. Zhang, Q., Zhou, A., Zhao, S., Suganthan, P.N., Liu, W., Tiwari, S.: Multiobjective optimization test instances for the CEC 2009 special session and competition. Tech. rep., University of Essex and Nanyang Technological University, CES-487 (2008)
25. Zhang, Q., Liu, W., Li, H.: The performance of a new version of MOEA/D on CEC09 unconstrained MOP test instances. In: Congress on Evolutionary Computation, pp. 203–208 (2009)
26. Zhao, S.Z., Suganthan, P.N., Zhang, Q.: Decomposition-based multiobjective evolutionary algorithm with an ensemble of neighborhood sizes. IEEE Trans. Evol. Comput. **16**(3), 442–446 (2012)

Using Hyper-Heuristic to Select Leader and Archiving Methods for Many-Objective Problems

Olacir R. Castro Jr. (✉) and Aurora Pozo

Computer Science's Department, Federal University of Paraná,
Curitiba, Paraná, Brazil
olacirjr@gmail.com, aurora@inf.ufpr.br

Abstract. Multi-objective Particle Swarm Optimization (MOPSO) is a promising meta-heuristic to solve Many-Objective Problems (MaOPs). Previous works have proposed different leader and archiving methods to tackle the challenges caused by the increase in the number of objectives, however, selecting the most appropriate components for a given problem is not a trivial task. Moreover, the algorithm can take advantage by using a variety of methods in different phases of the search. To deal with those issues, we adopt the use of hyper-heuristics, whose concept emerges for dynamically selecting components for effectively solving a problem. In this work, we use a simple hyper-heuristic to select leader and archiving methods during the search. Unlike other studies, our hyper-heuristic is guided by the R_2 indicator due to its good measuring characteristics and low computational cost. Experimental studies were conducted to validate the new algorithm where its performance is compared to its components individually and to the state-of-the-art MOEA/D-DRA algorithm. The results show that the new algorithm is robust, presenting good results in different situations.

Keywords: MOPSO · Many-objective · Hyper-heuristics · Leader selection · Archiving · R_2 indicator

1 Introduction

A promising meta-heuristic to deal with Many-Objective Problems (MaOPs) is the Multi-Objective Particle Swarm Optimization (MOPSO), a multi-objective version of the well-known Particle Swarm Optimization (PSO) [10]. PSO is a stochastic meta-heuristic based on the movement of bird flocks looking for food, created to optimize non-linear functions.

A MOPSO that has presented good results in the literature and is used as base in many works [2], [5] is the Speed-constrained Multi-objective PSO (SMPSO) [15]. SMPSO is a MOPSO that restricts the velocity of the particles to prevent erratic movements.

Despite the good results obtained by SMPSO using different leader [5] and archiving [2] methods, the correct choice of these operators is hard because they are problem-dependent. The difficulty in selecting operators, which is as well usual

© Springer International Publishing Switzerland 2015
A. Gaspar-Cunha et al. (Eds.): EMO 2015, Part I, LNCS 9018, pp. 109–123, 2015.
DOI: 10.1007/978-3-319-15934-8_8

to other meta-heuristics, leads to the development of a methodology called hyper-heuristic.

A hyper-heuristic can be used as a high-level methodology to dynamically select low-level components, producing automatically a suitable combination to solve a given problem [4].

In this work we investigate two hypotheses: h_1 selecting good combinations between leader and archiving methods improve the results obtained by MOPSO; h_2 R_2 [9] is an appropriate indicator to guide a hyper-heuristic.

To investigate these hypotheses, we design a new MOPSO algorithm called H-MOPSO that uses a simple hyper-heuristic to dynamically select a combination of leader and archiving methods guided by the R_2 indicator.

Few works on the literature apply hyper-heuristics on multi-objective optimizers, and this work, as far as we known, is the first one that uses a hyper-heuristic to select leader and archiving methods in a MOPSO.

Empirical analyses are conducted to observe the behavior of H-MOPSO when faced to many-objective scenarios. The experiments use the DTLZ many-objective family of benchmarking problems [6]. A first study is conducted to validate the new algorithm where its performance is compared to its low-level heuristics employed separately. Additionally, a second study is performed to compare it to the state-of-the-art MOEA/D-DRA [19] algorithm.

The remainder of this paper is organized as follows: Section 2 presents the background concepts used in this work. Section 3 describes H-MOPSO, and Section 4, the empirical study conducted to assess the performance of the new H-MOPSO. Finally, the conclusions are presented in Section 5.

2   Background

This section presents some background concepts used in this paper. Section 2.1 surveys PSO and MOPSO, as well as the SMPSO algorithm. Some characteristics of hyper-heuristics are summarized in Section 2.2.

2.1   MOPSO

Particle Swarm Optimization (PSO) [10] is a stochastic meta-heuristic created to optimize nonlinear functions based on the movement of bird flocks looking for food. In this method a swarm (population) of particles (solutions) moves across the search space (evolve) guided by personal and social leaders.

Unlike in single objective optimization, a Multi-Objective Problem (MOP) requires the simultaneous optimization of two or more objective functions. The objective functions are usually in conflict, so these problems do not present only one optimal solution, but a set of them. This set of solutions is called Pareto optimal set.

In a MOP, a solution is said to dominate another if it is not worse in any objective and is strictly better in at least one. Solutions in the Pareto set are non-dominated by any other solution of the feasible solution space and the image of these solutions in the objective space is called Pareto front [13].

To expand a PSO algorithm into a Multi Objective PSO (MOPSO), usually two main modifications are made: the creation of an external archive (or repository) to store the non-dominated solutions, and the use of a leader selection method to select a global leader for the particles among a set of equally good solutions according to some criterion.

When the external archive becomes full, an archiving strategy is needed to prune it and keep it on a predefined size, discarding some non-dominated solutions based on some criterion. This criterion has great impact in the quality of the solutions generated in the search, especially in many-objectives due to the large portion of the population that becomes non-dominated. There are many approaches in the literature to manage the repository, and a comparison among some of them is done in [2].

As in MOPs there is no single optimal solution, a leader selection method is also needed, and this method have impact on the quality of the solutions as well. A comparison between some of the leader selection methods available in the literature is presented in [5].

Other aspect that has been observed in a MOPSO is that in some conditions the velocity of the particles can become too high, generating erratic movements towards the limits of the decision space. To avoid such situations the Speed-constrained Multi-objective PSO (SMPSO) [15] algorithm presents a velocity constriction mechanism based on a factor χ that varies based on the values of the influence factor of the social (C_1) and personal (C_2) leaders. In SMPSO, the Crowding Distance (CD) metric [7] is used in both, leader and archiving methods. Due to its good results in the literature, SMPSO is an algorithm frequently used as reference [2], [5].

2.2   Hyper Heuristics

The task of choosing appropriate parameters or algorithms to solve an optimization problem is often hard. In this context, the hyper-heuristic approach emerges as a high-level methodology which, when given a particular problem instance or class of instances, and a number of low-level heuristics or components, automatically produces an adequate combination of these components to effectively solve the given problem [4].

An iteration of a hyper-heuristic can be subdivided into two parts: *heuristic selection* and *move acceptance*. Some heuristic selection methods generate online score(s) for each heuristic based on their performances. Then these values are processed and/or combined in a systematic manner to select the heuristic to be applied to the candidate solution at each step. A valid manner to implement it is to use a roulette-wheel (score proportionate) strategy to associate a probability with each heuristic that is computed by dividing each individual score by the total score. Then, a heuristic is randomly selected based on these probabilition. A high score generates a higher probability of being selected [4], [1].

The acceptance strategy is an important component of any local search heuristic. This strategy can be either deterministic or non-deterministic. Deterministic methods make the same decision for acceptance regardless of the decision point during the search using given current and new candidate solutions(s). A non-deterministic approach might generate a different decision for the same input. The decision process in most non-deterministic move acceptance methods requires additional parameters, such as the current iteration number [4].

In this first study, we use a simple hyper-heuristic. Our selection mechanism consists of a roulette-wheel where the probability of each low-level heuristic is updated according to the difference in the R_2 value obtained before and after the application of the selected heuristic (score). We use the deterministic move acceptance strategy Improving and Equal (IE) due to its simplicity and the good results presented in [1]. In IE, a new solution is accepted if it improves or maintains the previous score value.

3   Hyper-MOPSO

As seen in previous works [2], [5], leader and archiving methods have a significant impact in the optimization process of a MOPSO, however the methods are more or less suitable to different problems and there is no single method that excels in all the problems.

In this work, we consider combinations of leader and archiving methods as low-level heuristics and select them during the search based in the R_2 quality indicator in order to achieve a good performance in all the optimization problems at hand.

The R_2 indicator [9] was originally proposed to assess the relative quality of two approximation sets. This indicator evaluates the desired aspects of a Pareto front approximation and presents a low computational cost [3]. In this work we use Tchebycheff as utility function in the R_2 calculation.

The hyper-heuristic proposed here works on the following way: until the repository is full, the search is conducted using a regular SMPSO to achieve a good diversity and fill the repository quickly. Once the repository is full, the leader selection and archiving methods are selected by a roulette wheel that starts with equal probability of choosing any of the low-level heuristics (leader/archiving) available.

This roulette is updated at each iteration based in the R_2 indicator. In a given iteration, if the selected heuristic improves or maintains the quality of the set of solutions in the repository (measured by the R_2 indicator), then the probability of the selected heuristic is increased in the roulette, the probabilities of the other heuristics are decreased and the solution is accepted.

However, if the quality of the solutions decreases, the heuristic used have its probability decreased, while the probabilities of the others are increased and the solution is rejected (a copy of the repository is restored).

This simple hyper-heuristic rewards or punishes the heuristics increasing or decreasing their participation in the search. A small minimum probability (calibrated before the experiments) is considered for the heuristics so they are not completely removed from the search, as it can present bad results in some stage, but could perform well ahead in the search procedure.

Algorithm 1. H-MOPSO

```
initialize(particles)
repository=initializeRepository(particles)
roulette=initializeRoulette();
t = 0
while t < tmax do
    if repository is full then
        | selectLeaderArchiver(roulette)
    end if
    else
        | selectLeaderArchiver(CD-CD)
    end if
    for each particle in the repository do
        selectGlobalLeader(particle, repository)
        updatePosition(particle)
        mutation(particle)
        evaluation(particle)
        updatePersonalLeader(particle)
    end for
    if repository is full then
        repositoryPrevious=repository
        repository=updateRepository(particles)
        R₂=calculateR₂(repository)
        R₂Ant=calculateR₂(repositoryPrevious)
        roulette=updateRoulette(roulette, R₂Ant, R₂)
        if R₂Ant < R₂ then
            | repository=repositoryPrevious
        end if
    end if
    else
        | repository=updateRepository(particles)
    end if
    t + +
end while
return repository
```

Algorithm 1 shows a pseudocode of H-MOPSO, where firstly the particles are randomly initialized. Then, the repository is initialized with the non-dominated solutions from the population. At next, the roulette is initialized with equal probabilities for all the low-level heuristics.

In the main loop of the algorithm, if the repository is (or was at some point) full, the leader and archiving method is selected based on the probabilities of the roulette, otherwise is used the original SMPSO until the repository becomes full (CD-CD combination).

Then for each particle, the procedures are the same as the SMPSO algorithm, where the particle selects a global leader from the repository, update its position based in its own previous position and in the position of the personal and global leaders. After, if the particle is selected for mutation, it is mutated. Following the particle is evaluated (objective vector calculated regarding new position) and updates its personal best.

Returning to the main loop, if the repository is (or was) full, a copy of the actual repository is done and the repository is updated with the new solutions. Then the R_2 of both, the actual and the previous repositories are calculated and the probabilities on the roulette are updated based in the performance of the selected heuristic. At next, if the solutions in the new repository have a worse R_2 value than the previous, the old repository replaces the new (IE acceptance). Case the repository was not full it is updated normally using the CD archiver. Finally the iteration counter is incremented. The main loop repeats *tmax* times, and the repository (best non-dominated solutions found so far) is returned as result of the search process.

In this procedure, the heuristics with better performance are chosen more often, and the algorithm can adapt itself to prefer the heuristics that improve its results according to the problem at hand.

4   Empirical Study

In this section, firstly the parameters used in all experiments are shown in Section 4.1. Then, Section 4.2 presents a comparative study between H-MOPSO and its nine low-level heuristics to assess the hyper-heuristic capacity of selecting viable components along the search. A comparative study between H-MOPSO and MOEA/D-DRA, a state-of-the-art algorithm from the literature, is shown in Section 4.3.

4.1   Experimental Setup

The parameters used in this study for the SMPSO and derivatives follows the same used in [15]. For MOEA/D-DRA we followed the same parameters used in [19]. In the hyper-heuristic, the initial probabilities are the same for all the low-level heuristics, the minimum probabilities of each heuristic was set to 0.5% to prevent it to be removed from the search, and the increment (or decrement) in the probabilities of the selected heuristic is set to one tenth of the initial probabilities. The move acceptance criterion used is Improving and Equal (IE).

In both experimental studies the stop criterion was the number of iterations, set to 100, or its equivalent in fitness evaluations (10100). The number of particles and the size of the repository are also set to 100.

In the first experimental study, in order to assess if the hyper-heuristic is selecting the low-level heuristics properly, we compare its performances using the R_2 indicator, the same used to guide the hyper-heuristic.

In the second experimental study, we used two indicators to assess the quality of the fronts generated by each algorithm: the first of them is a modification of the Inverted Generational Distance (IGD) known as IGD_p [17], which indicates if the solution set found by the algorithm is well distributed over the true discretized

Pareto front. The second indicator is the Hypervolume [18] that measures the size of the portion of the objective space dominated by a solution set.

In both studies, the entire DTLZ [6] family of well-known multi-objective problems was used (DTLZ1 - DTLZ7). These problems can scale both in number of objectives and decision variables, also the true Pareto optimal front is known.

The results measured with the quality indicators in 30 independent runs of the algorithms are submitted to the Kruskal-Wallis test [11] at 5% significance level. Moreover, to summarize the results, the averages of the 30 runs of each algorithm are submitted to the Friedman test [8], also at 5% significance level.

These results are presented as tables showing the indicator values achieved for the compared algorithms in the form of the ranks used by the statistical test applied. The number in parentheses indicates the final classification of the algorithms, where smaller rank values are better. The algorithm with the better rank is highlighted in bold font.

When calculating the final ranks, in case of statistical tie (algorithms presenting no statistical difference), the final rank of each of the tied algorithms is equal to the average of the ranks that would be assigned to them. Two algorithms are considered statistically equal if the difference between their ranks is smaller than the critical difference.

4.2 H-MOPSO vs. Low-Level Heuristics

This section presents the results obtained by the algorithms H-MOPSO and its nine low-level heuristics, consisting of combinations between the three archivers (Crowding Distance (CD) [7], Ideal [2] and Multilevel Grid Archiving (MGA) [12]) and the three leader selection methods (Crowding Distance (CD) [7], Sigma [14], NWSum [16]).

Table 1 show the Kruskal-Wallis ranks of the R_2 results achieved by the algorithms. In this table, the critical difference is 73.03. From the results presented, for few objectives (two and three), H-MOPSO presents the general better performance, being among the better ranks in all the instances, except in DTLZ4 for two objectives. SMPSO presented good results as well, outperforming H-MOPSO in DTLZ4 for two objectives, and being tied with it in six instances.

For five objectives, H-MOPSO also presented the better general results, outperforming all the low-level heuristics in all problems, except for DTLZ4, where it statistically tied with SMPSO and NWSum-CD.

For ten objectives, again H-MOPSO had overall good results, losing only in DTLZ2 and DTLZ4. In DTLZ2, SMPSO and Sigma-CD had the better results, however in DTLZ4 only SMPSO achieved the smaller ranking.

For fifteen objectives, H-MOPSO had the better rankings in all problems, except for DTLZ4 where it lost, and in DTLZ2, where it tied with SMPSO and Sigma-CD. The better performance in DTLZ4 was achieved by SMPSO.

For twenty objectives, H-MOPSO also had the better performance in most problems, losing in DTLZ2 and DTLZ4, and being tied with CD-Ideal and NWSum-Ideal in DTLZ7. SMPSO had the better results in DTLZ2 and DTLZ4.

Table 1. Kruskal-Wallis ranks for the R_2 indicator

Obj.	Algorithms	DTLZ1	DTLZ2	DTLZ3	DTLZ4	DTLZ5	DTLZ6	DTLZ7
2	H-MOPSO	**17.70 (1.50)**	**30.23 (1.50)**	**20.70 (1.00)**	44.70 (2.50)	**17.43 (1.50)**	**34.23 (1.50)**	**16.93 (2.00)**
	SMPSO	**44.10 (1.50)**	**30.77 (1.50)**	107.70 (4.00)	**16.30 (1.50)**	**43.57 (1.50)**	**54.37 (1.50)**	**44.40 (2.00)**
	CD-Ideal	216.90 (7.00)	267.97 (9.50)	202.00 (8.00)	256.07 (8.00)	263.00 (9.50)	192.00 (6.50)	263.00 (8.50)
	CD-MGA	141.67 (6.00)	208.50 (6.50)	171.03 (6.00)	200.63 (8.00)	221.53 (8.00)	129.07 (6.50)	210.93 (7.50)
	NWSum-CD	161.10 (6.50)	136.97 (6.00)	118.20 (4.00)	105.07 (3.50)	133.73 (5.50)	169.13 (6.50)	89.83 (2.50)
	NWSum-Ideal	159.63 (6.50)	178.37 (6.00)	112.70 (4.00)	119.87 (4.00)	171.40 (6.00)	171.47 (6.50)	177.43 (6.50)
	NWSum-MGA	174.67 (6.50)	142.00 (6.00)	98.73 (4.00)	115.40 (3.50)	140.83 (5.50)	173.17 (6.50)	121.67 (4.50)
	Sigma-CD	185.30 (6.50)	136.47 (6.00)	226.10 (8.00)	212.37 (8.00)	140.03 (5.50)	191.07 (6.50)	201.00 (7.50)
	Sigma-Ideal	193.07 (6.50)	187.23 (6.00)	214.90 (8.00)	217.63 (8.00)	187.83 (6.00)	196.67 (6.50)	173.43 (6.50)
	Sigma-MGA	210.87 (6.50)	186.50 (6.00)	232.93 (8.00)	216.97 (8.00)	185.63 (6.00)	193.83 (6.50)	206.37 (7.50)
3	H-MOPSO	**15.60 (1.50)**	**15.63 (2.00)**	**16.80 (1.00)**	**21.27 (2.00)**	**25.77 (1.50)**	**32.33 (1.50)**	**20.40 (1.00)**
	SMPSO	68.53 (3.00)	63.03 (3.00)	126.13 (4.50)	55.20 (2.50)	**35.23 (1.50)**	79.73 (2.50)	108.93 (4.50)
	CD-Ideal	188.77 (7.50)	265.27 (8.00)	176.20 (5.50)	252.00 (8.00)	270.10 (9.50)	174.73 (6.50)	164.83 (5.50)
	CD-MGA	159.97 (5.50)	218.60 (8.00)	144.60 (4.50)	224.40 (7.50)	215.47 (7.00)	113.10 (4.50)	169.17 (5.50)
	NWSum-CD	115.07 (4.00)	115.93 (3.50)	117.83 (4.50)	82.00 (2.50)	148.10 (6.00)	201.43 (7.00)	246.90 (9.00)
	NWSum-Ideal	136.13 (5.00)	198.10 (8.00)	111.07 (4.50)	178.63 (6.50)	178.10 (6.00)	212.40 (7.00)	221.60 (8.00)
	NWSum-MGA	115.03 (4.00)	121.07 (3.50)	103.37 (4.50)	112.90 (3.50)	171.27 (6.00)	201.63 (7.00)	228.53 (8.00)
	Sigma-CD	193.33 (7.50)	68.33 (3.00)	222.10 (8.50)	189.70 (7.50)	117.57 (5.50)	146.10 (6.00)	125.43 (4.50)
	Sigma-Ideal	257.40 (8.50)	213.83 (8.00)	253.17 (9.00)	192.17 (7.50)	179.43 (6.00)	170.97 (6.50)	112.73 (4.50)
	Sigma-MGA	255.17 (8.50)	225.20 (8.00)	233.73 (8.50)	196.73 (7.50)	169.97 (6.00)	177.57 (6.50)	106.47 (4.50)
5	H-MOPSO	**16.73 (1.00)**	**18.67 (2.00)**	**23.13 (1.00)**	**18.83 (2.00)**	**38.93 (2.50)**	**28.43 (2.00)**	**20.83 (2.00)**
	SMPSO	160.27 (5.00)	103.17 (3.50)	130.00 (4.50)	**53.97 (2.00)**	109.73 (4.50)	99.10 (3.00)	220.77 (8.00)
	CD-Ideal	104.00 (4.50)	256.40 (8.50)	116.60 (4.50)	193.90 (6.50)	284.23 (9.50)	107.27 (4.00)	143.13 (4.50)
	CD-MGA	119.13 (4.50)	234.60 (8.00)	120.80 (4.50)	171.17 (6.50)	165.07 (6.00)	69.97 (2.50)	154.70 (6.00)
	NWSum-CD	121.67 (4.50)	69.63 (3.00)	131.90 (4.50)	**83.10 (2.00)**	178.77 (6.00)	203.33 (7.50)	251.67 (8.50)
	NWSum-Ideal	114.60 (4.50)	177.80 (7.00)	105.57 (4.50)	267.83 (10.00)	236.00 (8.00)	249.60 (8.50)	210.67 (7.50)
	NWSum-MGA	137.47 (4.50)	130.27 (4.00)	135.47 (4.50)	189.10 (6.50)	134.03 (5.00)	183.57 (7.50)	238.73 (8.50)
	Sigma-CD	231.87 (8.50)	61.20 (3.00)	244.93 (9.00)	173.87 (6.50)	84.67 (3.00)	161.17 (6.00)	116.07 (4.00)
	Sigma-Ideal	253.53 (9.00)	213.50 (8.00)	243.00 (9.00)	179.57 (6.50)	163.07 (6.00)	175.73 (6.50)	75.07 (3.00)
	Sigma-MGA	245.73 (9.00)	239.77 (8.00)	253.60 (9.00)	173.67 (6.50)	110.50 (4.50)	226.83 (7.50)	73.37 (3.00)
10	H-MOPSO	**19.67 (1.00)**	79.57 (3.00)	**21.07 (1.00)**	155.53 (5.50)	**26.53 (2.00)**	**45.40 (1.50)**	**19.97 (2.00)**
	SMPSO	258.50 (8.50)	**21.30 (2.50)**	263.17 (8.50)	**15.53 (1.50)**	96.80 (4.00)	110.83 (4.50)	229.30 (8.00)
	CD-Ideal	119.60 (4.00)	246.93 (8.50)	97.80 (4.00)	227.50 (8.00)	273.63 (9.00)	177.93 (6.00)	117.67 (4.00)
	CD-MGA	116.80 (4.00)	238.33 (8.50)	109.47 (4.00)	165.57 (6.00)	150.37 (5.50)	119.43 (5.00)	153.60 (5.50)
	NWSum-CD	96.73 (4.00)	93.27 (3.50)	136.77 (4.00)	77.50 (3.00)	164.37 (5.50)	172.97 (6.00)	201.87 (7.50)
	NWSum-Ideal	106.03 (4.00)	165.00 (5.50)	111.10 (4.00)	280.47 (9.00)	216.00 (7.50)	215.13 (7.00)	173.63 (7.00)
	NWSum-MGA	107.70 (4.00)	145.70 (5.00)	105.63 (4.00)	238.13 (8.50)	92.40 (4.00)	155.40 (6.00)	236.03 (8.00)
	Sigma-CD	208.47 (8.50)	**54.97 (2.50)**	226.60 (8.50)	110.73 (4.50)	132.60 (5.00)	204.40 (7.00)	210.10 (7.50)
	Sigma-Ideal	243.20 (8.50)	215.63 (7.50)	219.00 (8.50)	114.03 (4.50)	151.27 (5.50)	147.83 (6.00)	77.80 (2.50)
	Sigma-MGA	228.30 (8.50)	244.30 (8.50)	214.40 (8.50)	120.00 (4.50)	201.03 (7.00)	155.67 (6.00)	85.03 (3.00)
15	H-MOPSO	**22.60 (2.00)**	79.43 (2.50)	**19.07 (1.00)**	166.33 (5.50)	**29.83 (1.50)**	**25.00 (1.00)**	**18.73 (1.50)**
	SMPSO	259.80 (8.50)	**24.23 (2.50)**	281.30 (9.00)	**15.50 (1.50)**	109.77 (4.50)	116.73 (5.00)	220.77 (7.00)
	CD-Ideal	128.13 (4.00)	260.90 (8.50)	103.50 (4.00)	207.43 (7.50)	266.53 (9.00)	159.40 (5.50)	62.63 (2.00)
	CD-MGA	114.33 (4.00)	243.83 (8.50)	97.40 (4.00)	182.43 (7.00)	178.90 (6.50)	103.43 (5.00)	151.60 (6.50)
	NWSum-CD	112.10 (4.00)	95.40 (3.50)	134.70 (4.50)	97.27 (4.00)	164.40 (5.50)	206.60 (7.00)	172.47 (7.00)
	NWSum-Ideal	92.93 (3.50)	160.13 (5.50)	101.80 (4.00)	280.97 (9.50)	225.57 (8.00)	263.40 (9.50)	96.90 (3.00)
	NWSum-MGA	93.77 (3.50)	161.33 (6.00)	111.33 (4.00)	248.73 (8.50)	97.00 (3.50)	151.63 (5.50)	185.27 (7.00)
	Sigma-CD	208.27 (8.50)	**48.73 (2.50)**	207.33 (7.50)	76.30 (3.00)	104.90 (4.00)	160.50 (5.50)	204.27 (7.00)
	Sigma-Ideal	235.87 (8.50)	196.93 (7.50)	223.00 (8.50)	121.23 (4.50)	196.23 (7.50)	171.37 (5.50)	176.13 (7.00)
	Sigma-MGA	237.20 (8.50)	234.07 (8.00)	225.57 (8.50)	108.80 (4.00)	131.87 (5.00)	146.93 (5.50)	216.23 (7.00)
20	H-MOPSO	**16.53 (1.00)**	81.37 (2.50)	**21.30 (1.00)**	163.30 (5.50)	**20.70 (1.00)**	**22.27 (1.50)**	**15.53 (2.00)**
	SMPSO	272.03 (8.50)	**18.87 (2.00)**	268.13 (8.50)	**15.50 (1.50)**	94.90 (4.00)	123.93 (5.50)	194.27 (7.00)
	CD-Ideal	107.40 (4.00)	256.20 (8.50)	113.70 (4.00)	196.43 (7.50)	242.50 (8.50)	184.40 (6.00)	**46.23 (2.00)**
	CD-MGA	106.33 (4.00)	238.07 (8.50)	94.40 (4.00)	189.03 (7.00)	183.13 (7.00)	91.23 (3.50)	175.67 (6.50)
	NWSum-CD	123.33 (4.00)	97.03 (4.00)	121.77 (4.00)	118.73 (4.50)	141.83 (5.00)	183.30 (6.00)	153.20 (6.00)
	NWSum-Ideal	91.90 (4.00)	160.13 (5.50)	101.03 (4.00)	278.87 (9.50)	231.60 (8.50)	261.70 (10.00)	**75.27 (2.00)**
	NWSum-MGA	112.97 (4.00)	161.13 (5.50)	121.00 (4.00)	249.33 (8.50)	146.77 (5.00)	146.07 (5.50)	173.17 (6.50)
	Sigma-CD	217.73 (8.50)	50.20 (2.50)	220.03 (8.50)	83.83 (3.00)	107.90 (4.00)	163.60 (5.50)	161.00 (6.00)
	Sigma-Ideal	231.50 (8.50)	201.80 (7.50)	218.20 (8.50)	99.50 (4.00)	148.40 (5.00)	178.33 (6.00)	245.30 (8.00)
	Sigma-MGA	225.27 (8.50)	240.20 (8.50)	225.43 (8.50)	110.47 (4.00)	187.27 (7.00)	150.17 (5.50)	265.37 (9.00)

Table 2. Friedman overall ranks for the R_2 indicator

H-MOPSO	SMPSO	CD-Ideal	CD-MGA	NWSum-CD	NWSum-Ideal	NWSum-MGA	Sigma-CD	Sigma-Ideal	Sigma-MGA
61.0 (1.0)	171.0 (4.5)	302.0 (7.5)	226.0 (5.0)	198.0 (4.5)	255.0 (6.0)	227.0 (5.0)	211.0 (4.5)	324.0 (8.5)	335.0 (8.5)

Regarding the performance of H-MOPSO per problem, in DTLZ1 it had alone the better rankings in all numbers of objectives, except for two, where it statistically tied with SMPSO. In DTLZ2, it was among the better performances in all cases, except for ten and twenty objectives. In DTLZ3, H-MOPSO had the better rankings alone in all cases.

DTLZ4 was the problem where H-MOPSO had the worst performance, being among the better algorithms only for three and five objectives. In DTLZ5, it shared the better rank with SMPSO for two and three objectives, and outperformed all the other algorithms for five objectives onwards.

In DTLZ6, H-MOPSO also shared the better ranking with SMPSO for two objectives, but outperformed all the other algorithms in the remaining cases. In DTLZ7, H-MOPSO shared the better results with SMPSO for two objectives, for twenty objectives, it statistically tied with CD-Ideal and NWSum-Ideal, outperforming the other algorithms in the remaining instances.

In general H-MOPSO stood among the better algorithms in all the problems, except for some instances of DTLZ2 and DTLZ4. DTLZ2 is an easy problem that does not impose challenges to convergence or diversity, in this case the algorithms converge easily to the front, and increase the R_2 value by improving the diversity. In this problem H-MOPSO was outperformed only by SMPSO and Sigma-CD that are two algorithms characterized by generating fronts with high diversity.

DTLZ4 is a hard problem by presenting diversity challenge. In this problem H-MOPSO had its worst performance, while SMPSO performed very well due to its high diversity characteristic.

Despite of the good results of SMPSO in few objectives and problems that demand higher diversity, it performs very badly in many-objective problems that present convergence challenge like DTLZ1 and DTLZ3. The generality obtained by H-MOPSO due to its hyper-heuristic is an advantage in such cases.

Due to the high amount of data presented in Table 1, it is hard to take overall conclusions from the performance of the algorithms over all the instances, hence we present Table 2 where is shown the Friedman ranks obtained for the overall analysis of the algorithms. In this test, the average of the 30 independent runs of each subproblem (problem/objective number) is considered. The test is performed with the 42 subproblems for each algorithm. The critical difference in this table is 90.78.

In the results summarized on this table, H-MOPSO achieves the lower Friedman rank and the lower final rank, which indicates that even not presenting the better individual result in all cases, it is a robust algorithm capable of obtaining good results in most of the cases.

Those results indicate that in general the proposed hyper-heuristic is able to properly select low-level heuristics to lead the search to regions that enhance its indicator values. The problems where H-MOPSO did not achieved the better performances in all cases (DTLZ2 and DTLZ4), are problems where the diversity characteristic is preferred over a balance of convergence and diversity, hence SMPSO in general obtained good results.

4.3   H-MOPSO vs. MOEA/D-DRA

In this section, we compare our proposed algorithm with MOEA/D-DRA [19], a state-of-the-art algorithm winner of the CEC 2009 MOEA contest. Since we are comparing the performance of the algorithms, and not their compliance with the R_2 metric, the popular indicators IGD_p and Hypervolume are used.

Tables 3 and 4 present the IGD_p and Hypervolume results achieved for both algorithms respectively. Those results are shown in the form of the average ranks used by the Kruskal-Wallis test, and the critical difference in both tables is 8.83.

Table 3. Kruskal-Wallis ranks for the IGD_p indicator

Obj.	Algorithms	DTLZ1	DTLZ2	DTLZ3	DTLZ4	DTLZ5	DTLZ6	DTLZ7
2	H-MOPSO	**16.90 (1.00)**	**15.50 (1.00)**	**15.50 (1.00)**	**17.70 (1.00)**	**15.50 (1.00)**	**16.50 (1.00)**	**15.50 (1.00)**
	MOEA/D-DRA	44.10 (2.00)	45.50 (2.00)	45.50 (2.00)	43.30 (2.00)	45.50 (2.00)	44.50 (2.00)	45.50 (2.00)
3	H-MOPSO	**16.57 (1.00)**	**15.50 (1.00)**	**15.80 (1.00)**	**22.07 (1.00)**	**15.50 (1.00)**	**17.43 (1.00)**	**15.50 (1.00)**
	MOEA/D-DRA	44.43 (2.00)	45.50 (2.00)	45.20 (2.00)	38.93 (2.00)	45.50 (2.00)	43.57 (2.00)	45.50 (2.00)
5	H-MOPSO	**19.93 (1.00)**	**15.83 (1.00)**	**15.63 (1.00)**	**19.90 (1.00)**	45.50 (2.00)	43.63 (2.00)	**16.20 (1.00)**
	MOEA/D-DRA	41.07 (2.00)	45.17 (2.00)	45.37 (2.00)	41.10 (2.00)	**15.50 (1.00)**	**17.37 (1.00)**	44.80 (2.00)
8	H-MOPSO	**19.73 (1.00)**	35.30 (2.00)	**16.57 (1.00)**	**16.50 (1.00)**	45.50 (2.00)	40.97 (2.00)	**16.57 (1.00)**
	MOEA/D-DRA	41.27 (2.00)	**25.70 (1.00)**	44.43 (2.00)	44.50 (2.00)	**15.50 (1.00)**	**20.03 (1.00)**	44.43 (2.00)
9	H-MOPSO	**17.37 (1.00)**	36.37 (2.00)	**15.50 (1.00)**	**16.23 (1.00)**	45.50 (2.00)	40.67 (2.00)	**44.57 (2.00)**
	MOEA/D-DRA	43.63 (2.00)	**24.63 (1.00)**	45.50 (2.00)	44.77 (2.00)	**15.50 (1.00)**	**20.33 (1.00)**	**16.43 (1.00)**

Table 4. Kruskal-Wallis ranks for the Hypervolume indicator

Obj.	Algorithms	DTLZ1	DTLZ2	DTLZ3	DTLZ4	DTLZ5	DTLZ6	DTLZ7
2	H-MOPSO	**15.80 (1.00)**	**15.53 (1.00)**	**15.50 (1.00)**	**16.73 (1.00)**	**15.50 (1.00)**	**16.50 (1.00)**	**15.50 (1.00)**
	MOEA/D-DRA	45.20 (2.00)	45.47 (2.00)	45.50 (2.00)	44.27 (2.00)	45.50 (2.00)	44.50 (2.00)	45.50 (2.00)
3	H-MOPSO	**15.92 (1.00)**	**16.03 (1.00)**	**15.87 (1.00)**	**17.10 (1.00)**	**15.50 (1.00)**	**17.33 (1.00)**	**15.50 (1.00)**
	MOEA/D-DRA	45.08 (2.00)	44.97 (2.00)	45.13 (2.00)	43.90 (2.00)	45.50 (2.00)	43.67 (2.00)	45.50 (2.00)
5	H-MOPSO	**18.10 (1.00)**	**15.63 (1.00)**	**15.60 (1.00)**	**15.63 (1.00)**	**15.50 (1.00)**	**18.00 (1.00)**	**15.50 (1.00)**
	MOEA/D-DRA	42.90 (2.00)	45.37 (2.00)	45.40 (2.00)	45.37 (2.00)	45.50 (2.00)	43.00 (2.00)	45.50 (2.00)
8	H-MOPSO	**16.67 (1.00)**	43.90 (2.00)	**15.53 (1.00)**	**34.67 (1.50)**	**15.50 (1.00)**	**27.20 (1.50)**	**15.50 (1.00)**
	MOEA/D-DRA	44.33 (2.00)	**17.10 (1.00)**	45.47 (2.00)	**26.33 (1.50)**	45.50 (2.00)	**33.80 (1.50)**	45.50 (2.00)
9	H-MOPSO	**15.60 (1.00)**	44.73 (2.00)	**15.50 (1.00)**	38.03 (2.00)	**15.50 (1.00)**	**17.57 (1.00)**	**15.50 (1.00)**
	MOEA/D-DRA	45.40 (2.00)	**16.27 (1.00)**	45.50 (2.00)	**22.97 (1.00)**	45.50 (2.00)	43.43 (2.00)	45.50 (2.00)

From the results presented, H-MOPSO presents outstanding performance for two and three objectives, where it outperforms MOEA/D-DRA in all cases according to both metrics.

Considering the IGD_p for five objectives onwards, H-MOPSO outperformed MOEA/D-DRA in most of the cases, losing only on the problems DTLZ5 and DTLZ6, for eight and nine objectives on DTLZ2 and for nine objectives on DTLZ7.

According to the hypervolume, H-MOPSO had better results in all cases, except for eight and nine objectives on problem DTLZ2 and for nine objectives on problem DTLZ4. For eight objectives on problems DTLZ4 and DTLZ6 the results of both algorithms statistically tied.

Summarizing these results, despite of the very good performance of H-MOPSO for low number of objectives, its search ability is deteriorated by the increase of the number of objectives. MOEA/D-DRA on the other hand uses decomposition instead of Pareto dominance, hence has an advantage on many-objective scenarios.

Here we can see that IGD_p and Hypervolume present controversial results for problems DTLZ5 and DTLZ6 for five objectives onwards. This can be explained by the fact that indicators not aware of the shape of the true Pareto front like Hypervolume and R_2 may present misleading results in problems with special fronts under some circumstances. Due to this fact, we assume that H-MOPSO was outperformed by MOEA/D-DRA on these problems.

These results can confirm the robustness of H-MOPSO, being able to outperform a state-of-the-art algorithm in most of the cases tested.

4.4   Analysis of the Probabilities

In this section, we analyze the behavior of the proposed hyper-heuristic through the probabilities of choosing each of its low-level heuristic along the search. Here we only show two representative cases due to lack of space.

Figures 1 and 2 present this behavior, where each line in the figures, represents the average probability of choosing one low-level heuristic in a given iteration. The probabilities in each iteration are averaged over the 30 runs, to show a pattern.

In these figures the number of iterations is different, because it is only shown the iterations in which the probabilities were updated, being ignored the first iterations where the SMPSO algorithm was used until the repository is full.

In a usual case, represented by Figure 1, after some iterations, a few low-level heuristics presenting better performances have its probabilities increased, while the others have its probabilities decreased, with a gap between the groups.

Another common case is represented by Figure 2, where the differences among the probabilities are smaller, hence there is no gap and some low-level heuristics are a little higher, while others are a little lower. In this case it is common to some low-level heuristics that presented bad performances, begin to perform better and overcome the previous best combinations.

To indicate the compliance of the selection of the low-level heuristics with the probabilities, we observed the number of times (in the 30 executions) that

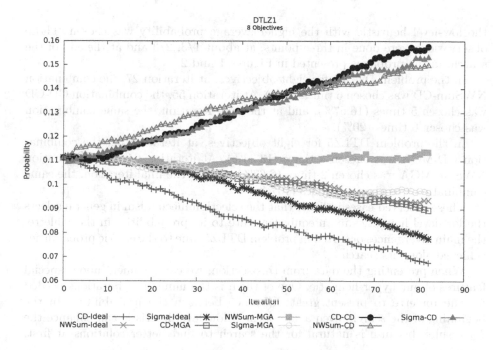

Fig. 1. Average probabilities of DTLZ1 for 8 objectives

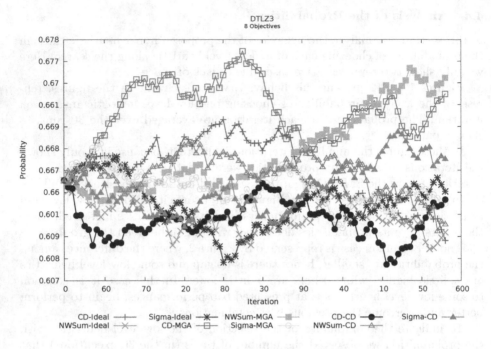

Fig. 2. Average probabilities of DTLZ5 for 8 objectives

the low-level heuristic with the higher average probability was chosen. Those observations were done in three points, at about 1/3, 2/3 and at the end of the search, in the problems presented in Figures 1 and 2.

In the problem DTLZ1 for eight objectives, at iteration 27, the combination NWSum-CD was chosen 6 times (20%). At iteration 55, the combination CD-CD was chosen 5 times (16.67%), and at the final iteration, the same combination was chosen 6 times (20%).

In the problem DTLZ5 for eight objectives, at iteration 32, the combination CD-MGA was chosen 4 times (13.33%). At iteration 64, the combination NWSum-MGA was chosen 2 times (6.67%), and at the final iteration, the same combination was chosen 3 times (10%).

These observed cases indicate that the selection mechanism in general selects the low-level heuristics in an equivalent rate to its probabilities in the roulette, the main divergences occurred in problem DTLZ5, due to the erratic probabilities achieved along its search.

When presenting the data from this section, we can comment about special features of our hyper-heuristic. One of them is the number of iterations needed for the roulette to present great difference between the probabilities. In the beginning of the search, even a sub-optimal low-level heuristic can enhance the R_2 results, because is natural for the search to find better solutions at first, however it can take several iterations to the roulette increase the probabilities of the good heuristics.

Other problem noted is related to the roulette mechanism. When using a high number of low-level heuristics, if they present great differences in performance, as in Figure 1, better heuristics are selected more frequently and the chances of getting better results are higher. However when the differences in the probabilities are smaller, as in Figure 2, sub-optimal heuristics are selected frequently, hence sub-optimal results are expected with the hyper-heuristic.

Another problem observed is regarding the R_2 indicator that in some problems with special Pareto shapes, can lead to misleading results. Examples of these results can be noted in the previous section, where IGD_p and hypervolume presented controversial results. The focus of future works will be addressing such problems.

5   Conclusion

This work presented H-MOPSO, a new MOPSO algorithm based on hyper-heuristic to select good leader and archiving methods. H-MOPSO uses a hyper heuristic guided by the R_2 performance indicator, which is a fast indicator that evaluates desired aspects of a Pareto front approximation. There are few works on the literature that apply hyper-heuristics on multi-objectives optimizers, and this work, as far as we known, is the first using a hyper-heuristic to select leader and archiving methods in a MOPSO.

An experimental study was conducted to compare the performance of H-MOPSO algorithm to the low-level heuristics used within it. This experimental study used the well-known DTLZ family of benchmark problems with 2, 3, 5, 10, 15 and 20 objectives. To this end, the solutions obtained from the different algorithms were compared using the R_2 indicator. The results from this first study indicate that the proposed hyper-heuristic is robust, presenting good performance in most of the cases.

These good results lead to a second experimental study, where we compare H-MOPSO with a state-of-the-art algorithm from the literature called MOEA/D-DRA. This second study also used the well-known DTLZ family of benchmark problems for 2, 3, 5, 8 and 9 objectives, and the outputs of the algorithms were evaluated using the IGD_p and the Hypervolume indicators. These results indicate that H-MOPSO is a robust and competitive meta-heuristic, outperforming MOEA/D-DRA in most of the cases.

The results obtained in these two experimental studies were used to validate two hypotheses: h_1 stated that selecting good combinations between leader and archiving methods is able to improve the results of a MOPSO. The results presented indicate that h_1 is confirmed, due to the number of cases in which H-MOPSO outperformed the low-level heuristics.

The other hypothesis investigated is h_2 that states that R_2 is an appropriate indicator to guide a hyper-heuristic. From the results presented, this hypothesis can be partially confirmed. In most of the cases, with regular Pareto shapes,

R_2 was able to efficiently measure the quality of the solutions, however in problems with special Pareto shapes, R_2 might present misleading results, leading the hyper-heuristic to select sub-optimal low-level heuristics.

Despite the good results obtained by H-MOPSO in the experimental studies conducted, its hyper-heuristic still presents some problems to be treated, like the large number of iterations needed for the hyper-heuristic to effectively work, the high usage of sub-optimal low-level heuristics in some circumstances, and the misleading results obtained by R_2 in some special conditions.

To address those problems, future works will focus on more effective hyper-heuristics, as well as using other and/or more quality indicators to better assess the real situation of the search. Another possible future direction is applying the hyper-heuristic to control or choose different parameters or operators of the MOPSO algorithm.

References

1. Bilgin, B., Özcan, E., Korkmaz, E.: An experimental study on hyper-heuristics and exam timetabling. In: Burke, E., Rudová, H. (eds.) PATAT 2007. LNCS, vol. 3867, pp. 394–412. Springer, Heidelberg (2007)
2. Britto, A., Pozo, A.: Using archiving methods to control convergence and diversity for many-objective problems in particle swarm optimization. In: 2012 IEEE Congress on Evolutionary Computation (CEC), pp. 1–8, June 2012
3. Brockhoff, D., Wagner, T., Trautmann, H.: On the properties of the R2 indicator. In: Proceedings of the 14th Annual Conference on Genetic and Evolutionary Computation, GECCO 2012, pp. 465–472. ACM, New York (2012). http://doi.acm.org/10.1145/2330163.2330230
4. Burke, E.K., Gendreau, M., Hyde, M., Kendall, G., Ochoa, G., Ozcan, E., Qu, R.: Hyper-heuristics: a survey of the state of the art. J. Oper. Res. Soc. **64**(12), 1695–1724 (2013)
5. Castro, Jr., O.R., Britto, A., Pozo, A.: A comparison of methods for leader selection in many-objective problems. In: IEEE Congress on Evolutionary Computation, pp. 1–8, June 2012
6. Deb, K., Thiele, L., Laumanns, M., Zitzler, E.: Scalable multi-objective optimization test problems. In: Proceedings of the 2002 Congress on Evolutionary Computation, CEC 2002, vol. 1, pp. 825–830, May 2002
7. Deb, K., Agrawal, S., Pratap, A., Meyarivan, T.: A fast elitist non-dominated sorting genetic algorithm for multi-objective optimisation: NSGA-II. In: Deb, K., Rudolph, G., Lutton, E., Merelo, J.J., Schoenauer, M., Schwefel, H.-P., Yao, X. (eds.) PPSN 2000. LNCS, vol. 1917, pp. 849–858. Springer, Heidelberg (2000)
8. Demsar, J.: Statistical comparisons of classifiers over multiple data sets. Journal of Machine Learning Research **7**, 1–30 (2006)
9. Hansen, M.P., Jaszkiewicz, A.: Evaluating the quality of approximations to the non-dominated set. Tech. Rep. IMM-REP-1998-7, Technical University of Denmark, March 1998
10. Kennedy, J., Eberhart, R.: Particle swarm optimization. In: Proceedings of IEEE International Conference on Neural Networks, vol. 4, pp. 1942–1948, November/December 1995

11. Kruskal, W.H., Wallis, W.A.: Use of ranks in one-criterion variance analysis. Journal of the American Statistical Association **47**(260), 583–621 (1952)
12. Laumanns, M., Zenklusen, R.: Stochastic convergence of random search methods to fixed size pareto front approximations. European Journal of Operational Research **213**(2), 414–421 (2011)
13. von Lücken, C., Barán, B., Brizuela, C.: A survey on multi-objective evolutionary algorithms for many-objective problems. Computational Optimization and Applications **58**(3), 707–756 (2014). http://dx.doi.org/10.1007/s10589-014-9644-1
14. Mostaghim, S., Teich, J.: Strategies for finding good local guides in multi-objective particle swarm optimization (MOPSO). In: Proceedings of the 2003 IEEE Swarm Intelligence Symposium, SIS 2003, pp. 26–33, April 2003
15. Nebro, A., Durillo, J., Garcia-Nieto, J., Coello Coello, C.A., Luna, F., Alba, E.: SMPSO: A new PSO-based metaheuristic for multi-objective optimization. In: Computational Intelligence in Multi-Criteria Decision-Making, pp. 66–73, March 2009
16. Padhye, N., Branke, J., Mostaghim, S.: Empirical comparison of MOPSO methods: guide selection and diversity preservation. In: Proceedings of the Eleventh Congress on Evolutionary Computation, CEC 2009, pp. 2516–2523. IEEE Press, Piscataway (2009)
17. Schutze, O., Esquivel, X., Lara, A., Coello Coello, C.A.: Using the averaged hausdorff distance as a performance measure in evolutionary multiobjective optimization. IEEE Transactions on Evolutionary Computation **16**(4), 504–522 (2012)
18. While, L., Bradstreet, L., Barone, L.: A fast way of calculating exact hypervolumes. IEEE Transactions on Evolutionary Computation **16**(1), 86–95 (2012)
19. Zhang, Q., Liu, W., Li, H.: The performance of a new version of MOEA/D on CEC 2009 unconstrained MOP test instances. In: IEEE Congress on Evolutionary Computation, CEC09, pp. 203–208, May 2009

Algorithms

Adaptive Reference Vector Generation for Inverse Model Based Evolutionary Multiobjective Optimization with Degenerate and Disconnected Pareto Fronts

Ran Cheng[1], Yaochu Jin[1,3](✉), and Kaname Narukawa[2]

[1] Department of Computing, University of Surrey, Guildford, Surrey GU2 7XH, UK
{r.cheng,yaochu.jin}@surrey.ac.uk
[2] Honda Research Institute Europe GmbH, 63073 Offenbach am Main, Germany
[3] College of Information Sciences and Technology, Donghua University,
Shanghai 201620, China

Abstract. Inverse model based multiobjective evolutionary algorithm aims to sample candidate solutions directly in the objective space, which makes it easier to control the diversity of non-dominated solutions in multiobjective optimization. To facilitate the process of inverse modeling, the objective space is partitioned into several subregions by predefining a set of reference vectors. In the previous work, the reference vectors are uniformly distributed in the objective space. Uniformly distributed reference vectors, however, may not be efficient for problems that have nonuniform or disconnected Pareto fronts. To address this issue, an adaptive reference vector generation strategy is proposed in this work. The basic idea of the proposed strategy is to adaptively adjust the reference vectors according to the distribution of the candidate solutions in the objective space. The proposed strategy consists of two phases in the search procedure. In the first phase, the adaptive strategy promotes the population diversity for better exploration, while in the second phase, the strategy focused on convergence for better exploitation. To assess the performance of the proposed strategy, empirical simulations are carried out on two DTLZ benchmark problems, namely, DTLZ5 and DTLZ7, which have a degenerate and a disconnected Pareto front, respectively. Our results show that the proposed adaptive reference vector strategy is promising in tacking multiobjective optimization problems whose Pareto front is disconnected.

Keywords: Multiobjective optimization · Model based evolutionary optimization · Inverse modeling · Reference vectors

1 Introduction

A multiobjective optimization problem (MOP) involves several conflicting objectives to be optimized simultaneously. Without loss of generality, an MOP can be formulated as follows:

© Springer International Publishing Switzerland 2015
A. Gaspar-Cunha et al. (Eds.): EMO 2015, Part I, LNCS 9018, pp. 127–140, 2015.
DOI: 10.1007/978-3-319-15934-8_9

$$\min \quad \boldsymbol{f}(\boldsymbol{x}) = (f_1(\boldsymbol{x}), f_2(\boldsymbol{x}), ..., f_m(\boldsymbol{x}))$$
$$\text{s.t.} \quad \boldsymbol{x} \in X, \quad \boldsymbol{f} \in Y \tag{1}$$

where $X \subset \mathbb{R}^n$ is the decision space and $\boldsymbol{x} = (x_1, x_2, ..., x_n) \in X$ is the decision vector, $Y \subset \mathbb{R}^m$ is the objective space and $\boldsymbol{f} \in Y$ is the objective vector, which is composed of m objective functions $f_1(\boldsymbol{x})$, $f_2(\boldsymbol{x})$,...,$f_m(\boldsymbol{x})$ that map $\boldsymbol{x}$ from X to Y. Due to the conflicting nature of the objectives, it is impossible to optimize all the objectives with one single solution. Consequently, there exists a set of optimal solutions, termed as Pareto optimal solutions, that trade-off between different objectives. The Pareto optimal solutions are often called the Pareto set in the decision space and image formed by the Pareto optimal solutions in the objective is termed Pareto front.

To obtain the Pareto optimal solutions, various multiobjective evolutionary algorithms (MOEAs) have been proposed, e.g. the elitist non-dominated sorting algorithm, known as NSGA-II [5], the decomposition based algorithm, called MOEA/D [14], among many others [16]. Most traditional MOEAs often require a high degree of diversity in storing the non-dominated solutions found so far in the current population or in an external archive. By contrast, model-based MOEAs [11,12,15] can alleviate the requirement on solution diversity by focusing on the construction of a probabilistic model in the decision space during the search. Such model based MOEAs, however, still rely on the use of a solution set, such as an archive, to represent the obtained non-dominated solutions. Another line of research that aims to alleviate the requirement on diversity is to build a regression model to represent the solutions obtained in the final generation by the optimizer [7,9], which can be used to generate new solutions after the optimization process is complete, thereby enhancing the diversity of the final solutions. Inspired by the ideas in this line of research, a multiobjective evolutionary algorithm using Gaussian process based inverse modeling (IM-MOEA) has been proposed [2].

In IM-MOEA, an inverse model that maps candidate solutions in the objective space onto the decision space is built during the optimization. To facilitate the inverse modeling, the objective space is partitioned into several subregions using predefined *reference vectors*. By associating each candidate solution with a particular reference vector, a number of inverse models are built for each subregion by using the candidate solutions relating to this subregion as training data. In the previous work of IM-MOEA, the reference vectors are uniformly generated by means of the canonical simplex-lattice design method [3]. This method for generating reference vectors works well for MOPs with a continuous and uniform Pareto front. However, for some MOPs with a nonuniform or disconnected Pareto front, the predefined, uniformly distributed reference vectors may result in low efficiency, as some reference vectors may not be associated with any candidate solutions, thus causing a waste of computational resource. To tackle this problem, here we present an adaptive reference vector generation strategy for IM-MOEA. The proposed strategy is able to adapt the distribution of the reference vectors to the distribution of the candidate solutions in the objective space.

In the following, we first briefly introduce the recently proposed IM-MOEA in Section 2. Then the adaptive reference vector generation strategy is described in Section 3. Section 4 presents experimental results for assessing the performance of the proposed adaptive strategy. Finally, conclusion is drawn in Section 5.

2   IM-MOEA

Traditional EDAs aim to estimate the distribution of the candidate solutions in the decision space, while the models in IM-MOEA are built to represent the inverse mapping from the objective space to the decision space. With the inverse models thus built, evenly distributed candidate solutions can be directly sampled in the objective space and then mapped onto the decision space.

Considering that the estimation of the entire inverse mapping from the m-dimensional objective space to the n-dimensional decision space can be technically difficult, the multivariate inverse model is decomposed into a number of univariate regression models:

$$P(X|Y) \approx \prod_{i=1}^{n} (P(x_i|f_j) + \epsilon_{j,i}), \tag{2}$$

where $j = 1, 2, ..., m$, $i = 1, 2, ..., n$, $P(x_i|f_j)$ is an univariate model that represents the inverse mapping from objective f_j to decision variable x_i, and $\epsilon_{j,i}$ is an error term. For convenience, it is assumed that $\epsilon_{j,i} \sim \mathcal{N}(0, (\sigma_n)^2)$ can be captured by additive Gaussian noise. Consequently, each univariate model together with the error term is realized using Gaussian process [13], which has the advantage of modeling both the global regularity and the local randomness in the distribution of the non-dominated solutions during the search. It is worth noting that although the decomposition strategy does not take into account the variable linkages explicitly, in our algorithm, a *random grouping method* has been adopted to implicitly learn the correlations between different decision variables by relating a number of decision variables with each objective. For example, for a three-objective MOP, if the group size is 2, three groups of models will be generated, each containing two univariate models. The reader is referred to [2].

In order to facilitate the inverse modelling, some pre-defined uniformly distributed reference vectors are used to partition the objective space into a number of subregions. To generate these reference vectors, the uniformly distributed points are firstly generated on a unit hyperplane and then mapped to a unit hypersphere, as shown in Fig. 1. With the pre-defined reference vectors (or subregions), the entire population can be partitioned into a number of subpopulations by associating the candidate solutions with different reference vectors. To associate each candidate solution with a particular reference vector, the angle (in the objective space) between each candidate solution and each reference vector is calculated, and a candidate solution is associated with a reference vector if

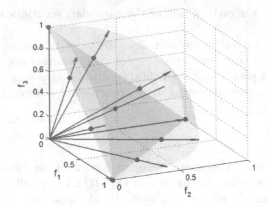

Fig. 1. An example of how to generate a number of 15 uniformly distributed reference vectors in a 3-objective space

and only if the angle between the candidate solution and the reference vector is smallest among all reference vectors.

Partitioning a population using reference vectors in the objective space was first suggested in [10], which has also been adopted in a few other recently proposed algorithms such as NSGA-III [4,8]. However, the population partition strategy in NSGA-III is to use reference points distributed on a unit hyperplane in the objective space to guide the convergence of the population, and as a consequence, each individual in the population is expected to converge to a corresponding reference point. By contrast, our method is motivated to partition the actual objective space by setting a number of reference vectors, and around each reference vector, a subpopulation is maintained in the subregion defined by this reference vector. In each subregion, promising candidate solutions are selected using non-dominated sorting and crowd distance [5]. Inverse models are then built using the selected candidate solutions as the training data. Therefore, reproduction is operated in each subregion by sampling the inverse models built for this region. At the end of each generation, the offspring generated in each subregion is combined together to create the parent population for the next generation.

As shown in Fig. 2, the main operations of IM-MOEA, i.e., non-dominated sorting, selection, inverse modeling and reproduction, are all carried out within each subpopulation once the entire population is partitioned. Therefore, the reference vectors, which directly determine how the population is partitioned, play a central role in IM-MOEA.

3   Adaptive Reference Vector Generation

The adoption of uniformly distributed reference vectors in the original IM-MOEA is based on the implicit assumption that the Pareto front of the MOP

Adaptive Reference Vector Generation 131

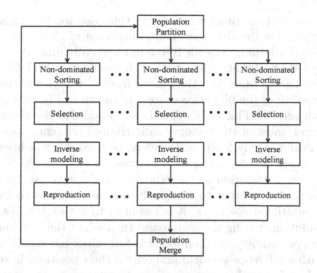

Fig. 2. The framework of the IM-MOEA

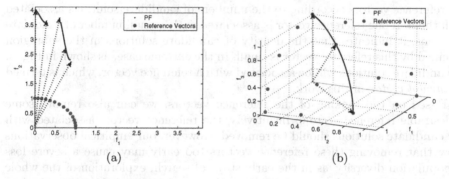

Fig. 3. Examples where there exist *invalid* reference vectors: (a) only 5 out of 15 reference vectors are covered by the Pareto front of 2-objective DTLZ7; (b) only 3 out of 15 reference vectors are covered by the Pareto front of 3-objective DTLZ5

is uniformly distributed in the whole objective space. This assumption may be impractical for many real-world MOPs. In this work, without the loss of generality, we use two benchmark functions in the DTLZ test suite [6], namely, DTLZ5 and DTLZ7, as examples to examine the effectiveness of the proposed strategy for adaptively generating reference vectors.

DTLZ7 is a typical MOP with a disconnected Pareto front consisting of 2^{m-1} disconnected segments, where m is the objective number. For example, as shown in Fig. 3(a), the Pareto front of a 2-objective DTLZ7 consists of two Pareto optimal regions. Moreover, both segments of the Pareto front distribute in the left part of the objective space, resulting in a large region in the objective space that has no Pareto optimal solutions. In this case, for 15 reference vectors uniformly distributed in the objective space, only 5 out of 15 are associated with the Pareto optimal solutions, leaving 10 reference vectors unused.

Another typical MOP that suffers from such problems is DTLZ5. This MOP has a degenerate Pareto front, i.e., the Pareto front is always a curve regardless of the dimensionality of the objective space. As shown in Fig. 3(b), the Pareto front of the 3-objective DTLZ5 is a curve in the middle of the objective space. In this case, again, most of the uniformly distributed reference vectors are not in use (in this example only 3 out of 15 reference vectors), which will give rise to considerable waste of computational resources.

In practice, the distribution of the true Pareto front is usually not known beforehand. Therefore, in order to effectively use all reference vectors and the associated computational resources, it is essential to detect the distribution of the candidate solutions during the search and then adapt the distribution of the reference vectors accordingly. As mentioned before, since one candidate solution is associated with a reference vector if and only if their positions in terms of the angle between them in the objective space are closest, the density of the solutions in a subregion can be easily estimated by counting the number of candidate solutions associated with each reference vector. In this way, we are able to rank the reference vectors according to the numbers of candidate solutions associated with them. If one reference vector is associated with a small number of candidate solutions, it indicates that the density of candidate solutions in the subregion specified by this reference vector is small. In the extreme case, as shown in Fig. 3, no candidate solutions will be associated with a reference vector, which is termed an *invalid vector* in this work.

Based on the ranking of the reference vectors, we can also remove some undesirable reference vectors. Intuitively, the reference vectors associated with few candidate solutions should be removed. However, our empirical observations show that removing these reference vectors too early may cause a severe loss of population diversity, as in the early stage of search, exploration of the whole search landscape can be more important than exploitation, whereas in the late stage of the search, exploitation is more desirable. In order to adapt the reference vectors to the different preferences at different search stages, we divide the search into two phases: exploration phase and exploitation phase. In the exploration phase, since the reference vectors are expected to be widespread, the reference vectors associated with too many candidate solutions are preferentially removed. By contrast, in the exploitation phase, reference vectors associated few candidate solutions are removed since a high density of reference vectors in the subregions to be exploited can accelerate convergence.

To maintain a relatively stable distribution of the reference vectors, only one reference vector, i.e., the one is ranked first or last, will be removed in each generation. Meanwhile, a new reference vector is randomly generated to replace the removed one. In this way, in each generation, the extreme reference vector will be replaced with a new, randomly generated reference vector. The procedure of the adaptive reference vector generation strategy is summarized in Algorithm 1.

Algorithm 1. The pseudo code of the adaptive reference vector generation strategy.

Require:current generation t, max generation max_t, the current reference vectors $V(t)$, current population $P(t)$
Ensure:the adapted reference vectors $V(t+1)$

1: randomly generate a new reference vector v_n;
2: calculate the numbers of candidate solutions in $P(t)$ that associated with each reference vector in $V(t)$;
3: **if** $t < \theta * max\_t$ **then**
4:    /*exploration phase*/
5:    remove the reference vector in $V(t)$ which is associated with the *maximal* number of candidate solutions and replace it with v_n;
6: **else**
7:    /*exploitation phase*/
8:    remove the reference vector in $V(t)$ which is associated with the *minimal* number of candidate solutions and replace it with v_n;
9: **end if**
10: $V(t+1) = V(t)$;

It can be seen that a control parameter θ is introduced to determine at which generation the exploration stage is switched to exploitation. In Section 4, some preliminary empirical studies have been carried out to investigate the influence of parameter θ on the search performance.

4   Simulation Results

To assess the performance of the proposed adaptive reference vector generation strategy, IM-MOEA with the adaptive strategy, denoted as A-IM-MOEA hereafter, is compared with the original IM-MOEA on four three-objective DTLZ benchmark MOPs, including DTLZ1, DTLZ2, DTLZ5 and DTLZ7. The first two MOPs (DTLZ1, DTLZ2) have a uniformly distributed Pareto front, while the other two MOPs (DTLZ5, DTLZ7) have a degenerate and a disconnected Pareto front, respectively, refer to Fig. 3. The specific settings of these four MOPs follow the recommendations in [6].

In IM-MOEA and the proposed adaptive A-IM-MOEA, there are three parameters to be specified: the population size, the number of reference vectors, denoted as K_r hereafter, the group size (for random grouping). The population size is set to 150 in all the experiments. To investigate the sensitivity of the proposed adaptive strategy to K_r, different settings ($K_r = 10$, $K_r = 15$ and $K_r = 28$) have been used in the comparisons with IM-MOEA. In addition, it is worth noting that the setting of the group size is dependent on the number of decision variables. Since the numbers of decision variables of three-objective DTLZ1, DTLZ2, DTLZ5 and DTLZ7 (7, 11, 11 and 21, respectively) are small, the group size is simply set to 1.

The inverted generational distance (IGD) [1] is used as the performance indicator in the performance comparisons:

$$IGD(P^*, P) = \frac{\sum_{v \in P^*} d(v, P)}{|P^*|}, \tag{3}$$

where P^* is a set of uniformly distributed solutions along the true Pareto front, and P is an approximation, $d(v, P)$ is the minimum Euclidean distance from the point v to P. The IGD metric is able to measure both diversity and convergence of P if $|P^*|$ is large enough, and a smaller IGD value indicates a better performance. In this work, a number of 500 uniformly distributed points are selected for each benchmark MOP to be P^*.

Table 1. Statistical results of IGD values obtained by A-IM-MOEA and IM-MOEA (mean values in the first line and standard deviations in the second line). If one result is statistically significantly better than the other one, it is highlighted.

K_r	Algorithm	DTLZ1	DTLZ2	DTLZ5	DTLZ7
10	A-IM-MOEA	5.91E-02	6.21E-02	**6.05E-03**	**6.98E-02**
		1.37E-02	3.83E-03	**6.24E-04**	**1.01E-02**
	IM-MOEA	4.97E-02	**5.67E-02**	1.77E-02	2.07E-01
		9.60E-03	**1.54E-03**	1.94E-03	2.30E-02
15	A-IM-MOEA	**3.83E-02**	5.44E-02	**4.77E-03**	**6.00E-02**
		6.19E-03	1.84E-03	**6.23E-04**	**7.07E-03**
	IM-MOEA	5.04E-02	**5.06E-02**	1.50E-02	1.18E-01
		1.71E-02	**7.54E-04**	1.56E-03	8.75E-03
28	A-IM-MOEA	**4.55E-02**	5.18E-02	**4.04E-03**	**5.46E-02**
		1.12E-02	2.40E-03	**5.53E-04**	**3.79E-03**
	IM-MOEA	4.76E-02	4.88E-02	1.06E-02	8.96E-02
		1.07E-02	5.63E-04	1.12E-03	5.03E-03
$+/\approx/-$		2 / 1 / 0	0 / 2 / 1	3 / 0 / 0	3 / 0 / 0

The statistical results presented in this section are obtained from 20 independent runs. In each independent run, a maximum of 50000 fitness evaluations is used as a termination criterion for DTLZ2, DTLZ5 and DTLZ7. For DTLZ1, 150000 fitness evaluations are used. To compare the results obtained by A-IM-MOEA and IM-MOEA, the Wilcoxon rank sum test is adopted at a significance level of 0.05. As a result of the Wilcoxon rank sum test, "+" means that the IGD values obtained by A-IM-MOEA are statistically significantly smaller than those obtained by IM-MOEA; "−" means that the IGD values obtained by A-IM-MOEA are statistically significantly larger than those obtained by IM-MOEA; and "$\approx$" means that there is no statistically significant difference between the IGD values obtained by A-IM-MOEA and IM-MOEA.

Before comparing the performance of A-IM-MOEA and IM-MOEA, some investigations regarding the setting of θ will be conducted. *theta* determines the generation at which the exploration search stage is to be switched to exploitation,

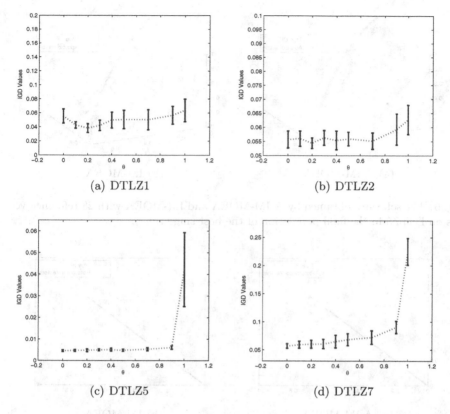

Fig. 4. The statistical results of the IGD values obtained by A-IM-MOEA with 15 reference vectors and different settings of θ. In this figure, error bars are used to present the mean and standard deviation.

refer to Algorithm 1. As shown in Fig. 4, different settings of θ may have different impacts on different benchmark problems. For DTLZ5 and DTLZ7, which have a degenerate and a disconnected PF, respectively, it seems that the performance of A-IM-MOEA is relatively insensitive to the settings of θ, as long as it is not larger than 0.7. This might be due to the fact that DTLZ5 and DTLZ7 are uni-modal, exploration has no significant effect on the search performance. Exploitation, which mainly contributes to convergence, can be important as the Pareto fronts of these two MOPs are not uniformly distributed in the objective space. By contrast, since the fitness landscape of DTLZ1 contains a large number of local optima, sufficient exploration becomes more important. It can be seen in Fig. 4 (a) that when θ is between 0.1 and 0.3, the standard deviation of IGD is smaller compared to that in other settings, which implies a more stable performance of A-IM-MOEA. Among the four benchmark MOPs, DTLZ2 is uni-modal and has a uniformly distribute Pareto front. Therefore, the performance of A-IM-MOEA is not very sensitive to the settings of θ either.

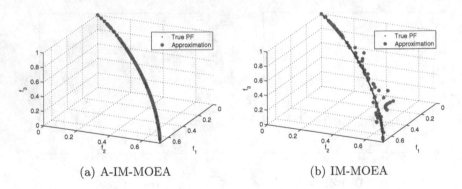

(a) A-IM-MOEA   (b) IM-MOEA

Fig. 5. The solutions obtained by A-IM-MOEA and IM-MOEA with 28 reference vectors on DTLZ5 in the final population of the best singe run.

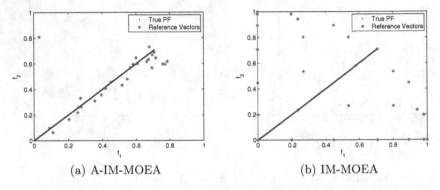

(a) A-IM-MOEA   (b) IM-MOEA

Fig. 6. The solutions obtained by A-IM-MOEA and IM-MOEA with 28 reference vectors on DTLZ5 in the final population of the best singe run. To ease the observations, the points are mapping into a 2-D (f_1 and f_2) plane from the 3-D objective space.

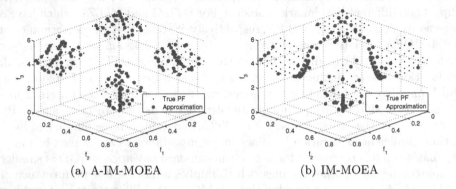

(a) A-IM-MOEA   (b) IM-MOEA

Fig. 7. The solutions obtained by A-IM-MOEA and IM-MOEA with 28 reference vectors on DTLZ7 in the final population of the best single run.

Based on the empirical investigations on the setting of θ, we use $\theta = 0.2$ for all experiments for comparing A-IM-MOEA and IM-MOEA. The statistical results obtained by A-IM-MOEA and IM-MOEA are summarized in Table 1. It can be seen that A-IM-MOEA significantly outperforms IM-MOEA on DTLZ5 and DTLZ7, regardless of the number of reference vectors. As evident from Fig. 5, the solutions obtained by A-IM-MOEA show significantly better convergence. It is because the reference vectors in A-IM-MOEA have been successfully adapted, thus increasing the sampling density around the true Pareto front rather than the entire objective space. To verify this statement, the reference vectors are plotted together with the true Pareto front. To better visualize the adapted distribution of the reference vectors, the points are mapping into a 2-D (f_1 and f_2) plane, as shown in Fig. 6. It can be seen that the reference vectors in A-IM-MOEA are mostly distribute around the true Pareto front, whilst the reference vectors in IM-MOEA, without any adaption, still uniformly distributed in the entire objective space.

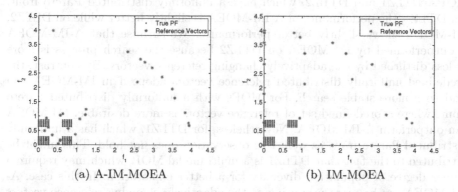

| | (a) A-IM-MOEA | (b) IM-MOEA |

Fig. 8. The solutions obtained by A-IM-MOEA and IM-MOEA with 28 reference vectors on DTLZ7 in the final population of the best singe run. The points are mapping into a 2-D (f_1 and f_2) plane from the 3-D objective space for better visualization.

Similar observations can be made from the results on DTLZ7 as well. As evident from Fig. 7, the solutions obtained by A-IM-MOEA show a promising distribution, while most of the solutions obtained by IM-MOEA distribute on the edges of the true Pareto front consisting of four disconnected piecewise segments. In addition to disconnection, the Pareto front of DTLZ7 shows significant bias on the m-th objective, thus resulting the distribution of the Pareto front centralized close to the third axis (f_3) in a 3-D objective space. These properties raise considerable difficulties for IM-MOEA which adopts a uniformly distributed reference vectors. By contrast, the adaptive reference vector generation strategy adopted in A-IM-MOEA has significantly better efficiency, as indicated in Fig. 8.

From the statistical results in Table 1, another interesting observation is the comparable performance of A-IM-MOEA and IM-MOEA on the other two

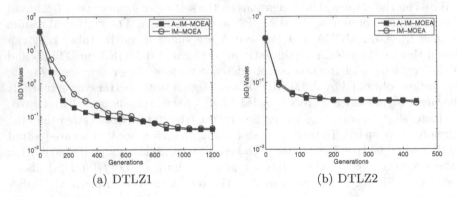

(a) DTLZ1   (b) DTLZ2

Fig. 9. The convergence profiles of the IGD values in the best single run with 15 reference vectors of A-IM-MOEA and IM-MOEA respectively.

MOPs, DTLZ1 and DTLZ2, which have a uniformly distributed Pareto front. On DTLZ1, the performance of A-IM-MOEA is slightly better while on DLTZ2, IM-MOEA shows slightly better performance. We surmise that A-IM-MOEA is outperformed by IM-MOEA on DTLZ2 because the search process is more or less disturbed by the adaptively changing reference vectors. By contrast, the predefined uniformly distributed reference vectors adopted in IM-MOEA can lead to a more stable search. For MOPs with a uniformly distributed Pareto front, where a predefined set of reference vectors is more desirable, IM-MOEA can outperform A-IM-MOEA. Nevertheless, for DTLZ1, which has a uniformly distributed Pareto front as well, this observation does not hold. This might be attributed to the fact that DTLZ1 is a multi-modal MOP, which may require a higher degree of population diversity for a better exploration. In this case, A-IM-MOEA can be more promising as the adaptively changing reference vectors can generate higher population diversity than the predefined reference vectors. As shown in Fig. 9, the adaptively changing reference vectors have enhanced the convergence speed of A-IM-MOEA in the exploration stage on DTLZ1. However, on DTLZ2, which is a uni-modal MOP, the convergence profiles of A-IM-MOEA and IM-MOEA show little difference.

5   Conclusion

An adaptive reference vector generation strategy is proposed in this paper, which has shown to be promising on two MOPs having a discrete or non-uniform Pareto front. In addition, the MOEA using the proposed strategy performs comparably with the one using uniformly distributed reference vectors on MOPs having a uniform Pareto front distributed in the whole objective space. In addition, an interesting observation is that the adaptive reference vectors are able to generate better population diversity to enhance the performance of IM-MOEA on multi-modal MOPs like DTLZ1.

In the future, the performance of the proposed adaptive strategy for generating reference vectors will be further assessed on additional MOPs. For example, it can be interesting to see how it performs on constrained MOPs, where the Pareto front is irregular as well. The mechanism for switching between exploration and exploitation also needs further examination.

Acknowledgments. This work was supported in part by the Honda Research Institute Europe GmbH and the Joint Research Fund for Overseas Chinese, Hong Kong and Macao Scholars of the National Natural Science Foundation of China (Grant No. 61428302).

References

1. Bosman, P.A., Thierens, D.: The balance between proximity and diversity in multiobjective evolutionary algorithms. IEEE Transactions on Evolutionary Computation **7**(2), 174–188 (2003)
2. Cheng, R., Jin, Y., Narukawa, K., Sendhoff, B.: A multiobjective evoltuionary algorithm using Gaussian process based inverse modeling. IEEE Transactions on Evolutionary Computation (accepted in 2015)
3. Cornell, J.A.: Experiments with mixtures: designs, models, and the analysis of mixture data. John Wiley & Sons (2011)
4. Deb, K., Jain, H.: An evolutionary many-objective optimization algorithm using reference-point based non-dominated sorting approach, part I: Solving problems with box constraints. IEEE Transactions on Evolutionary Computation **18**(4), 577–601 (2013)
5. Deb, K., Pratap, A., Agarwal, S., Meyarivan, T.: A fast and elitist multiobjective genetic algorithm: NSGA-II. IEEE Transactions on Evolutionary Computation **6**(2), 182–197 (2002)
6. Deb, K., Thiele, L., Laumanns, M., Zitzler, E.: Scalable multi-objective optimization test problems. In: Proceedings of the Congress on Evolutionary Computation, pp. 825–830. IEEE (2002)
7. Giagkiozis, I., Fleming, P.J.: Pareto front estimation for decision making. Evolutionary Computation (accepted in 2014)
8. Jain, H., Deb, K.: An evolutionary many-objective optimization algorithm using reference-point based non-dominated sorting approach, part II: Handling constraints and extending to an adaptive approach. IEEE Transactions on Evolutionary Computation **18**(4), 602–622 (2013)
9. Jin, Y., Sendhoff, B.: Connectedness, regularity and the success of local search in evolutionary multi-objective optimization. In: Proceedings of the IEEE Congress on Evolutionary Computation, vol. 3, pp. 1910–1917. IEEE (2003)
10. Liu, H.L., Gu, F., Zhang, Q.: Decomposition of a multiobjective optimization problem into a number of simple multiobjective subproblems. IEEE Transactions on Evolutionary Computation **18**(3), 2450–455 (2014)
11. Martí, L., García, J., Berlanga, A., Coello Coello, C., Molina, J.M.: On current model-building methods for multi-objective estimation of distribution algorithms: Shortcommings and directions for improvement. Department of Informatics, Universidad Carlos III de Madrid, Madrid, Spain, Technical Report GIAA2010E001 (2010)

12. Okabe, T., Jin, Y., Sendoff, B., Olhofer, M.: Voronoi-based estimation of distribution algorithm for multi-objective optimization. In: Proceedings of the IEEE Congress on Evolutionary Computation, vol. 2, pp. 1594–1601. IEEE (2004)
13. Rasmussen, C.E.: Gaussian processes for machine learning. MIT Press (2006)
14. Zhang, Q., Li, H.: MOEA/D: A multiobjective evolutionary algorithm based on decomposition. IEEE Transactions on Evolutionary Computation 11(6), 712–731 (2007)
15. Zhang, Q., Zhou, A., Jin, Y.: RM-MEDA: A regularity model-based multiobjective estimation of distribution algorithm. IEEE Transactions on Evolutionary Computation 12(1), 41–63 (2008)
16. Zhou, A., Qu, B.Y., Li, H., Zhao, S.Z., Suganthan, P.N., Zhang, Q.: Multiobjective evolutionary algorithms: A survey of the state of the art. Swarm and Evolutionary Computation 1(1), 32–49 (2011)

MOEA/PC: Multiobjective Evolutionary Algorithm Based on Polar Coordinates

Roman Denysiuk[1][✉], Lino Costa[2], Isabel Espírito Santo[2], and José C. Matos[3]

[1] Algoritmi R&D Center, University of Minho, Braga, Portugal
roman.denysiuk@algoritmi.uminho.pt
[2] Department of Production and Systems Engineering, University of Minho, Braga, Portugal
{lac,iapinho}@dps.uminho.pt
[3] Institute for Sustainability and Innovation in Structural Engineering, University of Minho, Braga, Portugal
jmatos@civil.uminho.pt

Abstract. The need to perform the search in the objective space constitutes one of the fundamental differences between multiobjective and single-objective optimization. The performance of any multiobjective evolutionary algorithm (MOEA) is strongly related to the efficacy of its selection mechanism. The population convergence and diversity are two different but equally important goals that must be ensured by the selection mechanism. Despite the equal importance of the two goals, the convergence is often used as the first sorting criterion, whereas the diversity is considered as the second one. In some cases, this can lead to a poor performance, as a severe loss of diversity occurs.

This paper suggests a selection mechanism to guide the search in the objective space focusing on maintaining the population diversity. For this purpose, the objective space is divided into a set of grids using polar coordinates. A proper distribution of the population is ensured by maintaining individuals in corresponding grids. Eventual similarities between individuals belonging to neighboring grids are explored. The convergence is ensured by minimizing the distances from individuals in the population to a reference point. The experimental results show that the proposed approach can solve a set of problems producing competitive performance when compared with state-of-the-art algorithms. The ability of the proposed selection to maintain diversity during the evolution appears to be indispensable for dealing with some problems, allowing to produce significantly better results than other considered approaches relying on different selection strategies.

1 Introduction

Evolutionary algorithms have gained popularity as a powerful tool for solving multiobjective optimization problems (MOPs) [1], [2]. They draw inspiration from the process of natural evolution to iteratively evolve to a better set of potential solutions. An important driving force behind evolution is natural selection. It is the one process that is responsible for the evolution of adaptations

A. Gaspar-Cunha et al. (Eds.): EMO 2015, Part I, LNCS 9018, pp. 141–155, 2015.
DOI: 10.1007/978-3-319-15934-8_10

of species to their environment. Natural selection leads to evolutionary change when individuals best suited for their environment are more likely to survive and reproduce, transferring useful genetic characteristics from parents to their offspring.

A common approach to simulate natural selection in MOEAs consists in assigning fitness values to individuals and sampling the population according to these values. The fitness values reflect the individuals quality in the problem environment and are the basis for selection. Nowadays, one can distinguish three major trends to the fitness assignment, which are dominance-, scalarizing- and indicator-based strategies. Dominance-based approaches [3], [4] calculate an individual's fitness on the basis of the Pareto dominance relation. Until recently, it has been probably the most commonly used approach that is usually combined with some diversity maintenance techniques. Scalarizing-based approaches [5], [6], [7] use traditional mathematical techniques based on the aggregation of multiple objectives into a single parameterized objective to assign scalar fitness values to population members. In turn, indicator-based approaches [8], [9], [10], which are a relatively recent trend, employ performance indicators for fitness assignment.

Regardless of the working principles, the selection mechanism must ensure the convergence to the Pareto set and the diversity of obtained solutions. The convergence and diversity are equally important goals that are somewhat conflicting in nature. Despite their equal importance, during selection many existing MOEAs explicitly or implicitly put more emphasis on the convergence. This is often the case in dominance-based approaches, which use the Pareto dominance relation first and the diversity as the second criterion [4], [11], [12]. Although scalarizing- and indicator-based approaches assign a scalar fitness value that reflects an aggregated quality with respect to the convergence and diversity, the necessary diversity may not be ensured in some cases. As a result of prioritizing the convergence, a severe loss of diversity can occur resulting in a poor performance on some problems.

In this work, we use the idea of grid division of the objective space using polar coordinates [13] to develop a new framework for solving MOPs. The main feature of the proposed approach is the selection mechanism, which does not rely on any of the aforementioned fitness assignment strategies. Its strengths are related to the particular ability of promoting the population diversity. This is achieved by dividing the objective space into a set of grids using polar coordinates and maintaining an individual in each grid, regardless of how good solutions with respect to the convergence appear in the population. For each grid, MOEA/PC seeks a solution minimizing the euclidean distance to a reference point.

The remainder of this paper is organized as follows. Section 2 describes the proposed multiobjective evolutionary algorithm. Section 3 provides the study of different parameter settings for the algorithm and discusses the results of comparison study with state-of-the-art approaches. Section 4 concludes the work and outlines some possible future work.

2   Algorithm Design

This section presents a multiobjective evolutionary algorithm based on polar coordinates (MOEA/PC). First, some general concepts are introduced. Then, the framework of the proposed algorithm is described.

2.1   Preliminaries

This paper considers an optimization problem of the form:

$$\underset{x \in \Omega}{\text{minimize: }} \boldsymbol{f}(\boldsymbol{x}) = (f_1(\boldsymbol{x}), f_2(\boldsymbol{x}), \ldots, f_m(\boldsymbol{x}))^{\text{T}}, \tag{1}$$

where m is the number of objectives, n is the number of variables, $\Omega = \{\boldsymbol{x} \in \mathbb{R}^n : lb_i \leq x_i \leq ub_i, i = 1, \ldots, n\}$ is the feasible decision space, lb_i and ub_i are the lower and upper bounds of the i-th variable.

In our work, we divide the objective space into a set of grids, $G = \{\boldsymbol{g}^1, \ldots, \boldsymbol{g}^{n_{grids}}\}$. The idea of grid devisions for two and tree-dimensional cases is shown in Figure 1. The number of grids, n_{grids}, corresponds to the number of solutions maintained in the population, where one population member is assigned to a single grid. Within the limits of the given grid, an assigned individual attempts to minimize the distance to a reference point. A reference point, $\boldsymbol{z} = (z_1, \ldots, z_m)^{\text{T}}$, is given by the lowest values found during the search for each objective. Each grid is defined by the vector $\boldsymbol{g} = (g_1, \ldots, g_{m-1})$, with components corresponding to angles in the first quadrant ($0 \leq g_i < \pi/2, i = 1, \ldots, m - 1$).

For the 2-dimensional case, to obtain 10 grids, we divide the right angle into 10 identical angles of size $\Delta\theta = \frac{\pi/2}{10} = 0.1571$. The set of 10 grids is given as:

$\boldsymbol{g}^1 = (0)$	$\boldsymbol{g}^2 = (0.1571)$	$\boldsymbol{g}^3 = (0.3142)$	$\boldsymbol{g}^4 = (0.4712)$	$\boldsymbol{g}^5 = (0.6283)$
$\boldsymbol{g}^6 = (0.7854)$	$\boldsymbol{g}^7 = (0.9425)$	$\boldsymbol{g}^8 = (1.0996)$	$\boldsymbol{g}^9 = (1.2566)$	$\boldsymbol{g}^{10} = (1.4137)$

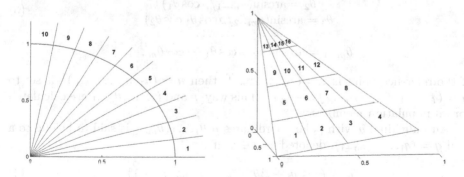

Fig. 1. Two and three-dimensional grid divisions of the objective space using polar coordinates

A vector belongs to the first grid if its polar angle is $0 \leq \theta < 0.1571$. A vector belongs to the second grid if its polar angle is $0.1571 \leq \theta < 0.3142$ and so on.

For the 3-dimensional case, to obtain 16 grids, the two right angles must be divided into 4 identical angles of size $\Delta\theta = \frac{\pi/2}{4} = 0.3927$. The set of 16 grids is given as:

$g^1 = (0,0)$	$g^2 = (0.3927,0)$	$g^3 = (0.7854,0)$	$g^4 = (1.1781,0)$
$g^5 = (0,0.3927)$	$g^6 = (0.3927,0.3927)$	$g^7 = (0.7854,0.3927)$	$g^8 = (1.1781,0.3927)$
$g^9 = (0,0.7854)$	$g^{10} = (0.3927,0.7854)$	$g^{11} = (0.7854,0.7854)$	$g^{12} = (1.1781,0.7854)$
$g^{13} = (0,1.1781)$	$g^{14} = (0.3927,1.1781)$	$g^{15} = (0.7854,1.1781)$	$g^{16} = (1.1781,1.1781)$

A vector belongs to the first grid if its polar angles are $0 \leq \theta_1 < 0.3927 \wedge 0 \leq \theta_2 < 0.3927$. A vector belongs to the second grid if its polar angles are $0.3927 \leq \theta_1 < 0.7854 \wedge 0 \leq \theta_2 < 0.3927$ and so on.

For an m-dimensional case, to generate n_{grids}, the $m-1$ right angles must be divided into $n_{div} = {}^{m-1}\!\sqrt{n_{grids}}$ identical angles of size $\Delta\theta = \frac{\pi/2}{n_{div}}$.

An individual in the population, a, is represented by the tuple of the form $\{x, f, \rho, \theta\}$, where x and f are the decision and the objective vectors, whereas ρ, θ are the polar coordinates.

For a vector $u = (u_1, u_2, \ldots, u_m)^{\mathrm{T}}$, its polar coordinates are expressed as:

$$
\begin{aligned}
u_1 &= \rho \cos\theta_1 \cos\theta_2 \cos\theta_3 \cdots \cos\theta_{m-1} \\
u_2 &= \rho \cos\theta_1 \cos\theta_2 \cos\theta_3 \cdots \sin\theta_{m-1} \\
u_3 &= \rho \cos\theta_1 \cos\theta_2 \cdots \sin\theta_{m-2} \\
&\ldots\ldots\ldots\ldots\ldots \\
u_m &= \rho \sin\theta_1
\end{aligned}
\tag{2}
$$

where ρ is the radius, θ_i is the polar angle $(i = 1, \ldots, m-1)$. They can be calculated as:

$$
\begin{aligned}
\rho &= \sqrt{u_1^2 + u_2^2 + \cdots + u_m^2} \\
\theta_1 &= \arcsin(u_m/\rho) \\
\theta_2 &= \arcsin(u_{m-1}/\rho\cos\theta_1) \\
\theta_3 &= \arcsin(u_{m-2}/\rho\cos\theta_1\cos\theta_2) \\
&\ldots\ldots\ldots\ldots\ldots \\
\theta_{m-1} &= \arcsin(u_2/\rho\cos\theta_1\cdots\cos\theta_{m-2}).
\end{aligned}
\tag{3}
$$

If the reference point is $z = (z_1, \ldots, z_m)^{\mathrm{T}}$, then $u = (u_1, u_2, \ldots, u_m)^{\mathrm{T}}$ is set to $u = (f_1 - z_1, f_2 - z_2, \ldots, f_m - z_m)^{\mathrm{T}}$. This way, ρ and $\theta_i, \ldots, \theta_{m-1}$ are calculated for all population members.

An individual a with polar coordinates $\rho, \theta_1, \ldots, \theta_{m-1}$ is said to belong to a grid $g = (g_1, \ldots, g_{m-1})$, denoted as $a \in g$, if

$$
g_i \leq \theta_i \leq g_i + \Delta\theta \quad \forall i \in \{1, \ldots, m-1\}.
\tag{4}
$$

If (4) is not fulfilled, then it is said that a does not belong to g, denoted as $a \notin g$.

2.2 Proposed Framework

MOEA/PC works as follows:
Input:

- CR - crossover probability;
- F - scaling factor;
- p_m - mutation probability;
- η_m - mutation distribution index;
- δ - probability for mating pool;
- T - neighborhood size;
- μ - population size;
- $maxEval$ - maximum number of function evaluations.

Output:

- $\{x^1, \ldots, x^{n_{grids}}\}$ - approximation to the Pareto set;
- $\{f(x^1), \ldots, f(x^{n_{grids}})\}$ - approximation to the Pareto front.

Step 1 Initialization
 Step 1.1 Compute a set of grids $G = \{g^1, \ldots, g^{n_{grids}}\}$.
 Step 1.2 For each grid, select T closest grids.
 Step 1.3 For each grid, randomly generate an individual.
 Step 1.4 Initialize a reference point, z.
 Step 1.5 For each individual, compute polar coordinates.
Step 2 Mating selection
 Step 2.1 Uniformly at random select a grid (say the i-th grid is selected).
 Step 2.2 With probability δ, select two different individuals $r1$ and $r2$ from the neighborhood of the i-th grid, whereas with probability $(1 - \delta)$ these individuals are selected from the whole set of grids.
Step 3 Variation
 Step 3.1 Generate a candidate, c, using a differential evolution (DE) operator.
 Step 3.2 Apply polynomial mutation on the candidate.
 Step 3.3 Repair the candidate.
Step 4 Update
 Step 4.1 For each $j \in \{1, \ldots, m\}$, if $f_j(x^c) < z_j$, then set $z_j = f_j(x^c)$.
 Step 4.2 Calculate polar coordinates for c.
 Step 4.3 Calculate polar coordinates for individual in each grid, if z was updated in Step 4.1.
Step 5 Environmental selection
 Step 5.1 Find a grid, say g^k, such that $c \in g^k$.
 Step 5.2 Swap a^k and c, if one of the following conditions is true:
  1. $a^k \notin g^k \ \wedge \ \exists j \in \{1, \ldots, m\} : f_j(x^c) < f_j(x^h)$;
  2. $a^k \in g^k \ \wedge \ \rho^c < \rho^k$.
 Step 5.3 If a^k and c were swapped, go to Step 5.1.
Step 6 If the maximum number of function evaluations is reached, then stop. Otherwise, go to **Step 2**.

In **Step 1**, after the grids are generated, the set of closest grids is selected for each grid. Since a grid is given as an $(m-1)$-dimensional vector, this is done by computing the Euclidean distance between grids and sorting according to the computed values.

In **Step 3**, the DE operator works as follows:

$$x_j^c = \begin{cases} x_j^i + F \times (x_j^{r1} - x_j^{r2}) & \text{with probability } CR \\ x_j^i & \text{with probability } 1 - CR \end{cases} \quad \forall j \in \{1, \ldots, n\}. \tag{5}$$

The polynomial mutation is performed as follows:

$$x_j^c = \begin{cases} x_j^c + \sigma_j \times (ub_j - lb_j) & \text{with probability } p_m \\ x_j^c & \text{with probability } 1 - p_m \end{cases} \quad \forall j \in \{1, \ldots, n\}, \tag{6}$$

where

$$\sigma_j = \begin{cases} (2u_j)^{1/(1+\eta_m)} - 1 & \text{if } u_j \leq 0.5 \\ 1 - (2 - 2u_j)^{1/(1+\eta_m)} & \text{otherwise} \end{cases} \quad \forall j \in \{1, \ldots, n\}, \tag{7}$$

and $u_j \in [0, 1]$ is a uniform random number.

To ensure the feasibility, the candidate is repaired as:

$$x_j^c = \min\{\max\{x_j^c, lb_j\}, ub_j\} \quad \forall j \in \{1, \ldots, n\}. \tag{8}$$

In **Step 5**, the survival process for the candidate and population members is designed to push out the worst individual from the population and ensure the distribution of the population according to the grid, which can be perturbed if a reference point have been updated in **Step 4**. The candidate enter into the population replacing the individual in a certain grid, whereas the latter becomes the candidate. The process continues until the candidate cannot enter into the population.

3 Performance Assessment

This section presents and discusses the results of experimental study carried out to investigated the performance of MOEA/PC. The experiments are divided into two parts. The first one estimates the effects of some parameter settings for the proposed algorithm. The second one compares the performance of MOEA/PC with state-of-the-art algorithms.

3.1 Experimental Setup

MOEA/PC is implemented in C++. Its performance is compared with those produced by MOEA/D [14], GDE3 [15] and IBEA [8], which are used within the jMetal [16] framework. The performance of the algorithms is studied on a set of challenging problems proposed in [14] and [17], which in the following will be referred as the LZ09 and LGZ test suites, respectively.

Table 1. Parameter settings for the algorithms

MOEA/PC	MOEA/D	GDE3	IBEA
$CR = 1.0$	$CR = 1.0$	$CR = 1.0$	$\eta_c = 20$
$F = 0.5$	$F = 0.5$	$F = 0.5$	$p_c = 0.9$
$\eta_m = 20$	$\eta_m = 20$	$\eta_m = 20$	$\eta_m = 20$
$p_m = 1/n$	$p_m = 1/n$	$p_m = 1/n$	$p_m = 1/n$
	$\delta = 0.9$		
	$T = 20$		

The quality of approximation sets [18] returned by the algorithms is evaluated using the inverted generational distance (IGD) indicator [19]. To calculate the IGD indicator, 1,000 uniformly distributed points along the Pareto front are generated for each problem.

For each algorithm, 30 independent runs are performed on each problem with a population size of $\mu = 300$, running for 1.5×10^5 and 3×10^5 function evaluations on LZ09 and LGZ problems, respectively. The other parameter settings used in comparative study are shown in Table 1.

All the parameters are chosen to guarantee a fair comparison between the algorithm. The values of δ and T for MOEA/D are used according to the original paper, whereas the effects of different values of δ and T are investigated in the first part of experimental study and chosen based on the obtained results.

3.2 Parametrization

In MOEA/PC, δ and T are two major control parameters. To examine the sensitivity of the proposed algorithm to these parameters, we carry out experiments for $\delta \in \{0.7, 0.8, 0.9, 1.0\}$ and $T \in \{10, 20, 30, 40\}$. Figure 2 shows the results of the IGD indicator on the LZ09 test suite. From this figure, we can see that the performance does not vary significantly on F1, F6, F7, whereas on the majority of problems very small values of δ and large values of T lead to a poorer performance. This can be due to that MOEA/PC with too small δ and too large T is poor at exploitation. Table 2 shows the mean ranks of the median IGD values achieved by the algorithm with different settings of δ and T. From the table, we can see that the algorithm has the best rank for $\delta = 0.8$ and $T = 20$.

Figure 3 illustrates the median values of IGD on the LGZ test suite. From the figure, we can see that MOEA/PC is less sensitive on problems F1, F4, F6 and F7, whereas the performance significantly deteriorates on the other problems for $\delta = 0.7$. Similarly to the LZ09 problems, MOEA/PC with too small δ performs poorly in terms of exploitation on the LGZ problems. Nevertheless, too large T is not so critical on these problems. Table 3 presents the mean ranks of the median values of IGD achieved by the algorithm with different settings on the LGZ test suite. From the table, we can see that the algorithm ranks the best on this set of problems having $\delta = 1.0$ and $T = 30$.

Thus, the results indicate that problems with different characteristics may require a careful setting of control parameters. Table 4 summarizes the results

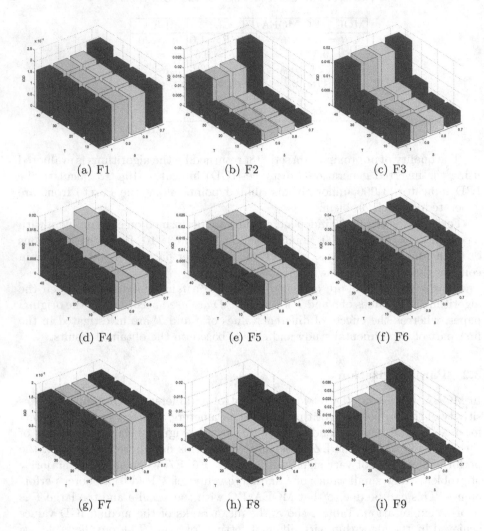

(a) F1   (b) F2   (c) F3

(d) F4   (e) F5   (f) F6

(g) F7   (h) F8   (i) F9

Fig. 2. Results for different parameter settings on the LZ09 test suite

Table 2. Mean ranks for different parameter settings on the LZ09 test suite

T \ δ	10	20	30	40
0.7	5.67	4.78	7.56	9.33
0.8	6.44	4.33	7.44	8.11
0.9	8.22	7.11	7.44	9.78
1.0	14.67	12.78	11.11	11.22

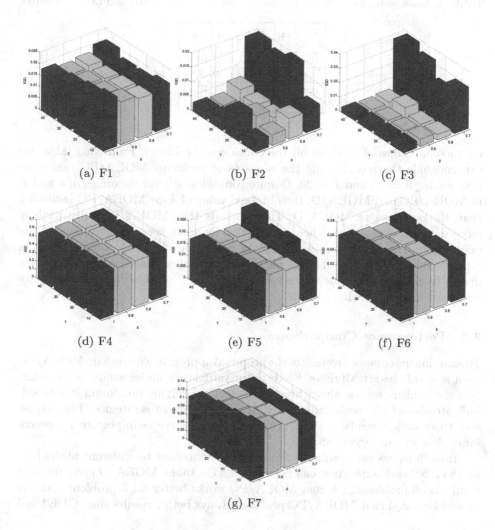

(a) F1   (b) F2   (c) F3

(d) F4   (e) F5   (f) F6

(g) F7

Fig. 3. Results for different parameter settings on the LGZ test suite

Table 3. Mean ranks for different parameter settings on the LGZ test suite

T / δ	10	20	30	40
0.7	10.29	7.86	10.14	12.43
0.8	10.43	5.86	7.57	12.00
0.9	9.71	4.14	4.86	10.57
1.0	13.43	5.43	3.00	8.29

Table 4. Mean ranks for different parameter settings on the LZ09 and LGZ test suites

δ \ T	10	20	30	40
0.7	7.69	6.13	8.69	10.69
0.8	8.19	5.00	7.50	9.81
0.9	8.88	5.81	6.31	10.13
1.0	14.13	9.56	7.56	9.94

for median values of IGD on all the considered problems. From this table, we can conclude that considering the two sets of problems MOEA/PC ranks the best having $\delta = 0.8$ and $T = 20$. Despite somewhat similar meanings of δ and T in MOEA/PC and MOEA/D, the obtained value of δ for MOEA/PC is smaller than the one used in MOEA/D. This suggests that MOEA/PC must slightly more focus on exploration when its selection forces to keep diversity among the population. Though it should be kept in mind that the final obtained rank is influenced by the number of problems in each test suite. Based on the conclusions drawn from these experiments we use $\delta = 0.8$ and $T = 20$ in comparative study discussed in the following.

3.3 Performance Comparison

To examine the competitiveness of the proposed approach, we compare MOEA/PC with state-of-the-art MOEAs. Since the main feature under study is the selection mechanism, we use algorithms with selections relying on: dominance-based, indicator-based and scalarizing-based fitness assignment strategies. They represent three major trends in MOEAs and, therefore, serve as important references for evaluating our algorithm.

Table 5 shows the results in terms of IGD obtained by different algorithms on the LZ09 test suite. As it can be seen from the table, MOEA/D gives the best results on 6 problems, whereas MOEA/PC works better on 3 problems. Also it should be noted that MOEA/PC provides always better results than GDE3 and

Table 5. Median and interquartile range of the IGD indicator on the LZ09 test suite

	MOEA/PC	MOEA/D	GDE3	IBEA
F1	1.64e-03 (8.9e-06)	1.31e-03 (7.7e-06)	2.19e-03 (1.0e-04)	6.77e-03 (7.4e-04)
F2	3.85e-03 (4.4e-04)	2.75e-03 (2.8e-04)	4.23e-02 (5.2e-03)	1.12e-01 (1.5e-02)
F3	4.41e-03 (1.6e-03)	2.69e-03 (4.2e-03)	3.66e-02 (2.7e-03)	5.48e-02 (2.9e-02)
F4	8.66e-03 (2.9e-03)	6.46e-03 (9.9e-03)	3.62e-02 (3.4e-03)	7.14e-02 (4.3e-02)
F5	1.06e-02 (3.2e-03)	1.22e-02 (5.4e-03)	3.53e-02 (4.4e-03)	3.96e-02 (1.5e-02)
F6	3.52e-02 (5.8e-04)	4.69e-02 (8.9e-03)	1.14e-01 (2.3e-02)	5.33e-01 (4.6e-02)
F7	1.64e-03 (3.2e-05)	1.34e-03 (2.1e-05)	3.94e-01 (0.0e+00)	1.98e-01 (9.1e-02)
F8	5.05e-03 (9.5e-03)	1.71e-03 (1.8e-03)	3.94e-01 (0.0e+00)	2.07e-01 (4.5e-02)
F9	3.02e-03 (9.8e-04)	3.85e-03 (2.1e-03)	4.50e-02 (7.8e-03)	1.20e-01 (6.7e-02)

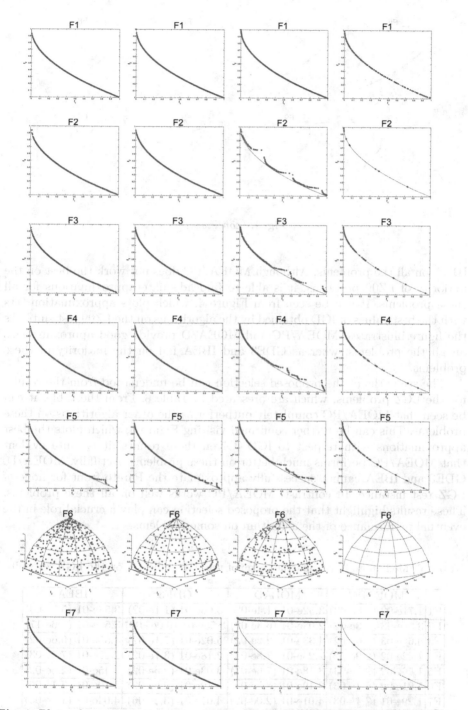

Fig. 4. Plots of approximation sets with the best IGD values for the LZ09 test suite obtained by MOEA/PC, MOEA/D, GDE3 and IBEA, left to right, respectively

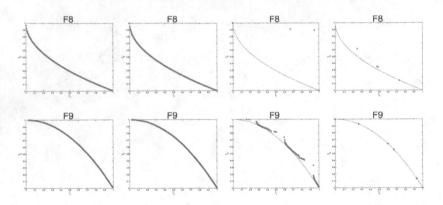

Fig. 4. (*continued*)

IBEA on all the problems. Although MOEA/PC does not work the best on the majority of LZ09 problems, it is able to find adequate approximations for all these problems. It can be seen from Figure 4, which plots approximation sets with the best values of IGD obtained by the algorithms on the LZ09 test suite. As the figure illustrates, MOEA/PC and MOEA/D provide good approximations on all the problems, whereas GDE3 and IBEA fail on the majority of these problems.

The strengths of the proposed selection can be understood from the results for the LGZ problems, which are presented in Table 6. From the table, it can be seen that MOEA/PC completely outperforms the other algorithms on these problems. This can be further confirmed visiting Figure 5, which plots the best approximations with respect to IGD. Given these results, it is quite evident that MOEA/PC performs much better on these problems. Actually, MOEA/D, GDE3 and IBEA cannot successfully approximate the Pareto front for none of LGZ test instance. In contrast, MOEA/PC works well on all these problems. These results highlight that the proposed selection can play a crucial role in the eventual performance of the algorithm on some problems.

Table 6. Median and interquartile range of the IGD indicator on the LGZ test suite

	MOEAPC	MOEAD	GDE3	IBEA
F1	1.71e-02 (2.8e-03)	3.52e-01 (1.4e-02)	3.60e-01 (1.1e-02)	3.58e-01 (5.7e-03)
F2	3.06e-03 (1.3e-02)	2.08e-01 (6.3e-02)	3.55e-01 (0.0e+00)	3.55e-01 (5.6e-17)
F3	8.64e-03 (2.6e-02)	4.33e-01 (2.2e-02)	5.02e-01 (7.1e-02)	4.55e-01 (6.6e-02)
F4	1.42e-02 (2.9e-02)	2.25e-01 (7.8e-03)	2.18e-01 (3.7e-02)	2.22e-01 (7.4e-03)
F5	1.59e-02 (3.0e-03)	3.18e-01 (7.6e-03)	3.03e-01 (2.6e-02)	2.45e-01 (2.6e-02)
F6	5.91e-02 (6.1e-03)	3.19e-01 (8.5e-08)	3.19e-01 (2.6e-02)	3.19e-01 (7.1e-06)
F7	1.29e-01 (7.1e-03)	4.01e-01 (2.0e-02)	4.01e-01 (3.2e-06)	4.04e-01 (1.0e-03)

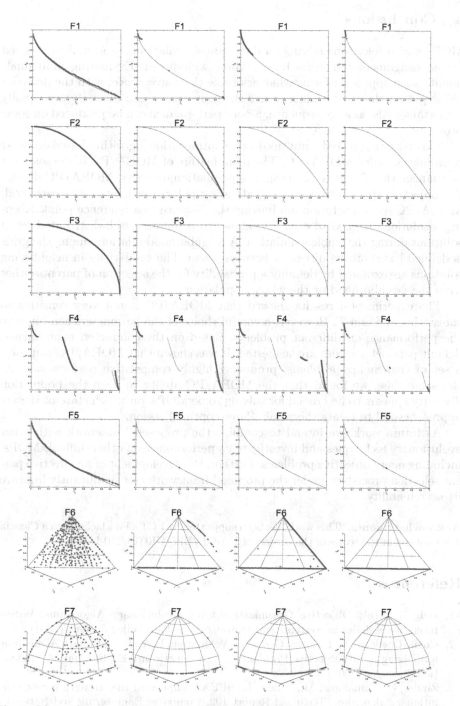

Fig. 5. Plots of approximation sets with the best IGD values for the LGZ test suite obtained by MOEA/PC, MOEA/D, GDE3 and IBEA, left to right, respectively

4   Conclusions

MOEAs with selections relying on dominance-, scalarizing- and indicator-based fitness assignment strategies have become widespread. Evaluating individuals quality such approaches somewhat prioritize the convergence, with the diversity being put on the second place. Since the convergence and diversity are equally important goals, as a consequence, a poor performance can be produced on some problems.

This paper suggested a multiobjective optimization algorithm based on polar coordinates, called MOEA/PC. The main feature of MOEA/PC is the selection mechanism that forces to maintain the population diversity. MOEA/PC divides the objective space into a set of grids using polar coordinates. For each grid, MOEA/PC seeks a solution minimizing the distance to a reference point. Keeping a solution assigned to each grid allows to maintain a well-distributed set of solutions during the whole simulation. A neighborhood relation among the grids is defined based on the distances between them. The similarities in neighboring solutions are explored by defining a probability for the selection of parents either from the neighborhood or the whole population.

The experimental results showed that MOEA/PC is not very sensitive to the setting of δ and T. However, a careful choice of the parameters can improve the performance on different problems. Based on the conducted experiments, default parameter values are suggested. It was shown that MOEA/PC can solve a set of challenging problems, producing highly competitive performance. At the same time, we found that the MOEA/PC ability to keep the population diversity appears to be crucial for solving some MOPs, on which state-of-the-art approaches fail to locate the whole Pareto optimal region.

As future work, we intend to combine the proposed framework with other evolutionary techniques and investigate its performance on other difficult MOPs, including many-objective problems. Further, the development of an effective parent selection procedure under the proposed framework can significantly improve its search ability.

Acknowledgments. This work has been supported by FCT - Fundação para a Ciência e Tecnologia in the scope of the project: PEst-OE/EEI/UI0319/2014.

References

1. Deb, K.: Multi-Objective Optimization using Evolutionary Algorithms. Wiley-Interscience Series in Systems and Optimization. John Wiley & Sons (2001)
2. Coello Coello, C.A., Lamont, G.B., Van Veldhuizen, D.A.: Evolutionary Algorithms for Solving Multi-Objective Problems, 2 edn. Genetic and Evolutionary Computation. Springer (2007)
3. Zitzler, E., Laumanns, M., Thiele, L.: SPEA2: Improving the strength pareto evolutionary algorithm. Technical Report 103, Computer Engineering and Networks Laboratory (TIK), ETH Zurich, Zurich, Switzerland (2001)

4. Deb, K., Pratap, A., Agarwal, S., Meyarivan, T.: A fast and elitist multiobjective genetic algorithm: NSGA-II. IEEE Transactions on Evolutionary Computation **6**(2), 182–197 (2002)

5. Hughes, E.J.: Multiple single objective Pareto sampling. In: Proceedings of the IEEE Congress on Evolutionary Computation, CEC 2003, pp. 2678–2684 (2003)

6. Ishibuchi, H., Doi, T., Nojima, Y.: Incorporation of scalarizing fitness functions into evolutionary multiobjective optimization algorithms. In: Runarsson, T.P., Beyer, H.-G., Burke, E.K., Merelo-Guervós, J.J., Whitley, L.D., Yao, X. (eds.) PPSN 2006. LNCS, vol. 4193, pp. 493–502. Springer, Heidelberg (2006)

7. Zhang, Q., Li, H.: MOEA/D: A multiobjective evolutionary algorithm based on decomposition. IEEE Transactions on Evolutionary Computation **11**(6), 712–731 (2007)

8. Zitzler, E., Künzli, S.: Indicator-based selection in multiobjective search. In: Yao, X., et al. (eds.) PPSN 2004. LNCS, vol. 3242, pp. 832–842. Springer, Heidelberg (2004)

9. Beume, N., Naujoks, B., Emmerich, M.: SMS-EMOA: Multiobjective selection based on dominated hypervolume. European Journal of Operational Research **181**(3), 1653–1669 (2007)

10. Rodríguez Villalobos, C.A., Coello Coello, C.A.: A new multi-objective evolutionary algorithm based on a performance assessment indicator. In: Proceedings of the Genetic and Evolutionary Computation Conference, GECCO 2012, pp. 505–512 (2012)

11. Denysiuk, R., Costa, L., Espírito Santo, I.: Many-objective optimization using differential evolution with variable-wise mutation restriction. In: Proceedings of the Conference on Genetic and Evolutionary Computation, GECCO 2013, pp. 591–598 (2013)

12. Denysiuk, R., Costa, L., Espírito Santo, I.: Clustering-based selection for evolutionary many-objective optimization. In: Bartz-Beielstein, T., Branke, J., Filipič, B., Smith, J. (eds.) PPSN 2014. LNCS, vol. 8672, pp. 538–547. Springer, Heidelberg (2014)

13. Kuang, D., Zheng, J.: Strategies based on polar coordinates to keep diversity in multi-objective genetic algorithm. In: Proceedings of the IEEE Congress on Evolutionary Computation, CEC 2005, pp. 1276–1281 (2005)

14. Li, H., Zhang, Q.: Multiobjective optimization problems with complicated Pareto sets, MOEA/D and NSGA-II. IEEE Transactions on Evolutionary Computation **13**(2), 284–302 (2009)

15. Kukkonen, S., Lampinen, J.: GDE3: the third evolution step of generalized differential evolution. In: Proceedings of the IEEE Congress on Evolutionary Computation, CEC 2005, 443–450 (2005)

16. Durillo, J.J., Nebro, A.J.: jMetal: A Java framework for multi-objective optimization. Advances in Engineering Software **42**(10), 760–771 (2011)

17. Liu, H.L., Gu, F., Zhang, Q.: Decomposition of a multiobjective optimization problem into a number of simple multiobjective subproblems. IEEE Transactions on Evolutionary Computation **18**(3), 450–455 (2014)

18. Zitzler, E., Thiele, L., Laumanns, M., Fonseca, C.M., Grunert da Fonseca, V.: Performance assessment of multiobjective optimizers: An analysis and review. IEEE Transactions on Evolutionary Computation **7**(2), 117–132 (2003)

19. Bosman, P.A.N., Thierens, D.: The balance between proximity and diversity in multiobjective evolutionary algorithms. IEEE Transactions on Evolutionary Computation **7**(2), 174–188 (2003)

GD-MOEA: A New Multi-Objective Evolutionary Algorithm Based on the Generational Distance Indicator

Adriana Menchaca-Mendez[✉] and Carlos A. Coello Coello

CINVESTAV-IPN (Evolutionary Computation Group), Departamento de
Computación, 07300 Mexico D.F., Mexico
adriana.menchacamendez@gmail.com, ccoello@cs.cinvestav.mx

Abstract. In this paper, we propose a new selection mechanism for
Multi-Objective Evolutionary Algorithms (MOEAs), which is based on
the generational distance indicator and uses a technique that relies on
Euclidean distances to maintain diversity in the population (in objective
function space). Our proposed selecion mechanism is incorporated into
a MOEA which adopts the operators of NSGA-II (crossover and mutation) to generate new individuals. The new MOEA is called "Generational
Distance - Multi-Objective Evolutionary Algorithm (GD-MOEA)." Our
GD-MOEA is validated using standard test problems taken from the specialized literature, having three to six objective functions. GD-MOEA is
compared with respect to MOEA/D using Penalty Boundary Intersection
(PBI), which is based on decomposition, and to SMS-EMOA-HYPE (a
version of SMS-EMOA that uses a fitness assignment scheme based on
the use of an approximation of the hypervolume indicator). Our preliminary results indicate that if we consider both quality in the solutions and
the running time required to generate them, our GD-MOEA is a good
alternative to solve multi-objective optimization problems having both
low dimensionality and high dimensionality in objective function space.

1 Introduction

Many real-world applications involve the solution of problems that have multiple (conflicting) objective functions which have to be simultaneously optimized.
These are the so-called "Multi-objective Optimization Problems (MOPs)". Since
their objective functions are in conflict with each other, the notion of optimality refers to finding the best possible trade-offs among the objective functions.
Consequently, there is no single optimal solution but a set of solutions, which is
called *Pareto optimal set*, whose image is known as the *Pareto front*. Since the
use of mathematical programming techniques to solve MOPs has several limitations, the use of evolutionary algorithms has become very popular in this area
in recent years, giving rise to the so-called Multi-Objective Evolutionary Algorithms (MOEAs) [6]. MOEAs have two main goals: (i) To find solutions that

The second author acknowledges support from CONACyT project no. 221551.

A. Gaspar-Cunha et al. (Eds.): EMO 2015, Part I, LNCS 9018, pp. 156–170, 2015.
DOI: 10.1007/978-3-319-15934-8_11

are, as close as possible, to the true Pareto front, and, (ii) to produce solutions that are spread along the Pareto front as uniformly as possible.

There are different indicators to assess the quality of the approximation of the Pareto optimal set generated by a MOEA, e.g., error ratio, generational distance, inverted generational distance, spacing, hypervolume, $R2$-indicator, Δ_p-indicator, ϵ-indicator, two set coverage, etc. [6]. However, very few performance indicators are "Pareto Compliant".[1] In recent years, MOEAs based on indicators have become popular because the use of Pareto-based selection has several limitations. Perhaps, the most remarkable is its poor scalability regarding the number of objective functions of a MOP.[2]

MOEAs based on the hypervolume indicator (I_H) have been relatively popular (see for example [2,15–17,28]) mainly because I_H is the only unary indicator which is known to be "Pareto compliant" [29]. However, I_H has an important disadvantage: its high computational cost (the problem of computing I_H is *NP-hard* [3]). Therefore, this type of MOEAs is impractical when we want to solve MOPs having four or more objective functions. On the other hand, after the study on the properties of the $R2$-indicator (I_{R2}) presented by Brockhoff et al. [4], a number of proposals of MOEAs based on I_{R2} have been introduced [13,21,22,26]. Although I_{R2}-based MOEAs can solve MOPs with many objective functions at an affordable computational cost, this type of algorithms also has an important disadvantage: They need to generate a set of well-distributed convex weights and this task becomes more difficult as we increase the number of objective functions. The same applies to the well-known MOEA/D [27] which decomposes the MOP into N scalar optimization subproblems and solves them simultaneously using an evolutionary algorithm. Recently, the Δ_p-indicator (I_{Δ_p}) was introduced [20] and some MOEAs based on it have already been proposed [10,12,19]. The Δ_p-indicator is composed of slight modifications of two well-known indicators: generational distance (I_{GD}) [23] and inverted generational distance (I_{IGD}) [5]. It is well-known that for computing I_{GD} and I_{IGD}, it is necesary to know the true Pareto front. Therefore, the most important disadvantage of MOEAs based on I_{Δ_p} is perhaps that they need a reference set which must contain well-distributed solutions. Not being able to produce a good reference set could produce a diversity loss in the population which might cause that the algorithm cannot generate the complete Pareto front, or that it generates poorly distributed solutions. In extreme cases, the lack of an appropriate reference set could prevent convergence.

In this paper, we propose a new MOEA based on I_{GD} and we use the technique proposed in [18], which is based on Euclidean distances, to maintain diversity in objective function space. The idea is to use the non-dominated set produced at

---

[1] Let Ω be the set of all feasible solutions and $\mathcal{A}$ and $\mathcal{B}$ two approximations of the Pareto optimal set, such that, $\mathcal{A} \preceq \mathcal{B}$ denotes that every point $\mathbf{b} \in \mathcal{B}$ is weakly dominated by at least one point $\mathbf{a} \in \mathcal{A}$. An indicator $I : \Omega \to \mathbb{R}$ is **Pareto compliant** if for all $\mathcal{A}, \mathcal{B} \in \Omega : \mathcal{A} \preceq \mathcal{B} \Rightarrow I(\mathcal{A}) \geq I(\mathcal{B})$, assuming that greater indicator values correspond to higher quality.

[2] The quick increase in the number of non-dominated solutions as we increase the number of objective functions, rapidly dilutes the effect of the selection mechanism of a MOEA [11].

each generation as a reference set to calculate I_{GD}, even if it is not well-distributed, since at the beginning, the aim is to achieve convergence to the true Pareto front. Then, when we have produced many non-dominated solutions, the aim will be to improve their distribution. In this way, we can address the disadvantages of MOEAs based on I_H, I_{R2} and I_{Δ_p}: Our new selection mechanism has linear complexity with respect to the number of objective functions because computing the I_{GD} and maintaining diversity by means of computing Euclidean distances have linear complexity with respect to the number of objective functions. Further, it is not necessary to generate a set of well-distributed convex weights, and also, it is not necessary to generate a well-distributed reference set.

The remainder of this paper is organized as follows. Section 2 describes the generational distance indicator. The technique to maintain diversity in the population is described in Section 3. Our proposal is presented in Section 4. The experimental validation and the results obtained are shown in Section 5. Finally, we provide our conclusions and some possible paths for future work in Section 6.

2   Generational Distance Indicator

The generational distance indicator (I_{GD}) reports how far, on average, $\mathcal{A}$ is from $\mathcal{PF}$ [7,24,25], where $\mathcal{PF}$ is the true Pareto front and $\mathcal{A}$ is an approximation of the true Pareto front. I_{GD} is Pareto non-compliant and it is defined as:

$$I_{GD} = \frac{1}{|\mathcal{A}|} \left(\sum_{i=1}^{|\mathcal{A}|} d_i^p \right)^{\frac{1}{p}} \tag{1}$$

where $|\mathcal{A}|$ is the number of vectors in $\mathcal{A}$, $p = 2$ and d_i is the Euclidean phenotypic distance between each member, i, of $\mathcal{A}$ and the closest member in $\mathcal{PF}$ to that member, i. If $I_{GD} = 0$, $\mathcal{A} \subseteq \mathcal{PF}$. Figure 1 shows how this indicator is calculated.

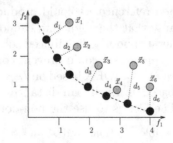

Fig. 1. The black points are the reference set. The approximation of the Pareto optimal set, $\mathcal{A}$, is composed by all point in gray. d_i is the Euclidean distance between x_i and its closest neighbor in $\mathcal{PF}$. Therefore: $I_{GD} = \frac{1}{6} \left(d_1^2 + d_2^2 + d_3^2 + d_4^2 + d_5^2 + d_6^2 \right)^{\frac{1}{2}}$.

3 A Distribution Technique Based on Euclidean Distances

In [18], Menchaca and Coello proposed a technique based on Euclidean distances to improve the diversity in objective function space. This technique works as follows: Let's suppose that we have already a set of non-dominated solutions which we call "S". If we want to improve its diversity using another set of non-dominated solutions which is called "B", then, the solutions in B compete with the solutions in S, considering that the size of S is fixed, as follows: For each solution $x \in B$, we obtain its nearest neighbor from S, x_{near}, and we choose a random individual from S, x_{random} such that $x_{near} \neq x_{random}$, and then, these three solutions compete to survive. First, x competes with x_{random}, if the Euclidean distance from x to its nearest neighbor in S is greater than the Euclidean distance from x_{random} to its nearest neighbor in S, x replaces x_{random}. If x loses the competition, x competes with its nearest neighbor to survive. If the Euclidean distance from x to its nearest neighbor in S (without considering

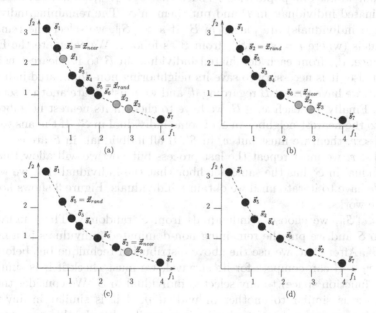

(a) (b) (c) (d)

Fig. 2. All points in black are in S and all points in gray are in B. In (a), we consider solution x_1, its nearest neighbor in S is s_2 and we choose s_5 as a random solution. First, x_1 competes with s_5 and s_5 loses because the distance from x_1 to s_2 is greater than the distance from s_5 to s_6; therefore, x_1 replaces s_5. In (b), we consider solution x_2, its nearest neighbor in S is s_6 and we choose s_2 as a random solution. First, x_2 and s_2 compete and x_2 wins because the distance from x_2 to s_6 is greater than the distance from s_2 to s_5; therefore, x_2 replaces s_2. Finally, in (c), we consider solution x_3, its nearest neighbor in S is s_2 and we choose s_5 as a random solution. First, x_3 competes with s_5 and s_5 wins because the distance from x_3 to s_2 is less than the distance from s_5 to s_3; therefore, x_3 competes with s_2 and x_3 wins because the distance from x_3 to x_6 is greater than the distance from s_2 to s_6. Thus, x_3 replaces s_2. In (d), we can see the new S.

x_{near}) is greater than the Euclidean distance from x_{near} to its nearest neighbor in $\mathbb{S}$, then x replaces x_{near}.

The authors mentioned that x_{near} is used with the idea of improving the diversity locally. If we move x_{near} to x, do we increase the Euclidean distance from x_{near} to its nearest neighbor in S? And, x_{rand} is used to avoid that solutions in unexplored regions are eliminated, e.g., if x and x_{near} are in an unexplored region, it is not good to delete one of the two solutions. Figure 2 illustrates how this technique works.

4   Our Proposal

In this work, we propose a new selection mechanism for MOEAs. The idea is to use I_{GD} as a convergence strategy and to use the above distribution technique to maintain diversity in the population when many (even all) solutions are non-dominated. Our selection mechanism works as follows: If we want to select s individuals of a population $\mathcal{P}$, such that $s < \|\mathcal{P}\|$, we have to obtain the non-dominated individuals in $\mathcal{P}$ and put them in S. The remaining individuals (dominated individuals) are placed in $\mathcal{B}$. If $s > \|S\|$, we select the remaining r individuals (where $r = s - \|S\|$) from $\mathcal{B}$ as follows: We calculate the Euclidean distance, d_i, from each dominated individual in $\mathcal{B}$ to its nearest neighbor in S, and also, it is necessary to save its neighboring non-dominated individual. After that, we have to sort $\mathcal{B}$ regarding d_i and we must create another set called "$S' = \emptyset$". Finally, for each $x_i \in \mathcal{B}$, we have to check if its nearest neighbor in S is equal to the nearest neighbor in S of some individual in S'. If the answer is no and $\|S'\| < r$, then, we must put x_i in S'. If all individuals in $\mathcal{B}$ are considered and $\|S'\| < r$, we must repeat the last process but now we will allow that only one individual in S' has the same neighbor that the individual that we want to select. We have to iterate until we obtain r individuals. Figure 3 shows how this procedure works.

If $s < \|S\|$, we choose s individuals from S randomly. These individuals remain in S and we put the remaining non-dominated individuals in a new set called "$\mathcal{B}$". After that, we use the above distribution technique but before each solution $x_i \in \mathcal{B}$ can compete for its survival, we must check if it is similar (in objective function space) to any selected individual in S. We consider that one individual x is similar to another individual y, if it is similar in any objective function: $x.f_i - y.f_i < \epsilon$, where ϵ is a small value. In this way, we avoid that weakly non-dominated individuals are selected. If we do not apply this constraint, we can obtain many weakly Pareto optimal solutions, which could prevent convergence. Figure 4 shows how this diversity technique works. The complete selection process is shown in Algorithm 1.

4.1   Generational Distance - Multi-Objective Evolutionary Algorithm (GD-MOEA)

In order to validate our selection mechanism, we designed a multi-objective evolutionary algorithm which uses the operators of NSGA-II (crossover and mutation)

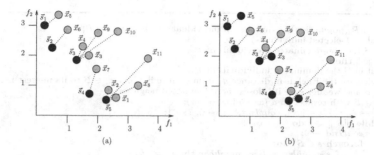

(a)   (b)

Fig. 3. Let's assume that we want to select eight individuals from the population. In (a), the non-dominated individuals are identified (black points) and then $\mathcal{S} = \{s_1, s_2, s_3, s_4, s_5\}$. After that, we calculate d_i for each dominated individual (gray points), we store its nearest neighbor in $\mathcal{S}$ and we sort them with respect to d_i, such that $x_i.d_i \leq x_{i+1}.d_{i+1}$. In (b), we proceed to select the remaining 3 individuals. First, we select individual x_1 ($\mathcal{S}' = \{x_1\}$). After that, individual x_2 is considered but it is not selected because its nearest neighbor in $\mathcal{S}$ is the same that the nearest neighbor of x_1. Then, we consider individual x_3 and we select it ($\mathcal{S}' = \{x_1, x_3\}$). Finally, individual x_4 is considered and it is not selected because its nearest neighbor is the same that the nearest neighbor of individual x_3. Then, we consider individual x_5 and we select it ($\mathcal{S}' = \{x_1, x_3, x_5\}$). Therefore, the selected individuals are $\mathcal{S} \cup \mathcal{S}'$ (black points).

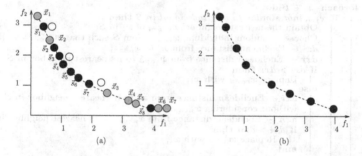

(a)   (b)

Fig. 4. Let's assume that we want to select eight individuals from the population. In (a), we select randomly eight non-dominated individuals and we put them in $\mathcal{S}$ (black points), and the remaining non-dominated individuals are placed in $\mathcal{B}$ (gray points). After that, we apply the distribution technique described in Section 3. (b) shows the final $\mathcal{S}$ and we can see that although x_6 and x_7 could replace individual s_8, they were not selected because they are similar to s_8.

to create new individuals. This is because our main aim is to validate the effect of our proposed selection mechanism comparing it with respect to other two selection mechanisms: The first is based on decomposition and the second one is based on the approximation of the hypervolume indicator. For this sake, we used the following MOEAs: MOEA/D [27] (using PBI to decompose the MOP) and SMS-EMOA-HYPE (a version of SMS-EMOA [2] that uses a fitness assignment based

---

Input : $\mathcal{P}$ (population), s (number of individuals to choose $s < \|\mathcal{P}\|$).
Output: $\mathcal{S}$ (selected individuals).
Put in $\mathcal{S}$ the non-dominated individuals of $\mathcal{P}$;
if $s > \|\mathcal{S}\|$ **then**
    Put in $\mathcal{B}$ the dominated individuals of $\mathcal{P}$;
    Calculate the Euclidean distance d_i from each individual $\boldsymbol{x}_i \in \mathcal{B}$ to its nearest neighbor in $\mathcal{S}$ and we also save its closest non-dominated neighbor;
    Sort $\mathcal{B}$ with respect to d (ascending order);
    $\mathcal{S}' \leftarrow \emptyset$, $r \leftarrow s - \|\mathcal{S}\|$, $contIndAux \leftarrow 0$, $i \leftarrow 1$;
    while $\|\mathcal{S}'\| < r$ **do**
        $contInd \leftarrow 0$;
        foreach $s \in \mathcal{S}'$ **do**
            if $s.neighbor = \mathcal{B}.\boldsymbol{x}_i.neighbor$ **then**
             | $contInd \leftarrow contInd + 1$;
            end
        end
        if $contInd \leq contIndAux$ **then**
        | Put $\mathcal{B}.\boldsymbol{x}_i$ in $\mathcal{S}'$;
        end
        repeat
        | $i \leftarrow i + 1$;
        until $\mathcal{B}.\boldsymbol{x}_i \notin \mathcal{S}'$;
        if $i = \|\mathcal{B}\|$ **then**
        | $i \leftarrow 0$, $contIndAux \leftarrow contIndAux + 1$;
        end
    end
    $\mathcal{S} \leftarrow \mathcal{S} \cup \mathcal{S}'$;
else
    if $s < \|\mathcal{S}\|$ **then**
        Choose randomly $\|\mathcal{S}\| - s$ individuals of $\mathcal{S}$ and put them in a new set called $\mathcal{B}$;
        foreach $\boldsymbol{x}_i \in \mathcal{B}$ **do**
            if $\boldsymbol{x}_i$ *is not similar to any individual in* $\mathcal{S}$ **then**
                Obtain the nearest neighbor, $\boldsymbol{x}_{near}$, of $\boldsymbol{x}_i$ in $\mathcal{S}$;
                Choose a random individual, $\boldsymbol{x}_{rand}$, from $\mathcal{S}$ such that $\boldsymbol{x}_{near} \neq \boldsymbol{x}_{rand}$;
                $dxi \leftarrow$ Euclidean distance from $\boldsymbol{x}_i$ to $\boldsymbol{x}_{near}$;
                $dxr \leftarrow$ Euclidean distance from $\boldsymbol{x}_{rand}$ to its nearest neighbor in $\mathcal{S}$;
                if $dxi > dxr$ **then**
                | Replace $\boldsymbol{x}_{rand}$ with $\boldsymbol{x}_i$;
                else
                    $dxi \leftarrow$ Euclidean distance from $\boldsymbol{x}_i$ to its nearest neighbor in $\mathcal{S}$ without considering $\boldsymbol{x}_{near}$;
                    $dxn \leftarrow$ Euclidean distance from $\boldsymbol{x}_{near}$ to its nearest neighbor in $\mathcal{S}$;
                    if $dxi > dxn$ **then**
                    | Replace $\boldsymbol{x}_{near}$ with $\boldsymbol{x}_i$;
                    end
                end
            end
        end
    end
end
return $\mathcal{S}$;

Algorithm 1: I_{GD}-Selection

on the approximation of the hypervolume indicator, proposed in [1]). Since these MOEAs use the same operators as our proposed approach to create new individuals, the comparison is fair.

Our proposed MOEA is called **"Generational Distance - Multi-Objective Evolutionary Algorithm (GD-MOEA)"** and it works as follows. First, it creates an initial population of size P. After that, it creates P new individuals and it combines the population of parents and offspring to obtain a population of size $2P$. Then, we use the proposed selection mechanism to choose the P individuals that will take part of the following generation. Finally, this process is repeated for a (pre-defined) number of generations.

5   Experimental Results

As mentioned before, we compare our proposed GD-MOEA with respect to MOEA/D and SMS-EMOA-HYPE. In the case of MOEA/D, we generated the convex weights using the technique proposed in [8] and after that, we applied clustering (k-means) to obtain a specific number of weights. In the case of SMS-EMOA-HYPE, we used the source code of HyPE available in the public domain [1] adopting 10^4 as our number of samples to assign fitness in the original SMS-EMOA.[3]

For our experiments, we used seven problems taken from the Deb-Thiele-Laumanns-Zitzler (DTLZ) test suite [9]. We used $k = 5$ for DTLZ1, DTLZ3 and DTLZ6 and $k = 10$ for the remaining test problems. Also, we used seven problems taken from the WFG toolkit [14], with $k\_factor = 2$ and $l\_factor = 10$. For each test problem, we performed 30 independent runs. For all three algorithms, we adopted the parameters suggested by the authors of NSGA-II: $p_c = 0.9$ (crossover probability), $p_m = 1/n$ (mutation probability), where n is the number of decision variables. We also used $\eta_c = 15$ and $\eta_m = 20$, respectively. We performed a maximum of 50,000 fitness function evaluations (in this case, we used a population size of 100 individuals and we iterated for 500 generations).

5.1   Performance Indicators

We adopted only the hypervolume indicator (I_H) to validate our results because it rewards both convergence towards the Pareto front as well as the maximum spread of the solutions obtained. To calculate the hypervolume indicator, we normalized the approximations of the Pareto optimal set, generated by the MOEAs, and $y_{ref} = [y_1, \cdots, y_k]$ such that $y_i = 1.1$ is used as our reference point. The normalization was performed considering all approximations generated by the different MOEAs (i.e., we put, in one set, all the non-dominated solutions found and from this set we calculate the maximum and minimum for each objective function).

5.2   Discussion of Results

Table 1 shows the results with respect to I_H as well as the results of the statistical analysis that we made to validate our experiments, for which we used Wilcoxon's rank sum. In Table 1, we can see that our proposed GD-MOEA outperformed MOEA/D in forty-three problems and in all cases we can reject the

---

[3] The source code of the three algorithms (MOEA/D, SMS-EMOA-HYPE and GD-MOEA) is avaialble the first author upon request. For MOEA/D, we used the source code available in the MOEA/D webpage.

null hypothesis (medians are equal). Only in thirteen problems our proposed approach was outperformed by MOEA/D. With respect to SMS-EMOA-HYPE, we can see that our GD-MOEA was outperformed in forty-nine problems. Only in six problems our GD-MOEA outperformed SMS-EMOA-HYPE and in one case they had a similar behavior (we cannot reject the null hypothesis). However, it is important to analyze the running time required by the three algorithms. In Table 2, we can see that MOEA/D is better than our GD-MOEA because, in the worst case, MOEA/D required 1.8199 seconds to find the approximation of the Pareto optimal set while our GD-MOEA required 2.6672 seconds, i.e., MOEA/D is 1.46 times faster than our GD-MOEA. In the case of SMS-EMOA-HYPE, we can see that it required 445.7333 seconds in the worst case, i.e., our GD-MOEA is 167.11 times faster than SMS-EMOA-HYPE. Therefore, we can say that our GD-MOEA is a good choice to solve MOPs having both low dimensionality and high dimensionality in objective function space, if we consider both quality in the approximation of the Pareto set and running time.

Finally, we will present a brief study on the effect of the population size on the performance of our approach. We know that if we increase the number of objective functions, we should increase the population size as well. However, algorithms such as SMS-EMOA cannot be used with large population sizes, because its running time rapidly increases (in the worst case, it needs to calculate P times the contribution to the hypervolume, where P is the population size, in order to decide which individual will be removed). In the case of MOEA/D and our GD-MOEA, it is indeed possible to increase the population size. In order to study the behavior of these two MOEAs, we adopted a population size of 300 individuals. Table 3 shows the results and we can see in (a) that in seven problems both algorithms have a similar behavior because we cannot reject the null hypothesis. In twelve cases, MOEA/D outperformed our proposed GD-MOEA and in thirty-seven cases our proposed GD-MOEA outperformed MOEA/D. With respect to the running time, we can see in (b) that in the worst case, MOEA/D required 4.7445 seconds and our proposed GD-MOEA required 11.4934 seconds, i.e., MOEA/D is 2.42 times faster than our proposed GD-MOEA. Therefore, we can say that our proposed GD-MOEA has a better performance than MOEA/D in most cases, while requiring a higher (but not significantly long) running time.

An interesting thing is that our proposed GD-MOEA had serious difficulties to solve DTLZ1, DTLZ6 and WFG1 with more than three objective functions when we used a population size of 100 individuals. However, when we increased the population size, our proposed GD-MOEA was able to obtain better results, and kept a good behavior, in general, for all the problems considered in our study.

6 Conclusions and Future Work

We have proposed a new selection mechanism based on the generational distance indicator (I_{GD}) and a technique based on Euclidean distances to improve the

Table 1. Results obtained in the DTLZ and WFG test problems. We compare our proposed GD-MOEA with respect to MOEA/D and SMS-EMOA-HYPE, using the hypervolume indicator (I_H). We show average values over 30 independent runs. The values in parentheses correspond to the standard deviations. The third column of each table shows the results of the statistical analysis applied to our experiments using Wilcoxon's rank sum. P is the probability of observing the given result (the null hypothesis is true). Small values of P cast doubt on the validity of the null hypothesis. $H = 0$ indicates that the null hypothesis ("medians are equal") cannot be rejected at the 5% level. $H = 1$ indicates that the null hypothesis can be rejected at the 5% level.

f	moead I_H	gd-moea I_H	$P(H)$	sms-emoa-hype I_H	gd-moea I_H	$P(H)$
DTLZ1(3)	1.0710(0.003)	**1.0842(0.005)**	0.00(1)	**1.1011(0.006)**	1.0842(0.005)	0.00(1)
DTLZ2(3)	0.7102(0.000)	**0.7150(0.006)**	0.00(1)	**0.7435(0.002)**	0.7150(0.006)	0.00(1)
DTLZ3(3)	1.3130(0.001)	**1.3276(0.003)**	0.00(1)	**1.3299(0.000)**	1.3276(0.003)	0.00(1)
DTLZ4(3)	0.8191(0.000)	**0.8344(0.008)**	0.00(1)	**0.8639(0.002)**	0.8344(0.008)	0.00(1)
DTLZ5(3)	0.2467(0.001)	**0.2620(0.003)**	0.00(1)	**0.2654(0.000)**	0.2620(0.003)	0.00(1)
DTLZ6(3)	1.0002(0.011)	**1.1060(0.012)**	0.00(1)	1.1048(0.014)	**1.1060(0.012)**	0.92(0)
DTLZ7(3)	0.4472(0.026)	**0.5055(0.061)**	0.00(1)	**0.5389(0.034)**	0.5055(0.061)	0.00(1)
DTLZ1(4)	1.1858(0.005)	**1.2077(0.199)**	0.00(1)	**1.2586(0.057)**	1.2077(0.199)	0.43(0)
DTLZ2(4)	0.8603(0.001)	**0.9030(0.013)**	0.00(1)	**1.0086(0.003)**	0.9030(0.013)	0.00(1)
DTLZ3(4)	1.4556(0.001)	**1.4605(0.006)**	0.00(1)	**1.4636(0.000)**	1.4605(0.006)	0.00(1)
DTLZ4(4)	0.8589(0.001)	**0.9017(0.015)**	0.00(1)	**1.0143(0.004)**	0.9017(0.015)	0.00(1)
DTLZ5(4)	0.8553(0.022)	**0.9122(0.031)**	0.00(1)	**0.9842(0.002)**	0.9122(0.031)	0.00(1)
DTLZ6(4)	**1.0244(0.013)**	0.7536(0.087)	0.00(1)	**1.0561(0.018)**	0.7536(0.087)	0.00(1)
DTLZ7(4)	0.3389(0.008)	**0.5288(0.036)**	0.00(1)	0.5270(0.037)	**0.5288(0.036)**	0.95(1)
DTLZ1(5)	**1.2463(0.011)**	0.4102(0.490)	0.00(1)	**1.2371(0.348)**	0.4102(0.490)	0.00(1)
DTLZ2(5)	0.9483(0.003)	**1.0500(0.027)**	0.00(1)	**1.2614(0.005)**	1.0500(0.027)	0.00(1)
DTLZ3(5)	1.5818(0.008)	**1.5933(0.014)**	0.00(1)	**1.6089(0.000)**	1.5933(0.014)	0.00(1)
DTLZ4(5)	0.9299(0.003)	**1.0280(0.017)**	0.00(1)	**1.2537(0.005)**	1.0280(0.017)	0.00(1)
DTLZ5(5)	1.0353(0.027)	**1.1208(0.104)**	0.00(1)	**1.3005(0.003)**	1.1208(0.104)	0.00(1)
DTLZ6(5)	**1.2919(0.021)**	0.7747(0.104)	0.00(1)	**1.4414(0.009)**	0.7747(0.104)	0.00(1)
DTLZ7(5)	0.0967(0.067)	**0.4426(0.045)**	0.00(1)	**0.4964(0.049)**	0.4426(0.045)	0.00(1)
DTLZ1(6)	**1.3056(0.013)**	0.0149(0.048)	0.00(1)	**1.5000(0.235)**	0.0149(0.048)	0.00(1)
DTLZ2(6)	0.9712(0.011)	**1.1841(0.029)**	0.00(1)	**1.5413(0.005)**	1.1841(0.029)	0.00(1)
DTLZ3(6)	**1.7610(0.004)**	1.7133(0.052)	0.00(1)	**1.7711(0.000)**	1.7133(0.052)	0.00(1)
DTLZ4(6)	0.9559(0.005)	**1.1805(0.044)**	0.00(1)	**1.5446(0.005)**	1.1805(0.044)	0.00(1)
DTLZ5(6)	0.7277(0.014)	**0.8356(0.037)**	0.00(1)	**1.0777(0.010)**	0.8356(0.037)	0.00(1)
DTLZ6(6)	**1.3298(0.037)**	0.5249(0.069)	0.00(1)	**1.6348(0.010)**	0.5249(0.069)	0.00(1)
DTLZ7(6)	0.0194(0.004)	**0.5360(0.066)**	0.00(1)	0.4480(0.125)	**0.5360(0.066)**	0.00(1)
WFG1(3)	**0.9183(0.017)**	0.8572(0.046)	0.00(1)	**1.0174(0.068)**	0.8572(0.046)	0.00(1)
WFG2(3)	0.1539(0.202)	**0.5245(0.128)**	0.00(1)	**0.6506(0.055)**	0.5245(0.128)	0.00(1)
WFG3(3)	0.4989(0.026)	**0.5982(0.008)**	0.00(1)	**0.6061(0.007)**	0.5982(0.008)	0.00(1)
WFG4(3)	0.5943(0.013)	**0.6473(0.008)**	0.00(1)	**0.7023(0.005)**	0.6473(0.008)	0.00(1)
WFG5(3)	0.4710(0.010)	**0.5286(0.004)**	0.00(1)	**0.5370(0.003)**	0.5286(0.004)	0.00(1)
WFG6(3)	0.4548(0.007)	**0.5245(0.008)**	0.00(1)	**0.5475(0.004)**	0.5245(0.008)	0.00(1)
WFG7(3)	0.4933(0.056)	**0.6528(0.012)**	0.00(1)	0.5617(0.029)	**0.6528(0.012)**	0.00(1)
WFG1(4)	**1.1040(0.058)**	0.6312(0.110)	0.00(1)	**1.1404(0.026)**	0.6312(0.110)	0.00(1)
WFG2(4)	0.0030(0.016)	**0.1481(0.179)**	0.00(1)	**0.4888(0.227)**	0.1481(0.179)	0.00(1)
WFG3(4)	0.2872(0.034)	**0.4502(0.025)**	0.00(1)	**0.5343(0.016)**	0.4502(0.025)	0.00(1)
WFG4(4)	0.6492(0.026)	**0.7987(0.015)**	0.00(1)	**0.9308(0.008)**	0.7987(0.015)	0.00(1)
WFG5(4)	0.3672(0.015)	**0.5232(0.006)**	0.00(1)	**0.5586(0.005)**	0.5232(0.006)	0.00(1)
WFG6(4)	0.2887(0.016)	**0.3580(0.044)**	0.00(1)	**0.5654(0.010)**	0.3580(0.044)	0.00(1)
WFG7(4)	0.2887(0.036)	**0.6944(0.017)**	0.00(1)	0.4220(0.032)	**0.6944(0.017)**	0.00(1)
WFG1(5)	**1.2195(0.063)**	0.4916(0.037)	0.00(1)	**1.2517(0.027)**	0.4916(0.037)	0.00(1)
WFG2(5)	0.0105(0.033)	**0.1194(0.122)**	0.00(1)	**0.4422(0.241)**	0.1194(0.122)	0.00(1)
WFG3(5)	0.1508(0.038)	**0.2609(0.062)**	0.00(1)	**0.4852(0.027)**	0.2609(0.062)	0.00(1)
WFG4(5)	0.6399(0.024)	**0.8698(0.029)**	0.00(1)	**1.1159(0.019)**	0.8698(0.029)	0.00(1)
WFG5(5)	0.2401(0.014)	**0.4614(0.018)**	0.00(1)	**0.5798(0.010)**	0.4614(0.018)	0.00(1)
WFG6(5)	**0.2408(0.016)**	0.2149(0.054)	0.00(1)	**0.5222(0.022)**	0.2149(0.054)	0.00(1)
WFG7(5)	0.2149(0.014)	**0.6888(0.019)**	0.00(1)	0.3138(0.022)	**0.6888(0.019)**	0.00(1)
WFG1(6)	**1.1466(0.022)**	0.5648(0.041)	0.00(1)	**1.3539(0.029)**	0.5648(0.041)	0.00(1)
WFG2(6)	0.0094(0.034)	**0.1403(0.153)**	0.00(1)	**0.4834(0.239)**	0.1403(0.153)	0.00(1)
WFG3(6)	**0.0993(0.044)**	0.0962(0.044)	0.58(0)	**0.4375(0.034)**	0.0962(0.044)	0.00(1)
WFG4(6)	0.5947(0.029)	**0.9251(0.035)**	0.00(1)	**1.2800(0.025)**	0.9251(0.035)	0.00(1)
WFG5(6)	0.1613(0.017)	**0.3368(0.044)**	0.00(1)	**0.5823(0.015)**	0.3368(0.044)	0.00(1)
WFG6(6)	**0.2273(0.021)**	0.1651(0.049)	0.00(1)	**0.5142(0.033)**	0.1651(0.049)	0.00(1)
WFG7(6)	0.1842(0.014)	**0.6745(0.031)**	0.00(1)	0.2735(0.018)	**0.6745(0.031)**	0.00(1)

Table 2. Time required (in seconds) by MOEA/D, SMS-EMOA-HYPE and our proposed GD-MOEA for the test problems adopted. All algorithms were compiled using the GNU C compiler and they were executed on a computer with a 2.66GHz processor and 4GB in RAM.

f	moead time	gd-moea time	sms-emoa-hype time	gd-moea time
DTLZ1(3)	**0.4993(0.016)**	0.7695(0.010)	47.0000(2.620)	**0.7695(0.010)**
DTLZ2(3)	**0.5783(0.010)**	1.1593(0.014)	106.1333(4.105)	**1.1593(0.014)**
DTLZ3(3)	**0.5195(0.012)**	0.6366(0.026)	135.9667(21.629)	**0.6366(0.026)**
DTLZ4(3)	**0.6037(0.008)**	1.2361(0.082)	107.1667(3.822)	**1.2361(0.082)**
DTLZ5(3)	**0.5922(0.007)**	1.0845(0.015)	64.3333(5.430)	**1.0845(0.015)**
DTLZ6(3)	**0.5007(0.018)**	0.8883(0.035)	59.0667(9.747)	**0.8883(0.035)**
DTLZ7(3)	**0.5397(0.008)**	0.9039(0.038)	98.4333(9.106)	**0.9039(0.038)**
DTLZ1(4)	**0.5230(0.008)**	0.8578(0.033)	59.6667(3.280)	**0.8578(0.033)**
DTLZ2(4)	**0.6147(0.012)**	1.1623(0.014)	156.0333(6.555)	**1.1623(0.014)**
DTLZ3(4)	**0.5533(0.020)**	0.8227(0.047)	165.9333(18.995)	**0.8227(0.047)**
DTLZ4(4)	**0.6440(0.011)**	1.2127(0.021)	157.2667(9.602)	**1.2127(0.021)**
DTLZ5(4)	**0.6128(0.009)**	1.1188(0.023)	143.1667(4.796)	**1.1188(0.023)**
DTLZ6(4)	**0.5351(0.010)**	1.2979(0.019)	129.1000(7.648)	**1.2979(0.019)**
DTLZ7(4)	**0.5860(0.008)**	0.9282(0.016)	185.6667(16.067)	**0.9282(0.016)**
DTLZ1(5)	**0.5532(0.005)**	0.9387(0.033)	79.1333(5.632)	**0.9387(0.033)**
DTLZ2(5)	**0.6453(0.010)**	1.1378(0.030)	188.3333(8.231)	**1.1378(0.030)**
DTLZ3(5)	**0.5785(0.012)**	1.0330(0.072)	177.1000(24.347)	**1.0330(0.072)**
DTLZ4(5)	**0.6949(0.004)**	1.2085(0.022)	190.5000(6.845)	**1.2085(0.022)**
DTLZ5(5)	**0.6455(0.004)**	1.1452(0.030)	229.3333(14.328)	**1.1452(0.030)**
DTLZ6(5)	**0.5784(0.008)**	1.6253(0.033)	225.3667(11.056)	**1.6253(0.033)**
DTLZ7(5)	**0.6289(0.004)**	0.9783(0.068)	296.9333(23.678)	**0.9783(0.068)**
DTLZ1(6)	**0.5816(0.011)**	1.4877(0.288)	98.9333(6.904)	**1.4877(0.288)**
DTLZ2(6)	**0.6750(0.003)**	1.1846(0.122)	233.3667(11.182)	**1.1846(0.122)**
DTLZ3(6)	**0.6162(0.017)**	1.5827(0.185)	185.3000(22.371)	**1.5827(0.185)**
DTLZ4(6)	**0.7485(0.003)**	1.2092(0.024)	234.6333(10.581)	**1.2092(0.024)**
DTLZ5(6)	**0.6683(0.011)**	1.1497(0.024)	336.9000(18.293)	**1.1497(0.024)**
DTLZ6(6)	**0.6308(0.006)**	1.6625(0.024)	340.4333(16.669)	**1.6625(0.024)**
DTLZ7(6)	**0.6589(0.012)**	1.0392(0.026)	377.9000(42.232)	**1.0392(0.026)**
WFG1(3)	**1.1427(0.019)**	1.5656(0.027)	147.0000(3.670)	**1.5656(0.027)**
WFG2(3)	**0.9272(0.024)**	1.3674(0.018)	98.4333(6.786)	**1.3674(0.018)**
WFG3(3)	**0.9738(0.018)**	1.4281(0.017)	148.7333(3.941)	**1.4281(0.017)**
WFG4(3)	**0.9919(0.007)**	2.2732(0.061)	107.5000(4.233)	**2.2732(0.061)**
WFG5(3)	**0.9594(0.007)**	2.1035(0.031)	153.0667(8.246)	**2.1035(0.031)**
WFG6(3)	**0.9478(0.010)**	1.5356(0.007)	168.9333(8.330)	**1.5356(0.007)**
WFG7(3)	**1.1988(0.026)**	2.3854(0.020)	151.5667(6.530)	**2.3854(0.020)**
WFG1(4)	**1.1697(0.017)**	1.4167(0.017)	233.7333(8.434)	**1.4167(0.017)**
WFG2(4)	**0.9473(0.021)**	1.4471(0.016)	170.6333(11.232)	**1.4471(0.016)**
WFG3(4)	**1.0207(0.011)**	1.3199(0.034)	247.1000(7.939)	**1.3199(0.034)**
WFG4(4)	**1.0258(0.009)**	2.4373(0.017)	157.8333(6.455)	**2.4373(0.017)**
WFG5(4)	**0.9848(0.009)**	1.8332(0.080)	206.7667(19.689)	**1.8332(0.080)**
WFG6(4)	**0.9774(0.007)**	1.2582(0.010)	216.9667(17.647)	**1.2582(0.010)**
WFG7(4)	**1.2529(0.014)**	2.4961(0.021)	252.1333(8.429)	**2.4961(0.021)**
WFG1(5)	**1.2474(0.015)**	1.5476(0.027)	335.0667(7.607)	**1.5476(0.027)**
WFG2(5)	**1.0083(0.020)**	1.5561(0.015)	269.5667(20.717)	**1.5561(0.015)**
WFG3(5)	**1.0908(0.010)**	1.4337(0.018)	378.1667(6.362)	**1.4337(0.018)**
WFG4(5)	**1.1067(0.005)**	2.5754(0.015)	220.6667(13.553)	**2.5754(0.015)**
WFG5(5)	**1.0683(0.006)**	1.4752(0.033)	276.2000(31.841)	**1.4752(0.033)**
WFG6(5)	**1.0342(0.024)**	1.3588(0.013)	274.2667(47.308)	**1.3588(0.013)**
WFG7(5)	**1.4166(0.021)**	2.6519(0.051)	358.9667(10.005)	**2.6519(0.051)**
WFG1(6)	**1.3214(0.012)**	1.6343(0.022)	383.8000(42.576)	**1.6343(0.022)**
WFG2(6)	**1.0430(0.021)**	1.6106(0.019)	377.4333(29.319)	**1.6106(0.019)**
WFG3(6)	**1.1115(0.011)**	1.5232(0.037)	445.7333(46.018)	**1.5232(0.037)**
WFG4(6)	**1.1695(0.009)**	2.6444(0.049)	316.2000(12.098)	**2.6444(0.049)**
WFG5(6)	**1.1185(0.009)**	1.4109(0.027)	246.7000(6.435)	**1.4109(0.027)**
WFG6(6)	**1.0602(0.024)**	1.4250(0.013)	259.2333(5.024)	**1.4250(0.013)**
WFG7(6)	**1.8199(0.145)**	2.6672(0.072)	408.2667(40.609)	**2.6672(0.072)**

Table 3. In (a), we show the results obtained in the DTLZ and WFG test problems using a population size of 300 individuals. We compare our GD-MOEA with respect to MOEA/D, using the hypervolume indicator (I_H). We show average values over 30 independent runs. The values in parentheses correspond to the standard deviations. The third column of each table shows the results of the statistical analysis applied to our experiments using Wilcoxon's rank sum. P is the probability of observing the given result (the null hypothesis is true). Small values of P cast doubt on the validity of the null hypothesis. $H = 0$ indicates that the null hypothesis ("medians are equal") cannot be rejected at the 5% level. $H = 1$ indicates that the null hypothesis can be rejected at the 5% level. In (b), we show the time required by MOEA/D and our proposed GD-MOEA for the test problems adopted in seconds. All algorithms were compiled using the GNU C compiler and they were executed on a computer with a 2.66GHz processor having 4GB in RAM.

f	moead I_H	gd-moea I_H	$P(H)$	moead time	gd-moea time
DTLZ1 (3)	1.0395 (0.001)	1.0265 (0.008)	0.00 (1)	1.5296 (0.023)	4.4234 (0.095)
DTLZ2 (3)	0.8847 (0.000)	0.8834 (0.004)	0.34 (0)	1.7870 (0.011)	5.5353 (0.143)
DTLZ3 (3)	1.3307 (0.000)	1.3308 (0.001)	0.00 (1)	1.5982 (0.042)	3.2699 (0.176)
DTLZ4 (3)	0.7786 (0.000)	0.7717 (0.007)	0.00 (1)	bf 1.7937 (0.015)	5.4645 (0.099)
DTLZ5 (3)	0.2612 (0.000)	0.2666 (0.002)	0.00 (1)	1.7661 (0.039)	5.4361 (0.058)
DTLZ6 (3)	1.0933 (0.005)	1.0813 (0.022)	0.00 (1)	1.4588 (0.028)	4.3155 (0.044)
DTLZ7 (3)	0.6397 (0.001)	0.5852 (0.080)	0.00 (1)	1.5896 (0.021)	4.8026 (0.146)
DTLZ1 (4)	1.3880 (0.003)	1.4575 (0.002)	0.00 (1)	1.6032 (0.024)	4.9574 (0.090)
DTLZ2 (4)	0.9845 (0.000)	0.9906 (0.046)	0.00 (1)	1.8629 (0.008)	5.0777 (0.096)
DTLZ3 (4)	1.4639 (0.000)	1.4641 (0.000)	0.00 (1)	1.7703 (0.041)	4.1087 (0.200)
DTLZ4 (4)	1.0157 (0.001)	1.0201 (0.013)	0.17 (0)	1.9052 (0.018)	5.1225 (0.033)
DTLZ5 (4)	1.0040 (0.002)	1.0405 (0.009)	0.00 (1)	1.8249 (0.038)	4.0589 (0.047)
DTLZ6 (4)	1.3351 (0.003)	1.2511 (0.025)	0.00 (1)	1.5590 (0.029)	5.9208 (0.049)
DTLZ7 (4)	0.4997 (0.002)	0.6487 (0.053)	0.00 (1)	1.7103 (0.024)	4.9728 (0.039)
DTLZ1 (5)	1.5813 (0.003)	1.6104 (0.000)	0.00 (1)	1.6717 (0.019)	5.4628 (0.091)
DTLZ2 (5)	1.1373 (0.002)	1.1504 (0.022)	0.00 (1)	1.9844 (0.016)	5.0243 (0.110)
DTLZ3 (5)	1.6102 (0.000)	1.6104 (0.000)	0.00 (1)	1.8143 (0.042)	5.1946 (0.265)
DTLZ4 (5)	1.1464 (0.002)	1.1500 (0.041)	0.25 (0)	2.0551 (0.015)	5.0612 (0.033)
DTLZ5 (5)	1.2309 (0.002)	1.2474 (0.015)	0.00 (1)	2.0464 (0.115)	4.0763 (0.043)
DTLZ6 (5)	1.4478 (0.005)	1.2493 (0.036)	0.00 (1)	1.6522 (0.043)	7.0208 (0.111)
DTLZ7 (5)	0.3545 (0.009)	0.6688 (0.026)	0.00 (1)	1.8392 (0.014)	5.1262 (0.057)
DTLZ1 (6)	1.7573 (0.001)	1.7716 (0.000)	0.00 (1)	1.7966 (0.040)	6.0618 (0.096)
DTLZ2 (6)	1.1549 (0.004)	1.2616 (0.040)	0.00 (1)	2.0712 (0.019)	5.1637 (0.063)
DTLZ3 (6)	1.7697 (0.000)	1.7694 (0.004)	0.00 (1)	1.9080 (0.023)	6.5666 (0.679)
DTLZ4 (6)	1.1329 (0.003)	1.3010 (0.054)	0.00 (1)	2.2384 (0.016)	5.4694 (0.084)
DTLZ5 (6)	1.2176 (0.007)	1.3123 (0.019)	0.00 (1)	2.0917 (0.108)	4.4795 (0.047)
DTLZ6 (6)	1.6242 (0.005)	1.3955 (0.049)	0.00 (1)	1.7541 (0.032)	7.9697 (0.137)
DTLZ7 (6)	0.0458 (0.009)	0.5867 (0.029)	0.00 (1)	1.9459 (0.024)	5.6369 (0.101)
WFG1 (3)	0.7547 (0.002)	0.6978 (0.016)	0.00 (1)	3.3185 (0.042)	5.4840 (0.082)
WFG2 (3)	0.7263 (0.079)	0.9019 (0.034)	0.00 (1)	2.8354 (0.129)	6.2730 (0.055)
WFG3 (3)	0.6044 (0.012)	0.6241 (0.004)	0.00 (1)	2.8724 (0.052)	5.5970 (0.044)
WFG4 (3)	0.7233 (0.006)	0.7528 (0.004)	0.00 (1)	2.9580 (0.041)	10.9987 (0.068)
WFG5 (3)	0.5484 (0.004)	0.5673 (0.003)	0.00 (1)	2.8350 (0.028)	8.8325 (0.235)
WFG6 (3)	0.5130 (0.004)	0.5444 (0.005)	0.00 (1)	2.8116 (0.022)	5.8362 (0.035)
WFG7 (3)	0.7685 (0.011)	0.7498 (0.009)	0.00 (1)	3.4059 (0.025)	10.2300 (0.120)
WFG1 (4)	0.2787 (0.018)	0.4129 (0.033)	0.00 (1)	3.4881 (0.038)	5.5105 (0.027)
WFG2 (4)	0.9008 (0.162)	0.9788 (0.078)	0.13 (0)	2.8905 (0.109)	6.5906 (0.056)
WFG3 (4)	0.4771 (0.017)	0.5514 (0.013)	0.00 (1)	3.1342 (0.059)	5.5723 (0.038)
WFG4 (4)	0.8773 (0.015)	0.9835 (0.010)	0.00 (1)	3.0843 (0.018)	11.4934 (0.153)
WFG5 (4)	0.5234 (0.008)	0.5960 (0.006)	0.00 (1)	3.0682 (0.033)	6.2124 (0.030)
WFG6 (4)	0.3364 (0.012)	0.4274 (0.032)	0.00 (1)	2.9030 (0.013)	5.6417 (0.035)
WFG7 (4)	0.7698 (0.036)	0.8675 (0.015)	0.00 (1)	3.8373 (0.054)	9.0491 (0.171)
WFG1 (5)	0.0533 (0.032)	0.4287 (0.020)	0.00 (1)	3.7985 (0.056)	6.2184 (0.037)
WFG2 (5)	0.9970 (0.134)	1.0072 (0.162)	0.55 (0)	3.0944 (0.083)	7.0006 (0.068)
WFG3 (5)	0.4882 (0.027)	0.5041 (0.043)	0.29 (0)	3.2425 (0.060)	6.1288 (0.060)
WFG4 (5)	0.9023 (0.035)	1.0894 (0.017)	0.00 (1)	3.3310 (0.017)	11.4864 (0.050)
WFG5 (5)	0.5085 (0.008)	0.6029 (0.020)	0.00 (1)	3.1917 (0.021)	5.7056 (0.046)
WFG6 (5)	0.2608 (0.012)	0.2939 (0.042)	0.00 (1)	3.0811 (0.011)	6.1855 (0.032)
WFG7 (5)	0.3798 (0.059)	0.8928 (0.020)	0.00 (1)	4.4487 (0.087)	8.7778 (0.227)
WFG1 (6)	0.0000 (0.000)	0.4658 (0.020)	0.00 (1)	4.0216 (0.077)	6.6320 (0.020)
WFG2 (6)	1.2803 (0.199)	1.1827 (0.224)	0.06 (0)	3.1562 (0.066)	7.1867 (0.063)
WFG3 (6)	0.5012 (0.055)	0.3781 (0.080)	0.00 (1)	3.3847 (0.074)	6.4182 (0.042)
WFG4 (6)	0.8909 (0.022)	1.1262 (0.038)	0.00 (1)	3.4896 (0.063)	11.1641 (0.074)
WFG5 (6)	0.5136 (0.011)	0.6429 (0.031)	0.00 (1)	3.3490 (0.015)	5.9626 (0.065)
WFG6 (6)	0.2565 (0.019)	0.2046 (0.034)	0.00 (1)	3.1610 (0.019)	6.4419 (0.036)
WFG7 (6)	0.2596 (0.009)	0.7964 (0.032)	0.00 (1)	4.7445 (0.088)	8.1669 (0.314)

(a)   (b)

diversity of the population. Our idea is to use I_{GD} as a convergence strategy and, when having many non-dominated individuals, to switch to the use of a technique to maintain diversity. However, it is important to be careful in both cases. When using I_{GD} only as a convergence strategy, if we choose the individuals with low values of d_i without considering if we have already selected individuals close to a particular non-dominated individual, then, we can have difficulties, e.g., in MOPs with disconnected Pareto fronts such as DTLZ7 (in this case, we will only obtain some portions of the Pareto front). If we only use the distribution technique based on Euclidean distances without considering if the individual which will compete is similar to another individual in one objective function, then, we can obtain many weakly Pareto points and this could prevent us from converging to the true Pareto front.

Our preliminary results indicate that our proposed GD-MOEA is a good option to solve MOPs having both low and high dimensionality in objective function space, if we consider both quality in the solutions and running time required to obtain them. Our proposed approach is able to obtain better results than MOEA/D, in most cases, and MOEA/D is only 1.46 times faster than our GD-MOEA when we use a population size of 100 individuals and, it is only 2.4224 times faster when we use a population size of 300 individuals. Although, SMS-EMOA-HYPE is better, in most cases, than GD-MOEA in terms of the quality of the solutions generated, it requires up to 167.11 times more computational time than our proposed approach.

As part of our future work, we want to improve our proposed selection mechanism so that it can deal (in a better way) with problems in which many weakly Pareto optimal solutions are generated such as DTLZ1 and DTLZ3. Also, we want to use other indicators to conduct an in-depth study, e.g., we could use the two set coverage indicator to measure convergence and the spacing indicator to assess the quality of the distribution of solutions generated by our proposed approach. This is because SMS-EMOA-HYPE maximizes the hypervolume indicator, and therefore, it has advantages over other two MOEAs when we use the hypervolume indicator to assess our results (evidently, it is expected that SMS-EMOA-HYPE will have better hypervolume values than other MOEAs, since this is precisely the value that it aims to maximize).

References

1. Bader, J., Zitzler, E.: HypE: An Algorithm for Fast Hypervolume-Based Many-Objective Optimization. Evolutionary Computation 19(1), 45–76 (2011)
2. Beume, N., Naujoks, B., Emmerich, M.: SMS-EMOA: Multiobjective selection based on dominated hypervolume. European Journal of Operational Research 181(3), 1653–1669 (2007)
3. Bringmann, K., Friedrich, T.: Approximating the least hypervolume contributor: NP-hard in general, but fast in practice. Theoretical Computer Science 425, 104–116 (2012)
4. Brockhoff, D., Wagner, T., Trautmann, H.: On the properties of the R2 indicator. In: 2012 Genetic and Evolutionary Computation Conference (GECCO 2012), Philadelphia, USA, pp. 465–472. ACM Press, July 2012. ISBN: 978-1-4503-1177-9

5. Coello Coello, C.A., Cortés, N.C.: Solving Multiobjective OptimizationProblems using an Artificial Immune System. Genetic Programming and Evolvable Machines 6(2), 163–190 (2005)
6. Coello Coello, C.A., Lamont, G.B., Van Veldhuizen, D.A.: Evolutionary Algorithms for Solving Multi-Objective Problems, 2nd edn. Springer, New York (2007). ISBN 978-0-387-33254-3
7. Coello Coello, C.A., Van Veldhuizen, D.A., Lamont, G.B.: Evolutionary Algorithms for Solving Multi-Objective Problems. Kluwer Academic Publishers, New York (2002). ISBN 0-3064-6762-3
8. Das, I., Dennis, J.E.: Normal-boundary intersection: A new method for generating the pareto surface in nonlinear multicriteria optimization problems. SIAM J. on Optimization 8(3), 631–657 (1998)
9. Deb, K., Thiele, L., Laumanns, M., Zitzler, E.: Scalable test problems for evolutionary multiobjective optimization. In: Abraham, A., Jain, L., Goldberg, R. (eds.) Evolutionary Multiobjective Optimization. Theoretical Advances and Applications, pp. 105–145. Springer, London (2005)
10. Domínguez-Medina, C., Rudolph, G., Schüetze, O., Trautmann, H.: Evenly spaced pareto fronts of quad-objective problems using PSA partitioning technique. In: 2013 IEEE Congress on Evolutionary Computation (CEC 2013), Cancún, México, June 20–23, pp. 3190–3197. IEEE Press (2013). ISBN 978-1-4799-0454-9
11. Farina, M., Amato, P.: On the optimal solution definition for many-criteria optimization problems. In: Proceedings of the NAFIPS-FLINT International Conference 2002, Piscataway, New Jersey, pp. 233–238. IEEE Service Center, June 2002
12. Gerstl, K., Rudolph, G., Schütze, O., Trautmann, H.: Finding evenly spaced fronts for multiobjective control via averaging hausdorff-measure. In: The 2011 8th International Conference on Electrical Engineering, Computer Science and Automatic Control (CCE 2011), Mérida, Yucatán, México, pp. 975–980. IEEE Press, October 2011
13. Gómez, R.H., Coello Coello, C.A.: MOMBI: a new metaheuristic for many-objective optimization based on the $R2$ indicator. In: 2013 IEEE Congress on Evolutionary Computation (CEC 2013), Cancún, México, June 20–23, pp. 2488–2495. IEEE Press (2013). ISBN 978-1-4799-0454-9
14. Huband, S., Hingston, P., Barone, L., While, L.: A Review of Multiobjective Test Problems and a Scalable Test Problem Toolkit. IEEE Transactions on Evolutionary Computation 10(5), 477–506 (2006)
15. Igel, C., Hansen, N., Roth, S.: Covariance Matrix Adaptation for Multi-objective Optimization. Evolutionary Computation 15(1), 1–28 (2007)
16. Knowles, J., Corne, D.: Properties of an Adaptive Archiving Algorithm for Storing Nondominated Vectors. IEEE Transactions on Evolutionary Computation 7(2), 100–116 (April 2003)
17. Menchaca-Mendez, A., Coello Coello, C.A.: A new selection mechanism based on hypervolume and its locality property. In: 2013 IEEE Congress on Evolutionary Computation (CEC 2013), Cancún, México, June 20–23, pp. 924–931. IEEE Press (2013)
18. Menchaca-Mendez, A., Coello Coello, C.A.: MD-MOEA : a new MOEA based on the maximin fitness function and euclidean distances between solutions. In: 2014 IEEE Congress on Evolutionary Computation (CEC 2014), Beijing, China, July 6–11, pp. 2148–2155. IEEE Press (2014). ISBN 978-1-4799-1488-3

19. Rodríguez Villalobos, C.A., Coello Coello, C.A.: A new multi-objective evolutionary algorithm based on a performance assessment indicator. In: 2012 Genetic and Evolutionary Computation Conference (GECCO 2012), Philadelphia, USA, pp. 505–512. ACM Press, July 2012. ISBN: 978-1-4503-1177-9

20. Schütze, O., Esquivel, X., Lara, A., Coello, C.A.: Coello. Using the Averaged Hausdorff Distance as a Performance Measure in Evolutionary Multiobjective Optimization. IEEE Transactions on Evolutionary Computation **16**(4), 504–522 (2012)

21. Trautmann, H., Wagner, T., Brockhoff, D.: R2-EMOA: focused multiobjective search using R2-indicator-based selection. In: Nicosia, G., Pardalos, P. (eds.) LION 7. LNCS, vol. 7997, pp. 70–74. Springer, Heidelberg (2013)

22. Phan, D.H., Suzuki, J.: R2-IBEA: R2 Indicator based evolutionary algorithm for multiobjective optimization. In: 2013 IEEE Congress on Evolutionary Computation (CEC 2013), Cancún, México, June 20–23, pp. 1836–1845. IEEE Press (2013). ISBN 978-1-4799-0454-9

23. Van Veldhuizen, D.A.: Multiobjective Evolutionary Algorithms: Classifications, Analyses, and New Innovations. PhD thesis, Department of Electrical and Computer Engineering. Graduate School of Engineering. Air Force Institute of Technology, Wright-Patterson AFB, Ohio, May 1999

24. Van Veldhuizen, D.A., Lamont, G.B.: Evolutionary computation and convergence to a pareto front. In: Koza, J.R. (ed.) Late Breaking Papers at the Genetic Programming 1998 Conference, Stanford University, California, pp. 221–228. Stanford University Bookstore (1998)

25. Van Veldhuizen, D.A., Lamont, G.B.: Multiobjective evolutionary algorithm test suites. In: Carroll, J., Haddad, H., Oppenheim, D., Bryant, B., Lamont, G.B. (eds.) Proceedings of the 1999 ACM Symposium on Applied Computing. San Antonio, Texas, pp. 351–357. ACM (1999)

26. Wagner, T., Trautmann, H., Brockhoff, D.: Preference articulation by means of the $R2$ indicator. In: Purshouse, R.C., Fleming, P.J., Fonseca, C.M., Greco, S., Shaw, J. (eds.) EMO 2013. LNCS, vol. 7811, pp. 81–95. Springer, Heidelberg (2013)

27. Zhang, Q., Li, H.: MOEA/D: A Multiobjective Evolutionary Algorithm Based on Decomposition. IEEE Transactions on Evolutionary Computation **11**(6), 712–731 (December 2007)

28. Zitzler, E., Künzli, S.: Indicator-based selection in multiobjective search. In: Yao, X., Burke, E.K., Lozano, J.A., Smith, J., Merelo-Guervós, J.J., Bullinaria, J.A., Rowe, J.E., Tiño, P., Kabán, A., Schwefel, H.-P. (eds.) PPSN 2004. LNCS, vol. 3242, pp. 832–842. Springer, Heidelberg (2004)

29. Zitzler, E., Thiele, L., Laumanns, M., Fonseca, C.M., Grunert da Fonseca, V.: Performance Assessment of Multiobjective Optimizers: An Analysis and Review. IEEE Transactions on Evolutionary Computation **7**(2), 117–132 (2003)

Experiments on Local Search for Bi-objective Unconstrained Binary Quadratic Programming

Arnaud Liefooghe[1]([✉]), Sébastien Verel[2], Luís Paquete[3], and Jin-Kao Hao[4]

[1] CRIStAL, UMR CNRS 9189, Université Lille 1 — Inria Lille-Nord Europe,
Villeneuve d'ascq, France
arnaud.liefooghe@univ-lille1.fr
[2] LISIC, Université du Littoral Côte d'Opale, Calais, France
verel@lisic.univ-littoral.fr
[3] CISUC, Department of Informatics Engineering, University of Coimbra,
Coimbra, Portugal
paquete@dei.uc.pt
[4] LERIA, Université d'Angers, Angers, France
jin-kao.hao@univ-angers.fr

Abstract. This article reports an experimental analysis on stochastic local search for approximating the Pareto set of bi-objective unconstrained binary quadratic programming problems. First, we investigate two scalarizing strategies that iteratively identify a high-quality solution for a sequence of sub-problems. Each sub-problem is based on a static or adaptive definition of weighted-sum aggregation coefficients, and is addressed by means of a state-of-the-art single-objective tabu search procedure. Next, we design a Pareto local search that iteratively improves a set of solutions based on a neighborhood structure and on the Pareto dominance relation. At last, we hybridize both classes of algorithms by combining a scalarizing and a Pareto local search in a sequential way. A comprehensive experimental analysis reveals the high performance of the proposed approaches, which substantially improve upon previous best-known solutions. Moreover, the obtained results show the superiority of the hybrid algorithm over non-hybrid ones in terms of solution quality, while requiring a competitive computational cost. In addition, a number of structural properties of the problem instances allow us to explain the main difficulties that the different classes of local search algorithms have to face.

1 Introduction

The unconstrained binary quadratic programming (UBQP) problem is one of the most challenging problem from single-objective combinatorial optimization [11]. Given a collection of n items such that each pair of items is associated with a profit value that can be positive, negative or zero, the UBQP problem seeks a subset of items that maximizes the sum of their paired values. The value of a pair is summed up only if the two corresponding items are selected. From a computational point-of-view, a feasible solution to a UBQP instance can be represented as a binary string of size n. Each position from the binary string maps to a particular variable that indicates whether the corresponding item is included

© Springer International Publishing Switzerland 2015
A. Gaspar-Cunha et al. (Eds.): EMO 2015, Part I, LNCS 9018, pp. 171–186, 2015.
DOI: 10.1007/978-3-319-15934-8_12

in the subset of selected items or not. Beyond its theoretical significance [8], the utility of UBQP has been demonstrated on a wide variety of application fields [11]. Furthermore, a number of NP-hard combinatorial optimization problems can be recast as UBQP problems, such as graph coloring, max-cut, set packing, set partitioning, or maximum clique, among others [11]. The single-objective UBQP problem has received a growing interest in recent years [9,11], and a multi-objective extension of UBQP has been proposed recently [12].

In this paper, we focus on bi-objective UBQP, where two profit values are associated with each pair of items. By optimizing both sums of profit values simultaneously, we can improve the descriptive power of the conventional single-objective UBQP problem, and provide a more general formulation. However, as for many problems from multi-objective combinatorial optimization, the bi-objective UBQP problem raises several difficulties for heuristics design. In particular, the number of optimal solutions can be very large [12], and determining whether a candidate solution is optimal is NP-complete, even in the single-objective case [8]. For these reasons, we design and experiment with multi-objective stochastic local search algorithms, and measure their efficiency and their effectiveness on instances with different dimensions and correlation degrees between the objective function values. Furthermore, we analyze the problem structure to learn more about those difficulties, and to improve the design of algorithms. The contributions of the paper are two-fold.

(i) We characterize the features of small-size, enumerable bi-objective UBQP instances. More particularly, we analyze the number of global and local optimal solutions, based on scalarizing functions and on the Pareto dominance relation; and we examine the connectedness between optimal solutions.

(ii) We design and analyze local search algorithms for bi-objective UBQP, including two scalarizing approaches, a Pareto-based approach, and a hybrid approach combining these two complementary search strategies. The designed algorithms substantially improve over the previous attempts in solving large-size bi-objective UBQP instances [12]. More importantly, our experimental analysis allows us to better understand how the performance of these classes of algorithms relates to the structural properties of the search space, explaining the high efficiency and effectiveness of the proposed approaches.

The remainder of the paper is organized as follows. In Section 2, we present the bi-objective UBQP problem. In Section 3, we study the characteristics of small-size instances. In Section 4, we introduce four stochastic local search algorithms for bi-objective UBQP. In Section 5, we analyze the performance of these algorithms on a set of large-size bi-objective UBQP instances. In Section 6, we finally conclude the paper and discuss further research directions.

2   Bi-objective UBQP

This section presents the problem formulation, some definitions related to multi-objective combinatorial optimization, and the problem instances that are investigated in the paper.

2.1   Problem Formulation

The bi-objective UBQP (bUBQP) problem can be formalized as follows [12].

$$\max \ f_1(x) = \sum_{i=1}^{n} \sum_{j=1}^{n} q_1^{ij} x_i x_j$$

$$\max \ f_2(x) = \sum_{i=1}^{n} \sum_{j=1}^{n} q_2^{ij} x_i x_j \tag{1}$$

$$\text{subject to} \ \ x \in \{0,1\}^n$$

where (f_1, f_2) is the pair of objective functions to be maximized, n is the number of items, $Q_1 = (q_1^{ij})$ and $Q_2 = (q_2^{ij})$ are both an $n \times n$ matrix of constant values, either positive, negative or zero. As in the single-objective case, the solution space $X = \{0,1\}^n$ is defined on binary strings of size n; its size is then 2^n.

2.2   Definitions

We denote by $Z \subseteq \mathbb{R}^2$ the feasible region in the *objective space*, *i.e.* the image of feasible solutions when using the maximizing function vector $f = (f_1, f_2)$ such that $Z = f(X)$. The Pareto dominance relation is defined as follows. A solution $x \in X$ is dominated by a solution $x' \in X$, denoted as $x \prec x'$, if $f_k(x) \leq f_k(x')$ for all $k \in \{1, 2\}$, with at least one strict inequality. If neither $x \nprec x'$ nor $x' \nprec x$ holds, then both solutions are *mutually non-dominated*. A solution $x \in X$ is *Pareto optimal* if there does not exist any other solution $x' \in X$ such that $x \prec x'$. The set of all Pareto optimal solutions is the *Pareto set*, and its mapping in the objective space is the *Pareto front*. One of the most challenging issues in multi-objective optimization is to identify a minimal complete Pareto set, *i.e.* one Pareto optimal solution mapping to each point from the Pareto front. Since the bUBQP problem is both NP-hard and intractable [12], approximate algorithms like stochastic local search are well suited to identify a *Pareto set approximation*.

2.3   Problem Instances

Following [12], the definition of each bUBQP objective function is based on a matrix Q_k, $k \in \{1, 2\}$. As in the single-objective UBQP instances available in the OR-lib [3], non-zero matrix integer values are randomly generated following a uniform distribution in $[-100, +100]$. The density $d \in [0, 1]$ gives the expected proportion of non-zero entries in the matrix. Following a Bernoulli distribution of parameter d, a given entry at position (i, j) is set to zero on *both* matrices, *i.e.* $q_1^{ij} = q_2^{ij} = 0$. Moreover, we define a correlation coefficient ρ between the data contained in the two matrices. The positive (respectively negative) data correlation decreases (respectively increases) the degree of conflict between the objective function values. The generation of correlated data follows a multivariate uniform distribution of dimension 2 [12]. As reported in Fig. 1(a), the coefficient ρ allows to tune the correlation between objective function values with a high accuracy. The considered bUBQP problem instances as well as an instance generator are available at the following URL: http://mocobench.sf.net/.

3   Characteristics of Small-Size bUBQP Instances

Thereafter, we study the impact of the density d and of the objective correlation ρ on the number of Pareto optimal solutions, Pareto local optimal solutions, supported solutions, and on the connectedness property of bUBQP instances. More particularly, we consider a density $d \in \{0.2, 0.4, 0.6, 0.8, 1.0\}$ and an objective correlation $\rho \in \{-0.9, -0.7, -0.4, -0.2, 0.0, +0.2, +0.4, +0.7, +0.9\}$. The problem size is set to $n = 18$ in order to enumerate the solution space exhaustively. For each parameter setting, 30 independently generated random instances are considered. Experimental results are given in Fig. 1(b–f). In the following, we provide a detailed analysis of these statistics.

3.1   Pareto Optimal Solutions

Fig. 1(b) shows the proportion of Pareto optimal solutions in the solution space. Interestingly, the density d does not affect the size of the Pareto set. However, the objective correlation ρ modifies the number of Pareto optimal solutions to several orders of magnitude. Indeed, almost 0.05% of the solution space correspond to non-dominated solutions for conflicting objectives ($\rho = -0.9$), whereas this number drops to less than 0.003% for correlated objective ($\rho = +0.9$). As a consequence, the larger the objective correlation ρ, the lower the cardinality of the Pareto set. This means that an algorithm is expected to take more time to identify the whole Pareto set when the objectives are in conflict.

3.2   Supported Solutions

In multi-objective optimization, *scalarizing* approaches consist in transforming the original problem into a single-objective one by means of an aggregation of the objective function values. A typical example is the weighted-sum scalarizing function [6] that can be defined as follows.

$$g_\lambda(x) = \lambda_1 \cdot f_1(x) + \lambda_2 \cdot f_2(x) \tag{2}$$

where $x \in X$ is a candidate solution, and $\lambda = (\lambda_1, \lambda_2)$, such that $\lambda_1, \lambda_2 \geq 0$, is a weighting coefficient vector. Supported solutions are non-dominated solutions which are optimal with respect to a weighted-sum aggregation of the objective functions. Their corresponding objective vectors are located on the boundary of the convex hull of the Pareto front [6]. On the contrary, non-supported solutions are not optimal for any setting of the weighting coefficient vector λ. In order to explain the ability of scalarizing multi-objective optimization approaches to identify a large portion of Pareto optimal solutions, we should put the problem-related properties in relation with the proportion of *supported* solutions. Fig. 1(c) shows the proportion of supported solutions in the Pareto set. Once again, the matrix density d has a very small influence. However, when the objective correlation increases, and despite the absolute number of supported solution actually gets lower, the proportion of supported solutions on the Pareto set increases.

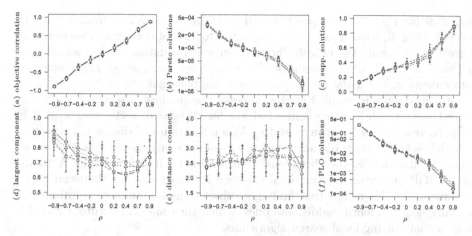

Fig. 1. (a) Spearman correlation coefficient between the objective function values, (b) ratio of the number of Pareto optimal solutions to the solution space size, (c) ratio of the number of supported solutions to the Pareto set size, (d) ratio of the size of the largest connected component of the Pareto graph for Hamming distance 1 to the Pareto set size, (e) minimal Hamming distance to connect the Pareto graph, and (f) ratio of Pareto local optimal solutions to the solution space size, with respect to the objective correlation ρ, for $d = 0.2$ (○), $d = 0.4$ (□), $d = 0.6$ (◇), $d = 0.8$ (△) and $d = 1.0$ (▽). For each parameter setting, average values and confidence intervals (with a significance level of 10^{-2}) are reported over 30 independently generated random instances. The problem size is $n = 18$. Notice the log-scale on the y-axis for (b) and (f).

For highly correlated objectives ($\rho = +0.9$), nearly all Pareto optimal solutions are supported (this is even the case for some of the instances). On the contrary, for conflicting objectives ($\rho = -0.9$), only 15% of Pareto optimal solutions are supported. By putting this property in relation with algorithm design, we can assume that scalarizing approaches should be more suited to approximate the Pareto set of bUBQP instances with correlated objectives.

3.3   Connectedness

In the following, we describe some properties related to the *connectedness* of the Pareto set [7]. We follow the definition of *k-Pareto graph* from [16]. The k-Pareto graph is a graph $PG_k = (V, E)$, where each vertex in V corresponds to a Pareto optimal solution, and there is an edge $e_{ij} \in E$ between two nodes i and j only if the shortest distance between solutions x_i and $x_j \in X$, with respect to a given neighborhood, is below a bound k. For bUBQP, we adopt the Hamming distance on binary strings. This corresponds to the number of moves performed with the *bit-flip* neighborhood operator. Fig. 1(d) shows the ratio between the size of the largest connected component in the 1-Pareto graph ($PG_{k=1}$) and the size of the Pareto set. The objective correlation ρ has a clear impact on this feature. Indeed, the proportion of Pareto optimal solutions in the largest connected component decreases from $\rho = -0.9$ to $\rho = 0.4$, and then slightly increases from $\rho = 0.4$ to $\rho = 0.9$. Overall, we can expect to reach 50% to 95% of the whole Pareto set by iteratively exploring the neighborhood of an approximation set initialized with at least one non-dominated solution. However, when there are several connected components in the 1-Pareto graph, it may happen that the distance between those components is small. Fig. 1(e) reports the smallest distance k such that the k-Pareto graph becomes connected, *i.e.* for all pairs of vertices $x_i, x_j \in V$ in PG_k, there is a path between x_i and x_j. When this minimal distance k is around 9, which is the average distance between random solutions for $n = 18$, we can conclude that the distance between Pareto optimal solutions is large. Actually, for bUBQP instances, this minimal distance is clearly smaller (between 2 and 3 in average). This means that finding a subset of non-dominated solutions can actually help to identify additional ones, which then may constitute a valuable asset for initializing local search algorithms.

3.4   Pareto Local Optimal Solutions

In Fig. 1(f), we report the proportion of Pareto Local Optimal (PLO) solutions [14] in the solutions space. A solution $x \in X$ is a PLO with respect to a neighborhood structure $\mathcal{N}$ if there does not exist any neighboring solution $x' \in \mathcal{N}(x)$ such that $x \prec x'$. As above, the neighborhood structure is taken as the 1-*bit-flip*, which is directly related to a Hamming distance 1. Once again, the distribution d does not seem to affect the number of PLO. However, similar to the trend observed on the Pareto set cardinality, the objective correlation ρ modifies the number of PLO to several orders of magnitude, from 20% of the solution space for $\rho = -0.9$ to less than 0.02% for $\rho = +0.9$. Therefore, by assuming that the difficulty for Pareto-based local search gets higher when the number of PLO is large, the difficulty of bUBQP instances might increase with the degree of conflict between the objectives.

4   Local Search for bUBQP

In this section, we give the working principles of four stochastic local search algorithms for identifying a Pareto set approximation to bUBQP instances. We start by introducing the algorithmic components shared by the different approaches. Then, we present the search strategies of two scalarizing local search algorithms, one Pareto-based local search algorithm, as well as a hybrid approach where a scalarizing and a Pareto local search phases are sequentially applied.

4.1   Main Ingredients

Neighborhood Relation. Similarly to the previous analysis, the neighborhood structure of the proposed local search algorithms is based on the 1-*bit-flip* operator: Two feasible solutions are neighbors if they differ exactly on one variable. In other words, a given neighbor can be reached by changing the value of a binary variable to its complement from the current solution. The size of the 1-*bit-flip* neighborhood structure is equal to the problem size n. As in the single-objective UBQP, each bUBQP objective function can be evaluated incrementally. We follow the fast incremental evaluation procedure proposed in [9] to calculate the move gain of a given neighbor. For each objective, the whole set of neighbors is evaluated in linear time. As a consequence, the objective values of *all* neighboring solutions are evaluated in $\mathcal{O}(n)$ in the two-objective case.

Tabu Search. The tabu search algorithm proposed in [10] is reported to be one of the best-performing approaches for single-objective UBQP. In order to extend it to the multi-objective case, we consider a simple weighted-sum aggregation, as presented in Section 3.2, so that the initial objective vector values are (temporarily) transformed into a single scalar fitness value. Once the objective values of a given neighboring solution have been (incrementally) evaluated, we compute its scalar fitness value with respect to the weighted-sum problem (Eq. 2) for a given definition of the weighing coefficient vector. As a short-term memory, we maintain the tabu list as follows: Revisiting solutions is avoided within a certain number of iterations, called the *tabu tenure*. The tabu tenure of a given variable x_i is denoted by $tenure(i)$. Hence, variable x_i will *not* be flipped again for a number of $tenure(i)$ iterations. Following [9], we set the tabu tenure of a given variable x_i after it has been flipped as $tenure(i) = tt + rand(10)$, where tt is a user-given parameter and $rand(10)$ gives a random integer value in $[1, 10]$. From the set of neighbors produced by all non-tabu moves, we select the one with the best (highest) fitness value. However, all the neighbors are always evaluated, and a tabu move can still be selected if it produces a better solution than the current global best; this is called an *aspiration criterion* in tabu search [10]. The stopping condition is satisfied when no improvement has been performed within a given number of moves α, called the *improvement cutoff*. For more details on the tabu search algorithm for single-objective UBQP, the reader is referred to [10].

4.2 Scalarizing Local Search with Uniform Weights (SLS_{unif})

The first approach consists in solving different settings of the weighted-sum problem (Eq. 2) by means of multiple weighting coefficient vectors defined in a way that the whole region of the Pareto front is covered in the objective space. For solving each scalarizing sub-problem, any algorithm for the resulting single-objective problem version can potentially be applied. In our case, we use the tabu search algorithm detailed above as a (single-objective) solver. Let us consider a set of μ uniformly defined weighting coefficient vectors $(\lambda^0, \ldots, \lambda^i, \ldots, \lambda^{\mu-1})$, such that $\lambda_1^i = i/(\mu - 1)$ and $\lambda_2^i = 1 - \lambda_1^i$. Each weighting coefficient vector λ^i corresponds to a scalarizing sub-problem WS^i. We start by identifying a high-quality solution with respect to the first objective function, corresponding to the scalarizing sub-problem WS^0, associated with the weighting coefficient vector $\lambda^0 = (0, 1)$. The final solution is then used as a seeding solution for solving the next sub-problem WS^1. We iterate this principle, each time the initial solution for sub-problem WS^i being the one that is returned by the tabu search algorithm for the previous sub-problem WS^{i-1}. At last, in order to avoid a bias in the search process towards one objective, we re-run the same strategy by considering the reversed sequence of weighting coefficient vectors $(\lambda^{\mu-1}, \lambda^{\mu-2}, \ldots, \lambda^0)$. The algorithm outputs the union of non-dominated solutions generated during these two phases. The resulting scalarizing local search with uniform weights (SLS_{unif}) adapts the *"double two-phase local search"* from [15] to bUBQP by using the single-objective tabu search procedure for solving scalarizing sub-problems.

4.3   Dichotomic Scalarizing Local Search (SLS_{dicho})

Similarly, the second approach is based on solving a sequence of scalarizing sub-problems, by means of a weighted-sum aggregation function, with the single-objective tabu search algorithm. However, unlike SLS_{unif} that defines them *a priori*, the weighting coefficient vectors are now iteratively determined based on the solutions identified at previous steps. The resulting SLS_{dicho} approach follows the principles of *dichotomic search* from exact bi-objective optimization [1], and adapt them to a local search engine strategy, similarly to [5]. Notice that, by using any exact algorithm instead of the tabu search procedure, such a dichotomic search would output the (exact) set of supported solutions [6]. Unfortunately, this would require to solve an NP-hard problem for each scalarizing sub-problem. Indeed, each sub-problem corresponds to a single-objective UBQP instance.

We start by identifying a high-quality solution for each separate objective. Let x^1 (resp. x^2) be the approximate solution found by tabu search for objective f_1 (resp. f_2). Both solutions are then added to a sequence $U_F = \{x^1, x^2, \ldots\}$, arranged in the decreasing order of f_1-values. Next, at each step of the algorithm, we define a weighting coefficient vector $\lambda = \left(f_2(x^2) - f_2(x^1), f_1(x^1) - f_1(x^2)\right)$, corresponding to the sub-problem to be solved in the current iteration. It gives a search direction that is perpendicular to the segment defined by $f(x^1)$ and $f(x^2)$ in the objective space. Let x be the solution identified by tabu search for this definition of λ. If $f_1(x^1) > f_1(x) > f_1(x^2)$ and $f_2(x^2) > f_2(x) > f_2(x^1)$, then x is added to the sequence U_F. Otherwise, we remove x^1 from U_F and add it to an external set U_T. Following [5], for each scalarizing sub-problem, we use the solutions found in previous iterations to seed the search process of the current iteration. Based on preliminary experiments, both x^1 and x^2 are here used as an initial solution for two independent runs of the tabu search procedure based on λ. The SLS_{dicho} algorithm iterates this principle until U_F contains less than two elements, and returns the non-dominated solutions from $U_F \cup U_T$.

4.4   Pareto Local Search (PLS)

Let us now consider a Pareto approach based on a set of solutions and local search principles. In contrast to scalarizing approaches, the selection process is here directly based on the Pareto dominance relation. A typical example is the Pareto Local Search (PLS) algorithm [14]. An archive of mutually non-dominated solutions found so far is maintained in two different sets: V_F for non-dominated solutions whose neighborhood has not yet been explored, and V_T for solutions whose neighborhood has already been explored. These two sets are used in order to avoid a useless re-evaluation of a solution's neighborhood. The algorithm starts with a set of mutually non-dominated solutions to initialize V_F, typically a single random solution. At each iteration, one unvisited solution is chosen at random from V_F. All its neighboring solutions are (incrementally) evaluated and checked for insertion in the archive. The current solution is then discarded from V_F and added to V_T, and dominated solutions are removed from $V_F \cup V_T$. The algorithm stops once V_F is empty, *i.e.* all solutions from the archive are *visited*. PLS always terminates and returns a maximal Pareto local optimum set [14].

4.5   Two-Phase Local Search (TP-LS)

The final algorithm consists in a hybrid two-phase approach, where SLS_{dicho} and PLS are applied in a sequential way. It combines two fundamentally different and complementary search strategies: a scalarizing and a Pareto-based approach. In the first phase, SLS_{dicho} is applied to identify a set of approximate supported solutions, as described in Section 4.3. This set of mutually non-dominated solutions is then used to initialize the archive V_F of the PLS algorithm, and is further improved by exploring the neighborhood of its own content until no improvement is possible. Hence, contrary to the conventional PLS, the search process does not start with a single random solution, but with a set of good-quality solutions identified by a scalarizing approach. The performance of the designed two-phase local search algorithm (TP-LS) should be impacted by the connectedness property for the problem under consideration; the more connected the Pareto optimal solutions, the easier to identify new non-dominated solutions from identified ones. Notice that TP-LS shares similar principles with existing approaches proposed for other problem classes [4, 13, 15].

5   Experimental Analysis

5.1   Experimental Design

We conduct an experimental study on the influence of the problem size (n) and of the objective correlation (ρ) over the performance of the proposed local search algorithms for approximating the Pareto set of bUBQP problem instances. In addition, we consider the best-known approximation sets (best-known) identified by multiple variants of evolutionary and memetic algorithms proposed in [12].

We investigate the following instance parameter setting: a problem dimension $n \in \{1000, 2000, 3000, 4000, 5000\}$ and a correlation between the objective function values $\rho \subset \{-0.5, -0.2, 0.0, +0.2, +0.5\}$. The density of the matrices is set to $d = 0.8$. One instance, generated at random, is considered *per* parameter combination. This leads to a total of 25 problem instances. A set of 30 runs *per* instance is performed for each algorithm. All the algorithms start with a random solution. The tabu tenure tt is set to $n/150$ and the improvement cutoff α

is set to n. At last, for each phase of SLS_{unif}, $\mu = 101$ weighting coefficient vectors $(\lambda^0, \ldots, \lambda^i, \ldots, \lambda^{100})$ are uniformly defined as $\lambda_1^i = i/100$ and $\lambda_2^i = 1 - \lambda_1^i$.

Since all the algorithms have a natural stopping condition, we measure their performance in terms of approximation set quality *and* computational cost. For each instance, we examine the quality of the Pareto set approximations identified by the competing algorithms in terms of hypervolume and epsilon indicators [17]. First, we compute the *hypervolume relative deviation* (hypervolume) as $(hv(R) - hv(A))/hv(R)$, where $A \subseteq Z$ is an approximation set and R is a reference set. The reference set is the best-found approximation over all tested configurations for the instance under consideration. Let z_k^- (resp. z_k^+) be the worst (resp. best) value obtained over all approximation sets for objective f_k, the reference point $\bar{z} = (\bar{z}_1, \bar{z}_2)$ for the hypervolume calculation is set to $\bar{z}_k = z_k^- - (z_k^+ - z_k^-) \cdot 10^{-2}$, $k \in \{1, 2\}$. Additionally, the epsilon indicator (epsilon) gives the minimum multiplicative factor by which an approximation set has to be shifted in the objective space in order to weakly dominate the reference set. In both cases, a lower indicator-value is better.

5.2 Experimental Results

A summary of our computational results is presented in Table 1, following the presentation from [2]. The first line corresponds to the bUBQP instance with $\rho = -0.5$ and $n = 1000$, and reports the quality of the Pareto set approximation obtained by the different algorithms with respect to hypervolume. The average hypervolume relative deviation obtained by SLS_{unif}, $\text{SLS}_{\text{dicho}}$, PLS, TP-LS and best-known over the 30 executions is respectively 0.009, 0.006, 0.002, 0.000 and 0.031. The ranking obtained by means of a pairwise Wilcoxon signed-rank non-parametric statistical test gives the following order for this particular setting: (1) TP-LS, (2) PLS, (3) $\text{SLS}_{\text{dicho}}$, (4) SLS_{unif}, and (5) best-known. Complementarily, Fig. 2 shows the average indicator-values for a subset of instances (the error bars indicate the confidence interval within a significance level of 10^{-2}). The results from best-known are omitted for a better readability.

Clearly, all the local search algorithms investigated in the paper largely improve over the previous best-known approximation sets from [12]. Indeed, for the 25 bUBQP instances under investigation, best-known obtains the lowest rank for 23 of them in terms of hypervolume and epsilon. A simple approach like SLS_{unif} is able to obtain better best-known results in all the instances but one. Among the algorithms proposed in the paper, SLS_{unif} is repeatedly dominated by the others with respect to both indicators. The only notable exceptions are for instances with correlated objectives where SLS_{unif} performs better than PLS in terms of hypervolume (while it is slightly worse in terms of epsilon), and for large-size instances with conflicting objectives where PLS encounters more difficulties compared with other approaches. In both cases, the reason seems to be that the approximation sets identified by PLS badly covers the lexicographically optimal regions of the Pareto front. This is also the reason why $\text{SLS}_{\text{dicho}}$ outperforms PLS on more than half of the instances with respect to hypervolume, whereas the same only happens four times with respect to epsilon.

Table 1. Comparison of the competing local search algorithms and of the previous best-known approximation [12] with respect to the hypervolume relative deviation (**hypervolume**) and to the unary multiplicative epsilon indicator (**epsilon**). The first value stands for the number of algorithms that statistically outperform the one under consideration with respect to a pairwise Wilcoxon signed-rank non-parametric statistical test with a p-value of 10^{-2} by using a Bonferroni correction (lower is better). The number in brackets stands for the average indicator-value, rounded to 10^{-3} (lower is better). **Bold** ranking values correspond to the best-performing algorithm for the instance and the indicator under consideration.

ρ	n	SLS$_{\text{unif}}$		SLS$_{\text{dicho}}$		PLS		TP-LS		best-known [12]	
						hypervolume					
−0.5	1000	3	(0.009)	2	(0.006)	1	(0.002)	**0**	(0.000)	4	(0.031)
−0.2	1000	3	(0.008)	2	(0.006)	1	(0.003)	**0**	(0.000)	4	(0.023)
0.0	1000	3	(0.007)	2	(0.006)	1	(0.004)	**0**	(0.000)	4	(0.016)
0.2	1000	1	(0.005)	1	(0.005)	3	(0.006)	**0**	(0.001)	4	(0.008)
0.5	1000	2	(0.002)	3	(0.003)	4	(0.008)	**0**	(0.002)	**0**	(0.002)
−0.5	2000	3	(0.007)	2	(0.004)	1	(0.002)	**0**	(0.000)	4	(0.053)
−0.2	2000	3	(0.007)	2	(0.004)	1	(0.003)	**0**	(0.001)	4	(0.047)
0.0	2000	3	(0.007)	1	(0.005)	1	(0.005)	**0**	(0.001)	4	(0.041)
0.2	2000	2	(0.005)	1	(0.004)	3	(0.006)	**0**	(0.001)	4	(0.023)
0.5	2000	1	(0.003)	1	(0.003)	4	(0.010)	**0**	(0.002)	3	(0.006)
−0.5	3000	3	(0.007)	2	(0.003)	1	(0.002)	**0**	(0.000)	4	(0.083)
−0.2	3000	3	(0.007)	1	(0.003)	1	(0.003)	**0**	(0.001)	4	(0.068)
0.0	3000	3	(0.007)	1	(0.004)	2	(0.006)	**0**	(0.001)	4	(0.062)
0.2	3000	2	(0.006)	1	(0.004)	3	(0.007)	**0**	(0.001)	4	(0.037)
0.5	3000	2	(0.003)	1	(0.002)	3	(0.010)	**0**	(0.001)	3	(0.010)
−0.5	4000	3	(0.007)	1	(0.003)	2	(0.006)	**0**	(0.000)	4	(0.092)
−0.2	4000	3	(0.007)	1	(0.003)	2	(0.004)	**0**	(0.001)	4	(0.077)
0.0	4000	3	(0.007)	1	(0.003)	2	(0.005)	**0**	(0.001)	4	(0.093)
0.2	4000	2	(0.004)	1	(0.002)	3	(0.006)	**0**	(0.001)	4	(0.047)
0.5	4000	2	(0.003)	1	(0.002)	3	(0.008)	**0**	(0.001)	4	(0.014)
−0.5	5000	2	(0.007)	1	(0.002)	3	(0.020)	**0**	(0.000)	4	(0.141)
−0.2	5000	2	(0.007)	1	(0.003)	3	(0.008)	**0**	(0.001)	4	(0.130)
0.0	5000	3	(0.006)	1	(0.003)	2	(0.006)	**0**	(0.001)	4	(0.130)
0.2	5000	2	(0.005)	1	(0.003)	3	(0.007)	**0**	(0.001)	4	(0.094)
0.5	5000	2	(0.003)	1	(0.002)	3	(0.010)	**0**	(0.001)	4	(0.021)
						epsilon					
−0.5	1000	3	(1.013)	2	(1.009)	1	(1.003)	**0**	(1.001)	4	(1.015)
−0.2	1000	2	(1.011)	2	(1.009)	1	(1.004)	**0**	(1.001)	4	(1.014)
0.0	1000	2	(1.010)	2	(1.010)	1	(1.005)	**0**	(1.001)	2	(1.010)
+0.2	1000	2	(1.009)	3	(1.009)	1	(1.005)	**0**	(1.001)	2	(1.008)
+0.5	1000	3	(1.011)	3	(1.012)	2	(1.008)	**0**	(1.002)	1	(1.005)
−0.5	2000	3	(1.009)	2	(1.007)	1	(1.003)	**0**	(1.001)	4	(1.026)
−0.2	2000	2	(1.008)	2	(1.008)	1	(1.003)	**0**	(1.001)	4	(1.027)
0.0	2000	3	(1.009)	2	(1.007)	1	(1.005)	**0**	(1.001)	4	(1.025)
+0.2	2000	3	(1.009)	2	(1.008)	1	(1.004)	**0**	(1.001)	4	(1.019)
+0.5	2000	2	(1.011)	1	(1.009)	1	(1.008)	**0**	(1.002)	4	(1.014)
−0.5	3000	3	(1.009)	2	(1.005)	1	(1.003)	**0**	(1.000)	4	(1.051)
−0.2	3000	3	(1.009)	2	(1.006)	1	(1.003)	**0**	(1.001)	4	(1.039)
0.0	3000	3	(1.008)	1	(1.006)	1	(1.005)	**0**	(1.001)	4	(1.034)
+0.2	3000	3	(1.007)	2	(1.006)	1	(1.004)	**0**	(1.001)	4	(1.025)
+0.5	3000	2	(1.005)	1	(1.004)	1	(1.004)	**0**	(1.001)	4	(1.011)
−0.5	4000	3	(1.008)	1	(1.004)	2	(1.007)	**0**	(1.000)	4	(1.055)
−0.2	4000	3	(1.008)	2	(1.005)	1	(1.003)	**0**	(1.001)	4	(1.042)
0.0	4000	3	(1.008)	2	(1.005)	1	(1.004)	**0**	(1.001)	4	(1.059)
+0.2	4000	3	(1.006)	2	(1.004)	1	(1.003)	**0**	(1.001)	4	(1.033)
+0.5	4000	3	(1.005)	1	(1.003)	1	(1.003)	**0**	(1.001)	4	(1.020)
−0.5	5000	2	(1.008)	1	(1.004)	3	(1.021)	**0**	(1.000)	4	(1.074)
−0.2	5000	2	(1.008)	1	(1.004)	2	(1.007)	**0**	(1.001)	4	(1.090)
0.0	5000	3	(1.008)	1	(1.005)	1	(1.004)	**0**	(1.001)	4	(1.064)
+0.2	5000	3	(1.007)	1	(1.004)	1	(1.004)	**0**	(1.001)	4	(1.050)
+0.5	5000	2	(1.004)	1	(1.003)	2	(1.004)	**0**	(1.001)	4	(1.025)

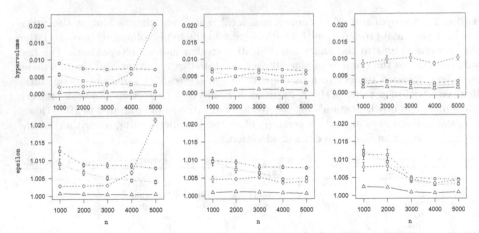

Fig. 2. Comparison of SLS$_{\text{unif}}$ (○), SLS$_{\text{dicho}}$ (□), PLS (◇) and TP-LS (△) with respect to hypervolume (top) and epsilon (bottom) for $\rho = -0.5$ (left), $\rho = 0.0$ (center), and $\rho = +0.5$ (right). A lower value is better.

Interestingly, the quality of the approximation sets identified by scalarizing approches (SLS$_{\text{unif}}$, SLS$_{\text{dicho}}$) slightly increases with the objective correlation ρ, as the proportion of supported solutions; see Fig. 1(c). By comparing SLS$_{\text{unif}}$ with SLS$_{\text{dicho}}$, the later is always at least as good as the former for all the instances we investigated but one ($\rho = +0.5$, $n = 1000$). The reason is that the SLS$_{\text{unif}}$ algorithm is limited on the approximation set size that it is able to identify (a fixed number of 101 weighting coefficient vectors times two phases, *i.e.* 202 solutions at most); see Fig. 3. On the contrary, SLS$_{\text{dicho}}$ adaptively determines a number of weighting coefficient vectors based on the solutions it iteratively identifies. As a consequence, it takes advantage of manipulating an unbounded approximation set, that allows SLS$_{\text{dicho}}$ to obtain better indicator-values overall. Still, as shown in Fig. 3, the number of solutions identified by both scalarizing approaches, which only seek for supported solutions, are lower than Pareto-based approaches by several orders of magnitude. However, the number of solutions found by all approaches reduces with the objective correlation ρ, as the number of Pareto optimal solutions reported in Section 3.1.

Overall, in terms of approximation quality, there is a clear advantage to TP-LS. Actually, hybridizing SLS$_{\text{dicho}}$ and PLS allows to obtain statistically better approximation sets, in terms of hypervolume and epsilon, than all the other competing algorithms, for all the instances. In particular, there is a substantial improvement of the indicator-values, showing that TP-LS is consistently able to identify a high-quality approximation set, which is very close to the reference set, especially for instances with conflicting objectives. Indeed, more than 99.67% of the best-found hypervolume is covered by the approximation set identified by TP-LS in the worst case. The epsilon indicator-value is always less than 1.003. The weakness of PLS in identifying good-quality lexicographical solutions seems to be overcome by initializing the archive with high-quality scalarizing solutions.

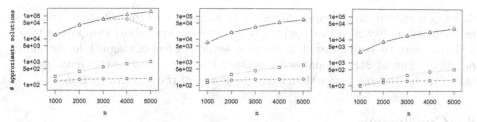

Fig. 3. Comparison of SLS$_{unif}$ ($\circ$), SLS$_{dicho}$ ($\square$), PLS ($\diamond$) and TP-LS ($\triangle$) with respect to the size of the approximation set found for $\rho = -0.5$ (left), $\rho = 0.0$ (center), and $\rho = +0.5$ (right). Notice the log-scale on the y-axis.

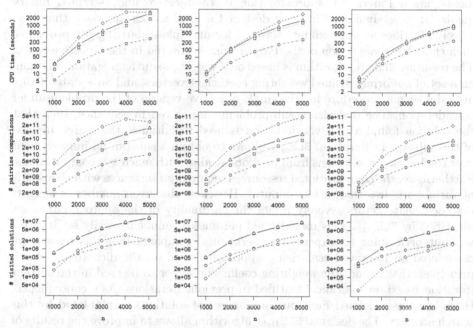

Fig. 4. Comparison of SLS$_{unif}$ ($\circ$), SLS$_{dicho}$ ($\square$), PLS ($\diamond$) and TP-LS ($\triangle$) with respect to to the CPU time (top), the number of comparisons (middle), and the number of visited solutions (bottom), for $\rho = -0.5$ (left), $\rho = 0.0$ (center), and $\rho = +0.5$ (right). Notice the log-scale on the y-axis.

This means that finding some non-dominated solutions can actually help to identify additional ones, as conjectured in Section 3.3.

More surprisingly, the hybrid TP-LS approach also allows to improve the performance of PLS in terms of computational time. As reported in Fig. 4, the running time of TP-LS is below the one of PLS for most of the instances. By analyzing this more carefully, TP-LS actually allows to generate much more candidate solutions, and as many potentially non-dominated solutions as PLS, by performing much less pairwise comparisons, particularly for large-size instances. As well, the overhead of TP-LS compared to only performing the first phase as in

SLS_{dicho} is almost insignificant. At last, Fig. 4 also reveals that the loss of SLS_{unif} in terms of quality is in fact counter-balanced by a very short computing time, which is lower than all other algorithms by several orders of magnitude. Actually, one single run of SLS_{unif} allows to improve the *aggregated* best-known results from [12] by running faster than each single run of the algorithms from [12].

6   Conclusions

In this paper, we designed and analyzed stochastic local search algorithms for identifying a Pareto set approximation in bi-objective unconstrained binary quadratic programming. First, we designed a local search approach that iteratively identifies an approximate solution for multiple scalarizing sub-problems, such that the whole region of the Pareto front is covered in the objective space. The resulting SLS_{unif} algorithm is based on a weighted-sum aggregation function, on a set of uniformly defined weighting coefficient vectors, and on a state-of-the-art tabu search procedure for the single-objective version of the problem under consideration. The scalarizing sub-problems are solved in sequence, such that the solution found at a given iteration is used to initialize the search process of the subsequent iteration. SLS_{unif} allows to obtain a substantial improvement over the best-know solutions from previous studies with much less computations. Furthermore, the computational ressources required for instances with thousands of variables is less than a few minutes. However, the given number of weighting coefficient vectors severely limits the cardinality of the approximation set identified by SLS_{unif}. This user-defined parameter cannot be easily set *a priori* without performing an expensive experimental campaign. For this reason, we considered an improved scalarizing approach, based on the dichotomic search principles, that defines the weighting coefficient vector to be used in the current iteration based on solutions identified in previous iterations. As a consequence, there is *no* user-defined limit on the number of solutions manipulated by this search strategy. The designed SLS_{dicho} algorithm allows to improve the results of SLS_{unif} in terms of quality, while inducing an extra cost in terms of computing time. Still, for these two scalarizing approaches, that both seek *supported* optimal solutions only, the cardinality of the obtained approximation set is less than those of Pareto-based approaches. In addition, we highlighted that the performance of scalarizing local search decreases with the degree of conflict between the objectives, following the proportional number of supported solutions. Next, a Pareto local search algorithm, that iteratively explores the neighborhood of an archive of mutually non-dominated solutions, obtains similar results in terms of quality indicator-values, while requiring even more computational resources. Indeed, the PLS strategy obtains larger approximation sets while exploring much less candidate solutions. The bottleneck of its effectiveness is the number of comparisons required to maintain the archive, while the bottleneck of its efficiency is a poor quality in finding strong solutions at the extremes of the Pareto front. At last, we designed a two-phase algorithm that applies SLS_{dicho} and PLS in a sequential way, the output of the former being the input of the latter. The TP-LS approach significantly surpasses the other algorithms over all the configurations we experimented. The computational overhead is negligible compared with the stand-alone SLS_{dicho} approach, TP–LS being able to generate as many solutions as PLS while performing much less comparisons. This hybrid algorithm profits from the closeness that exists between optimal solutions. Starting with a subset of approximate (supported) non-dominated solutions, TP–LS is able to identify additional ones, then improving the overall approximation set quality.

Extending the stochastic local search approaches investigated in the paper to unconstrained binary quadratic programming with more than two objectives would improve even more the expressive ability of the problem formulation. This would provide a more general unifying modeling and solution framework for multi-objective optimization that could potentially enable an efficient reformulation and resolution of a wide class of large-scale and NP-hard multi-objective combinatorial optimization problems.

References

1. Aneja, Y., Nair, K.: Bicriteria transportation problem. Manag. Sci. **25**(1) (1979)
2. Bader, J., Zitzler, E.: HypE: An algorithm for fast hypervolume-based many-objective optimization. Evol. Comp. **19**(1), 45–76 (2011)
3. Beasley, J.E.: OR-library: Distributing test problems by electronic mail. J. Oper. Res. Soc. **41**(11), 1069–1072 (1990)
4. Dubois-Lacoste, J., López-Ibáñez, M., Stützle, T.: A hybrid TP+PLS algorithm for bi-objective flow-shop scheduling problems. Comp. Oper. Res. **38**(8), 1219–1236 (2011)
5. Dubois-Lacoste, J., López-Ibáñez, M., Stützle, T.: Improving the anytime behavior of two-phase local search. Ann. Math. Artif. Intel. **61**(2), 125–154 (2011)
6. Ehrgott, M.: Multicriteria optimization, 2nd edn. Springer (2005)
7. Ehrgott, M., Klamroth, K.: Connectedness of efficient solutions in multiple criteria combinatorial optimization. Eur. J. Oper. Res. **97**(1), 159–166 (1997)
8. Garey, M.R., Johnson, D.S.: Computers and Intractability: A Guide to the Theory of NP-Completeness. W. H. Freeman & Co. Ltd (1979)
9. Glover, F., Hao, J.K.: Efficient evaluations for solving large 0-1 unconstrained quadratic optimisation problems. Int. J. Metaheuristics **1**(1), 3–10 (2010)
10. Glover, F., Lü, Z., Hao, J.K.: Diversification-driven tabu search for unconstrained binary quadratic problems. 4OR-Q. J. Oper. Res. **8**(3), 239–253 (2010)
11. Kochenberger, G., Hao, J.K., Glover, F., Lewis, M., Lü, Z., Wang, H., Wang, Y.: The unconstrained binary quadratic programming problem: a survey. J. Comb. Optim. **28**(1), 58–81 (2014)
12. Liefooghe, A., Verel, S., Hao, J.K.: A hybrid metaheuristic for multiobjective unconstrained binary quadratic programming. Appl. Soft. Comp. **16**, 10–19 (2014)
13. Lust, T., Teghem, J.: Two-phase Pareto local search for the biobjective traveling salesman problem. J. Heuristics **16**(3), 475–510 (2010)
14. Paquete, L., Schiavinotto, T., Stützle, T.: On local optima in multiobjective combinatorial optimization problems. Ann. Oper. Res. **156**(1), 83–97 (2007)

15. Paquete, L., Stützle, T.: A study of stochastic local search algorithms for the biobjective QAP with correlated flow matrices. Eur. J. Oper. Res. **169**(3), 943–959 (2006)
16. Paquete, L., Stützle, T.: Clusters of non-dominated solutions in multiobjective combinatorial optimization: an experimental analysis. In: Barichard, V., Ehrgott, M., Gandibleux, X., T'Kindt, V. (eds.) Multiobjective Programming and Goal Programming. LNEMS, vol. 618, pp. 69–77. Springer, Heidelberg (2009)
17. Zitzler, E., Thiele, L., Laumanns, M., Foneseca, C.M., Grunert da Fonseca, V.: Performance assessment of multiobjective optimizers: An analysis and review. IEEE Trans. Evol. Comput. **7**(2), 117–132 (2003)

A Bug in the Multiobjective Optimizer IBEA: Salutary Lessons for Code Release and a Performance Re-Assessment

Dimo Brockhoff(✉)

Inria Lille - Nord Europe, DOLPHIN Project-team, 59650 Villeneuve d'Ascq, France
dimo.brockhoff@inria.fr

Abstract. The Indicator-Based Evolutionary Algorithm (IBEA) is one of the first indicator-based multiobjective optimization algorithms and due to its wide availability in several algorithm packages is often used as a reference algorithm when benchmarking multiobjective optimizers. The original publication on IBEA proposes to use two specific variants: one based on the ε-indicator and one based on the hypervolume. Several experimental studies concluded that, surprisingly, the IBEA variant with the ε-indicator performs better than the one with the hypervolume—even if the (unary) hypervolume indicator itself is the quality measure used in the performance assessment. Recently, a small bug has been found in the hypervolume variant of IBEA with large implications on its performance. Here, we not only explain the bug in detail and correct it, but also present the (improved) results of the corrected version. Moreover, and probably even more important for the scientific community, we point out that this bug has been transferred to other than the original software package, discuss how this obscured the bug, and argue in favor of some simple, even obvious guidelines how the optimization community should deal with algorithm source codes, documentation, and the (natural) existence of bugs in the future.

1 Introduction

The Indicator-Based Evolutionary Algorithm (IBEA, [12]) is one of the first proposed indicator-based multiobjective optimization algorithms. Due to its simplicity, good performance, and wide availability in several algorithm packages such as PISA [3], Paradiseo [7], jMetal [5] or the MOEA Framework [6], IBEA is an often-used reference algorithm when benchmarking multiobjective optimizers.

The main idea behind IBEA is to employ in the calculation of a solution's fitness a binary quality indicator, which assigns two solution sets a scalar value indicating their relative quality. The original publication proposes to use two specific IBEA variants: one based on the additive ε-indicator, denoted IBEA$_{\varepsilon+}$ in the following, and one based on the hypervolume (denoted IBEA$_{HD}$; more details about the algorithm are provided in the following section). Several experimental studies concluded that, surprisingly, the IBEA variant with the ε-indicator performs better than the one with the hypervolume [1,2,10]—even if the (unary)

© Springer International Publishing Switzerland 2015
A. Gaspar-Cunha et al. (Eds.): EMO 2015, Part I, LNCS 9018, pp. 187–201, 2015.
DOI: 10.1007/978-3-319-15934-8_13

hypervolume indicator itself is the quality measure used in the performance assessment [12]. This led to the fact that most studies using IBEA use the version employing the ε-indicator.

Recently, a small bug has been reported in the hypervolume variant of IBEA in the Paradiseo [7] implementation which turned out to stem from its original PISA implementation [3] and which has some large implications on its performance. In the following, we not only explain the bug in detail and correct it, but also present the (improved) results of the corrected version on the same test problems as in the original publication [12]. As expected, the corrected version outperforms the buggy one with the exceptions of the discrete knapsack and network processor design problems and for a low number of objective functions where the two versions do not differ statistically significantly. On the ZDT6 problem, we furthermore show that the former version was not invariant under permutations of the objective functions while the corrected one is.

Moreover, we have seen that the same bug has been also present in other algorithm packages such as jMetal [5] and the MOEA framework [6]. Hence, we argue during the final part of the paper in favor of independent implementations, thorough testing, and a precise and honest documentation of algorithm packages within our community.

2   IBEA

The general Indicator-Based Evolutionary Algorithm (IBEA) as proposed by Zitzler and Künzli [12] is one of the very first multiobjective optimizers to integrate user preferences in a clear and mathematically sound way. The main contribution of IBEA was to open up a new research area on the design of multiobjective optimization algorithms which employ a so-called *quality indicator* in their (environmental) selection procedure.

Before we describe the original IBEA algorithm in more detail, let us mention that we consider, w.l.o.g., minimization problems here where the Pareto dominance relation $\prec$ is defined between solutions x^1 and x^2 as $x^1 \prec x^2$ if and only if $f_i(x^1) \leq f_i(x^2)$ for all objective functions $f_i : X \to Z$ ($1 \leq i \leq k$) and $f_i(x^1) < f_i(x^2)$ for at least one objective function. In this case, we also say x^1 dominates x^2. An m-ary quality indicator is furthermore a function $I : \Omega^m \to$ that maps m solution sets $X_1, \ldots, X_m$ from the set of all possible solutions ($X_1, \ldots, X_m \in \Omega = 2^X$) to a real number. Nowadays, mostly *unary quality indicators* such as the standard hypervolume indicator are used in both performance assessment and the definition of solution quality within the environmental selection. Instead, IBEA itself is based on *binary quality indicators* that map *two* solution sets to a real number.

To be more precise, the fitness of a solution x^1 in IBEA's population P is assigned by

$$F(x^1) = \sum_{x^2 \in P \setminus \{x^1\}} -e^{-I(\{x^2\}, \{x^1\})/(c \cdot \kappa)}$$

where $\kappa > 0$ is a parameter of the algorithm and $c = \max_{x^1, x^2 \in P} |I(x^1, x^2)|$ is the maximum indicator value between any two population members. This fitness assignment scheme of IBEA has the theoretical property that if a solution x^1 dominates solution x^2 then also $F(x^1) > F(x^2)$ as long as the chosen binary indicator I itself is "dominance preserving"[1] [12]. Note that in the following, we abuse the mathematical notation and write $I(x, y)$ instead of $I(\{x\}, \{y\})$ if x and y are single solutions. Examples of dominance preserving binary indicators are the binary hypervolume and the binary ε-indicator which, for that reason, have been proposed to be used in the original IBEA publication.

The binary (additive) ε-indicator assigns to two solution sets A and B the minimal objective value ε by which all solutions in A have to be improved (along each objective) in order to (weakly) dominate all solutions in B:

$$I_{\varepsilon+}(A, B) = \min_{\varepsilon} \left\{ \forall x^2 \in B \exists x^1 \in A : f_i(x^1) - \varepsilon \leq f_i(x^2) \text{ for } i \in \{1, \ldots, k\} \right\} .$$

The binary ε-indicator is negative if all solutions in B are dominated by at least one solution in A.

The binary hypervolume indicator used in [12], assigns to two solution sets A and B the "volume of the space that is dominated by B but not by A with respect to a predefined reference point" in objective space [12]:

$$I_{HD}(A, B) = \begin{cases} I_H(B) - I_H(A) & \text{if } \forall x^2 \in B \exists x^1 \in A : x^1 \prec x^2 \\ I_H(A + B) - I_H(A) & \text{else} \end{cases} \tag{1}$$

where $I_H(.)$ denotes the standard (unary) hypervolume proposed in [13] and the index "HD" stands for "hypervolume difference". Also $I_{HD}(A, B)$ is negative if all solutions in B are dominated by at least one solution in A. Note also that neither of the two binary indicators is symmetric, i.e., typically $I(A, B) \neq I(B, A)$ holds. The corresponding IBEA variants using the above defined hypervolume and ε-indicator are denoted IBEA$_{HD}$ and IBEA$_{\varepsilon+}$ respectively here.

Algorithm 1 shows the pseudo code of the entire IBEA procedure (copied and adapted from [12]). It starts with generating α solutions uniformly at random from the search space (Step 1). Then, IBEA follows the standard way of selecting solutions for mating (via a binary tournament with replacement, Step 5), generating new solutions from those selected solutions (via problem dependent crossover and mutation operators, Step 6), and environmental selection where the above described fitness assignment scheme is used (Step 2) to iteratively reduce the population back to the population size α by deleting the solutions with worst fitness successively (Step 3). Important to note is that the described adaptive version of IBEA scales both the objective values before computing the indicator values (Steps 2.1 and 2.2) and the indicator values themselves before to apply the above fitness assignment scheme (Steps 2.3, 2.4, and 3.3). Moreover, the calculation of the fitness is partially updated as soon as one solution

---

[1] A binary quality indicator I is called *dominance preserving* if for all solutions $x^1, x^2, x^3 \in X$ both $x^1 \prec x^2 \implies I(\{x^1\}, \{x^2\}) < I(\{x^2\}, \{x^1\})$ and $x^1 \prec x^2 \implies I(\{x^3\}, \{x^1\}) \geq I(\{x^3\}, \{x^2\})$ hold.

Algorithm 1. (Adaptive) IBEA as proposed in [12]

Input: α (population size)

N (maximum number of iterations)

κ (fitness scaling factor)

Output: A (Pareto set approximation)

Step 1: Initialization: Generate initial population P of size α at random; set iteration counter $m = 0$

Step 2: Fitness Assignment: First scale objective and indicator values; then use scaled values to assign fitness for each population member $x^1 \in P$:

1. Determine lower $(\underline{b_i} = \min_{x \in P} f_i(x))$ and upper bound $(\overline{b_i} = \max_{x \in P} f_i(x))$ of each objective function

2. Scale each objective to interval $[0, 1]$: $f_i'(x) = (f_i(x) - \underline{b_i})/(\overline{b_i} - \underline{b_i})$

3. Calculate indicator values $I(x^1, x^2)$ using the scaled objective values f_i' and determine the maximum absolute indicator value $c = \max_{x^1, x^2 \in P} |I(x^1, x^2)|$

4. For all $x^1 \in P$ set $F(x^1) = \sum_{x^2 \in P \setminus \{x^1\}} -e^{-I(x^2, x^1)/(c \cdot \kappa)}$

Step 3: Environmental Selection: Iterate the following three steps until the size of population P does not exceed α :

1. Choose an individual $x^* \in P$ with the smallest fitness value, i.e., $F(x^*) \leq F(x)$ for all $x \in P$

2. Remove x^* from the population

3. Update fitness values of all remaining individuals $x \in P$ as $F(x) = F(x) + e^{-I(x^*, x)/(c \cdot \kappa)}$

Step 4: Termination: If $m \geq N$ or another stopping criterion is fulfilled, stop and return the non-dominated solutions in P as A

Step 5: Mating Selection: Perform binary tournament selection with replacement on P in order to fill the temporary mating pool P'

Step 6: Variation: Apply recombination and mutation operators to the mating pool P' and add the resulting offspring to P. Increment iteration counter $(m = m + 1)$ and go to Step 2

---

is deleted during the environmental selection (Step 3.3). Finally, the algorithm terminates when the total number of iterations N are reached or another user-defined stopping criterion is reached (not implemented in the PISA version).

3   The Bug

The reported bug appeared in the hypervolume calculation of IBEA$_{HD}$, more precisely in the recursive "hypervolume by slicing objectives" technique used in the original PISA implementation, see line 33 of the original C code in Fig. 2. It is caused by a typo which misplaces the correct variable "a" by "b"—resulting in wrongly adding the volume of an objective space part to the indicator value $I(a, b)$ that is not dominated by either solution. Figure 1 shows an example where the bug not only miscalculates the binary hypervolume indicator values but also results in a different order of the two solutions with respect to their fitness. Note that the bug results only in erroneous decisions where the point on the left is wrongly preferred while the opposite never happens. Hence, it can be expected

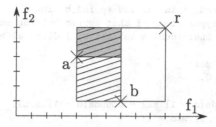

Fig. 1. Illustration of the impact of the bug on the comparison of two solutions with objective vectors a and b. The gray box corresponds to the true hypervolume dominated solely by objective vector a while the striped box shows the actual contribution computed by the original, buggy code. The buggy code considers a better than b while the corrected code considers b better. In both cases, the hypervolume solely dominated by b is computed correctly.

that the correction of this bug has an impact on the search performance of IBEA. It might even explain and counterbalance the impression of previous benchmarking studies that the hypervolume-based IBEA does surprisingly not perform as well as the ε-indicator-based version when the (unary) hypervolume indicator is used as performance measure. In the following, we will thus investigate the effect of the bug fix on the performance of IBEA$_{HD}$ extensively.

4   Concrete Implications for the Performance of IBEA

To investigate the concrete implications of the bug (and its correction) on the performance of IBEA, we rerun the experiments of the original IBEA paper by Zitzler and Künzli [12]. Before we have a closer look on the results, let us note that here, we can show the results for the 2-objective ZDT6 and EXPO2 problems, the 3-objective DTLZ2, DTLZ6, and EXPO3 problems, and on the 4-objective EXPO4 problem mentioned in the original publication—many of which had to be omitted in the original IBEA paper. As much as possible, we tried to use the same problem and algorithm parameters as in the original study.

4.1   Experimental Setup

All experiments were performed in PISA [3] with all modules downloaded from http://www.tik.ee.ethz.ch/sop/pisa in their August 2014 version. As problems, we chose the continuous ZDT6, DTLZ2, and DTLZ6 problems with 10, 12, and 22 variables and 2, 3, and 3 objective functions respectively as suggested in the original publications [4, 11]. In addition, we chose the discrete 2-objective 0-1-knapsack problem with 100 items [13] as well as the 2-, 3-, and 4-objective EXPO problem (network processor design) with standard PISA settings [9].

Together with IBEA$_{\varepsilon+}$ and the buggy and corrected IBEA$_{HD}$, we ran the PISA implementations of NSGA-II and SPEA2 as in [12]. All algorithms used a

```
 1 double calcHypInd(ind *p_ind_a, ind *p_ind_b, int d)
 2 /* calculates the hypervolume of that portion of the objective space that
 3    is dominated by individual a but not by individual b */
 4 {
 5    double a, b, r, max;
 6    double volume = 0;
 7
 8    r = rho * (bounds[d - 1].max - bounds[d - 1].min);
 9    max = bounds[d - 1].min + r;
10
11    a = p_ind_a->f[d - 1];
12    if (p_ind_b == NULL)
13        b = max;
14    else
15        b = p_ind_b->f[d - 1];
16
17    if (d == 1)
18    {
19        if (a < b)
20            volume = (b - a) / r;
21        else
22            volume = 0;
23    }
24    else
25    {
26        if (a < b)
27        {
28            volume = calcHypInd(p_ind_a, NULL, d - 1) * (b - a) / r;
29            volume += calcHypInd(p_ind_a, p_ind_b, d - 1) * (max - b) / r;
30        }
31        else
32        {
33            volume = calcHypInd(p_ind_a, p_ind_b, d - 1)
                       * (max - a) / r; \\ corrected version
                                   \\ original version: "* (max - b) / r;"
24        }
25    }
26
27    return (volume);
28 }
```

Fig. 2. The source code snippet of the PISA implementation of IBEA and the bug in line 33. For readability, the function name is shortened here.

population size of $\alpha = 100$, a binary tournament mating selection, and were run independently 30 times for 200 iterations each.

Regarding the variation operators, the continuous problems used SBX crossover and polynomial mutation with the PISA parameters set as individual_mutation_probability=1;   individual_recombination_probability=1; variable_ mutation_ probability=0.01; variable_recombination_probability=1;

variable_ swap_probability=0; eta_mutation=20; eta_recombination=20 and use_ symmetric_recombination=1. For the discrete knapsack problem, we used one-bit mutation and one-point-crossover with the PISA parameters mutation_probability =1 and recombination_probability=0.8. For EXPO, we used the standard PISA settings. For IBEA, we furthermore used $\kappa = 0.05$ as suggested in [12] and a reference point of $(2, \ldots, 2)$ for the internal normalized calculations of I_{HD}.

To compare the algorithms, the hypervolume and the additive ε-indicator were recorded every 50 iterations and computed relative to a reference set, obtained by joining all non-dominated solutions at this specific iteration over all algorithm runs. Before computing the indicators, the objective vectors had been normalized such that all non-dominated points at the investigated iteration over all algorithms defined the box $[1, 2]^k$. The reference point for the hypervolume indicator was chosen as $(2.1, \ldots, 2.1)$ as in [12].

4.2   Comparison Between Buggy and Corrected IBEA$_{HD}$

Figures 3 and 4 show the box plots of both the hypervolume and ε-indicator for the four algorithms NSGA-II, SPEA2, IBEA$_{\varepsilon+}$, and IBEA$_{HD}$ after 200 iterations—on the left for the buggy version of IBEA$_{HD}$ and on the right for the corrected version of IBEA$_{HD}$[2]. Figure 5 shows the direct comparison between the buggy IBEA$_{HD}$ and the corrected version on each problem after 50, 100, 150, and 200 iterations. All boxplots are drawn with Matlab and the ends of the notches correspond to "$q_2 - 1.57(q_3 - q_1)/\sqrt{n}$ and $q_2 + 1.57(q_3 - q_1)/\sqrt{n}$, where q_2 is the median, q_1 and q_3 the 25th and 75th percentiles [...], and n is the number of runs", thus indicating statistically significant medians "at the 5% significance level if their intervals do not overlap" (compare the Matlab documentation for further details).

Overall, three main observations can be made: As to the continuous problems, the bug fix has a positive effect on IBEA$_{HD}$ on the 3-objective problems DTLZ2 and DTLZ6 with respect to both indicators. The corrected IBEA$_{HD}$ now results in similar or better indicator values than IBEA$_{\varepsilon+}$. For the 2-objective ZDT6 problem, however, the effect is small and sometimes slightly detrimental (though not statistically significant as the boxplots' notches do overlap). We give a possible explanation for this behavior on ZDT6 in the following section.

As to the discrete problems, no positive effect of the bug fix can be reported. The results before and after the bug fix are similar and only very few statistically significant differences can be observed when looking at the notches in Fig. 5. The corrected IBEA$_{HD}$ version is not better than IBEA$_{\varepsilon+}$ with respect to the hypervolume indicator except for the 4-objective EXPO problem and a larger number of iterations. Exactly in these cases of the 4-objective EXPO problem,

---

[2] The reason for two sets of figures with four algorithms each instead of showing all five algorithms in a single plot is to see the effect of the bug fix directly. Because all results are relative to other algorithms, joining all algorithms alters the box plots (slightly) and thus makes comparisons with the original paper [12] more difficult. For a direct comparison of the buggy and corrected IBEA$_{HD}$, we provide Fig. 5.

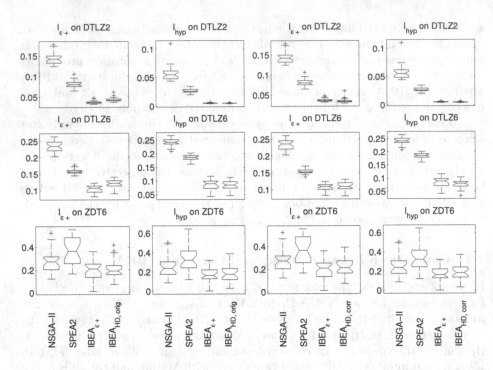

Fig. 3. Boxplots comparing the performance of NSGA-II, SPEA2, IBEA$_{\varepsilon+}$, and IBEA$_{HD}$ for different continuous problems (rows). The left two columns show the results for the original IBEA$_{HD}$ implementation, the two right columns the same results for the corrected IBEA$_{HD}$ version. Columns 1 and 3 show results for the additive ε-indicator; columns 2 and 4 results for the hypervolume indicator. All indicators are computed after 200 generations with respect to the reference set stemming from all algorithms of that plot.

on the other hand, the additive ε-indicator is significantly worse for the corrected IBEA$_{HD}$ version. It seems as if the discreteness of the problems and the comparatively small number of non-dominated solutions in the resulting populations do not allow the (correct) hypervolume fitness to be effective. It can be also noted that especially for the knapsack problem, the variance between runs is quite large, which means that the impact of the bug fix can only be rather small anyway as the results differ much more among the runs than among the algorithms.

Last, we see that the positive effect of the bug fix, at least for the selected problems, becomes larger with an increasing number of objective functions and more pronounced on the continuous problems with more function evaluations. Another fundamental improvement caused by the bug fix will be discussed in the next subsection.

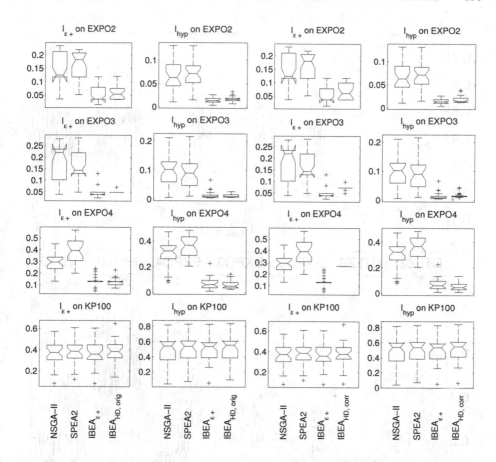

Fig. 4. Boxplots comparing the performance of NSGA-II, SPEA2, IBEA$_{\varepsilon+}$, and IBEA$_{HD}$ for some discrete problems (rows). The left two columns show the results for the original IBEA$_{HD}$ implementation, the two right columns the same results for the corrected IBEA$_{HD}$ version. Columns 1 and 3 show results for the additive ε-indicator; columns 2 and 4 results for the hypervolume indicator. All indicators are computed after 200 generations with respect to the reference set stemming from all algorithms of that plot.

4.3   Invariance With Respect to Objective Permutations

One additional observation, we can make when comparing the buggy and the corrected version of IBEA$_{HD}$, is that the corrected version is, as expected, invariant over a permutation of the objective functions, i.e., the performance is the same when we for example exchange the first and the second objective function. This invariance is a desired property of an optimization algorithm as it generalizes the statements we can make about the performance of the algorithm without actually testing it. The invariance properties of the hypervolume indicator give

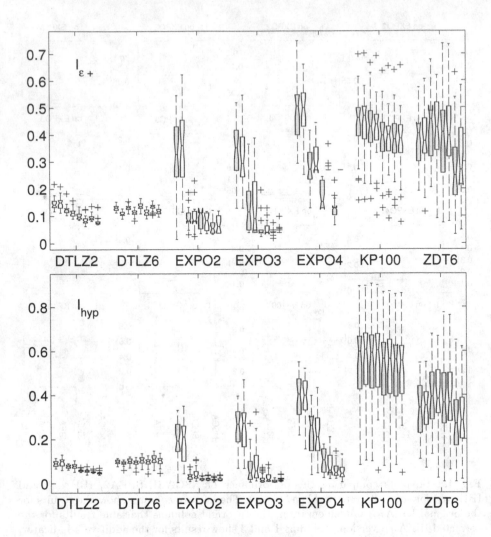

Fig. 5. Comparison via boxplots between the buggy and the corrected IBEA$_{HD}$ version for various problems and number of iterations. For each problem and from left to right, iterations 50, 50, 100, 100, 150, 150, 200, and 200; the left gray box corresponds to the buggy IBEA$_{HD}$ while the right white box corresponds to the corrected version. The top plot shows the additive ε-indicator values over 30 runs, the bottom plot the results for the hypervolume indicator.

us theoretically this invariance of IBEA$_{HD}$, but it turns out that the bug in the original PISA implementation resulted in an algorithm that is not invariant. To investigate this, we ran both the buggy and the corrected version of IBEA$_{HD}$ with a population size of 100 for 200 iterations on the ZDT6 problem—this time with an increased number of 100 variables to better see the effect. To be precise,

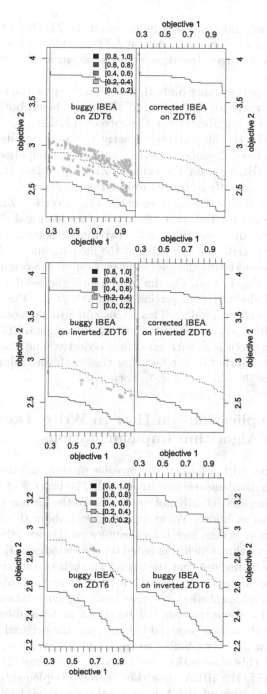

Fig. 6. Empirical attainment function plots after 200*D* function evaluations comparing the buggy and corrected IBEA*HD* on the ZDT6 problem with 100 variables (top) and on its inverted version (middle, after swapping the objectives again for comparison reasons). On the bottom plot, the results of the buggy IBEA*HD* on both problems.

we ran each algorithm independently 30 times on the ZDT6 problem and again, with the same initial random seeds, on the ZDT6 problem where the two objective functions are exchanged (we denote this problem as the "inverted" ZDT6 problem in the figures).

To compare the performance on both problems, we plot the empirical attainment functions [8] where for the "inverted" ZDT6 problem, both objectives are again swapped for comparability. For the corrected IBEA$_{HD}$, due to the same initial random seeds, the 30 runs are exactly the same, while for the buggy IBEA$_{HD}$, we see some differences. Figure 6 shows the comparison of the buggy and the corrected IBEA$_{HD}$ on the original ZDT6 problem (top) and on the "inverted" ZDT6 problem (middle), as well as the results of the buggy IBEA$_{HD}$ on ZDT6 against the buggy IBEA$_{HD}$ results on the "inverted" ZDT6 (bottom)— once again, the results for the corrected version are identical. Not only do the empirical attainment functions differ on the two functions for the buggy IBEA$_{HD}$ (Fig. 6, bottom), especially when looking at the median, but the buggy IBEA$_{HD}$ also significantly outperforms the corrected version on the original ZDT6 problem (gray areas in Fig. 6, top). On the other hand, the results are comparable or even in favor of the corrected version to the left of the Pareto front for the "inverted" ZDT6 (Fig. 6, middle). This let us come to the conclusion that the buggy version of IBEA$_{HD}$ exploits the fact that the original ZDT6 problem can be solved by finding solutions with good first objective and then moving along the axis, which is supported by the fact that the bug favors solutions on the left as mentioned in Sec. 3 (compare Fig. 6, bottom).

5   General Implications on How to Write, Document, and Distribute Algorithm Implementations

Let us end the paper with a look at the broader picture and the way we should deal with algorithm implementations in general. The process of finding, solving, and tracking the bug was actually not optimal from the author's personal point-of-view and occurred as follows. We were made aware about this bug by receiving an e-mail from Yann Semet and his colleagues at Thales who had problems debugging their algorithm which was based on the Paradiseo [7] implementation of IBEA. It quickly turned out that the reason for their suspicious results was the bug reported above. Moreover, it turned out that the bug was also present in the original PISA [3] implementation and that both implementations were actually the same (except for some renaming of functions and variables in Paradiseo). Unfortunately, the Paradiseo code did not mention the original implementation such that it was not directly possible to track the observed bug to its origin.

In the wake of this observation, we checked more carefully other software packages that provide the IBEA algorithm. Though implemented in Java while the PISA implementation of IBEA is in C, also for jMetal [5] the same code snippet of the PISA implementation was used without any reference to the original PISA code. Solely the implementation of IBEA in the MOEA framework [6] mentioned clearly where the code was coming from. It was furthermore easy to report the bug via the corresponding online bug-tracking system—a functionality that the other three frameworks (including PISA itself) either do not offer at the moment or that, in the case of Paradiseo, are not linked from the webpage, such that the bug had to be therefore reported by plain e-mail. Let us mention that, for all above software packages, the developers quickly replied to our bug report and the latest versions of the MOEA Framework (v2.3), jMetal (4.5.1), PISA (from October 13, 2014 on), and the Github repository of Paradiseo already contain the bug fix.

The discovery of the bug within the IBEA implementation in the software packages mentioned and the discovery that several implementations are just copies of the original code without references to it will hopefully have a lasting impact on how source code of optimization algorithms is written and distributed in our community. At least, we should always try to remind ourselves on the following two main aspects:

- Algorithm implementations, even if they are provided in big and well-known algorithm packages, should be always questioned and tested thoroughly. Simple unit tests and even more important independent implementations would have exposed the bug.
- Re-using code can also be beneficial in terms of a broader distribution of an algorithm due to reduced implementation times. But if code is copied, the original code basis should always be mentioned in the code for easier tracking of bugs over different software packages—independent of any (obvious) copyright issues one needs to adhere. To allow for easier tracking in the opposite direction and to distribute bug fixes to other packages, it is furthermore recommended that the original code is regularly checked for bug fixes and the secondary code updated accordingly.

Last, we would like to suggest reporting version numbers of the algorithms used in our papers, such as done frequently for example when the single-objective CMA-ES is used. This will, in case of a bug, make it much simpler to find out whether the reported results are trustworthy or not.

6 Conclusions

The correction of a bug in the hypervolume calculation of the multiobjective optimizer IBEA, uncovered since its first implementation in 2004, showed an important impact in the algorithm's search performance. The buggy implementation is not invariant against permutations of the objective functions and shows worse results especially for continuous problems and when the number of objective functions is high. On the tested discrete problems, the buggy and the corrected IBEA behave similar with the only observed significant worsening for the network processor design problem EXPO when the ε-indicator is used as performance measure. The bug might, furthermore, explain why the IBEA variant employing the ε-indicator was so far more often used in empirical studies than the one using the hypervolume indicator—and thus resulting in the wrong perception of algorithm performances.

Probably even more important than the correction of the bug was the observation that several algorithm frameworks such as Paradiseo and jMetal copied the original PISA code of the (buggy) IBEA without mentioning where the code was coming from. Let us be clear that—as long as no copyright is violated of course—having comparable algorithm implementations with the same performance, in general, is a big plus for comparing and applying algorithms in practice (for example by having platform independent implementations). But without a truly independent implementation of IBEA, it took almost 10 years to find such an important bug as the one discussed. Our community should therefore try to have (at least) *two independent implementations of the main algorithms available* (and *a thorough check that they do the same*). Copying code without referencing to the original source is furthermore detrimental as it is almost impossible to track bug fixes over different software packages. We should therefore also aim at *more visible links between our software packages* and more scientific and technical exchanges among their developers. Furthermore, we should aim at more (and the right) *unit tests* as the simple test described in Sec. 4.3 could have detected the bug earlier. However, also testing cannot detect all bugs: the unit tests in the MOEA Framework for example fully cover the package containing the bug. Since we have to live with the fact that our software will naturally contain bugs, it is therefore even more important that we provide easy ways to report them via *bugtrackers* and to always *mention version numbers* of the algorithms we use in our papers. Addressing the mentioned challenges in the future when it comes to algorithm implementations and distributions will hopefully allow the (multiobjective) optimization community to appear even stronger and more trustworthy to the outside.

Acknowledgments. This work was supported by the grant ANR-12-MONU-0009 (NumBBO) of the French National Research Agency. The author would also like to acknowledge the JSPS funded project "Global Research on the Framework of Evolutionary Solution Search to Accelerate Innovation". Special thanks go to Yann Semet and his colleagues from Thales who originally pointed out the bug in the Paradiseo implementation of IBEA. Many thanks also go to Beat Futterknecht, Benny Gächter, David Hadka, Arnaud Liefooghe, Antonio Nebro, and Lothar Thiele for their help with the software packages mentioned in the paper and to Anne Auger, Nikolaus Hansen, Arnaud Liefooghe and the anonymous reviewers for their fruitful comments on the paper.

References

1. Bader, J.: Hypervolume-Based Search For Multiobjective Optimization: Theory and Methods. Ph.D. thesis, ETH Zurich (2010)
2. Basseur, M., Burke, E.K.: Indicator-based multi-objective local search. In: Congress on Evolutionary Computation (CEC 2007), pp. 3100–3107 (2007)

3. Bleuler, S., Laumanns, M., Thiele, L., Zitzler, E.: PISA – a platform and programming language independent interface for search algorithms. In: Fonseca, C.M., Fleming, P.J., Zitzler, E., Deb, K., Thiele, L. (eds.) EMO 2003. LNCS, vol. 2632, pp. 494–508. Springer, Heidelberg (2003)
4. Deb, K., Thiele, L., Laumanns, M., Zitzler, E.: Scalable test problems for evolutionary multi-objective optimization. In: Evolutionary Multiobjective Optimization: Theoretical Advances and Applications, pp. 105–145 (2005)
5. Durillo, J.J., Nebro, A.J.: jMetal: a Java Framework for Multi-Objective Optimization. Advances in Engineering Software **42**, 760–771 (2011)
6. Hadka, D.: MOEA Framework (2014). http://www.moeaframework.org/
7. Liefooghe, A., Basseur, M., Jourdan, L., Talbi, E.-G.: ParadisEO-MOEO: a framework for evolutionary multi-objective optimization. In: Obayashi, S., Deb, K., Poloni, C., Hiroyasu, T., Murata, T. (eds.) EMO 2007. LNCS, vol. 4403, pp. 386–400. Springer, Heidelberg (2007)
8. López-Ibáñez, M., Paquete, L., Stützle, T.: Exploratory analysis of stochastic local search algorithms in biobjective optimization. In: Experimental Methods for the Analysis of Optimization Algorithms, pp. 209–222 (2010)
9. Thiele, L., Chakraborty, S., Gries, M., Künzli, S.: Design space exploration of network processor architectures. In: Network Processor Design 2002: Design Principles and Practices. Morgan Kaufmann (2002)
10. Wagner, T., Beume, N., Naujoks, B.: Pareto-, aggregation-, and indicator-based methods in many-objective optimization. In: Obayashi, S., Deb, K., Poloni, C., Hiroyasu, T., Murata, T. (eds.) EMO 2007. LNCS, vol. 4403, pp. 742–756. Springer, Heidelberg (2007)
11. Zitzler, E., Deb, K., Thiele, L.: Comparison of Multiobjective Evolutionary Algorithms: Empirical Results. Evolutionary Computation **8**(2), 173–195 (2000)
12. Zitzler, E., Künzli, S.: Indicator-based selection in multiobjective search. In: Yao, X., et al. (eds.) PPSN VIII. LNCS, vol. 3242, pp. 832–842. Springer, Heidelberg (2004)
13. Zitzler, E., Thiele, L.: Multiobjective Evolutionary Algorithms: A Comparative Case Study and the Strength Pareto Approach. IEEE Transactions on Evolutionary Computation **3**(4), 257–271 (1999)

A Knee-Based EMO Algorithm with an Efficient Method to Update Mobile Reference Points

Yu Setoguchi[1]($\boxtimes$), Kaname Narukawa[2], and Hisao Ishibuchi[1]

[1] Department of Computer Science and Intelligent Systems,
Graduate School of Engineering, Osaka Prefecture University, Sakai, Osaka, Japan
yu.setoguchi@ci.cs.osakafu-u.ac.jp, hisaoi@cs.osakafu-u.ac.jp
[2] Honda Research Institute Europe GmbH, 63073 Offenbach am Main, Germany
kaname.narukawa@honda-ri.de

Abstract. A number of evolutionary multi-objective optimization (EMO) algorithms have been proposed to search for non-dominated solutions around reference points that are usually assumed to be given by a decision maker (DM) based on his/her preference. However, setting the reference point needs a priori knowledge that the DM sometimes does not have. In order to obtain favorable solutions without a priori knowledge, "knee points" can be used. Some algorithms have already been proposed to obtain solutions around the knee points. TKR-NSGA-II is one of them. In this algorithm, the DM is supposed to specify the number of knee points as a parameter whereas such information is usually unknown. In this paper, we propose an EMO algorithm that does not require the DM to specify the number of knee points in advance. We demonstrate that the proposed method can efficiently find solutions around knee points.

Keywords: Knee point · Preference · Reference point · Decision maker

1 Introduction

Most of evolutionary multi-objective optimization (EMO) algorithms have been designed to approximate the entire Pareto front. However, if a decision maker (DM) faces a large number of solutions that approximate the entire Pareto front, it would be a difficult task for the DM to choose only a single solution among them. Some EMO algorithms [6,8] have been proposed to obtain solutions around preference regions that the DM specifies according to his/her preference. Those algorithms do not try to find a large number of solutions. As a result, the DM can decrease a burden to choose a final single solution. However, setting preference regions needs a priori knowledge that the DM sometimes does not have.

As a substitute for the preference information, "knee points" have attracted much attention [1–5,9]. If the Pareto front has a clear knee point, most DMs may prefer solutions around the knee point. This is because a small improvement of any objective from the knee point leads to a large deterioration of at least one

© Springer International Publishing Switzerland 2015
A. Gaspar-Cunha et al. (Eds.): EMO 2015, Part I, LNCS 9018, pp. 202–217, 2015.
DOI: 10.1007/978-3-319-15934-8_14

of the other objectives. Until now, some methods to identify knee points have been proposed [4,5,7,9]. For example, Das [4] proposed the normal boundary intersection method. As shown in Fig. 1, the distance is calculated between the Pareto optimal solutions and the line connecting the extreme points. In minimization problems, the knee point has the largest distance in a convex part.

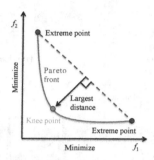

Fig. 1. Normal boundary intersection method

Trade-off information is also used to identify knee points [9]. The trade-off information can be calculated as follows:

$$T(x_i, x_j) = \frac{\sum_{k=1}^n \max\left[0, \frac{f_k(x_j)-f_k(x_i)}{f_k^{\max}-f_k^{\min}}\right]}{\sum_{k=1}^n \max\left[0, \frac{f_k(x_i)-f_k(x_j)}{f_k^{\max}-f_k^{\min}}\right]}, \tag{1}$$

where n is the number of objectives, $f_k(x_i)$ indicates the k-th objective value of the solution x_i, $f_k^{\max}$ and $f_k^{\min}$ correspond to the maximum and minimum values of the k-th objective in the population, respectively. Rachmawati and Srinivasan [9] used (2) to calculate a metric in terms of the trade-off for solutions:

$$\mu(x_i, S) = \min_{x_j \in S, x_i \nprec x_j, x_j \nprec x_i} T(x_i, x_j). \tag{2}$$

We note that x_i and x_j are non-dominated with each other and S indicates the set of solutions x_j in the population. $\mu(x_i, S)$ denotes the least amount of improvement per unit deterioration by substituting any solution x_j from S with x_i. The knee point achieves the largest value of the trade-off metric.

In Figs. 2(a) and 3(a), the red curve shows the Pareto front with four convex regions. It should be noted that we assume a minimization problem. The dashed line in Fig. 2(a) connects the extreme points. We also show the distance from the dashed line to the Pareto front and the trade-off metric calculated by (2) in Figs. 2(b) and 3(b), respectively. The largest value of each metric appears in the regions corresponding to the convex parts.

Recently, Bechikh et al. have proposed algorithms [1,2] that are designed to obtain solutions around knee points. In their algorithms, the DM specifies the number of knee points that he/she is searching for, whereas most DMs do not have information about the number of knee points especially in the case of real

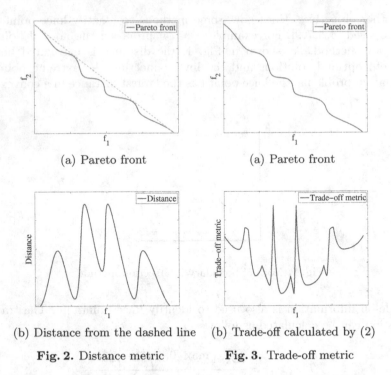

(a) Pareto front   (a) Pareto front

(b) Distance from the dashed line   (b) Trade-off calculated by (2)

Fig. 2. Distance metric   **Fig. 3.** Trade-off metric

world problems. As a result, the DM should examine several specifications of the number of knee points. In order to avoid such a situation, we propose an EMO algorithm that does not require the DM to specify the number of knee points. To demonstrate the performance of our proposed method, we apply it to two- and three-objective benchmark problems and a two-objective real world problem.

The rest of this paper is organized as follows. First, we explain a related study in Section 2. Then, we describe our proposed method in Section 3. Section 4 shows experimental results on benchmark problems and a real world problem, and finally we provide conclusions in Section 5. It should be noted that we consider minimization problems throughout this paper.

2   TKR-NSGA-II

Bechikh et al. have proposed TKR-NSGA-II [2], which is designed to obtain solutions around knee points. This algorithm uses the Reference point NSGA-II (R-NSGA-II) procedure [6] as the basic structure of the algorithm. R-NSGA-II is designed to obtain solutions around reference points. In R-NSGA-II, the DM specifies reference points in advance, while in TKR-NSGA-II, an algorithm updates reference points every generation. This algorithm is called Trade-off-based Mobile Reference Points Updating Strategy (T-MRPUS), which specifies knee-like points as reference points. This algorithm is shown in Algorithm 1.

T-MRPUS uses two parameters, namely N_k and ξ. N_k denotes the number of knee points that are searched for by the DM. ξ denotes the minimum distance between reference points. In this algorithm, extreme solutions and N_k knee-like points are specified as reference points by use of the parameter ξ and the trade-off metric, which is explained in Section 1. We explain the algorithm according to each step in Algorithm 1. First, extreme solutions on the first non-dominated front (FF) are specified as reference points. Then, the trade-off metric is calculated for each solution on the FF and the solutions are sorted in descending order of the trade-off metric. After that, we examine each solution in the sorted order. A solution is chosen as a reference point if it satisfies the following condition: its shortest distance to other reference points is larger than ξ in the normalized objective space. When N_k solutions are selected, the algorithm stops choosing solutions. When the final number of chosen solutions is smaller than N_k, T-MRPUS permits to choose a solution even if its shortest distance to other reference points is not larger than ξ in the normalized objective space.

Algorithm 1 . T-MRPUS

Input:
   FF: the first non-dominated front, N_k: the number of knees,
   n: the number of objectives, ξ: the minimum distance between reference points
Output:
   MRP: the set of mobile reference points

1: ES ← extreme_solutions(FF, n);
2: MRP ← ES;
3: **for** $i = 1$ **to** size(FF) **do**
4:     FF(i).trade-off ← evaluate_trade-off_metric(FF(i), FF);
5: **end for**
6: sorted_FF ← **sort**(FF, descend);
7: $j \leftarrow 1$, $k \leftarrow 1$;
8: **while** ($k <= N_k$) **and** ($j <=$ size(FF)) **do**
9:     **if** (**not** (is_ξ_duplicate(sorted_FF(j), MRP))) **then**
10:         **append** sorted_FF(i) **to** MRP;
11:         $k \leftarrow k + 1$;
12:     **end if**
13:     $j \leftarrow j + 1$;
14: **end while**
15: **if** (size(MRP) $< N_k + n$) **then**
16:     **for** $i = $ (size(MRP)+1) **to** $N_k + n$ **do**
17:         **append** sorted_FF(i) **to** MRP;
18:     **end for**
19: **end if**

The DM occasionally consumes a lot of time for specifying the parameter N_k because its appropriate value should be found by trial and error. We demonstrate the difficulty of the parameter setting. As an example, we use the DEB2DK problem [3]. This problem is a two-objective minimization problem and the number of knee points can be arbitrarily specified using a parameter K. In this example, we specify the number of knee points as four (A, B, C, and D in Fig. 4). Since we assume a situation where the number of knee points is unknown, first we specify $N_k = 2$ and $\xi = 0.1$. The obtained solutions are shown in Fig. 4(a). From Fig. 4(a), we can see that the solutions are distributed around the two extreme points and the two knee points B and C. From these solutions, the DM would wonder there are other knee points. Therefore we change the value of N_k from 2 to 10. The obtained solutions are shown in Fig. 4(b). In Fig. 4(b), the solutions are distributed not only around the two extreme points and the four knee points, but also around other regions such as concave parts of the Pareto front. From these solutions, the DM would assume that $N_k = 10$ is larger than the actual number of knee points. Therefore, the next step would be decreasing N_k. Like this, the DM should adjust the parameter N_k until he/she is satisfied with an obtained result. Moreover if the number of objectives is more than three, it is difficult to visualize solutions in the objective space. As a result, it is hard for the DM to find a proper value of N_k.

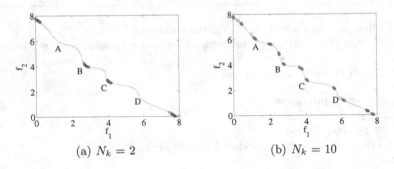

(a) $N_k = 2$ (b) $N_k = 10$

Fig. 4. Obtained solutions by TKR-NSGA-II on DEB2DK problem

3   Proposed Method

In this section, we explain our proposed method. There are two main differences between our proposed method and TKR-NSGA-II [2]. One is the basic structure of the algorithm, which corresponds to R-NSGA-II in TKR-NSGA-II. The other is a reference point selection strategy, which corresponds to T-MRPUS in TKR-NSGA-II. We explain the basic structure of the algorithm and the reference point selection strategy in our proposed method.

3.1   Basic Structure

TKR-NSGA-II uses R-NSGA-II as the basic structure of the algorithm, while our proposed method uses the Preference-based NSGA-II (P-NSGA-II) that has recently been proposed by Narukawa et al. [8]. P-NSGA-II uses a hyperplane and Gaussian functions to reflect the DM's preference. In this paper, we modify the selection of next-generation individuals in P-NSGA-II so as to realize a distribution that follows Gaussian functions. The modified P-NSGA-II is explained below.

Preference Configuration. In configuring a preference, the DM specifies the center and spread values of Gaussian functions that are lying on the hyperplane. Firstly, we set the hyperplane passing through $(1, 0, ..., 0)$, $(0, 1, ..., 0)$, ..., $(0, 0, ..., 1)$ for an n-objective minimization problem. In Fig. 5, the hyperplane for a two-objective minimization problem is depicted. Next, k n-dimensional Gaussian functions $p_i(\mathbf{v})$ are used to represent the preference of the DM with respect to a vector $\mathbf{v} = (v_1, v_2, ..., v_n)$ on the hyperplane (i.e., $v_1 + v_2 + ... + v_n = 1$) as follows:

$$p(\mathbf{v}) = \max_{i=1}^{k} p_i(\mathbf{v}), \tag{3}$$

$$p_i(\mathbf{v}) = \exp\{-\sum_{l=1}^{n} \frac{(v_l - w_l^i)^2}{(s_l^i)^2}\}, \ i = 1, 2, ..., k, \tag{4}$$

where k is the number of the preference regions, n is the number of objectives, $w_l^i \geq 0$ is the center (or the mean) and $s_l^i > 0$ is the spread (or the standard deviation) of the l-th element of the n-dimensional Gaussian function $p_i(\mathbf{v})$. The larger $p(\mathbf{v})$ is, the more $\mathbf{v}$ is preferred. The center vector $\mathbf{w}^i = (w_1^i, w_2^i, ..., w_n^i)$ is specified under the following condition:

$$w_1^i + w_2^i + ... + w_n^i = 1, w_l^i \geq 0 \tag{5}$$

For example, Fig. 5 depicts the preference of the DM for a two-objective optimization problem where a two-dimensional Gaussian function is specified by $\mathbf{w} = (w_1, w_2) = (0.5, 0.5)$ and $\mathbf{s} = (s_1, s_2) = (1.0, 1.0)$. As we can see in the figure, $\mathbf{v}_A$ is more preferred than $\mathbf{v}_B$ since $p(\mathbf{v}_A) > p(\mathbf{v}_B)$ holds in Fig. 5.

Preference Calculation for Solutions. In order to calculate how much the DM prefers a solution with an objective vector $\mathbf{f} = (f_1, f_2, ..., f_n)$, $\mathbf{f}$ is first shifted to $\mathbf{f}' = (f_1', f_2', ..., f_n')$, where

$$f_l' = f_l - f_l^{\min}, \ l = 1, 2, ..., n, \tag{6}$$

and $f_l^{\min}$ is the minimum value found so far for the l-th objective. Next, $\mathbf{f}'$ is normalized to $\mathbf{f}'' = (f_1'', f_2'', ..., f_n'')$ as follows:

$$f_l'' = \frac{f_l'}{f_l'^{\text{ext}}}, \ l = 1, 2, ..., n, \tag{7}$$

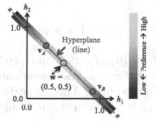

Fig. 5. Preference on a hyperplane (line) in two-objective optimization

where f'^{ext}_l is the l-th objective value of an extreme vector $\mathbf{f}'^{\text{ext}}_l$ for the l-th objective. The extreme vector $\mathbf{f}'^{\text{ext}}_l$ is identified by finding a vector that minimizes the following achievement scalarizing function ASF with a weight vector $\mathbf{d}^l = (d^l_1, d^l_2, ..., d^l_n)$ indicating the l-th objective direction (e.g., $\mathbf{d}^1 = (d^1_1, d^1_2, ..., d^1_n) = (1, 0, ..., 0)$):

$$ASF(\mathbf{f}', \mathbf{d}^l) = \max_{i=1}^{n} f'_i / d^l_i, \tag{8}$$

where $d^l_i = 0$ is replaced with $d^l_i = 10^{-6}$ for avoiding a division by zero. Then, the hyperplane passing through n extreme vectors is generated and $\mathbf{f}''$ is mapped to $\mathbf{v} = (v_1, v_2, ..., v_n)$ on the hyperplane as follows:

$$v_i = \frac{f''_i}{f''_1 + f''_2 + ... + f''_n}, \quad i = 1, 2, ..., n. \tag{9}$$

It should be noted that $v_1 + v_2 + ... + v_n = 1$ holds because $\mathbf{v}$ is on the hyperplane. These coordinate transformations are depicted in Fig. 6. Now the preference of the DM on the solution having the objective vector $\mathbf{f}$ is calculated as $p(\mathbf{v})$ in (3)-(4).

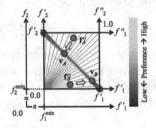

Fig. 6. Coordinate transformations of solutions

Selection Method. In the selection, first we choose extreme solutions that are needed for constructing a widespread hyperplane. Then, we pick up solutions according to their ranks in the same manner as in NSGA-II. Let us assume S_i is the set of solutions having rank i. Next, we consider the selection among S_L. The value of L is the minimum integer that makes $|S_1 \cup S_2 \cup ... \cup S_L| \geq \mu$ hold, where μ is the parent population size. In the selection among S_L, we use an allowable radius and a shortest distance.

The allowable radius is defined for a solution having an objective vector $\mathbf{f} = (f_1, f_2, ..., f_n)$ as follows:

$$r(\mathbf{v}) = 1 - p(\mathbf{v}), \tag{10}$$

where $\mathbf{v}$ is obtained from $\mathbf{f}$ by (6)-(7), (9), and $p(\mathbf{v})$ is calculated by (3)-(4). The shortest distance $d(\mathbf{v})$ is defined as the Euclidean distance of a solution to its nearest solution on the hyperplane. The allowable radius and the shortest distance are illustrated in Fig. 7, where a dashed line represents the preference function $p(\mathbf{v})$ in (3).

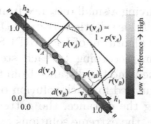

Fig. 7. Allowable radius and shortest distance for solutions $\mathbf{v}_A$ and $\mathbf{v}_B$

In the selection among S_L, first we randomly choose a solution from S_L and calculate its allowable radius and shortest distance. If $r(\mathbf{v}) > d(\mathbf{v})$ holds, the solution is removed. Otherwise, another solution is randomly selected prohibiting the same solution(s) to be chosen, and then applied to the same operation. If any solutions are not removed, the solution with the minimum value of $d(\mathbf{v})$ is removed. We iterate this procedure until $|S_1 \cup S_2 \cup ... \cup S_L| = \mu$ holds. By removing solutions whose shortest distances are smaller than the allowable radius, it is expected that the remaining solutions follow the distribution of $p(\mathbf{v})$ in (3).

We explained the procedure of P-NSGA-II. Next, we apply R-NSGA-II and P-NSGA-II to DTLZ2 problem with two objectives and compare the obtained solutions. In R-NSGA-II, a reference point is specified as $(0.7, 0.7)$. In P-NSGA-II, $\mathbf{w}$ and $\mathbf{s}$ are specified as $(0.5, 0.5)$ and $(0.5, 0.5)$, respectively. In both cases, the preference region is specified as the center of the Pareto front.In the case of R-NSGA-II, solutions are uniformly distributed around a reference point like Fig. 8(a). In the case of P-NSGA-II, the distribution of obtained solutions is similar to a Gaussian function as shown in Fig. 8(b), where they are densely distributed around the center of a preference. This distribution is useful when we search for solutions around knee points. This is because most DMs may prefer solutions that are dense around the knee point.

3.2   Reference Point Selection Strategy

For the selection of reference points, TKR-NSGA-II uses a selection strategy called T-MRPUS, which picks up solutions having a large trade-off metric as explained in Section 2, whereas our proposed method uses a selection strategy

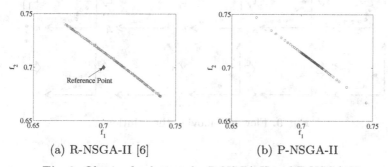

(a) R-NSGA-II [6]   (b) P-NSGA-II

Fig. 8. Obtained solutions by R-NSGA-II and P-NSGA-II

which eliminates solutions having a small distance from the hyperplane connecting extreme solutions. We call this selection strategy Distance-based Elimination MRPUS (DE-MRPUS). The algorithm of DE-MRPUS is shown in Algorithm 2.

We explain the algorithm according to each step in Algorithm 2. First, extreme solutions are identified and the distance from the hyperplane connecting the extreme solutions is calculated for each solution on the first non-dominated front (FF). After calculating the distance, the extreme solutions are removed from the FF. This is because, the extreme solutions are necessary just for calculating the distance in this algorithm. Next, the remaining solutions are sorted in ascending order of the distance. We examine each solution in the sorted order. A solution is removed if it satisfies the following condition: its shortest distance to other remaining FF solutions is not larger than ξ in the normalized objective space. ξ is a parameter which denotes the minimum distance between reference points. In Fig. 9, we show a flow of removing the solutions. Consequently, we may obtain knee-like points in each convex part as shown in Fig. 9.

After examining all solutions, we discriminate the remaining solutions based on distances that are calculated in the first step of this algorithm. For normalization, the distances of the remaining solutions are divided by the largest distance of the remaining solutions. If they have a larger normalized distance than a threshold T, they are specified as reference points. In this algorithm, the DM can control a distribution of solutions by specifying the threshold T. For example, when we use a large value as the threshold T, the DM only obtains solutions around knee points having a large distance from the hyperplane connecting extreme solutions. This result would be nice for the DM who wants to obtain solutions around knee points that have the largest distances from the hyperplane connecting extreme solutions.

In this algorithm, we can use other metrics such as the trade-off metric to identify knee-like points. In computational experiments, we examine the use of the trade-off metric in Fig. 3 as well as the distance metric in Fig. 2. When we use the trade-off metric, we call it Trade-off-based Elimination MRPUS (TE-MRPUS). Compositions of each algorithm are shown in Table 1.

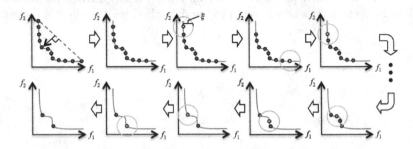

Fig. 9. Flow of removing the solutions

Algorithm 2 . DE-MRPUS

Input:
   FF: the first non-dominated front, T: the threshold to discriminate reference points, n: the number of objectives, ξ: the minimum distance between reference points
Output:
   MRP: the set of mobile reference points

 1: ES $\leftarrow$ extreme_solutions(FF, n);
 2: **for** $i = 1$ **to** size(FF) **do**
 3:     FF(i).distance $\leftarrow$ normal_boundary_intersection_method(FF(i), ES);
 4: **end for**
 5: **remove** ES **from** FF;
 6: sorted_FF $\leftarrow$ **sort**(FF, **ascend**);
 7: $j \leftarrow 1$;
 8: **while** ($j <=$ size(FF)) **do**
 9:     **if** (is_ξ_duplicate(sorted_FF(j), sorted_FF)) **then**
10:        **remove** sorted_FF(j) **from** sorted_FF;
11:     **else**
12:        $j \leftarrow j + 1$;
13:     **end if**
14: **end while**
15: **for** $i = 1$ **to** size(sorted_FF) **do**
16:     sorted_FF(i).normalized_distance $\leftarrow$ normalize(i, sorted_FF);
17:     **if** ($T <$ sorted_FF(i).normalized_distance) **then**
18:        **append** sorted_FF(i) **to** MRP;
19:     **end if**
20: **end for**

Table 1. Compositions of each algorithm

Algorithm	Basic structure	Reference point selection	Metric
TKR-NSGA-II [2]	R-NSGA-II [6]	T-MRPUS [2]	Trade-off [9]
Proposed method	P-NSGA-II	DE-MRPUS	Distance [4]
	P-NSGA-II	TE-MRPUS	Trade-off [9]

4   Computational Experiment

In this section, we apply our proposed method to two- and three-objective bench-mark problems and a two-objective real world problem. In order to examine the performance, we also apply TKR-NSGA-II [2] to those problems and compare the results. In all experiments, the population size is specified as 100. We use the SBX with a distribution index of 15 and the polynomial mutation with a distribution index of 20. The crossover probability is set to 0.5 and the mutation probability is set to $1/m$,where m is the number of decision variables. In this study, we use parameter values in Table 2 unless otherwise specified.

Table 2. Specification of parameters

Algorithm	Parameter values
TKR-NSGA-II [2]	$\xi = 0.1$, $\epsilon = 0.001$, N_k = actual number of knee points
Proposed method	$\xi = 0.1$, $\mathbf{s} = (0.5, ..., 0.5)$, $T = 0$

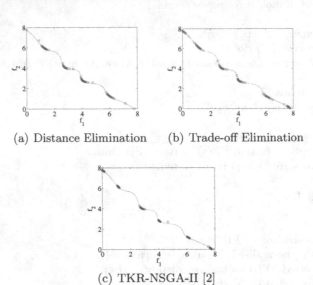

(a) Distance Elimination   (b) Trade-off Elimination

(c) TKR-NSGA-II [2]

Fig. 10. Obtained solutions on DEB2DK

4.1 Benchmark Problem

Branke et al. [3] have proposed three knee-based test problems, namely DO2DK, DEB2DK and DEB3DK problems to evaluate the performance of knee-based EMO algorithms. The DO2DK and the DEB2DK are two-objective problems and the DEB3DK is a three-objective problem. In all problems, the number of knee points can be arbitrarily specified using a parameter K: the number of knee points is specified as K^{n-1}, where n is the number of objectives. Rachmawati and Srinivasan [9] modified those problems and proposed DO2DK-1, DEB2DK-1 and DEB3DK-1 problems that require an ability to converge towards the Pareto front. In this paper, we only show the results on DEB2DK, DEB2DK-1, and DEB3DK problems due to the paper length limitation.

First, we consider the DEB2DK where the number of knee points is specified as four. The stopping criterion is specified as 200 generations. We show solutions obtained by Distance Elimination and Trade-off Elimination in Figs. 10(a) and 10(b), respectively. From Figs. 10(a) and 10(b), we can see that both approaches obtain solutions around knee points. Moreover we show solutions obtained by TKR-NSGA-II in Fig. 10(c). We can see that the distribution of solutions is similar to those obtained by Distance Elimination and Trade-off

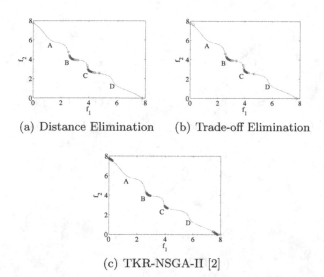

(a) Distance Elimination   (b) Trade-off Elimination

(c) TKR-NSGA-II [2]

Fig. 11. Change of distribution by T and N_k ($T = 0.8$ in (a) and (b), $N_k = 2$ in (c))

Elimination. However it should be noted that TKR-NSGA-II needs a trial-and-error adjustment for the parameter N_k.

Next, we change a value of the threshold T from 0 to 0.8 in our proposed method. When the threshold T is set to a large value, we only obtain solutions around knee points that have a large distance metric or trade-off metric as shown in Figs. 11(a) and 11(b). We also show solutions obtained by TKR-NSGA-II, where N_k is specified as 2 in Fig. 11(c). From Fig. 11(c), we can see that TKR-NSGA-II also obtains solutions around knee points that have a large trade-off metric. However for obtaining such a result, the DM should specify the number of knee points having a large trade-off metric, whereas the DM usually does not have such information. Moreover, we show solutions around the knee point B in Fig. 12. We can see that the distribution of obtained solutions is similar to a Gaussian function in our proposed method.

We also check an effect of the spread vector that specifies the Gaussian function. We specify the spread vector as $\mathbf{s} = (0.3, ..., 0.3)$, $\mathbf{s} = (0.5, ..., 0.5)$ and $\mathbf{s} = (0.7, ..., 0.7)$ and show the obtained solutions in Figs. 13(a), 13(b) and 13(c), respectively. In all experiments, Distance Elimination is used. We can see that when $\mathbf{s}$ is large, solutions are widely distributed around knee points, while when $\mathbf{s}$ is small, solutions are narrowly distributed around knee points.

Then, we solve the DEB2DK-1 where the number of knee points is specified as four. We specify the stopping criterion as 1000 generations. We show the obtained solutions in Fig. 14. From Fig. 14, we can see that solutions are converging towards the four knee points in our proposed method and TKR-NSGA-II.

Finally, we solve the DEB3DK where the number of knee points is specified as 1. The stopping criterion is specified as 300 generations. The obtained results

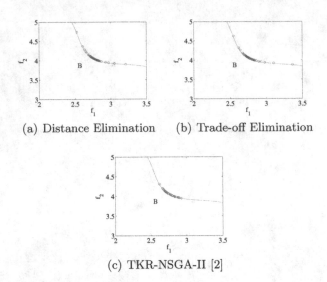

(a) Distance Elimination   (b) Trade-off Elimination

(c) TKR-NSGA-II [2]

Fig. 12. Solutions around the knee point B

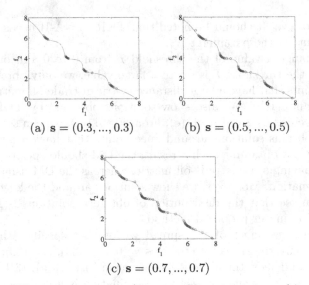

(a) $\mathbf{s} = (0.3, ..., 0.3)$ (b) $\mathbf{s} = (0.5, ..., 0.5)$

(c) $\mathbf{s} = (0.7, ..., 0.7)$

Fig. 13. Effect of $\mathbf{s}$ parameter (Distance Elimination is used in all cases)

are shown in Fig. 15. From Fig. 15, we can confirm that Distance Elimination and TKR-NSGA-II obtain solutions around the knee point existing in the center of the Pareto front. Trade-off Elimination obtains solutions not only around the knee point, but also around many other regions. It is not a good characteristic if the DM only wants to know information around knee points.

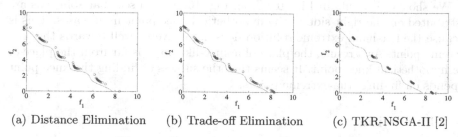

(a) Distance Elimination   (b) Trade-off Elimination   (c) TKR-NSGA-II [2]

Fig. 14. Obtained solutions on DEB2DK-1

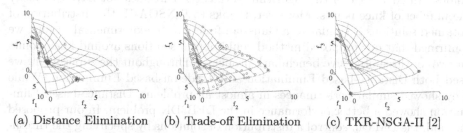

(a) Distance Elimination   (b) Trade-off Elimination   (c) TKR-NSGA-II [2]

Fig. 15. Obtained solutions on DEB3DK

4.2   Real World Problem

As a real world problem, we use a welded beam design problem [6]. This problem has two objectives and one knee point. One of the objectives is minimizing the cost of fabrication and the other is minimizing the end of deflection of the welded beam. The stopping criterion is specified as 500 generations. We specify the threshold T as 0.5 in our proposed method for a strong convergence towards the knee point. Since there is no easy way to calculate the Pareto front of this problem, we compare the result obtained by NSGA-II in which the stopping criterion is specified as 10000 generations. Among the solutions obtained by NSGA-II, we identify the solution having the largest distance and the solution having the largest trade-off metric, respectively. (The same solution is identified in both cases.) We call the solution a hypothetical knee point.

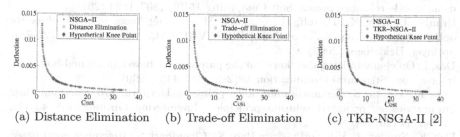

(a) Distance Elimination   (b) Trade-off Elimination   (c) TKR-NSGA-II [2]

Fig. 16. Obtained solutions on welded design problem ($T = 0.5$ in (a) and (b))

We show the results in Fig. 16. From Fig. 16, we can see that solutions are distributed on the right side of the hypothetical knee point in all cases. This is because the left-hand extreme solution does not converge well towards the true extreme point. As a result, the place of a knee-like point is far from the place of the hypothetical knee point. It seems that the success of finding the knee point depends on finding true extreme points.

5   Conclusion

In this paper, we proposed an EMO algorithm that obtains solutions around knee points. An advantage of the algorithm is that the DM does not have to specify the number of knee points. Moreover, thanks to P-NSGA-II, the distribution of obtained solutions is similar to a Gaussian function. In experimental results, we confirmed that our proposed method could obtain solutions around knee points on two- or three-objective benchmark problems. Throughout the experiment, we used both Distance-based Elimination and Trade-off-based Elimination. While they showed similar performances in almost all problems, Distance-based Elimination showed a better performance in the DEB3DK problem. In our proposed method, the DM can control a distribution of solutions by specifying the threshold T. In the case of a high threshold T, we only obtain solutions around knee points with a large distance or trade-off metric. Besides, we confirmed that a spread of solutions is controllable by the spread vector. In the case of a two-objective real world problem, solutions are distributed on the right side of the hypothetical knee point, because the extreme solution does not converge well.

As a future work, it would be interesting to try self-adjusting of the parameter ξ in order to decrease the burden of parameter setting. In this paper we used only two- or three-objective problems. Therefore, it would be also interesting to apply our proposed method to many-objective problems.

References

1. Bechikh, S., Said, L.B., Ghedira, K.: Searching for knee regions in multi-objecitve optimization using mobile reference points. In: Proc. of the 2010 ACM Symposium on Applied Computing, pp. 1118–1125. ACM (2010)
2. Bechikh, S., Said, L.B., Ghedira, K.: Searching for knee regions of the pareto front using mobile reference points. Soft Computing 15(9), 1807–1823 (2011)
3. Branke, J., Deb, K., Dierolf, H., Osswald, M.: Finding knees in multi-objective optimization. In: Yao, X., et al. (eds.) PPSN VIII. LNCS, vol. 3242, pp. 722–731. Springer, Heidelberg (2004)
4. Das, I.: On characterizing the "knee" of the pareto curve based on normal-boundary intersection. Structural Optimization 18(2–3), 107–115 (1999)
5. Deb, K., Gupta, S.: Understanding knee points in bicriteria problems and their implications as preferred solution principles. Engineering Optimization 43(11), 1175–1204 (2011)
6. Deb, K., Sundar, J., Udaya Bhaskara Rao, N., Chaudhuri, S.: Reference point based multi-objective optimization using evolutionary algorithms. International Journal of Computational Intelligence Research 2(3), 273–286 (2006)

7. Mattson, C.A., Mullur, A.A., Messac, A.: Smart pareto filter: Obtaining a minimal representation of multiobjective design space. Engineering Optimization **36**(6), 721–740 (2004)
8. Narukawa, K., Tanigaki, Y., Ishibuchi, H.: Evolutionary many-objective optimization using preference on hyperplane. In: Proc. of 2014 Conference Companion on Genetic and Evolutionary Computation Conference, pp. 91–92. ACM (2014)
9. Rachmawati, L., Srinivasan, D.: Multiobjective evolutionary algorithm with controllable focus on the knees of the pareto front. IEEE Trans. on Evolutionary Computation **13**(4), 810–824 (2009)

A Hybrid Algorithm for Stochastic Multiobjective Programming Problem

Zutong Wang[1](✉), Jiansheng Guo[1], Mingfa Zheng[2],
and Qifang He[3]

[1] College of Materiel Management and Safety Engineering,
Air Force Engineering University, Xi'an 710051, China
bravetom@163.com
[2] College of Science, Air Force Engineering University, Xi'an 710051, China
mingfazheng@126.com
[3] College of Information and Navigation,
Air Force Engineering University, Xi'an 710051, China

Abstract. The traditional approach in the solution of stochastic multiobjective programming problem involves transforming the original problem into a deterministic multiobjective programming problem. However, due to the complexity in practical application problems, the closed form of stochastic multiobjective programming problem is usually hard to obtain, and yet, there is surprisingly little literature that addresses this problem. The principal purpose of this paper is to propose a new hybrid algorithm to solve stochastic multiobjective programming problem efficiently, which is integrated with Latin Hypercube Sampling, Monte Carlo simulation, Support Vector Regression and Artificial Bee Colony algorithm. Several numerical examples are presented to illustrate the validity and performance of the hybrid algorithm. The results suggest that the proposed algorithm is very suitable for solving stochastic multiobjective programming problem.

Keywords: Stochastic programming · Multiobjective programming · Pareto efficient solution · Hybrid algorithm

1 Introduction

Many real-life problems require considering and optimizing multiple and conflicting objectives from the multiobjective optimization point of view, leading us into the area of multiobjective programming (MOP) probelm. The MOP problem in static environment with deterministic parameters has received much research interest [1][2][3]. However, since indeterminacy is inherent in most real cases, where observed phenomena are disturbed by indeterministic perturbations, the application of deterministic MOP methods to real-world problems

This work was supported by National Natural Science Foundation under Grant No. 71171199, and Natural Science Foundation of Shaanxi Province under Grant No. 2013JM1003.

© Springer International Publishing Switzerland 2015
A. Gaspar-Cunha et al. (Eds.): EMO 2015, Part I, LNCS 9018, pp. 218–232, 2015.
DOI: 10.1007/978-3-319-15934-8_15

often faces the difficulty that for a particular problem considered, the parameters involved take unknown or uncommensuarable values at the moment of making the decision. With the great improvement of probability theory, the probability distribution is widely adopted to depict such kind of indeterministic phenomenon in real-life MOP problem, which leads to the research field of stochastic MOP problem. The stochastic MOP modeling is widely used in many real-world decision making problems of management science, engineering, and technology, including distributed energy resources planning [4], network design [5], traveling salesman problem [6], capacitated arc routing problem [7], etc.

The review of these works shows that the stochastic MOP models can lead to very large scale problems, and the solution of such problems always involves introducing several equivalent deterministic models to remove the random ambiguity in original stochastic MOP problem, such as expected value model, minimum variance model, etc. In this paper, we use the valuation criteria of objective functions C in stochastic MOP problem to remove the random ambiguity and obtain the closed form of equivalent deterministic model, where C denotes the criteria of the specific valuation in practical application. Furthermore, we propose the definition of Pareto efficient solution in stochastic MOP problem based on criteria C.

However, due to the complex nature of real life problem, in most cases, the closed form of equivalent deterministic models in stochastic MOP problem is difficult to obtain. Under these circumstances, the methods based on approximation should be applied, such as Sample Average Approximation (SAA-N, where N is the sample size) method [8]. Though the sequence of SAA-N optimal values (N=1,2 ...) can converge almost surely to the true optimal value, it is prohibitively expensive when the problem to be solved is provided with complicated formulations and feasible set, such as NP-hard with many local extremums. In order to improve the computation efficiency, some hybrid algorithms using Monte Carlo simulation, artificial neural network (ANN) and genetic algorithm (GA) have been adopted for solving stochastic programming problems [9] [10]. However, since the traditional hybrid algorithms need to generate large-scale size of decision points through purely random sampling to obtain the desired precision in model approximation, and need long time to obtain the optimal solution in model optimization with GA, the computation cost of these hybrid algorithms is very time-expensive. A new powerful and efficient hybrid algorithm should be designed and applied to the stochastic MOP problem to reduce the computation cost and improve the computation accuracy. For this purpose, a new hybrid algorithm composed of Latin Hypercube Sampling (LHS), Monte Carlo simulation, Support Vector Regression (SVR) and Artificial Bee Colony (ABC) algorithm is built to obtain the Pareto efficient solutions in stochastic MOP problem in this paper.

In the hybrid algorithm presented, it is broken into four phases, that is, sample phase, simulation phase, approximation phase and optimization phase. The problem in sample phase is addressed using Latin hypercube sampling introduced by McKay et al [11], which is a very popular sampling method for use with computationally demanding models. It has been theoretically and experimentally proved that LHS is more precise and robust than traditional random sampling methods [12] [13]. The problem in simulation phase is addressed using Monte Carlo simulation method to calculate the specified valuation meanings of functions on the sample generated in sample phase. The problem in approximation phase is addressed using a new and very promising regression technique developed by Vapnik, Steven Golowich, and Alex Smola [14] in 1996, called support vector regression (SVR). The excellent performances of SVR in approximation have been obtained in [15]. The problem in optimization phase is addressed using ABC algorithm proposed by Karaboga in 2005 [16], which is a meta-heuristic bionic algorithm based on the intelligent foraging behavior of honey bees. It has been validated that its effectiveness and efficiency on algorithm performance are competitive to other optimization algorithms [17][18][19]. Since every phase in the hybrid algorithm is implanted with advanced methods, where using the LHS and Monte Carlo simulation for model data collection, SVR for model approximation, and ABC algorithm for model optimization, it is expected that it can reduce the computation cost and improve the computation accuracy greatly. The comparison result with traditional hybrid algorithm in a numerical example shows that this new algorithm is more precise and efficient.

The paper is organized in the following manner. In Section 2, the mathematical formulation of stochastic MOP problem is introduced, and three equivalent deterministic models are presented, that is, expected value model, minimum variance model, and α-optimistic value model. In Section 3, a new powerful hybrid algorithm is built for solving the stochastic MOP problem more efficiently, and a numerical example with many stochastic local minimums is provided to illustrate the solution of stochastic MOP problem using the hybrid algorithm in Section 4. Finally, a brief summary is given and some open points are stated for future research work in Section 5.

2   Mathematical Formulation

In this section, the mathematical description of a general stochastic MOP problem is presented first, and then three equivalent deterministic models are proposed to remove the random ambiguity in original stochastic MOP problem.

2.1   Description of Stochastic Multiobjective Programming Problem

Let us consider the stochastic MOP problem as follows:

$$\min_{x \in D} f(x, \xi) = (f_1(x, \xi_1), f_2(x, \xi_2), \cdots, f_p(x, \xi_n)) \tag{2.1}$$

where $x \in D$ is a vector of decision variables of the problem; $\xi_1, \xi_2, \cdots, \xi_p$ are random vectors whose components are random variables, $\xi_i = (\xi_{i1}, \xi_{i2}, \cdots, \xi_{in})$, defined on the probability space $(\Omega, \mathcal{F}, \mathrm{Pr})$; and the set of feasible solutions $D \subset R^n$ is crisp, nonempty and compact.

Since the objective function in model (2.1) becomes dependent not only on the solution, but also on a random influence, i.e., it becomes a random variable. There are no methods can compare the random variables directly, it needs to remove the random ambiguity in it before comparison. Most frequently, the practical aim is then to propose several equivalent deterministic models to optimize the specific valuation of the objective functions in model (2.1).

2.2 Equivalent Deterministic Models

Here, the general equivalent deterministic model of original stochastic MOP problem is described as follows:

$$\min_{x \in D} C[f(x, \xi)] = (C[f_1(x, \xi_1)], C[f_2(x, \xi_2)], \cdots, C[f_p(x, \xi_n)]) \qquad (2.2)$$

where C denotes the criteria of the specific valuation of the objective functions in model (2.1).

Different real-life problems call for different criteria of valuation to satisfy its need in practical application, when C denotes the expected value of the objective functions, the model (2.2) is called expected value model of stochastic MOP problem, and can be presented as follows:

$$\min_{x \in D} E[f(x, \xi)] = (E[f_1(x, \xi_1)], E[f_2(x, \xi_2)], \cdots, E[f_p(x, \xi_n)]) \qquad (2.3)$$

When C denotes the variance of the objective functions, the model (2.2) is called minimum variance model of stochastic MOP problem, and can be presented as follows:

$$\min_{x \in D} V[f(x, \xi)] = (V[f_1(x, \xi_1)], V[f_2(x, \xi_2)], \cdots, V[f_p(x, \xi_n)]) \qquad (2.4)$$

When C denotes the $\alpha-$optimistic value of the objective functions, the model (2.2) is called $\alpha-$optimistic value model of stochastic MOP problem, and can be presented as follows:

$$\min_{x \in D} f(x, \xi)_{sup}(\alpha) = (f_1(x, \xi_1)_{sup}(\alpha), f_2(x, \xi_2)_{sup}(\alpha), \cdots, f_p(x, \xi_n)_{sup}(\alpha))$$
$$(2.5)$$

Though the random ambiguity is removed, the objectives are usually in conflict in stochastic MOP problem, there is no optimal solution that simultaneously minimizes all the objective functions. In this case, we have to introduce the concept of Pareto efficient solution in stochastic MOP problem, which means that it is impossible to improve any one of objectives without sacrificing on one or more of the other objectives.

Definition 2.1 *A Pareto efficient solution* x^* *in model* (2.2) *is said to be C Pareto efficient to the stochastic MOP problem* (2.1), *where the feasible solution* x^* *is said to be a Pareto efficient solution of model* (2.2) *if there is no feasible solution* x *such that*

$$C[f_j(x, \xi_j)] \leq C[f_j(x^*, \xi_j)], j = 1, 2, \ldots, p$$

and $C[f_j(x, \xi_j)] < C[f_j(x^*, \xi_j)]$ *for at least one index j.*

3 Hybrid Intelligent Algorithm for Stochastic MOP Problem

To solve stochastic MOP problem with complicated closed form and feasible set, the most direct approach is to nest the iterative loops by performing complete Monte Carlo estimation (inner loop) for each optimization data request (outer loop). However, this can be prohibitively expensive, for this reason, the optimization techniques must be combined with hybrid algorithm. In this section, the Latin Hypercube Sampling (LHS), Monte Carlo simulation, Support Vector Regression (SVR) and Artificial Bee Colony (ABC) algorithm are integrated to design a powerful hybrid algorithm for solving stochastic MOP problem.

3.1 Design of Hybrid Intelligent Algorithm

(1) Generation of Sample Using LHS

The goal of sampling in stochastic MOP problem is to generate a matrix of experiments $X^n = (x_{ij})_{n \times k}$ from the feasible set where n is the number of experiments and k is the number of variables. As an extension of stratified-random procedure, Latin hypercube sampling has a long history and has shown its robustness capabilities in sample generation. The LHS involves sampling ns values from the prescribed distribution of each of k decision variables $X_1, X_2, \ldots, X_k$ in stochastic MOP problem. Unlike simple random sampling, LHS ensures a full coverage of the range of each variable by maximally stratifying each marginal distribution. As the information of design variables in stochastic MOP problem is always hard-available beforehand, it is usually assumed that all design variables follow uniform distribution.

(2) Computation Using Monte Carlo Simulation

Due to the complexity in stochastic MOP problem, it is hard to obtain the closed form of its deterministic objective functions, the Monte Carlo simulation is adopted to calculate the valuation of objective functions on the sample generated by LHS. In this paper, the Monte Carlo simulation is adopted to calculate the expected value, variation and $\alpha-$optimistic value of the objective functions respectively.

(3) Approximation Using ABC-SVR

After obtain the sample in feasible set and the corresponding valuation of objective functions, it needs to build a surrogate model to map the relationships between them, which can be considered as a regression process. Support Vector Regression (SVR) is a new regression method different from traditional/statistical ones, it minimizes the generalized error bound instead of the observed training error, so as to achieve the generalized performance. In this paper, we start our study on the basis of the SVM toolbox—LIBSVM directly [20], rather than discuss about the principle and algorithm of SVR. LIBSVM is a library for SVM; its goal is to let users can easily use SVM as a tool. Since the control parameters of SVR are very sensitive to its performance, a successful parameter selection is very important, especially the parameter γ in kernel function and the parameter c of cost. The artificial bee colony (ABC) algorithm is used to find the optimal control parameters aiming at the best regression accuracy, called ABC-SVR, whose regression performance will be tested in Section 4. The main steps of the ABC-SVR application in the stochastic MOP problem can be summarized as follows:

Step 1: Generate train data and test data based on the sample and its corresponding valuation of objective functions;

Step 2: Normalize the train data and test data to improve the regression ability of SVR;

Step 3: Denote the parameter γ in kernel function and the parameter c of cost as a food position (c, γ), and the regression accuracy as the nectar amount;

Step 4: Adopt ABC algorithm to find the optimal control parameters of SVR (optimal food position) on the normalized data;

Step 5: Use the optimal control parameters to build a SVR model and train it according to the train data;

Step 6: Employ the test data to validate the accuracy of trained SVR model. If the regression accuracy does not meet, return to Step 3 to change the ABC options until the desired accuracy is met.

(4) Optimization Using ABC Algorithm

Created by Karaboga [16], the artificial bee colony algorithm is a new population-based meta-heuristic method motivated by the intelligent foraging behaviors of honeybee swarm. There are three essential components in the basic ABC algorithm, respectively are, food source positions, nectar-amount, and three kinds of foraging bees (employed bees, onlookers, and scouts). Each food source position represents a feasible solution to the optimization problem being considered and the nectar-amount of a food source corresponds to the quality (fitness) of the solution being represented by the food source. Each kind of foraging bee performs one particular operation for generating new candidate food source positions. Employed bees are those bees which are searching the food around the food source in their memory currently; they are responsible for sharing the information about food sources with onlooker bees. Onlooker bees are those bees which are waiting in the hive for the information from the employed bees; they tend to choose good food source with more nectar-amount shared by the employed bees, and then further tap the foods around the selected food source. Scout bees are those bees which are carrying out random searches for discovering new food sources if the employed bees and onlookers cannot find a better neighboring food source. Thus, the ABC algorithm visualizes the employed and onlooker bees as performing the job of local search (exploitation), whereas the onlookers and scouts bees as performing the job of global search (exploration). Specifically, unlike real bee colonies, the ABC algorithm assumes that there is a one-to-one correspondence between the employed bees and the food sources, that is, the number of food sources (solution) is the same as the number of employed bees. The role conversion in the algorithm is activated when the bees cannot find a better food source, the employed bee of an abandoned food source becomes a scout bee, which will becomes an employed bee again after it finds a new food source.

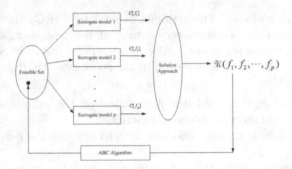

Fig. 1. Main idea of hybrid algorithm for stochastic MOP problem

Straightforwardly, the main idea of the proposed hybrid algorithm for stochastic MOP problem can be illustrated in Fig. 1. Firstly, the valuations of objective functions are obtained through the corresponding surrogate models built using ABC-SVR. Secondly, these valuations are integrated into one objective value using solution approaches which have been proved validated in deterministic MOP, such as linear weighted method, ideal point method, etc. Thirdly, the ABC algorithm finds the optimal solution in feasible set according to the integrated objective value, which is its optimization goal. According to Definition 2.1, the optimal solution ABC algorithm obtained is the Pareto efficient solution in stochastic MOP problem.

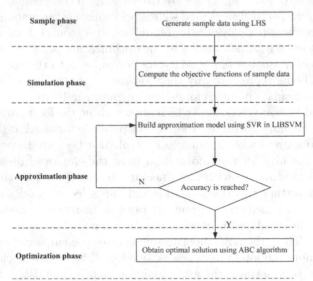

Fig. 2. Framework of hybrid algorithm for stochastic MOP problem

Specifically, the framework of the proposed hybrid algorithm is illustrated in Fig. 2, the main procedure can be summarized as follows:

Step 1: Generate sample data from feasible set as input data using LHS;

Step 2: Calculate corresponding objective values as the output data using Monte Carlo simulation;

Step 3: Build surrogate models for every objective function based on the approximation of input and output data using ABC-SVR;

Step 4: Send the employed bees to the food sources (solution), set it as the input data and determine the nectar amounts f_i (output data) using the SVR models built in Step 3 and the solution approaches such as linear weighted method, ideal point method, etc.;

Step 5: Calculate the fitness values of each solution fit_i and its corresponding probability values as follows:

$$fit_i = \begin{cases} 1/(1+f_i) & if \ f_i \geq 0 \\ 1 + abs(f_i) & if \ f_i < 0 \end{cases}$$

$$p_i = fit_i / \sum_{i=1}^{SN} fit_i$$

where $i = 1, 2, SN$; SN is the number of food sources;

Step 6: Send the onlooker bees to their food sources according to the probability values;

Step 7: Send the scouts to the search area if a food source could not be improved through "limit" trials, and replace it with a new randomly produced solution if the new solution is better;

Step 8: Memorize the best food source (solution) achieved so far;

Step 9: If a stopping criterion is met, then output the best food source, otherwise, go back to Step 4.

It can be seen that the hybrid algorithm not only applies classic sample method to generate sample data in feasible set, but also adopts advanced regression procedure to build surrogate model, and powerful optimization algorithm for solution improvement.

3.2 Performance Test

To test the performance of the proposed hybrid algorithm, two numerical examples are presented. The first one is a stochastic single objective problem with available closed form of expected value, which is presented as follows.

$$\begin{cases} \min & f(x, \xi) = \xi_1 x_1^2 + \xi_2 x_2^2 + \xi_3 x_3^2 \\ s.t. \\ & -1 \leq x_1, x_2, x_3 \leq 1 \end{cases} \tag{3.1}$$

where ξ_1, ξ_2, ξ_3 are random variables, and subject to uniform distribution $\mathscr{U}(0, 2)$, normal distribution $\mathscr{N}(1, 3)$, and exponential distribution $\mathscr{EXP}(1)$, respectively.

Due to the linearity characteristic of expected value in probability theory, we can deduce that

$$E[f(x, \xi)] = E[\xi_1]x_1^2 + E[\xi_2]x_2^2 + E[\xi_3]x_3^2$$

and the optimal solution is $(0, 0, 0)$, the corresponding optimal objective value is 0.

Numerically, let us solve (3.1) by using the hybrid algorithm proposed in this paper. The sample size in LHS is set as 500, and the number of expected value calculation cycle in Monte Carlo is set as 5000. Parameters set for the ABC algorithm are given in Table 1, and the maximum number of cycles in ABC is taken as 200. The optimal solution obtained is $(-0.0109, 0.0117, -0.0048)$, and the optimal approximated objective value is 2.1357E-4.

Table 1. Control parameters adopted in the ABC algorithm

Control parameters in ABC algorithm	
Colony size	40
Limit	100
Number of onlookers	Half of the colony size
Number of employed bees	Half of the colony size
Number of scouts	1

The convergence of hybrid algorithm in the solution (3.1) are shown in Fig. 3, from which it is easy to know that the value by simulation is almost equal to the true optimal value by performing over 450 sample data.

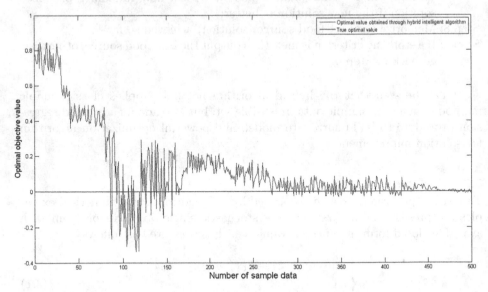

Fig. 3. Convergence to the true optimal value

The second one is a numerical example from [21], which has been solved by the traditional hybrid algorithm

$$\begin{cases} \min & E[\sqrt{(x_1 - \xi_1)^2 + (x_2 - \xi_2)^2 + (x_3 - \xi_3)^2}] \\ \text{s.t.} & \\ & x_1^2 + x_2^2 + x_3^2 \le 10 \end{cases} \quad (3.2)$$

where ξ_1, ξ_2, ξ_3 are random variables, subject to uniform distribution $\mathcal{U}(1,2)$, normal distribution $\mathcal{N}(3,1)$, and exponential distribution $\mathcal{EXP}(4)$, respectively. In the traditional hybrid algorithm, it needs to produce 2000 sample data to train the ANN, and 300 generations in the evolution of GA. While in the new algorithm we proposed, it just needs to produce 500 sample data, and go through 60 cycles. Programs are run independently for each algorithm in MATLAB R2010b (version of 7.11.0.584) on Intel(R) Core(TM) i3-2310M CPU @2.10GHz under Window XP environment. The obtained results are shown in Table 2. It is clear that the proposed algorithm is more precise and efficient than the traditional hybrid algorithm.

Table 2. Performance comparison

	Traditional algorithm	Proposed algorithm
Optimal solution	(1.1035,2.1693,2.0191)	(1.1469,2.3775,1.7375)
Minimum objective	3.56	3.34
Time cost	659 seconds	165 seconds

4   Application on A Theoretical Case

Here, a numerical example is provided to illustrate the proposed new hybrid algorithm. Assume that x_1, x_2 are two decision variables, and ξ_1, ξ_2 are random variables, subject to uniform distribution $\mathcal{U}(1,5)$ and normal distribution $\mathcal{N}(1,4)$, respectively. The problem under consideration is the following bi-objective programming problem involving random variables in the objective functions.

$$\begin{cases} \min_{x_1,x_2} f_1(x,\xi_1,\xi_2) = \xi_1 \sin^2(\xi_2 x_1) + \xi_1 \cos^2(\xi_2 x_2) \\ \min_{x_1,x_2} f_2(x,\xi_1,\xi_2) = \dfrac{(\xi_1 x_1 + \xi_2)^2 + (\xi_2 x_2 + \xi_1)^2}{10} \\ \text{subject to :} \\ \qquad -2 \le x_1, x_2 \le 2 \end{cases} \quad (4.1)$$

To obtain Pareto efficient solution in problem (4.1), the linear weighted method and ideal point method are adopted. Using the linear weighted method, the equivalent deterministic model can be presented as follows:

$$\begin{cases} \min_{x_1,x_2} C[f(x,\xi_1,\xi_2)] = \lambda_1 C[f_1(x,\xi_1,\xi_2)] + \lambda_2 C[f_2(x,\xi_1,\xi_2)] \\ \text{subject to :} \\ \qquad -2 \le x_1, x_2 \le 2 \end{cases} \quad (4.2)$$

where $\lambda_1, \lambda_2 > 0$, and $\lambda_1 + \lambda_2 = 1$.

Using the ideal point method, the equivalent deterministic model can be presented as follows:

$$
\begin{cases}
\min\limits_{x_1, x_2} C[f(x, \xi_1, \xi_2)] = \sqrt{(C[f_1(x, \xi_1, \xi_2)] - f_1^0)^2 + (C[f_2(x, \xi_1, \xi_2)] - f_2^0)^2} \\
\text{subject to :} \\
\qquad -2 \le x_1, x_2 \le 2
\end{cases} \tag{4.3}
$$

where f_1^0 and f_2^0 denote the optimal values of $C[f_1(x, \xi_1, \xi_2)]$ and $C[f_2(x, \xi_1, \xi_2)]$ without considering another objective, respectively.

Since C denotes the general meaning of valuation of random objective functions, three kinds of specific meaning and corresponding deterministic models are considered here, that is, the expected value model, the minimum variance model and the α-optiministic value model.

The expected value model is solved first using the hybrid algorithm proposed in Section 3. Firstly, using LHS to generate sample in the feasible set. As the information of (x_1, x_2) is unknown beforehand, it is assumed that decision variables (x_1, x_2) follow uniform distribution $\mathscr{U}(-2, 2)$. The sample size is set as 500. Then Monte Carlo simulation is adopted to calculate the expected value of $E[f_1(x, \xi_1, \xi_2)]$ and $E[f_2(x, \xi_1, \xi_2)]$ on the 500 sample points generated by LHS. The number of expected value calculation cycle is set as 5000. The first ten sample points and its corresponding objective values are shown in Table 3.

Table 3. The first ten sample points and corresponding objective values

Sample Point	Sample Data	$E[f_1(x, \xi_1, \xi_2)]$	$E[f_2(x, \xi_1, \xi_2)]$
SP1	(0.0012,-1.2514)	1.5118	1.5695
SP2	(-0.2989,-1.2339)	2.4112	1.4701
SP3	(-0.8486,-0.9854)	3.0114	1.6675
SP4	(-1.7186,1.8216)	3.0121	6.3310
SP5	(-1.5538,0.8206)	3.0125	3.9422
SP6	(-0.9030,1.7695)	3.0114	4.4770
SP7	(-0.9130,0.5722)	3.0456	2.3637
SP8	(1.9990,-1.9940)	3.0096	7.6842
SP9	(-1.5418,-0.4542)	3.1798	2.9080
SP10	(-0.2729,0.3027)	2.8948	1.6806

After obtain the sample in feasible set and the corresponding expected values of objective functions, we use ABC-SVR method to build the surrogate model. We take the 1-450 sample data as the train data, and the 451-500 sample data as the test data to validate the surrogate model. The accuracy of output prediction is used to represent the regression performance, and compare the regression performance of ABC-SVR with normal SVR using grid search method and BP artificial neural network. In the ABC-SVR, parameters set for the ABC algorithm are given in Table 1. The mean square error (MSE) and coefficient of determination R^2 are adopted to be regression performance index of these three methods, and the comparison results are shown in Table 4 and 5.

As shown in Table 4 and 5, the regression performance of ABC-SVR is best, and has achieved an ideal precision.

After obtaining the two surrogate models of $E[f_1(x, \xi_1, \xi_2)]$ and $E[f_2(x, \xi_1, \xi_2)]$, the ABC algorithm is adopted to find the optimal solution of expected value model using linear weighted method and ideal point method. In the ABC algorithm, control parameters set for the ABC algorithm are the same as shown in Table 1.

Using linear weighted method, three scenarios are considered here, they are $\lambda_1 = 0.7, \lambda_2 = 0.3$, $\lambda_1 = 0.5, \lambda_2 = 0.5$, $\lambda_1 = 0.3, \lambda_2 = 0.7$, respectively. The results obtained are shown in Table 4.4.

Table 4. Regression performance comparison of three methods (ABC-SVR / SVR / ANN) about $E[f_1(x, \xi_1, \xi_2)]$

Method	Parameters	Regression Performance	
		MSE	R^2
ABC-SVR	bestc=1.5340, bestg=3.5834	0.00012	0.99875
SVR	bestc=0.4682, bestg=0.3356	0.00356	0.98499
ANN	hidden layer number=10	0.01647	0.96522

Table 5. Regression performance comparison of three methods (ABC-SVR / SVR / ANN) about $E[f_2(x, \xi_1, \xi_2)]$

Method	Parameters	Regression Performance	
		MSE	R^2
ABC-SVR	bestc=0.7462,bestg=0.164	0.00065	0.9982
SVR	bestc=2.6284,bestg=0.1864	0.0017	0.9819
ANN	hidden layer number=10	0.0028	0.9712

Table 6. Results obtained in expected value model using linear weighted method

Scenarios	Results	
	$E[f_1(x, \xi_1, \xi_2)]$	1.3993
$\lambda_1 = 0.7, \lambda_2 = 0.3$	$E[f_2(x, \xi_1, \xi_2)]$	2.3562
	Optimal solution	(-0.0157,-2)
	$E[f_1(x, \xi_1, \xi_2)]$	1.5006
$\lambda_1 = 0.5, \lambda_2 = 0.5$	$E[f_2(x, \xi_1, \xi_2)]$	1.9588
	Optimal solution	(-0.0200,-1.7071)
	$E[f_1(x, \xi_1, \xi_2)]$	1.9742
$\lambda_1 = 0.3, \lambda_2 = 0.7$	$E[f_2(x, \xi_1, \xi_2)]$	1.3857
	Optimal solution	(-0.0236,-1.5080)

As shown in the Table 6, the results obtained are different in three scenarios, this is due to that different weights denote different importance of the objectives, higher weight implies higher importance. Therefore, the optimal solutions obtained are different.

Using ABC algorithm and ideal point method, we can obtain the minimum values of $E[f_1(x, \xi_1, \xi_2)]$ and $E[f_2(x, \xi_1, \xi_2)]$ on the feasible set without considering other objectives. They are 1.1196, 1.2716, respectively. Then, we can solve the

Table 7. Results obtained in expected value model using ideal point method

	Results
$E[f_1(x, \xi_1, \xi_2)]$	1.6053
$E[f_2(x, \xi_1, \xi_2)]$	1.7532
Optimal solution	(-0.0236,-1.5080)

Table 8. Results obtained in minimum variance model using linear weighted method

Scenarios		Results
	$V[f_1(x, \xi_1, \xi_2)]$	1.1581
$\lambda_1 = 0.7, \lambda_2 = 0.3$	$V[f_2(x, \xi_1, \xi_2)]$	2.3017
	Optimal solution	(0.1854,0.3297)
	$V[f_1(x, \xi_1, \xi_2)]$	1.1843
$\lambda_1 = 0.5, \lambda_2 = 0.5$	$V[f_2(x, \xi_1, \xi_2)]$	1.1596
	Optimal solution	(-0.1419,0.2198)
	$V[f_1(x, \xi_1, \xi_2)]$	1.2104
$\lambda_1 = 0.3, \lambda_2 = 0.7$	$V[f_2(x, \xi_1, \xi_2)]$	0.7798
	Optimal solution	(-0.1242,-0.1821)

Table 9. Results obtained in minimum variance model using ideal point method

	Results
$V[f_1(x, \xi_1, \xi_2)]$	1.1643
$V[f_2(x, \xi_1, \xi_2)]$	0.8005
Optimal solution	(-0.1668,-0.2527)

Table 10. Results obtained in α−optiministic value model using linear weighted method

Scenarios		Results
	$f_1(x, \xi_1, \xi_2)_{sup}(\alpha)$	0.2470
$\lambda_1 = 0.7, \lambda_2 = 0.3$	$f_2(x, \xi_1, \xi_2)_{sup}(\alpha)$	0.4366
	Optimal solution	(0.0012,-1.2514)
	$f_1(x, \xi_1, \xi_2)_{sup}(\alpha)$	0.7220
$\lambda_1 = 0.5, \lambda_2 = 0.5$	$f_2(x, \xi_1, \xi_2)_{sup}(\alpha)$	0.2138
	Optimal solution	(-0.2534,-1.6625)
	$f_1(x, \xi_1, \xi_2)_{sup}(\alpha)$	1.2862
$\lambda_1 = 0.3, \lambda_2 = 0.7$	$f_2(x, \xi_1, \xi_2)_{sup}(\alpha)$	0.1168
	Optimal solution	(-0.6105,-1.4748)

Table 11. Results obtained in α−optiministic value model using ideal point method

	Results
$f_1(x, \xi_1, \xi_2)_{sup}(\alpha)$	0.3876
$f_2(x, \xi_1, \xi_2)_{sup}(\alpha)$	0.3672
Optimal solution	(-0.0988,-1.2830)

expected value model using ideal point method with the same way in the solution of expected value model using linear weighted method. The results obtained are shown in Table 7.

Following the same procedure in the solution of expected value model illustrated above, the results obtained in the minimum variance model and the $\alpha-$optiministic value model are shown as follows, where α is set as 0.8.

As shown in the Table 6-11, the Pareto efficient solutions obtained in three deterministic models are different from each other, this is due to that the meanings applied to remove the random ambiguity are different. Therefore, in the practical application of stochastic MOP, it needs to specify the meanings of random objective functions first, then remove the random ambiguity in objectives using the meanings specified to obtain the equivalent deterministic MOP problem, and finally generate the Pareto efficient solutions under the meanings specified. Additionally, in the same deterministic MOP model, the solutions obtained using linear weighted method and ideal point method are Pareto efficient, this indicates that the solution approach and the new hybrid algorithm proposed in this paper are valid.

5 Conclusions

The general purpose of this study is to propose a powerful hybrid algorithm to address the difficulty that the closed form of converted deterministic model in practical stochastic MOP problem is usually hard to obtain, which is integrated with Latin Hypercube Sampling (LHS), Monte Carlo simulation, Support Vector Regression (SVR) and Artificial Bee Colony (ABC) algorithm. A numerical example was provided to illustrate the validity of the solution approach and the performance of the hybrid algorithm. Our study shows that, different criteria of the equivalent deterministic MOP model can result in different solutions obtained, and the hybrid algorithm built in this paper is efficient to solve stochastic MOP problem. In our view, to be studied in future, a new solution method is needed for treating stochastic MOP problem on the Pareto front directly.

References

1. Miettinen, K.: Nonlinear Multiobjective Optimization. International series in operations research & management science. Kluwer Academic Publishers (1999)
2. Ozlen, M., Azizoglu, M.: Multi-objective Integer Programming: A General Approach for Generating All Non-dominated Solutions. Euro. J. Oper. Res. **199**(1), 25–35 (2009)
3. Tarek, E.: Method of Centers Algorithm for Multi-objective Programming Problems. Act. Math. Sci. **29**(5), 1128–1142 (2009)
4. Arturo, A.R., Ault, G., Galloway, S.: Multi-objective Planning of Distributed Energy Resources: A Review of The State-of-The-Art. Ren. and Sus. Ene. Rev. **14**(5), 1353–1366 (2010)
5. Chen, A., Kim, J., Lee, S., Kim, Y.: Stochastic Multi-objective Models for Network Design Problem. Exp. Sys. with Appl. **37**(2), 1608–1619 (2010)

6. Luis, P., Thomas, S.: Design and Analysis of Stochastic Local Search for The Multi-objective Traveling Salesman Problem. Com. & Oper. Res. **36**(9), 2619–2631 (2009)
7. Fleury, G., Lacomme, P., Prins, C., Sevaux, M.: A Memetic algorithm for a bi-objective and stochastic CARP. In: Session : Multi Objective Combinatorial Optimization, The 6th Metaheuristics International Conference, MIC 2005, Vienna, Austria, August, 22–26 (2005)
8. Anton, J.K., Alexander, S., Tito, H.M.: The Sample Average Approximation Method for Stochastic Discrete Optimization. SIAM J. Opt. **12**(2), 479–502 (2002)
9. Zhao, R.Q., Liu, B.D.: Stochastic Programming Models for General Redundancy Optimization Problems. IEEE Tran. on Reliab. **52**(2), 181–191 (2003)
10. Zhou, J., Liu, B.D.: New Stochastic Models for Capacitated Location Allocation Problem. Com. & Ind. Eng. **45**(1), 111–125 (2003)
11. McKay, M.D., Beckman, R.J., Conover, W.J.: A Comparison of Three Methods for Selecting Values of Input Variables in The Analysis of Output from A Computer Code. Technometrics **42**(1), 55–61 (2000)
12. Stein, M.: Large Sample Properties of Simulations Using Latin Hypercube Sampling. Technometrics **29**(2), 143–151 (1987)
13. Giunta, A.A., McFarland, J.M., Swiler, L.P., Eldred, M.S.: The Promise and Peril of Uncertainty Quantification Using Response Surface Approximations. Struct. and Infra. Eng. **2**(3–4), 175–189 (2006)
14. Vapnik, V., Steven, E.G., Alex, S.: Support Vector Method for Function Approximation, Regression Estimation, and Signal Processing. Adv. in Neur. Inf. Pro. Sys., 281–287 (1997)
15. Drucker, H., Burges, C., Kaufman, L., Alex, S., Vapnik, V.: Support Vector Regression Machines. Adv. in Neur. Inf. Pro. Sys., 155–161 (1997)
16. Karaboga, D.: An Idea Based on Honey Bee Swarm for Numerical Optimization. Technical report, Erciyes University (2005)
17. Karaboga, D., Akay, B.: A Comparative Study of Artificial Bee Colony Algorithm. Appl. Math. Com. **214**(1), 108–132 (2009)
18. Wang, Z.T., Guo, J.S., Zheng, M.F., Wang, Y.: Uncertain Multiobjective Travelling Salesman Problem. Euro. J. Oper. Res. **241**(2), 478–489 (2014)
19. Guo, J.S., Wang, Z.T., Zheng, M.F., Wang, Y.: Uncertain multiobjective redundancy allocation problem of repairable systems based on artificial bee colony algorithm. Chin. J. Aero. (2014). DOI:http://dx.doi.org/10.1016/j.cja.2014.10.014
20. Chang, C.C., Lin, C.J.: Libsvm: A Library for Support Vector Machines. ACM Tran. on Int. Sys. and Tech. **2**(3), 27 (2011)
21. Liu, B.D., Zhao, R.Q., Wang, G.: Uncertain Programming with Application, fifth edn. Tsinghua University Publication (2012)

Parameter Tuning of MOEAs Using a Bilevel Optimization Approach

Martin Andersson[✉], Sunith Bandaru, Amos Ng, and Anna Syberfeldt

Virtual Systems Research Centre, University of Skövde, Skövde, Sweden
{martin.andersson,sunith.bandaru,amos.ng,anna.syberfeldt}@his.se

Abstract. The performance of an Evolutionary Algorithm (EA) can be greatly influenced by its parameters. The optimal parameter settings are also not necessarily the same across different problems. Finding the optimal set of parameters is therefore a difficult and often time-consuming task. This paper presents results of parameter tuning experiments on the NSGA-II and NSGA-III algorithms using the ZDT test problems. The aim is to gain new insights on the characteristics of the optimal parameter settings and to study if the parameters impose the same effect on both NSGA-II and NSGA-III. The experiments also aim at testing if the rule of thumb that the mutation probability should be set to one divided by the number of decision variables is a good heuristic on the ZDT problems. A comparison of the performance of NSGA-II and NSGA-III on the ZDT problems is also made.

Keywords: Parameter tuning · NSGA-II · NSGA-III · ZDT · Bilevel optimization · Multi-objective problems

1 Introduction

Real-world optimization problems are often formulated with multiple objectives and are therefore preferably solved using multi-objective evolutionary algorithms (MOEAs). Metaheuristics such as EAs involve a set of user-defined parameters that control various aspects of the algorithm. It is well-known [1,10] that these settings can greatly affect the search process and the overall performance of the algorithm. However, setting them for a particular problem is not always intuitive. A strategy that is often used is to choose parameter values that have been shown to be effective on similar problems. Some metaheuristics, such as evolutionary strategies (ES), come with their own heuristics or recommendations for choosing the parameters. Neither method guarantees maximal performance from the algorithm. This paper addresses this issue by using the idea of *optimal* parameters, similar in principle to the one proposed in [9]. The parameter setting problem can itself be viewed as an optimization problem in which the objective is to maximize the performance of the algorithm used on a particular problem. For single-objective problems, this performance indicator could be directly related to the best function value attained by the algorithm. Since this work considers

A. Gaspar-Cunha et al. (Eds.): EMO 2015, Part I, LNCS 9018, pp. 233–247, 2015.
DOI: 10.1007/978-3-319-15934-8_16

multi-objective problems, a commonly used performance indicator is the hyper-volume [12]. Thus, our formulation contains a multi-objective problem nested within a single-objective problem and resembles the following,

$$
\begin{aligned}
&\underset{\mathbf{p}}{\text{Maximize}} \quad HV(\mathbf{p}) \\
&\text{where,} \quad HV(\mathbf{p}) \text{ is the hypervolume of the non-dominated set} \\
&\qquad\quad \text{obtained by solving the following problem with parameters } \mathbf{p} \\
&\qquad\quad \underset{\mathbf{x}}{\text{Minimize}} \quad \{f_1(\mathbf{x}), f_2(\mathbf{x}), \ldots, f_M(\mathbf{x})\} \\
&\qquad\quad \text{Subject to} \quad g_j(\mathbf{x}) \geq 0 \,\forall\, j \in \{1, 2, \ldots, J\} \\
&\qquad\qquad\qquad\quad h_k(\mathbf{x}) = 0 \,\forall\, k \in \{1, 2, \ldots, K\} \\
&\qquad\qquad\qquad\quad \mathbf{x}_l \leq \mathbf{x} \leq \mathbf{x}_u
\end{aligned}
\tag{1}
$$

The algorithmic parameters of the lower-level optimization problem become the variables for the upper-level optimization problem.

Many real-world optimization problems are also designed to be scalable with respect to the variables. For example, consider a production line involving several machines in which their processing times have to be optimized for maximizing the overall throughput and minimizing the work in process of the line. Adding additional operations (machines) to such a line is equivalent to scaling the original optimization problem since the objectives remain the same. In such a situation, it is beneficial to study how the optimal parameter values for the algorithm change with respect to the problem size. Another important aspect is the computational cost. Objective functions in the real world are rarely analytical. In other words, evaluation of the objective functions may involve computationally expensive simulations. Studying the impact of the available computational budget on the optimal parameter values can lead to considerable savings in time and cost.

In order to illustrate the above ideas, two MOEAs, namely NSGA-II [4] and NSGA-III [3], are chosen with the ZDT test suite [13] to experimentally study the effects of problem size and available budget. NSGA-II and the ZDT test problems is a combination that is commonly used to assess the performance of a new metaheuristic. Finding the optimal parameters and the corresponding hypervolume for this combination will also allow a new metaheuristic to be compared against the NSGA-II best-case performance on the ZDT problems. Other test problems are not included in this study because that would reduce the number of experiments performed on each problem. This trade-off will allow for a more in-depth analysis of the ZDT problems.

It is worthwhile to mention here that the goal of this paper is not to find parameter settings that work across a range of problems, but to study how the optimal parameters vary for a given problem with the number of variables and budget size. In order to achieve this, several experiments will be performed on each problem to get multiple sets of optimized parameters. A secondary aim of this paper is to see how NSGA-III compares to NSGA-II in terms of performance and whether they use similar optimal parameter settings. Though NSGA-III was originally designed to handle many (> 3) objective problems, this paper will address how it performs against NSGA-II on the ZDT problems. Testing against problems with three or more objectives would be interesting but is out of scope of this paper and left for future work.

The rest of the paper is organized as follows. Section 2 introduces the parameter setting problem and related work. In Section 3 a description of the experimental design is provided. The experimental results appear in Section 4. The conclusions are summarized in Section 5.

2   Background

The problem of finding the optimal set of parameters for a particular problem can itself be formulated as an optimization problem and solved by an EA. This bilevel optimization approach is called a meta-EA [5]. Though computationally intensive, the approach is highly parallelizeable since replications of the optimizations, at both the upper and lower-level, are independent. A software framework was developed as part of this work that could efficiently distribute and run optimizations in parallel. This software together with a cluster of homogeneous commodity computers enabled the scope of the experiments to be extended well beyond what would have been feasible on a single computer.

One issue that has to be considered when testing different parameters is that they are usually not independent. This means that changing the parameters one by one may lead to to sub-optimal settings. Changing them simultaneously on the other hand will require a large number of experiments to be performed. It is therefore impractical to perform parameter tuning manually, even though there exist different techniques to overcome this problem to some extent. A detailed description and taxonomy of the available techniques can be found in [5].

2.1   Classification and Terminology

It is possible to distinguish three layers in parameter tuning: The application layer, the algorithm (lower) layer and the design (upper) layer [5]. The problem to be solved is located on the application layer and the metaheuristic to solve that problem is on the algorithm layer. On the design layer is the parameter tuner that tests different parameters for the metaheuristic on the algorithm layer. To avoid confusion, the quality of solutions for the problem on the application layer is called fitness while the quality of the parameters is called utility [5]. The classification that was proposed in [6] distinguishes between parameter tuning where the parameters are static and parameter control where they can change during the optimization.

Tuners can be divided into two main categories: iterative and non-iterative [5]. Non-iterative tuners generate all parameters at the start, usually in a systematic fashion. This allows the utility landscape to be modeled from the utility of the evaluated parameters. Iterative tuners, on the other hand, generate the parameters iteratively as the tuner progresses. This makes them more suitable for finding the (near-)optimal parameter vectors, because they can perform a search of the utility landscape.

2.2   Related Work

Using a bilevel optimization approach to do parameter tuning has been done before in the literature. In [1] a Genetic Algorithm (GA) was tuned on single objective sphere problem. The authors found that the GA using the optimized parameters to be significantly better than a GA with "standard" parameters.

Another example is [7] which used a GA to tune the parameters for a GA on a set of numerical test functions. The result were then validated on a image registration task, showing a small but statistical significant advantage to the tuned GA against a "standard" GA.

A more recent example is [9] in which the authors used NSGA-II to tune the parameters of Partical Swarm Optimization (PSO) and Differential Evolution (DE). The algorithms were tuned against both the precision and speed of convergence. It was found that in addition to finding good parameters, the approach could also extract relationships between parameters and the impact of a parameter on the quality criteria.

3   Experimental Design

The meta-EA approach only provides a single optimal parameter set $\mathbf{p}^*$ for each experiment, meaning that it does not provide much insight into the utility landscape. This paper will address this issue by running several different experiments on the same test problem. Two things will be varied for all test problems: the function evaluation budget and the number of decision variables (N).

3.1   Experimental Setup

The experiments involve four aspects that this paper studies, these are listed in Table 1. Each experimental setting is combination of these different aspects. Thus, in total there are 350 (2 MOEAs × 5 test problems × 7 budget sizes × 5 problem sizes) different experimental settings each of which is independently replicated 20 times. The outcome of each replication is the set of parameters with the best hypervolume. Therefore, each experimental setting produced 20 different sets of parameters.

Each experimental setting was a bilevel optimization with a function evaluation budget of 1000, using the parameters shown in the third column of Table 2. The budget was based on manually analyzing a small number of bilevel optimizations and identifying the fact that most of them stopped improving after about 500 evaluations. The MOEA being optimized, at the algorithm layer, was also independently replicated 20 times for each set of parameters being evaluated. The average hypervolume from these optimizations was then used as the utility of that set of parameters.

3.2   Experimental Settings

The following paragraphs explain each row of Table 1.

Table 1. Experimental settings and corresponding choices

Experimental setting	Experimental choices
MOEA	NSGA-II, NSGA-III
Test problem	ZDT1, ZDT2, ZDT3, ZDT4, ZDT6
Function evaluations	100, 500, 2000, 3500, 5000, 6500, 8000
Number of decision variables	2, 10, 20, 30, 40

MOEAs and Test Problems. The two tuned MOEAs, NSGA-II and NSGA-III, on the algorithm layer are both real-value coded. That is why the binary coded test problem ZDT5 was excluded from the experiments. All other ZDT test problems are used in this study. The reference point for the hypervolume calculations for ZDT{1, 2, 3, 6} is (11, 11) and (11, 1000) for ZDT4. The reason for the higher reference point on ZDT4 is that some of the optimizations failed to reach any solution within the (11, 11) reference point.

Both NSGA-II and NSGA-III use the SBX crossover operator and a polynomial mutation.

Function Evaluations. It has been argued that keeping the parameters static during an optimization is not optimal [6]. This would also indicate that it is advantageous to use different parameters for different function evaluation budgets, even though the parameters are static during the run. In order to test this, each experiment will be performed with different budget sizes.

Number of Decision Variables. A number of different rules of thumb have been proposed in the literature. For example, in a binary coded GA, the mutation rate should be proportional to the length of the chromosome [8], $p_m = 1/l$. For a real-value coded GA the length is substituted with the number of decision variables. Previous work has found this rule to be accurate on a single objective sphere problem [1]. This rule will be tested by varying the number of decision variables for each problem.

3.3 Meta-EA Parameters

At the design layer is a real-value coded meta-EA using the SBX crossover and a polynomial mutation. This introduces the problem of choosing a good set of parameters at the design layer as well. To avoid using yet another meta-EA to solve this problem, a full factorial experimental design was performed instead. The values for each parameter is shown in Table 2. To limit the runtime of these experiments only one test problem, ZDT1, was selected as the test problem on the application layer. To further limit the scope only the NSGA-II algorithm was used at the algorithm layer. The function evaluation budget for NSGA-II on the algorithm layer was 1000 and the number of replications were 10. On the design layer the function evaluation budget for the meta-EA was 250 with 10 replications. The parameters with the highest average hypervolume was then chosen as the set of parameters to use at the design layer for the rest of the experiments. The chosen parameters are shown in the third column in Table 2.

Table 2. Full factorial experimental design for meta-EA parameter settings

Meta-EA Parameter	Possible Values	Selected
Population Size	4, 8, 16	8
Mutation Probability	0.2, 0.4, 0.6, 0.8	0.4
Mutation Distribution Index	1, 2, 5, 10, 20, 40	1
Crossover Probability	0.2, 0.4, 0.6, 0.8	0.6
Crossover Distribution Index	1, 2, 5, 10, 20, 40	40

3.4   Parameters

NSGA-II and NSGA-III have very similar parameters, the only difference is that NSGA-III does not directly specify the population size. It is instead based on the number of reference points. The reference points are systematically created by placing them on a normalized hyperplane as described in [2]. To obtain the number of reference points created by this method the following equation is used:
$$H = \binom{M-1+divisions}{divisions}$$

1. Population size for NSGA-II (*pop*): An integer in the range [2, 300]. Upper bound determined by small scale experiments that showed all optimizations used a population size less than 300.
2. Divisions for NSGA-III (*divisions*): The number of divisions along each objective. The population size is set to exactly the number of reference points created by the divisions. An integer in the range [1, 299]. Upper bound set to 299 to get the same population size limits as for NSGA-II.
3. Mutation probability (p_m): The probability of random changes to the decision variables. A real-value in the range [0, 1].
4. Mutation Distribution Index (η_m): Index governing the proximity of the mutated child to its parent. A real-value in the range [0, 300]. Upper bound determined by small scale experiments that showed all optimizations used a η_m less than 300.
5. Crossover probability (p_c): The probability of creating offspring from parents. A real-value in the range [0, 1].
6. Crossover Distribution Index (η_c): Index governing the proximity of the mutated children to the parents. A real-value in the range [0, 300]. Upper bound determined by small scale experiments that showed all optimizations used a η_c less than 300.

The parameters of the optimization on the algorithm layer in Equation (1) become variables for the optimization on the design layer. Thus the variable vector $\mathbf{p}$ in Equation (1) is $\mathbf{p} = \{pop, p_m, \eta_m, p_c, \eta_c\}$ for NSGA-II and $\mathbf{p} = \{divisions, p_m, \eta_m, p_c, \eta_c\}$ for NSGA-III.

3.5 Performance Measure

The hypervolume measure is used to assess the performance of the EA at the algorithm layer. The hypervolume is the volume in objective space formed by a reference point and the Pareto front. The hypervolume is calculated using the technique described in [11], which also discusses the hypervolume measure in more detail. The advantage of the hypervolume measure is that provides single measure for both the convergence and spread of the solutions. The drawback is that it can be computationally expensive and that it can be sensitive to inclusion, or exclusion, of extremal points. Each EA keeps a Pareto archive of unlimited size that is used to calculated the hypervolume at the end of the optimization. So even though no limit was set for the archive size it is of course limited in practice by the available memory and running complexity of the hypervolume calculation. Neither proved to be a problem for the experiments in this study.

4   Experimental Results

This section will present the results from the experiments. Due to the large number of experiments conducted, totally 350, only a subset of all results can fit in this paper. The results for the most common problem size, 30, are shown in Table 3 and Table 4 for NSGA-II and NSGA-III respectively. The values are the median together with the standard deviation.

The experiments were run on a heterogeneous cluster of commodity hardware. In total there were 91 computers and the experiments took approximately 170 hours to complete.

4.1   Population Size

Most of the experiments found that a small population size was most optimal. Many found the smallest possible size, two, to be the best. Having a small population size increases the selection pressure since only a small amount of solutions survive each generation. Thus, allowing the optimization to advance more quickly. However, this comes at the cost of diversity among the solutions, but based on the results, the ZDT problems do not seem to require much diversity among the solutions. One reason the population size can be kept small is the fact that the hypervolume is calculated from the, unlimited, Pareto archive. Using the last generation to calculate the hypervolume would in most cases result in a smaller hypervolume, since fewer solutions would be used in the calculation. An exception to the small population size is experiments with $N = 2$. This is especially true when the budget size is 100. One reason for this might be that

Table 3. Optimal parameter values for NSGA-II with $N = 30$

	Budget	HV	pop	p_m	η_m	p_c	η_c
ZDT1	100	105.46 ± 2.62	2.0 ± 33.56	0.17 ± 0.03	0.04 ± 33.24	0.64 ± 0.24	178.27 ± 101.50
	500	118.42 ± 0.06	2.0 ± 0.0	0.07 ± 0.00	0.07 ± 0.21	0.43 ± 0.11	138.81 ± 88.95
	2000	120.62 ± 0.00	2.0 ± 0.0	0.04 ± 0.00	0.15 ± 0.41	0.49 ± 0.06	1.57 ± 25.64
	3500	120.66 ± 0.00	2.0 ± 0.0	0.03 ± 0.00	0.07 ± 0.55	0.74 ± 0.08	0.33 ± 0.19
	5000	120.66 ± 0.0	2.0 ± 0.0	0.02 ± 0.00	0.12 ± 0.45	0.94 ± 0.04	0.09 ± 0.08
	6500	120.66 ± 2.84	2.0 ± 0.0	0.02 ± 0.00	0.01 ± 0.46	0.99 ± 0.01	0.05 ± 0.07
	8000	120.66 ± 2.84	2.0 ± 0.0	0.01 ± 0.00	0.06 ± 0.35	0.99 ± 0.00	0.03 ± 0.05
ZDT2	100	92.19 ± 2.78	2.0 ± 2.83	0.16 ± 0.03	0.05 ± 65.34	0.83 ± 0.25	131.68 ± 91.13
	500	114.19 ± 4.27	2.0 ± 4.79	0.07 ± 0.01	0.08 ± 64.09	0.54 ± 0.18	149.06 ± 90.18
	2000	120.25 ± 0.00	2.0 ± 0.0	0.05 ± 0.00	0.30 ± 0.42	0.44 ± 0.07	254.41 ± 78.27
	3500	120.32 ± 0.00	2.0 ± 0.0	0.03 ± 0.00	0.07 ± 0.45	0.66 ± 0.07	0.31 ± 0.18
	5000	120.33 ± 0.00	2.0 ± 0.0	0.02 ± 0.00	0.05 ± 0.71	0.93 ± 0.06	0.05 ± 0.07
	6500	120.33 ± 2.84	2.0 ± 0.0	0.02 ± 0.00	0.20 ± 0.84	0.99 ± 0.02	0.01 ± 0.03
	8000	120.33 ± 1.42	2.0 ± 0.0	0.02 ± 0.00	0.07 ± 0.67	0.99 ± 0.00	0.01 ± 0.02
ZDT3	100	110.31 ± 1.58	2.0 ± 1.74	0.17 ± 0.13	0.06 ± 20.89	0.77 ± 0.22	107.82 ± 85.04
	500	125.72 ± 0.20	2.0 ± 0.0	0.07 ± 0.01	0.05 ± 0.34	0.59 ± 0.13	93.57 ± 63.37
	2000	128.69 ± 0.00	2.0 ± 0.0	0.05 ± 0.01	1.48 ± 0.86	0.62 ± 0.09	207.58 ± 127.42
	3500	128.76 ± 0.00	2.0 ± 0.0	0.03 ± 0.00	0.31 ± 0.88	0.85 ± 0.07	0.16 ± 0.11
	5000	128.77 ± 0.00	2.0 ± 0.6	0.02 ± 0.00	0.63 ± 2.90	0.98 ± 0.02	0.03 ± 0.62
	6500	128.77 ± 0.00	3.5 ± 1.92	0.02 ± 0.00	0.05 ± 1.17	0.99 ± 0.00	1.92 ± 2.40
	8000	128.77 ± 0.0	6.0 ± 2.70	0.02 ± 0.01	0.25 ± 39.46	0.99 ± 0.02	3.27 ± 7.72
ZDT4	100	7874.47 ± 40.69	2.0 ± 0.86	0.10 ± 0.03	5.23 ± 97.64	0.98 ± 0.05	28.36 ± 100.76
	500	9534.32 ± 218.76	2.0 ± 4.15	0.05 ± 0.03	3.46 ± 130.49	0.94 ± 0.11	33.89 ± 81.29
	2000	10233.5 ± 230.31	33.5 ± 18.85	0.03 ± 0.01	67.14 ± 119.39	0.99 ± 0.07	41.67 ± 13.91
	3500	10644.55 ± 177.93	29.5 ± 33.41	0.02 ± 0.01	26.21 ± 132.47	0.99 ± 0.14	41.89 ± 33.14
	5000	10898.0 ± 147.19	4.0 ± 45.68	0.02 ± 0.01	7.73 ± 125.00	0.99 ± 0.16	39.21 ± 58.90
	6500	10945.1 ± 99.07	3.0 ± 44.47	0.02 ± 0.01	9.14 ± 95.20	0.94 ± 0.20	26.79 ± 16.83
	8000	10960.95 ± 99.50	4.5 ± 62.07	0.02 ± 0.01	10.35 ± 115.74	0.99 ± 0.14	33.41 ± 15.60
ZDT6	100	43.81 ± 0.34	2.0 ± 0.0	0.18 ± 0.03	0.03 ± 0.13	0.73 ± 0.16	142.08 ± 103.52
	500	70.17 ± 0.41	2.0 ± 0.0	0.07 ± 0.01	0.01 ± 0.22	0.53 ± 0.18	79.68 ± 91.39
	2000	106.75 ± 0.19	2.0 ± 0.0	0.05 ± 0.00	0.02 ± 0.15	0.40 ± 0.07	128.93 ± 77.69
	3500	114.81 ± 0.04	2.0 ± 0.0	0.05 ± 0.00	0.11 ± 0.27	0.32 ± 0.05	223.53 ± 86.08
	5000	116.17 ± 0.01	2.0 ± 0.0	0.05 ± 0.00	0.10 ± 0.41	0.28 ± 0.05	209.40 ± 89.19
	6500	116.37 ± 0.00	2.0 ± 0.0	0.05 ± 0.00	0.08 ± 0.43	0.30 ± 0.05	16.30 ± 42.55
	8000	116.40 ± 0.00	2.0 ± 0.0	0.04 ± 0.00	0.24 ± 0.87	0.38 ± 0.05	3.42 ± 35.33

Table 4. Optimal parameter values for NSGA-III with $N = 30$

	Budget	HV	$divisions$	p_m	η_m	p_c	η_c
ZDT1	100	105.02 ± 3.98	1.0 ± 88.43	0.20 ± 0.14	0.08 ± 48.35	0.83 ± 0.27	151.25 ± 100.56
	500	118.30 ± 0.06	1.0 ± 0.0	0.07 ± 0.01	0.06 ± 0.22	0.88 ± 0.11	95.01 ± 98.37
	2000	120.62 ± 0.00	1.0 ± 0.0	0.04 ± 0.00	0.15 ± 0.45	0.97 ± 0.05	1.60 ± 68.94
	3500	120.66 ± 0.00	1.0 ± 0.0	0.02 ± 0.00	0.20 ± 0.87	0.99 ± 0.00	0.23 ± 0.21
	5000	120.66 ± 2.84	1.0 ± 0.49	0.02 ± 0.00	0.18 ± 0.76	0.99 ± 0.02	0.32 ± 0.34
	6500	120.66 ± 0.0	2.0 ± 0.43	0.02 ± 0.00	0.08 ± 0.34	0.99 ± 0.01	0.27 ± 0.20
	8000	120.66 ± 2.84	2.0 ± 0.0	0.02 ± 0.00	0.46 ± 0.84	0.99 ± 0.00	0.22 ± 0.16
ZDT2	100	91.46 ± 3.43	1.0 ± 64.72	0.17 ± 0.03	0.01 ± 60.35	0.92 ± 0.24	148.35 ± 96.71
	500	114.01 ± 0.14	1.0 ± 0.0	0.06 ± 0.00	0.00 ± 0.09	0.92 ± 0.06	142.13 ± 78.08
	2000	120.24 ± 0.00	1.0 ± 0.0	0.05 ± 0.00	0.11 ± 0.50	0.94 ± 0.09	230.93 ± 78.70
	3500	120.32 ± 0.00	1.0 ± 0.0	0.03 ± 0.00	0.07 ± 0.54	0.99 ± 0.02	0.32 ± 0.17
	5000	120.33 ± 0.00	1.0 ± 0.21	0.02 ± 0.00	0.11 ± 1.15	0.99 ± 0.00	0.11 ± 0.19
	6500	120.33 ± 2.84	2.0 ± 0.45	0.02 ± 0.00	0.45 ± 2.65	0.99 ± 0.01	0.16 ± 0.23
	8000	120.33 ± 2.84	2.0 ± 0.35	0.02 ± 0.00	0.43 ± 1.02	0.99 ± 0.00	0.13 ± 0.14
ZDT3	100	109.38 ± 5.39	1.0 ± 106.74	0.19 ± 0.25	0.13 ± 109.57	0.83 ± 0.34	157.28 ± 86.58
	500	125.57 ± 0.11	1.0 ± 0.0	0.08 ± 0.01	0.24 ± 0.28	0.95 ± 0.15	82.53 ± 56.98
	2000	128.68 ± 0.00	1.0 ± 0.0	0.05 ± 0.01	0.60 ± 0.75	0.97 ± 0.06	11.93 ± 91.16
	3500	128.76 ± 0.00	1.0 ± 0.55	0.02 ± 0.01	0.51 ± 1.86	0.99 ± 0.05	0.23 ± 12.31
	5000	128.77 ± 0.00	2.0 ± 0.80	0.02 ± 0.01	1.22 ± 14.54	0.99 ± 0.03	0.85 ± 2.68
	6500	128.77 ± 0.00	2.0 ± 1.04	0.03 ± 0.01	17.44 ± 40.01	0.99 ± 0.00	0.01 ± 2.43
	8000	128.77 ± 0.00	4.0 ± 2.03	0.02 ± 0.01	0.39 ± 86.62	0.99 ± 0.01	5.81 ± 27.95
ZDT4	100	7808.50 ± 103.39	1.0 ± 0.92	0.10 ± 0.11	6.64 ± 111.21	0.98 ± 0.14	42.87 ± 108.69
	500	9117.17 ± 253.37	8.5 ± 4.40	0.10 ± 0.05	264.02 ± 129.77	0.99 ± 0.01	97.43 ± 107.31
	2000	10420.05 ± 217.01	13.5 ± 15.15	0.03 ± 0.01	32.66 ± 129.23	0.99 ± 0.00	42.41 ± 30.90
	3500	10662.05 ± 173.42	22.5 ± 24.28	0.02 ± 0.01	124.65 ± 140.48	0.99 ± 0.03	33.41 ± 35.91
	5000	10893.45 ± 144.74	2.0 ± 33.95	0.02 ± 0.01	7.56 ± 139.79	0.99 ± 0.06	43.74 ± 17.90
	6500	10942.75 ± 117.76	1.0 ± 42.92	0.02 ± 0.01	8.81 ± 130.25	0.99 ± 0.11	44.83 ± 16.98
	8000	10959.35 ± 98.00	1.0 ± 48.80	0.02 ± 0.01	9.50 ± 113.74	0.99 ± 0.09	39.72 ± 17.57
ZDT6	100	43.41 ± 3.90	1.0 ± 97.06	0.19 ± 0.09	0.15 ± 66.85	0.92 ± 0.31	110.77 ± 81.40
	500	69.93 ± 0.45	1.0 ± 0.0	0.08 ± 0.01	0.03 ± 0.18	0.89 ± 0.16	82.35 ± 79.93
	2000	106.27 ± 0.21	1.0 ± 0.0	0.05 ± 0.00	0.07 ± 0.13	0.69 ± 0.16	144.05 ± 73.31
	3500	114.63 ± 0.08	1.0 ± 0.0	0.05 ± 0.00	0.14 ± 0.33	0.58 ± 0.12	167.48 ± 88.15
	5000	116.13 ± 0.01	1.0 ± 0.0	0.05 ± 0.00	0.08 ± 0.28	0.58 ± 0.07	161.48 ± 77.15
	6500	116.36 ± 0.00	1.0 ± 0.0	0.05 ± 0.00	0.16 ± 0.92	0.61 ± 0.08	23.18 ± 94.26
	8000	116.40 ± 0.00	1.0 ± 0.0	0.05 ± 0.00	0.25 ± 0.68	0.78 ± 0.11	6.02 ± 2.21

a random search, which both NSGA-II and NSGA-III degenerates to when the population size is greater or equal to the budget, has about the same performance when the problem is easy to solve and the function evaluation budget is very limited.

The median values for the population size parameter using NSGA-II are shown in Table 5. The same observations can be made for the NSGA-III and are not shown for that reason.

4.2 Mutation Probability: $1/N$ Rule of Thumb

Each experiment was run with five different values for the number of decision variables. This was done to test the accuracy of the rule of thumb that the mutation rate should be set to one divided by the number decision variables. Since each experiment was also run with different function evaluation budgets, it is also possible to see if that had any affect on the mutation probability. The usefulness of this evaluation is limited by the small number of problems used in this paper and no generalization can be made how this rule works on other problems. The mutation probabilities are also only from the best set of parameters found. Therefore, this evaluation does not test the accuracy of this rule of thumb for sub-optimal sets of parameters.

The experiment results can be divided into two groups based on the relationship between the mutation probability and the number of variables. ZDT{1, 2, 3} is in one group and ZDT{4, 6} is in the other. The first group start with a relatively high mutation probability for two variables, which then decreases and is kept almost constant for 10, 20, 30 and 40 variables. The second group has a more gradual decrease in the mutation probability. The median values from two problems are shown here, ZDT1 in Figure 1 and ZDT4 in Figure 2.

On ZDT1 the rule slightly overestimates the mutation probability for budget sizes greater than 500 when N is 10 and 20 because the optimized mutation probability does not change much between N 10 and 40. It also underestimates for all N when the budget size is less than 2000. On ZDT4 the rule overestimates when N is 2 and the budget size is greater than 500. It also underestimates for all N when the budget size is 100. For all other cases the rule matches well with the data.

Based on these results the rule of thumb is able to estimate good values for the mutation probability, especially for larger budgets, on the ZDT test problems.

4.3 Mutation Probability vs. Budget Sizes

A trend that can be observed throughout all experiments is that mutation probability decreases as the function evaluation budget increases. The trend is most prominent on ZDT{1, 2, 3} and less so on ZDT{4, 6}. Figure 5 and 6 shows the

Table 5. Experimental results for population sizes in NSGA-II

	N	100	500	2000	3500	5000	6500	8000
ZDT1	2	79.5 ± 121.99	11.0 ± 3.02	12.0 ± 5.83	15.0 ± 6.70	15.5 ± 6.27	20.0 ± 4.48	22.0 ± 6.66
	10	2.0 ± 0.0	2.0 ± 0.0	2.0 ± 0.0	2.0 ± 0.0	2.0 ± 0.0	2.0 ± 0.0	2.0 ± 0.8
	20	2.0 ± 0.0	2.0 ± 0.0	2.0 ± 0.0	2.0 ± 0.0	2.0 ± 0.0	2.0 ± 0.0	2.0 ± 0.0
	30	2.0 ± 33.56	2.0 ± 0.0	2.0 ± 0.0	2.0 ± 0.0	2.0 ± 0.0	2.0 ± 0.0	2.0 ± 0.0
	40	2.0 ± 0.0	2.0 ± 0.0	2.0 ± 0.0	2.0 ± 0.0	2.0 ± 0.0	2.0 ± 0.0	2.0 ± 0.0
ZDT2	2	2.0 ± 100.89	6.0 ± 1.65	17.5 ± 2.24	18.5 ± 6.49	22.0 ± 4.43	26.5 ± 4.02	28.0 ± 4.57
	10	2.0 ± 3.26	2.0 ± 0.0	2.0 ± 0.0	2.0 ± 0.0	2.0 ± 0.0	2.0 ± 0.0	2.0 ± 0.43
	20	2.0 ± 59.58	2.0 ± 0.0	2.0 ± 0.0	2.0 ± 0.0	2.0 ± 0.0	2.0 ± 0.0	2.0 ± 0.0
	30	2.0 ± 2.83	2.0 ± 4.79	2.0 ± 0.0	2.0 ± 0.0	2.0 ± 0.0	2.0 ± 0.0	2.0 ± 0.0
	40	2.0 ± 1.74	2.0 ± 0.0	2.0 ± 0.0	2.0 ± 0.0	2.0 ± 0.0	2.0 ± 0.0	2.0 ± 0.0
ZDT3	2	4.0 ± 119.75	4.5 ± 2.53	14.0 ± 5.20	32.0 ± 8.46	37.5 ± 7.40	48.5 ± 24.04	56.0 ± 34.45
	10	2.0 ± 3.31	2.0 ± 0.0	4.0 ± 1.90	9.0 ± 2.71	11.0 ± 1.10	11.5 ± 0.78	11.0 ± 3.28
	20	2.0 ± 2.71	2.0 ± 0.0	2.0 ± 0.73	2.0 ± 1.69	4.5 ± 2.53	9.0 ± 3.28	10.0 ± 2.25
	30	2.0 ± 1.74	2.0 ± 0.0	2.0 ± 0.0	2.0 ± 0.0	2.0 ± 0.6	3.5 ± 1.92	6.0 ± 2.70
	40	2.0 ± 0.0	2.0 ± 0.0	2.0 ± 0.0	2.0 ± 0.0	2.0 ± 0.43	2.0 ± 0.0	2.0 ± 0.87
ZDT4	2	3.0 ± 110.51	2.0 ± 15.87	8.0 ± 4.01	8.0 ± 2.47	12.0 ± 3.46	12.0 ± 2.78	17.0 ± 5.01
	10	2.5 ± 1.81	2.0 ± 8.92	3.0 ± 30.32	3.0 ± 28.57	5.0 ± 56.69	6.0 ± 35.50	11.0 ± 85.31
	20	2.5 ± 1.57	12.0 ± 6.07	4.0 ± 23.04	4.0 ± 34.12	4.0 ± 50.59	6.5 ± 60.39	6.0 ± 50.64
	30	2.0 ± 0.86	2.0 ± 4.15	33.5 ± 18.85	29.5 ± 33.41	4.0 ± 45.68	3.0 ± 44.47	4.5 ± 62.07
	40	2.5 ± 0.97	2.0 ± 2.62	15.0 ± 16.91	2.5 ± 28.88	77.0 ± 40.10	4.0 ± 38.47	6.0 ± 59.35
ZDT6	2	2.0 ± 79.04	2.0 ± 0.53	11.0 ± 2.94	20.0 ± 5.35	23.5 ± 4.63	26.0 ± 3.62	29.0 ± 5.31
	10	2.0 ± 0.0	2.0 ± 0.0	2.0 ± 0.0	2.0 ± 0.0	2.0 ± 0.0	2.0 ± 0.6	2.0 ± 1.39
	20	2.0 ± 0.0	2.0 ± 0.0	2.0 ± 0.0	2.0 ± 0.0	2.0 ± 0.0	2.0 ± 0.0	2.0 ± 0.0
	30	2.0 ± 0.0	2.0 ± 0.0	2.0 ± 0.0	2.0 ± 0.0	2.0 ± 0.0	2.0 ± 0.0	2.0 ± 0.0
	40	2.0 ± 0.0	2.0 ± 0.0	2.0 ± 0.0	2.0 ± 0.0	2.0 ± 0.0	2.0 ± 0.0	2.0 ± 0.0

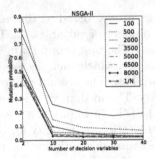
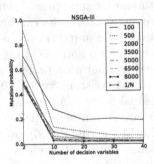

Fig. 1. Applicability of 1/N rule of thumb for p_m on ZDT1

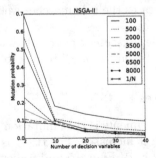
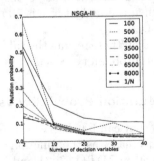

Fig. 2. Applicability of 1/N rule of thumb for p_m on ZDT4

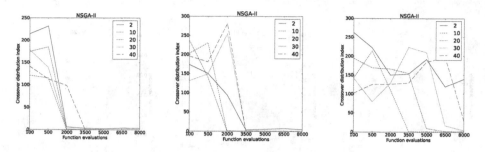

Fig. 3. Trends with budget size and η_c on ZDT1, ZDT2 and ZDT6

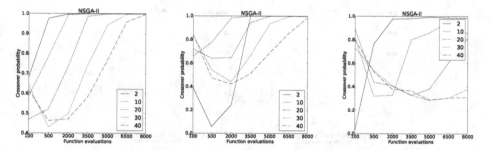

Fig. 4. Trends with budget size and p_c on ZDT1, ZDT2 and ZDT6

median values for NSGA-II on ZDT1 and ZDT6. The results for NSGA-III are similar, but they are not shown here because of space limitations.

4.4 Budget Size, p_c and η_c

A trend how p_c changes with respect to the budget size can be observed on all problems except ZDT4, although it is most clear on ZDT{1, 2, 3}. For small budgets p_c is relatively high. As the budget size increases p_c first falls and then rises, approaching a value of one. The point at which it starts to rise is related to the number of decision variables. Another trend is that there is a point at which an increase of the budget size causes a sharp fall of η_c. One explanation for why p_c is high and η_c is low for large budget sizes is based on the fact that most experiments use a population size of two. The two individuals are pushed apart by the crowding distance and with a large enough budget they will end up at each of the two extreme values. The rest of the non-dominated front is then filled by crossing these two solutions, and since they are at opposite extremes a low η_c is preferred. These trends are shown for NSGA-II on ZDT1, ZDT2 and ZDT6 in Figure 3 and Figure 4, the values are the median.

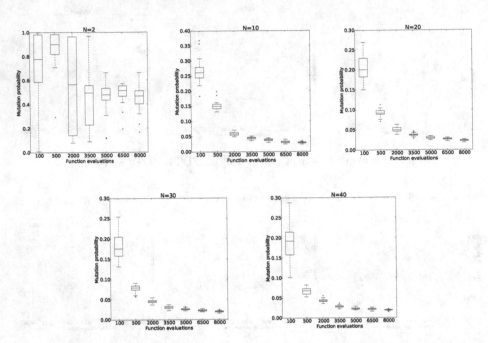

Fig. 5. Mutation probabilities with varying budget sizes for NSGA-II on ZDT1

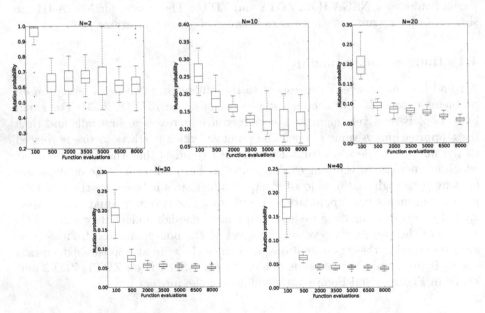

Fig. 6. Mutation probabilities with varying budget sizes for NSGA-II on ZDT6

Table 6. Hypervolume results for NSGA-II and NSGA-III

	N	ZDT1		ZDT2		ZDT3	
		NSGA-II	NSGA-III	NSGA-II	NSGA-III	NSGA-II	NSGA-III
100	2	119.68	119.43	**118.58**	116.73	126.41	125.58
	10	**115.35**	111.60	107.64	105.04	119.34	117.46
	20	**109.27**	106.38	95.78	96.77	**112.78**	108.35
	30	104.92	103.35	91.39	90.65	**110.06**	104.97
	40	**103.06**	101.24	87.85	86.87	**108.01**	102.54
500	2	120.65	**120.66**	120.32	**120.32**	128.76	**128.76**
	10	**120.53**	120.51	**120.11**	120.08	**128.50**	128.46
	20	**119.77**	119.69	**117.87**	117.77	**127.41**	127.27
	30	**118.43**	118.29	113.22	114.00	**125.71**	125.55
	40	**116.88**	116.64	110.72	109.39	**123.96**	122.75
2000	2	120.66	**120.66**	120.33	120.33	128.77	128.77
	10	120.66	120.66	**120.33**	120.32	**128.77**	128.77
	20	**120.65**	120.65	**120.31**	120.31	**128.74**	128.74
	30	120.62	120.62	**120.25**	120.24	**128.69**	128.68
	40	**120.55**	120.55	**120.14**	120.12	**128.61**	128.60
3500	2	120.66	120.66	120.33	120.33	128.77	128.77
	10	**120.66**	120.66	120.33	120.33	**128.77**	128.77
	20	**120.66**	120.66	**120.33**	120.33	**128.77**	128.77
	30	**120.66**	120.66	**120.32**	120.32	**128.76**	128.76
	40	**120.65**	120.65	120.31	120.31	128.75	128.75
5000	2	120.66	120.66	120.33	120.33	128.77	128.77
	10	120.66	120.66	120.33	120.33	**128.77**	128.77
	20	**120.66**	120.66	120.33	120.33	128.77	**128.77**
	30	**120.66**	120.66	**120.33**	120.33	**128.77**	128.77
	40	**120.66**	120.66	**120.32**	120.32	**128.77**	128.77
6500	2	120.66	**120.66**	120.33	120.33	128.77	128.77
	10	120.66	120.66	**120.33**	120.33	**128.77**	128.77
	20	120.66	120.66	**120.33**	120.33	128.77	128.77
	30	**120.66**	120.66	120.33	120.33	128.77	128.77
	40	120.66	120.66	**120.33**	120.33	**128.77**	128.77
8000	2	120.66	**120.66**	120.33	120.33	128.77	**128.77**
	10	120.66	120.66	120.33	120.33	128.77	**128.77**
	20	120.66	120.66	120.33	120.33	128.77	**128.77**
	30	120.66	120.66	**120.33**	120.33	128.77	128.77
	40	**120.66**	120.66	120.33	120.33	**128.77**	128.77

	N	ZDT4		ZDT6	
		NSGA-II	NSGA-III	NSGA-II	NSGA-III
100	2	**10993.44**	10990.83	109.83	106.26
	10	**10461.61**	10411.53	63.63	61.04
	20	**9251.83**	9192.64	49.29	47.53
	30	**7889.13**	7781.64	**43.82**	41.48
	40	**6391.02**	6289.29	40.93	40.29
500	2	10999.46	10999.46	116.41	116.41
	10	10849.93	10828.18	103.26	102.69
	20	10221.37	10247.87	82.07	81.78
	30	**9447.10**	9273.20	70.23	70.03
	40	8524.69	8408.60	**68.34**	63.00
2000	2	10999.7	10999.69	116.42	**116.42**
	10	10970.92	10973.84	**116.35**	116.34
	20	10778.65	10793.25	113.59	113.37
	30	10416.11	10425.08	**106.76**	106.31
	40	9968.78	9899.27	99.29	98.87
3500	2	10999.7	10999.7	116.43	116.43
	10	10991.61	10984.64	**116.41**	116.41
	20	10896.30	10911.16	**116.24**	116.21
	30	10656.69	10666.82	114.80	114.65
	40	10402.90	10438.98	**111.48**	111.19
5000	2	10999.7	10999.7	116.43	116.43
	10	10993.38	10993.26	**116.42**	116.41
	20	10931.79	10946.89	**116.40**	116.39
	30	10779.01	10780.29	**116.17**	116.13
	40	10498.14	10557.02	**115.19**	115.03
6500	2	10999.7	10999.7	116.43	116.43
	10	**10997.87**	10994.22	116.42	116.42
	20	10945.27	10956.11	**116.41**	116.41
	30	10896.20	10849.03	**116.37**	116.36
	40	10787.64	10712.04	**116.11**	116.06
8000	2	10999.7	10999.7	116.43	116.43
	10	10995.99	10998.15	**116.42**	116.42
	20	10977.54	10966.47	**116.41**	116.41
	30	10807.09	10808.35	116.40	116.40
	40	10752.03	10840.82	**116.34**	116.32

4.5   Hypervolume Comparisons Between NSGA-II and NSGA-III

The mean hypervolume values for both NSGA-II and NSGA-III are shown in Table 6. It is not possible, due to the number of experiments performed, to include all parameter settings used to obtain the hypervolume results. A subset of all the parameter settings and their corresponding hypervolumes are presented in Table 3 and Table 4. A Welch-t test with a significance of 5% is performed to determine if the two samples, NSGA-II and NSGA-III, are statistical different. If the null hypothesis can be rejected, the greater hypervolume is shown in bold.

The difference in hypervolume between NSGA-II and NSGA-III is for the most part small. However, for some of the experiments, NSGA-II is slightly better. NSGA-III is statistically better on some experiments but the difference is too small to be concluded as significant.

To summarize, NSGA-II is found to be marginally better than NSGA-III on the ZDT problems. Both NSGA-II and NSGA-III can find solutions very close to the Pareto front for ZDT{1, 2, 3, 6}. The most difficult problem is ZDT4, for which with $N > 10$ none of algorithms could reach the maximum theoretical hypervolume within 8000 evaluations.

5   Conclusions and Further Work

This paper utilized a bilevel optimization framework to find optimal parameter values for two different MOEAs, namely NSGA-II and NSGA-III, for maximal performance on the ZDT test suite. Both the number of decision variables and the function evaluation budgets were simultaneously varied to determine how they affect the optimal parameter settings for the respective algorithm. This made it possible to test the rule of thumb that the mutation probability should be set to $1/N$. The results show that, on the ZDT test problems, this rule is a good heuristic.

The experiments also made it possible to see what affect the different function evaluation budgets has on the optimized parameters. An important observation was that the optimal mutation probability is not only dependent on the number of decision variables but also on the available budget size. Specifically, it was observed that the optimal mutation probability decreases with increasing budget.

It was also clear from the results that the ZDT test problems do not require much diversity in the population because most experiments found the optimal population size to be less than 10, often close to the minimum of just two individuals. This also indicates that a parameter controlling the elitism should have been included in the experiments.

Another aim of this paper was to compare the performance between NSGA-II and NSGA-III on the ZDT test problems. From the results, it is possible to discern a slight advantage with NSGA-II over NSGA-III on the ZDT problems. As far as the optimal parameter values are concerned, it was observed that the differences are small.

Extending these experiments to scale the number of objectives instead of the number of decision variables would be interesting, and is intended as future work.

Since these results, as well as other earlier work, indicate that it is sub-optimal to keep parameter settings static during the run, it would be be worthwhile to modify an EA, on the algorithm layer, to be able to use multiple sets of parameters during an optimization. This would allow a meta-EA to tune multiple sets of parameters at different intervals of the optimization, instead of being limited to a single set throughout the optimization run.

References

1. Bäck, T.: Parallel optimization of evolutionary algorithms. In: Davidor, Y., Männer, R., Schwefel, H.-P. (eds.) PPSN 1994. LNCS, vol. 866, pp. 418–427. Springer, Heidelberg (1994)
2. Das, I., Dennis, J.: Normal-boundary intersection: A new method for generating the pareto surface in nonlinear multicriteria optimization problems. SIAM Journal on Optimization **8**(3), 631–657 (1998)
3. Deb, K., Jain, H.: An evolutionary many-objective optimization algorithm using reference-point-based nondominated sorting approach, part i: Solving problems with box constraints. IEEE Transactions on Evolutionary Computation **18**(4), 577–601 (2014)
4. Deb, K., Pratap, A., Agarwal, S., Meyarivan, T.: A fast and elitist multiobjective genetic algorithm: NSGA-II. IEEE Transactions on Evolutionary Computation **6**(2), 182–197 (2002)
5. Eiben, A.E., Smit, S.K.: Parameter tuning for configuring and analyzing evolutionary algorithms. Swarm and Evolutionary Computation **1**(1), 19–31 (2011)
6. Eiben, A., Hinterding, R., Michalewicz, Z.: Parameter control in evolutionary algorithms. IEEE Transactions on Evolutionary Computation **3**(2), 124–141 (1999)
7. Grefenstette, J.: Optimization of control parameters for genetic algorithms. IEEE Transactions on Systems, Man and Cybernetics **16**(1), 122–128 (1986)
8. Mühlenbein, H.: How genetic algorithms really work: mutation and hillclimbing. In: PPSN, pp. 15–26 (1992)
9. Ugolotti, R., Cagnoni, S.: Analysis of evolutionary algorithms using multi-objective parameter tuning. In: Proceedings of the 2014 Conference on Genetic and Evolutionary Computation, GECCO 2014, pp. 1343–1350. ACM, New York (2014)
10. Wessing, S., Beume, N., Rudolph, G., Naujoks, B.: Parameter tuning boosts performance of variation operators in multiobjective optimization. In: Schaefer, R., Cotta, C., Kołodziej, J., Rudolph, G. (eds.) PPSN XI. LNCS, vol. 6238, pp. 728–737. Springer, Heidelberg (2010)
11. While, L., Bradstreet, L., Barone, L.: A fast way of calculating exact hypervolumes. IEEE Transactions on Evolutionary Computation **16**(1), 86–95 (2012)
12. Zitzler, E.: Evolutionary Algorithms for Multiobjective Optimization: Methods and Applications. Ph.D. thesis, Shaker Verlag (1999)
13. Zitzler, E., Deb, K., Thiele, L.: Comparison of multiobjective evolutionary algorithms: Empirical results. Evol. Comput. **8**(2), 173–195 (2000)

Pareto Adaptive Scalarising Functions for Decomposition Based Algorithms

Rui Wang$^{1(\boxtimes)}$, Qingfu Zhang^2, and Tao Zhang^1

^{1} College of Information Systems and Management, National University of Defense Technology, Changsha 410073, Hunan, People's Republic of China
ruiwangnudt@gmail.com
^{2} Department of Computer Science, City University of Hong Kong, and the University of Essex, Colchester, UK

Abstract. Decomposition based algorithms have become increasingly popular for solving multi-objective problems. However, the effect of scalarising functions in decomposition based algorithms is under-explored. This study analyses the search behaviour of a family of frequently used scalarising functions— the L_p weighted approaches, and identifies that the p value corresponds to a trade-off between the L_p approach's search ability and its robustness on Pareto front geometries. That is, as the p value increases, the search ability of the L_p approach decreases whereas its robustness on Pareto front geometry increases. Based on this observation, we propose to use Pareto adaptive scalarising functions in decomposition based algorithms, where the p value is adaptively fine-tuned based on an estimation of the Pareto front shape. MOEA/D using Pareto adaptive scalarising functions (MOEA/D-par) is tested on a set of problems (with up to seven objectives) encompassing three basic Pareto front geometries, i.e., convex, concave and linear, and is shown to outperform MOEA/D using Chebyshev function on all the test problems.

Keywords: Multi-objective optimization · Evolutionary computation · Decomposition · Scalarising function · Pareto adaptive

1 Introduction

Multi-objective optimisation problems (MOPs) arise in many disciplines such as engineering, finance, logistics and control systems [1], where multiple objectives must be simultaneously optimised. Often objectives in a MOP are in competition with each other, and thus, the optimal solution set of MOPs is not a single solution but comprises of a set of trade-off solutions. Multi-objective evolutionary algorithms (MOEAs) are well suited for solving MOPs since (i) their population-based nature leads naturally to the generation of an approximate trade-off surface in a single run; and (ii) they tend to be robust to underlying objective function characteristics.

During the last two decades, a variety of MOEA approaches has been proposed. These approaches can be categorised into three main classes: Pareto-dominance or modified dominance based algorithms, e.g., MOGA [2], NSGA-II [3],

© Springer International Publishing Switzerland 2015
A. Gaspar-Cunha et al. (Eds.): EMO 2015, Part I, LNCS 9018, pp. 248–262, 2015.
DOI: 10.1007/978-3-319-15934-8_17

PICEA-g [4]; Performance indicator based algorithms, e.g., IBEA[5], HypE [6]; and decomposition based algorithms, e.g., CMOGA [7], MSOPS [8], MOEA/D [9]. Amongst these approaches, decomposition based algorithms become increasingly popular recently. Decomposition based algorithms decompose a MOP into a set of single objective problems by means of weighted scalarising functions, or a set of simple MOPs [10,11] and optimise them in a collaborative manner. Compared with the other two types of algorithms, decomposition based algorithms have a number of advantages such as high search ability for combinatorial optimisation, computational efficiency on fitness evaluation and high compatibility with local search [9,12–14]. The seminal decomposition based MOEA, i.e., MOEA/D [9], that popularised this method, has been used in many real-world applications [15]. Despite these advantages, the performance of decomposition based algorithms is arguably dependent on the specification of weights[16] and scalarising functions [17]. The choice of suitable weights and scalarising functions is typically problem-dependent and therefore is difficult if no information about the problem characteristics is known before the search proceeds.

Regarding the choice of weights, we have known that when the Pareto front geometry of a MOP is known *a priori*, an optimal distribution of weights for certain scalarising function can be identified [16,18]. Otherwise, a suitable set of weights can be configured adaptively. A number of effective methods have been proposed for this purpose, for example, co-evolving weights with solutions [19,20], using Pareto adaptive weights [21], adjusting weights adaptively based on an estimation of Pareto front geometry [22–24]. Regarding the choice of scalarising functions, although we have known: for example, the weighted Chevbshev is able to find solutions on both convex and non-convex regions whereas the weighted sum cannot [25]; the weighted sum can obtain better results than the weighted Chevbshev on multi-objective knapsack problems [9], this is still far from being well understood. It is in general unclear what the relation is between different scalarising functions; and how an appropriate scalarising function can be identified for a new problem. Towards a better understanding of the effect of scalarising functions in decomposition based algorithms as well as unlocking the aforementioned issues, in this study we analyse a family of frequently used scalarising functions, i.e., the L_p weighted approaches in terms of their search ability and their robustness on the Pareto front geometry. Moreover, based on the analysis, we propose to use Pareto adaptive L_p scalarising functions in decomposition based algorithms so as to enhance the algorithm's performance.

The remainder of this paper is organised as follows: in Section 2 some background knowledge about decomposition based approaches, is provided. Section 3 elaborates the effect of L_p scalarising functions and how to choose a suitable L_p scalarising function. Experiments and discussions are provided in Section 4. Finally, Section 5 concludes the paper and identities future studies.

2   Decomposition Approaches

Without loss of generality, a minimisation MOP is defined as,

$$\min_{\mathbf{x}} \mathbf{F}(\mathbf{x}) = (f_1(\mathbf{x}), f_2(\mathbf{x}), \cdots, f_m(\mathbf{x}))$$

$$\text{subject to } \mathbf{x} \in \Omega \tag{1}$$

where m is the number of objective functions (generally, $m > 2$); $\mathbf{x}$ is a *vector* in the *decision (variable) space* Ω. $\mathbb{R}^m$ is the *objective space*. $F : \Omega \to \mathbb{R}^m$ consists of m real-valued objective functions that are to be minimised.

Decomposition based approaches decompose a MOP into a set of single objective problems defined by means of scalarising functions with different weights. The optimal solution of each single objective problem corresponds to one Pareto optimal solution of a MOP [26]. The weight vector defines a search direction for the scalarising function. Diversified solutions can be obtained by employing different search directions.

A variety of scalarising functions can be used in decomposition based algorithms [26]. The weighted sum and the weighted Chebyshev from the family of weighted L_p scalarising functions are two of the most popular ones. Mathematically, the weighted L_p scalarising function can be written as,

$$g^{wd}(\mathbf{x}|\mathbf{w}, p) = \left(\sum_{i=1}^{m} \lambda_i (f_i(\mathbf{x}) - z_i^*)^p \right)^{\frac{1}{p}}, \ p > 0 \quad \lambda_i = (1/w_i)^p \tag{2}$$

where $\mathbf{z}^* = (z_1, z_2, \cdots, z_m)$ is the *ideal* point; $\mathbf{w} = (w_1, w_2, \cdots, w_m)^T$ is a weighting vector and $\sum_{i=1}^{m} w_i = 0, \ w_i \geq 0$; The $\mathbf{w}$ determines the search direction of the scalarising function. Note that whether the obtained Pareto optimal solution is along the search direction or not is also influenced by the Pareto front geometry [16,27]. The weighted sum and weighted Chebyshev are derived by setting $p = 1$ and $p \to \infty$, respectively.

In addition, decomposition based algorithms combine different objective function values into one scalar value. These objectives might have various units of measurement, and/or scaled disparately. It is therefore important to rescale different objectives to dimension-free units before aggregation. Typically, the normalisation procedure transforms an objective value f_i by

$$\overline{f}_i = \frac{f_i - z_i^*}{z_i^{nad} - z_i^*} \tag{3}$$

If the z_i^* and z_i^{nad} (the *nadir* point) are not available, the smallest and largest f_i of all non-dominated solutions found so far could be used instead.

3   The Choice of a Suitable L_p Scalarising Function

3.1   Analysis: Property of Different L_p Scalarising Functions

This section analyses the property of different L_p scalarising functions, that is, the trade-off between their search ability and their robustness on Pareto front

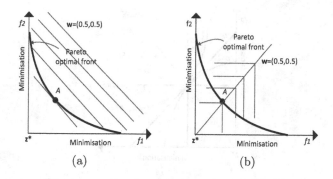

Fig. 1. Contour lines of the weighted sum (a) and weighted Chebyshev (b) scalarising functions

geometries [28]. Inspired by [29], we first look at two special cases, i.e., the weighted sum and the weighted Chebyshev. Fig. 1 shows contour lines of the two scalarising functions in a bi-objective case with *ideal* point at the origin and weight vector $\mathbf{w} = (0.5, 0.5)$. The objective space is divided into two sub-spaces by the contour line. Solutions in one sub-space are better than solutions on the contour line while solutions in the other sub-space are worse. Solutions that lie on the same contour line have the same scalar objective value. In Fig. 1, solution A is the optimal solution of $g^{wd} \left(\mathbf{x} | (0.5, 0.5), 1 \right)$ and $g^{wd} \left(\mathbf{x} | (0.5, 0.5), \infty \right)$.

The contour line of the weighted sum approach is a line, and the contour line of the weighted Chebyshev approach is a polygonal line (with vertical angle). According to the shape of the contour line we can observe that for the weighted sum approach the size of a better region equals to half of the whole objective space regardless of the number of objectives. This indicates that the probability of replacement of an existing solution by a newly generated solution always decreases from $\frac{1}{2}$ to 0 as the search progresses. The maximal probability of replacement (i.e., $\frac{1}{2}$) is not influenced by the number of objectives. In this sense, the search ability of the weighted sum approach is not affected by an increase in the number of objectives. With respect to the weighted Chebyshev function, a better region roughly equals to $\left(\frac{1}{2}^m \right)$ of the m-dimensional objective space. This indicates that the maximal probability of replacement is $\left(\frac{1}{2}^m \right)$. Compared with the weighted sum approach, the maximal probability of replacement significantly decreases as the number of objective increases. In other words, the search ability of Chebyshev scalarising function deteriorates as the number of objectives increases [28,30]. However, it is suspected that the search ability of the Chebysheve scalarising function is comparable to the Pareto-dominance relation as claimed in [28]. Our preliminary experiments show that compared with the Pareto dominance, solutions selected by the Chebyshev function are more likely to be closer to the *ideal* point [31]. Moreover, it has been widely demonstrated that decomposition based algorithms (even using random weighted Chebyshev functions) outperform Pareto-dominance based algorithms on many-objective problems [14,19].

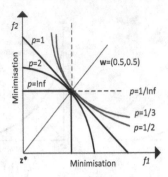

Fig. 2. Contour lines of the L_p scalarising function with different p values

Contour lines of the L_p scalarising functions with different p values are shown in Fig. 2. We can observe from the figure that the volume enclosed by the contour line and the *ideal* point decreases as p increases (a calculation of the volume can be referred to [28]). This indicates that as p increases, the probability of finding a better solution (measured by the L_p approach) decreases, that is, the search ability of the L_p scalarising function decreases. This observation is also experimentally demonstrated by applying MOEA/D with L_3, L_7 and L_∞ scalarising functions to solve the 4 objective WFG4 [32] problem whose Pareto optimal front is a hyper-sphere. Each of the algorithm instantiations is run for 31 independent runs. The mean hypervolume (HV) values and the generation distance (GD) values over generations are plotted in Fig. 3. We can clearly observe from Fig. 3 that MOEA/D with $p = 3$ performs the best, followed by $p = 7$, and then $p \to \infty$, i.e., the weighted Chebyshev function.

As previously mentioned, the weighted sum function may not be able to find all the Pareto optimal solutions in the case of non-convex PFs [26, p.79], whereas the Chebyshev scalarising function can find solutions in both convex and non-convex regions, see Fig. 4. Upon closer examination, we can imagine that all the

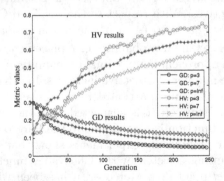

Fig. 3. (Colour online) The performance of MOEA/D using the $p = 3$, $p = 7$ weighted L_p scalsrising functions and the weighted Chebyshev function on the 4-objective WFG4 problem: the mean HV and GD values of over generations

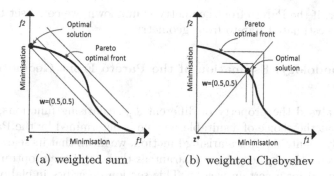

(a) weighted sum   (b) weighted Chebyshev

Fig. 4. Behaviours of the weighted sum (a) and Chebyshev (b) scalarising function on non-convex Pareto front

L_p scalarising functions except for the Chebyshev, face difficulties in searching for solutions in a non-convex region. To be more specific, a weighted L_p scalarising function can find solutions along certain search direction in a non-convex region only if the curvature of its contour line is larger than the curvature of the PF shape. Otherwise the selected scalarising function suffers from the non-convex geometry issue. Since the curvature of the Chebyshev function is ∞, it is able to find Pareto optimal solutions for any type of geometries. For example, assuming that the PF is a circle (quadratic) in the first quadrant, see Fig. 5. In order to find the Pareto optimal solution $\mathbf{x}$ along the search direction $(0.5, 0.5)$, the L_p with $p > 2$ should be used, e.g, $p = 3$.

Overall the search ability of a L_p scalarising function and its robustness on Pareto front geometries are a trade-off— the higher the search ability, the lower the robustness. If the Pareto front geometry is known *a priori*, we will be able to determine a suitable L_p scalarising function by taking into account the curvature of the Pareto front. For example, for a search direction $\mathbf{w}^j$, we can set the p value being larger than the curvature of the segmented Pareto front along

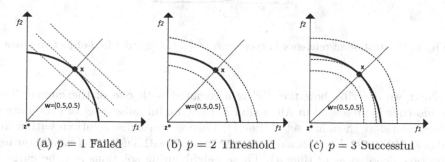

(a) $p = 1$ Failed   (b) $p = 2$ Threshold   (c) $p = 3$ Successful

Fig. 5. Searching the same solution using different L_p functions

$\mathbf{w}^j$. However, if the Pareto front geometry is unknown, we could set the p value based on the estimated Pareto front geometry.

3.2   Methodology: Estimation of the Pareto Front Geometry on Line

We have analysed the property of different L_p scalarising functions, and have identified that the choice of a suitable p value is determined by the Pareto front geometry. By a suitable L_p scalarising function, we mean that its search ability is maximised, and simultaneously, it guarantees that any Pareto optimal solution can be obtained for a certain weight. This section describes in elaborate detail how a suitable L_p scalarising function is determined. The key issue here is to effectively estimate the Pareto front geometry.

A number of methods are available in the literature for estimating the Pareto front geometry. Here, we borrow the idea from [21,33], that is, approximating the Pareto front using a family of reference curves:

$$\{(y_1)^\alpha + (y_2)^\alpha, \cdots, (y_m)^\alpha = 1; y_j \in (0,1], \alpha \in (0,\infty)\} \tag{4}$$

The family of curves as shown in Fig. 6 possesses the following properties: i) if $\alpha > 1$, the curve is a concave; ii) if $\alpha < 1$, the curve is convex; and iii) if $\alpha = 1$, the curve is linear.

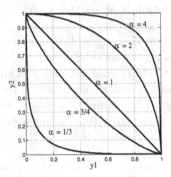

Fig. 6. Illustration of reference curves for $\alpha = \frac{1}{3}, \frac{3}{4}, 1, 2$ and 4 in 2-objective space

Next we describe how the PF is associated with one of the curves. The pseudo-code is presented in Algorithm 1. First we initialise a set of candidate p values, and store them in a set P (line 1); Then we normalise solutions within the range $[0, 1]$ (line 2). Next for a search direction $\mathbf{w}^j$, we identify its T neighbouring solutions, denoted as Q (line 4). These neighbouring solutions are the current solutions of the neighbouring problems. The parameter T is the same as the selection neighbourhood size in MOEA/D [9]. We compute the Eq. (5) for each candidate p value. The smaller the $h(p, Q)$, the better the solutions Q match the reference curve. The p is determined as the value that produces the second smallest $h(p, Q)$ (lines 5 and 6). The reason for not choosing the p associated with the minimal $h(p, Q)$ is that the curvature of L_p function is required to be larger than the curvature of this segmented PF shape, i.e. $p > \alpha$, see Fig. 5. In addition, we include a pre-defined large number, e.g., 1000, in the set P. If $h(Q, 1000)$ is found to be the minimal, p is set to ∞, i.e., the Chebyshev function is used instead (lines 7-9).

$$h(p, Q) = \sum_{\forall \mathbf{x}^k \in Q} \left(\sum_{i=1,...,m} \left(f_i\left(\mathbf{x}^k\right) \right)^p - 1 \right)^2, \quad p \in P \tag{5}$$

Algorithm 1. Selecting a suitable L_p scalarising function

Input: non-dominated solutions available Q, neighbourhood size, T
Output: p value.

1 Initialise the candidate L_p functions, e.g., $P = \{\frac{1}{2}, \frac{2}{3}, 1, 2, 3, ..., 10, 1000\}$;
2 Normalise solutions within the range $[0, 1]$;
3 **foreach** *search direction*, $\mathbf{w}^j$ **do**
4   Find the T neighbouring solutions, Q, of the search direction $\mathbf{w}^j$;
5   Compute the $h(p, Q)$ for each candidate p;
6   Find the second smallest $h(p, Q)$ and identify the corresponding p value;
7   **if** *p equals to a pre-defined large value in the* P **then**
8 | using the Chebyshev function instead, i.e., $p \leftarrow \infty$;
9   **end**
10 **end**

---

4 Experiments and Discussions

This section examines the effect of Pareto adaptive scalarising functions. We incorporate it into the state-of-the-art decomposition based algorithm, i.e., MOEA/D [9], and compare the derived algorithm, denoted as MOEA/D-par (see Algorithm 2), with MOEA/D using the Chebyshev scalarising functions.

4.1 Experimental Descriptions

Test Problems. Test problems used in this study are constructed by applying different shape functions provided in the WFG toolkit to the standard WFG4 benchmark problem, please refer to [19] for more details. The WFG41 has a concave Pareto optimal front. WFG42 has a convex Pareto optimal front. The Pareto optimal front of WFG43 is a hyperplane. The number decision variables of these problems is set to $n = 100$ wherein the WFG position parameter (k) and the distance parameter (l) are $\frac{m-1}{2}$ and $100 - k$, respectively. The Pareto optimal front of these problems has the same trade-off magnitudes, and it is within $[0, 2]$. These problems are invoked in 2-, 4- and 7-objective instances. Note that unless otherwise stated we use WFGn-Y to denote the problem WFGn with Y objectives.

Algorithm 2. MOEA/D using Pareto adaptive scalarising functions

Input: initial population, $S \leftarrow \{\mathbf{x}^1, \mathbf{x}^2, \cdots, \mathbf{x}^N\}$, initial weights,
$\qquad$ $W \leftarrow \{\mathbf{w}^1, \mathbf{w}^2, \cdots, \mathbf{w}^N\}$, selection neighbourhood size, T, replacement
$\qquad$ neighbourhood size, nr

Output: S

1 Initialise the L_k^p as the weighted sum style, i.e., $p_k \leftarrow 1, i \in \{1, 2, \cdots, N\}$;
2 Evaluate the objective function values of the initial S;
3 Update the *ideal* and *nadir* vectors, $\mathbf{z}^*$ and $\mathbf{z}^{nad}$;
4 Randomly assign each weight, $\mathbf{w}^i$ with a candidate solution, $\mathbf{x}^i$;
5 Calculate the Euclidean distance between weights, $\mathbf{w}^i$ and $\mathbf{w}^j$, $i, j \in 1, 2, \cdots, N$;
6 Find the T neighbouring weights $B(\mathbf{w}^i)$ of $\mathbf{w}^i$ based on the distance of weights
$\quad$ and identify the related neighbouring solutions Q of $\mathbf{x}^i$;
7 Set *iteration* $\leftarrow 0$, set *matingS* $\leftarrow \emptyset$;
8 **while** *the stopping criterion is not satisfied* **do**
9 $\quad$ **for** $i \leftarrow 1$ **to** N **do**
10 $\quad\quad$ **if** *rand* $< \delta$ **then**
11 $\quad\quad\quad$ | $\quad$ *matingS* $\leftarrow Q$;
12 $\quad\quad$ **else**
13 $\quad\quad\quad$ | $\quad$ *matingS* $\leftarrow S$;
14 $\quad\quad$ **end**
15 $\quad\quad$ Randomly select three solutions $\mathbf{x}^{r1}, \mathbf{x}^{r2}$ and $\mathbf{x}^{r3}$ from the mating pool,
$\quad\quad\quad$ *matingS*;
16 $\quad\quad$ Generate a new solution $\mathbf{x}^{new}$ by performing differential evolution (DE)
$\quad\quad\quad$ and polynomial mutation (PM) operators;
17 $\quad\quad$ Evaluate the objective value of $\mathbf{x}^{new}$, and update the ideal and nadir
$\quad\quad\quad$ vectors;
18 $\quad\quad$ **for** *each* $\mathbf{x}^k \in Q$ **do**
19 $\quad\quad\quad$ | $\quad$ Compare $g^{wd}(\mathbf{x}^{new}|\mathbf{w}^k, p^k)$ with $g^{wd}(\mathbf{x}^k|\mathbf{w}^k, p_k)$;
20 $\quad\quad$ **end**
21 $\quad\quad$ Replace no more than nr solutions in Q with $\mathbf{x}^{new}$ if $g^{wd}(\mathbf{x}^{new}|\mathbf{w}^k, p_k)$
$\quad\quad\quad$ is smaller;
22 $\quad$ **end**
23 $\quad$ Update the p_k value for each search direction using Algorithm 1;
24 **end**

General Parameters. The following parameters are set constant across all algorithm runs:

- *Algorithm runs and stopping criterion*: each algorithm is performed for 31 runs, each run for 25,000 function evaluations.
- *Population size*: $N = 200$ for bi-objective problems, 400 for 4-objective problems, and 700 for 7-objective problems.
- *DE and PM operators*: the DE control parameters are set as $F = 0.5$ and $CR = 0.9$. The mutation probability $pm = 1/n$ and its distribution index is set to be $\eta_m = 20$.

- *The initial candidate p values*: $p \in P = \{\frac{1}{2}, \frac{2}{3}, 1, 2, 3, 4, 5, 10, 1000\}$.
- *MOEA/D parameters*: the selection neighbourhood size is set to 10% of N, the replacement size (nr) is 10% of T.

4.2   Experimental Results

Median Attainment Surfaces. Plots of median attainment surfaces across the 31 runs of each algorithm are shown in Fig. 7. These allow visual inspection of performance in terms of the dual aims of proximity to and diversity across the global trade-off surface. The PF of each problem serves as a reference. From inspection of Fig. 7, the two algorithms appear to have comparable diversity performance while the MOEA/D-par has a clear better convergence performance than MOEA/D for all the three problems.

Comparison Results in Terms of the HV and C Metrics. Comparison results of MOEA/D-par with MOEA/D in terms of the HV and C metrics are presented in Table 1. A favourable HV value (larger, for a minimisation problem) implies good proximity with diversity. In our experimental studies, the reference point is set to $1.1 \times \mathbf{z}^{nad}$, i.e., $(2.2, 2.2, \cdots, 2.2)$. The C metric is a binary metric which provides information on convergence. For example, given two sets, A and B, $C(A, B)$ refers to the fraction of solutions in B that are dominated at least by one solution in A. $C(A, B) > C(B, A)$ indicates a better convergence of the A set. Moreover, the non-Parametric Wilcoxon-ranksum two-sided comparison procedure at the 95% confidence level is employed to compare the significance of difference between two algorithms.

From Table 1, we can clearly observe that MOEA/D-par performs better than MOEA/D for all problems in terms both the HV and C metrics. As the only difference between MOEA/D-par and MOEA/D lies in the use of Pareto adaptive scalarising functions, such results are able to confirm that provided a good estimation of the Pareto front geometry, the use of Pareto adaptive scalarising function is helpful, which can improve the performance of MOEA/D significantly for both bi- and many-objective problems (up to 7 objectives).

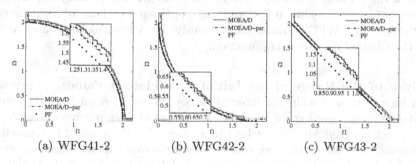

(a) WFG41-2   (b) WFG42-2   (c) WFG43-2

Fig. 7. (Color online) Attainment surfaces for the 2-objective WFG4X problems

Table 1. Comparison results of the HV and C metric values for the WFG4X problems. The symbol '$<$', '$=$' or '$>$' means MOEA/D is statistically worse, comparable or better than MOEA/D-par. A refers to MOEA/D, B refers to MOEA/D-par.

	$HV(A)$	$HV(B)$	$C(A,B)$	$C(B,A)$
WFG41-2	0.2982(0.0052) <	0.3138(0.0031)	0.0052(0.0104) <	0.9640(0.0426)
WFG42-2	0.7710(0.0047) <	0.7887(0.0038)	0.0984(0.0668) <	0.7846(0.0464)
WFG43-2	0.5286(0.0071) <	0.5460(0.0021)	0.0438(0.0876) <	0.7548(0.0404)
WFG41-4	0.4668(0.0097) <	0.5861(0.0028)	0.0020(0.0041) <	0.3495(0.0427)
WFG42-4	0.8828(0.0072) <	0.9100(0.0079)	0(0) <	0.1355(0.0642)
WFG43-4	0.7318(0.0218) <	0.8147(0.0194)	0.0044(0.0051) <	0.1118(0.0638)
WFG41-7	0.5319(0.0212) <	0.7204(0.0340)	0(0) <	0.0633(0.0368)
WFG42-7	0.9500(0.0069) <	0.9556(0.0011)	0.0263(0.0137) <	0.0551(0.0178)
WFG43-7	0.8112(0.0165) <	0.8884(0.0084)	0(0) <	0.0797(0.0239)

4.3 Experimental Discussions

This section investigates two issues, as part of a wider discussion for the use of Pareto adaptive scalarising functions. First, we examine the obtained p values in MOEA/D-par; Second, the range of the candidate p values.

Observation of the Obtained p Values. Empirical comparison results have demonstrated the benefits of using Pareto adaptive scalarising functions in MOEA/D. Here, we show the obtained p values for the search direction $\mathbf{w} = (0.5, 0.5)$ over generations, as an evidence of the superior performance of MOEA/D-par over MOEA/D. Due to the limited space, Fig. 8(a) only illustrates the obtained p values for WFG41-4. The Pareto optimal front of WFG41-4 is a hyper-sphere. This indicates that the threshold p value is 2, and thus, the obtained p value should be 3 provided on the considered candidate p values, i.e., $p \in P = \{\frac{1}{2}, \frac{2}{3}, 1, 2, 3, 4, 5, 10, 1000\}$. As is expected, it is observed from Fig. 8(a) that the p values gradually converge to 3. As previously analysed, the $L_{p=3}$ scalarising function is able to find all Pareto optimal solutions for a sphere type Pareto front, i.e., WFG41, and simultaneously, $L_{p=3}$ has a better search ability than the Chebyshev scalarising function.

Analysis of the Range of the Initial Candidate p Values. In principle p can be any value within the interval $(0, \infty]$. However, regarding the computational efficiency, we expect to shrink the range of p as much as possible. Of course, such a shrink should not lead to a severe deterioration of the algorithm performance. In this section, we conduct a simple experimental analysis on the effect of different p values so as to set an upper bound of the candidate p value.

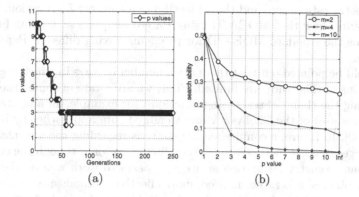

(a)   (b)

Fig. 8. (a)The obtained p values for WFG41-4 over generations; (b) The change of search ability of different L_p scalarising function in 2-, 4- and 10-objective problems

Let us consider a set of $1000 \times m$ points that are uniformly sampled from the hypercube $(0, 2]^m$, where m is the dimension of objective space. Also, consider the contour lines of the L_p scalarising functions along the direction of $\mathbf{w} = \{1/m, \cdots, 1/m\}$. Such contour lines intersect the point $\mathbf{x} = (1, \cdots, 1)$. Then we count the number of points that satisfy the condition $\sum_{i=1}^{m} (\mathbf{x}_i)^p < m$, indicating that $\mathbf{x}$ is better than solutions on the contour line of the $g^{wd}(\mathbf{x}|\mathbf{w}, p) = m$. The experiments are repeated for 100 times. The mean proportion of better points over p values varying from 1 to 10 are plotted in Fig. 8(b) for $m = 2, 4$ and 10-dimension spaces, respectively. From the figure, we find that the search ability of the L_p scalarising function decreases dramatically from $p = 1$ to $p = 5$ whereas slightly when $p > 5$. Moreover, the larger the problem dimension, the faster the decrease of the L_p search ability. The search ability of $L_{p=10}$ appears no significant advantage over the weighted Chebyshev function, in particular, in the 10-dimension space. Therefore, we tentatively recommend that despite the Chebyshev function, $p = 10$ might be considered as an upper bound for the candidate p values.

5   Conclusion

Decomposition based algorithms comprise a popular class of multi-objective evolutionary algorithms, and have been demonstrated to perform well when a suitable set of weighted scalarising functions are provided. The effect of weights, including methods for determining suitable weights, have been intensively studied. However, the effect of scalarising functions is far from being well understood. In this paper we study the properties of the family of L_p scalarising functions, and identify that the p value corresponds to a trade-off between the scalarising function's search ability and its robustness on Pareto front geometry. Moreover, we propose to use different Pareto adaptive scalarising functions along different search directions. A naive method is employed to perform an on line Pareto front geometry estimation, and thus, identifying a suitable L_p function. Experimental results show that MOEA/D using Pareto adaptive scalarising functions outperforms the standard MOEA/D for problems having different Pareto front geometries.

It should be pointed out that there are a number of ways in which the central contributions of this study are limited. First, we are aware of some other methods handling the choice of scalarising functions, for example, an adaptive use (a simultaneous use) of the Chebyshev and weighted sum approaches by Ichibuchi et al. [29,30]. In future, a comprehensive analysis regarding the advantages and disadvantages of these methods will be conducted. Second, though the employed Pareto front geometry estimation strategy appears to work well on the considered test problems, it is rather limited, more effective methods are required. As a start, it is non-trivial to investigate how a suitable set of neigobouring solutions should be chosen as this plays an important role for discontinuous Pareto front geometry estimation. Third, adaptation of scalarising functions accounts effectively varying the subproblems. As discussed in [34], the adaptation can lead to reduced convergence rates, and thus, the effect of adaptation of scalarising functions should be investigated further. Lastly, findings of this study are based on three basic continuous MOPs. It is also important to assess the performance of MOEA/D-par on problems having other complex geometries, other problem types, e.g. multi-objective combinatorial problems, and also, crucially, real-world problems.

Acknowledgments. Acknowledgment. This research was conducted in the Department of computer science, City University of Hong Kong, and the first author is grateful for the facilities and support provided by the University. The first author would also like to thank the National Natural Science Foundation of China (No. 61403404,61473241) and the National University of Defense Technology (No. JC 14-05-01) for their financial support.

References

1. Deb, K.: Multi-objective optimization using evolutionary algorithms. Wiley (2001)
2. Fonseca, C., Fleming, P.: Genetic Algorithms for Multiobjective Optimization: Formulation Discussion and Generalization. In: Proceedings of the 5th International Conference on Genetic Algorithms. Morgan Kaufmann Publishers Inc., pp. 416–423 (1993)
3. Deb, K., Pratap, A., Agarwal, S., Meyarivan, T.: A fast and elitist multiobjective genetic algorithm: NSGA-II. IEEE Transactions on Evolutionary Computation **6**(2), 182–197 (2002)
4. Wang, R., Purshouse, R., Fleming, P.: Preference-inspired Co-evolutionary Algorithms for Many-objective Optimisation. IEEE Transactions on Evolutionary Computation **17**(4), 474–494 (2013)
5. Zitzler, E., Künzli, S.: Indicator-Based Selection in Multiobjective Search. In: Yao, X., Burke, E.K., Lozano, J.A., Smith, J., Merelo-Guervós, J.J., Bullinaria, J.A., Rowe, J.E., Tiňo, P., Kabán, A., Schwefel, H.-P. (eds.) PPSN 2004. LNCS, vol. 3242, pp. 832–842. Springer, Heidelberg (2004)

6. Bader, J., Zitzler, E.: HypE: an algorithm for fast hypervolume-based many-objective optimization. Evolutionary Computation **19**, 45–76 (2011)
7. Murata, T., Ishibuchi, H., Gen, M.: Specification of Genetic Search Directions in Cellular Multi-objective Genetic Algorithms. In: Zitzler, E., Deb, K., Thiele, L., Coello Coello, C.A., Corne, D.W. (eds.) EMO 2001. LNCS, vol. 1993, pp. 82–95. Springer, Heidelberg (2001)
8. Hughes, E.: Multiple single objective pareto sampling. In: 2003 IEEE Congress on Evolutionary Computation (CEC), pp. 2678–2684. IEEE (2003)
9. Zhang, Q., Li, H.: MOEA/D: A Multiobjective Evolutionary Algorithm Based on Decomposition. IEEE Transactions on Evolutionary Computation **11**(6), 712–731 (2007)
10. Wang, R., Fleming, P., Purshouse, R.: General framework for localised multi-objective evolutionary algorithms. Information Sciences **258**(2), 29–53 (2014)
11. Liu, H.-L., Gu, F., Zhang, Q.: Decomposition of a Multiobjective Optimization Problem Into a Number of Simple Multiobjective Subproblems. IEEE Transactions on Evolutionary Computation **18**(3), 450–455 (2014)
12. Li, H., Zhang, Q.: Multiobjective Optimization Problems With Complicated Pareto Sets, MOEA/D and NSGA-II, IEEE Transactions on Evolutionary Computation **13**(2), 284–302 (2009)
13. Zhang, Q., Liu, W., Li, H.: The performance of a new version of MOEA/D on CEC09 unconstrained MOP test instances. In: 2009 IEEE Congress on Evolutionary Computation (CEC), pp. 203–208. IEEE (2009)
14. Ishibuchi, H., Sakane, Y., Tsukamoto, N., Nojima, Y.: Evolutionary many-objective optimization by NSGA-II and MOEA/D with large populations. In: Proceedings of the 2009 IEEE International Conference on Systems, Man, and Cybernetics, vol. 1, pp. 1820–1825. IEEE, San Antonio (2009)
15. Zhang, Q.: Research articles and applications related to MOEA/D. http://dces.essex.ac.uk/staff/zhang/webofmoead.html
16. Giagkiozis, I., Purshouse, R.C., Fleming, P.J.: Generalized Decomposition. In: Purshouse, R.C., Fleming, P.J., Fonseca, C.M., Greco, S., Shaw, J. (eds.) EMO 2013. LNCS, vol. 7811, pp. 428–442. Springer, Heidelberg (2013)
17. Ishibuchi, H., Akedo, N., Nojima, Y.: A Study on the Specification of a Scalarizing Function in MOEA/D for Many-Objective Knapsack Problems. In: Nicosia, G., Pardalos, P. (eds.) LION 7. LNCS, vol. 7997, pp. 231–246. Springer, Heidelberg (2013)
18. Giagkiozis, I., Purshouse, R., Fleming, P.: Generalized decomposition and cross entropy methods for many-objective optimization. Information Sciences, pp. 1–25 (2014) (in press)
19. Wang, R., Purshouse, R., Fleming, P.: Preference-inspired co-evolutionary algorithms using weight vectors. European Journal of Operational Research. http://dx.doi.org/10.1016/j.ejor.2014.05.019 (in press)
20. Wang, R., Purshouse, R., Fleming, P.: Preference-Inspired Co-Evolutionary Algorithm Using Weights for Many-objective Optimisation. In: GECCO 2013: Proceedings of the Genetic and Evolutionary Computation Conference, pp. 101–102. ACM, Amsterdam (2013)
21. Jiang, S., Cai, Z., Zhang, J., Ong, Y.: Multiobjective optimization by decomposition with Pareto-adaptive weight vectors. In: 2011 Seventh International Conference on Natural Computation (ICNC), pp. 1260–1264. IEEE (2011)
22. Li, H., Landa-Silva, D.: An adaptive evolutionary multi-objective approach based on simulated annealing. Evolutionary Computation **19**(4), 561–595 (2011)

23. Qi, Y., Ma, X., Liu, F., Jiao, L., Sun, J., Wu, J.: MOEA/D with Adaptive Weight Adjustment. Evolutionary Computation **22**(2), 231–264 (2013)
24. Derbel, B., Brockhoff, D., Liefooghe, A.: Force-Based Cooperative Search Directions in Evolutionary Multi-objective Optimization. In: Purshouse, R.C., Fleming, P.J., Fonseca, C.M., Greco, S., Shaw, J. (eds.) EMO 2013. LNCS, vol. 7811, pp. 383–397. Springer, Heidelberg (2013)
25. Jin, Y., Okabe, T., Sendhoff, B.: Adapting Weighted Aggregation for Multiobjective Evolution Strategies. In: Zitzler, E., Deb, K., Thiele, L., Coello Coello, C.A., Corne, D.W. (eds.) EMO 2001. LNCS, vol. 1993, pp. 96–110. Springer, Heidelberg (2001)
26. Miettinen, K.: Nonlinear multiobjective optimization. Springer (1999)
27. Derbel, B., Brockhoff, D., Liefooghe, A., Verel, S.: On the Impact of Multiobjective Scalarizing Functions. In: Bartz-Beielstein, T., Branke, J., Filipič, B., Smith, J. (eds.) PPSN 2014. LNCS, vol. 8672, pp. 548–558. Springer, Heidelberg (2014)
28. I. Giagkiozis, P. Fleming, Methods for multi-objective optimization: An analysis, Information Sciences 293, 338–350(2015)
29. Ishibuchi, H., Sakane, Y., Tsukamoto, N., Nojima, Y.: Adaptation of Scalarizing Functions in MOEA/D: An Adaptive Scalarizing Function-Based Multiobjective Evolutionary Algorithm. In: Ehrgott, M., Fonseca, C.M., Gandibleux, X., Hao, J.-K., Sevaux, M. (eds.) EMO 2009. LNCS, vol. 5467, pp. 438–452. Springer, Heidelberg (2009)
30. Ishibuchi, H., Sakane, Y., Tsukamoto, N., Nojima, Y.: Simultaneous use of different scalarizing functions in MOEA/D. In: GECCO 2010: Proceedings of the Genetic and Evolutionary Computation Conference. ACM, Portland, pp. 519–526 (2010)
31. Wang, R.: Towards understanding of selection strategies in many-objective optimisation., Research Report No. 1096, College of Information Systems and Management, National University of Defense Technology (November 2014)
32. Huband, S., Hingston, P., Barone, L., While, L.: A review of multiobjective test problems and a scalable test problem toolkit. IEEE Transactions on Evolutionary Computation **10**, 477–506 (2006)
33. Hernández-Díaz, A., Santana-Quintero, L., Coello Coello, C., Molina, J.: Pareto-adaptive ε-dominance. Evolutionary computation **15**(4), 493–517 (2007)
34. Giagkiozis, I., Purshouse, R.C., Fleming, P.J.: Towards Understanding the Cost of Adaptation in Decomposition-Based Optimization Algorithms. In: 2013 IEEE International Conference on Systems, Man, and Cybernetics (SMC), pp. 615–620. IEEE (2013)

A Bi-level Multiobjective PSO Algorithm

Pedro Carrasqueira[1]([⊠]), Maria João Alves[2], and Carlos Henggeler Antunes[3]

[1] INESC Coimbra, Coimbra, Portugal
pmcarrasqueira@net.sapo.pt
[2] Faculty of Economics, University of Coimbra / INESC Coimbra, Coimbra, Portugal
mjalves@fe.uc.pt
[3] Department of Electrical Engineering and Computers,
University of Coimbra / INESC Coimbra, Coimbra, Portugal
ch@deec.uc.pt

Abstract. Bi-level optimization represents a class of optimization problems with two decision levels: the upper level (leader) and the lower level (follower). Bi-level problems have been extensively studied for single objective problems, but there is few research in case of multiobjective problems in both levels. This case is herein studied using a multiobjective particle swarm optimization (MOPSO) based algorithm. To solve the bi-level multiobjective problem the algorithm searches for upper level Pareto optimal solutions. In every upper level search, the algorithm solves a lower level multiobjective problem in order to find a representative set of lower level Pareto optimal solutions for a fixed upper level vector of decision variables. The search in both levels is performed using the operators of a MOPSO algorithm. The proposed algorithm is able to solve bi-level multiobjective problems achieving solutions in the true Pareto optimal front or close to it.

1 Introduction

In many practical situations there are hierarchical decisions such that whenever a decision maker does not completely control all the variables of the problem. Suppose the case that a leader, the chief executive officer (CEO) of a company, intends to maximize global profits, and a follower, the division manager, is responsible for optimizing his/her resources. Each one controls different sets of variables and has his/her own goals, which are often different and may be conflicting. Two embedded optimization problems have to be considered to represent such situation as an optimization problem, each one relating to each decision maker, where the lower level problem acts as a constraint to the upper level one. This is a bi-level optimization problem. Here the CEO's level of decision is the upper level and the division manager's level of decision is the lower level. Hierarchical optimization structures appears naturally in management of decentralized organizations and in many aspects of policy making (e.g. transportation network design or energy pricing) [4]. In addition, the upper level and/or the lower level of a bilevel decision problem may involve multiple objectives. In our work, the bi-level problem with multiple objectives in both levels is considered

© Springer International Publishing Switzerland 2015
A. Gaspar-Cunha et al. (Eds.): EMO 2015, Part I, LNCS 9018, pp. 263–276, 2015.
DOI: 10.1007/978-3-319-15934-8_18

(BLMOP). To solve the BLMOP, a multiobjective particle swarm optimization (MOPSO) algorithm is proposed. MOPSO is a population based meta-heuristic which has obtained good results in a wide range of problems [7].

The existence of multiple Pareto optimal solutions at the lower level for a given upper level solution makes the resolution of the BLMOP a very difficult task. Recently some researchers have studied the BLMOP and proposed algorithms to tackle it.

Eichfelder [8], [9] presented new theoretical results and proposed an algorithm to solve the BLMOP with bi-objective problems in both levels and a single upper level variable. Nishizaki and Sakawa [14], Shi and Xia [16] and Abo-Sinna and Baky [3] proposed interactive algorithms to solve the BLMOP. Evolutionary algorithms have also been used to deal with BLMOP. Deb and Sinha [6] proposed an evolutionary algorithm (BLEMO) based on the NSGA-II multiobjective algorithm. The BLEMO algorithm has been used to solve the BLMOP, including problems with interdependent constraints in both levels. In [17], Sinha and Deb have considered additional rules to enhance the algorithm's ability to find the true Pareto front. Particle swarm optimization has also been used to solve bi-level problems with multiple objectives in one level or both levels. Halter and Mostaghim [11] proposed a particle swarm based algorithm to solve a real-world bi-level problem with a single objective in the upper level and three objectives in the lower level with linear constraints. Alves [1] proposed a bi-level MOPSO algorithm to solve bi-level linear problems with multiple objectives in the upper level and a single objective in the lower level. This algorithm was further improved in [2]. The algorithms in [1],[2] and [11] exploit the linearity of the objective functions and/or the constraints to solve the lower level problem. Zhang et al. [18] presented an algorithm based on multiobjective PSO to solve the BLMOP. In order to avoid premature convergence in the PSO, a crossover operator was later introduced in [19].

Most of these algorithms are devoted to specific cases such as requiring linearity, limiting the number of objectives or upper level variables. The algorithms developed in [6], [17], [18] and [19] are able to solve generic multiobjective bi-level problems.

The algorithm proposed in this paper (OMOPSO-BL) uses a MOPSO algorithm to solve the bi-level problem with multiple objectives in both levels. After each upper level search, OMOPSO-BL performs a lower level search in order to find a representative set of lower level Pareto optimal solutions for a fixed upper level vector of decision variables. This step is crucial and shall be thoroughly monitored in order to get feasible solutions to the upper level problem. The algorithm uses an upper level - lower level interactive process based on the one considered in BLEMO. To perform the MOPSO operations, the OMOPSO [15] algorithm has been selected due to the good performance it has demonstrated in comparison to other MOPSO approaches [7]. The proposed algorithm OMOPSO-BL is compared with the BLEMO algorithm in order to assess its competitiveness in some benchmark problems.

The remainder of this paper is organized as follows. In Section 2 the BLMOP formulation and some definitions are stated. The main principles of the BLEMO and OMOPSO-BL algorithms are presented in Section 3. In Section 4 some experimental results are shown. Section 5 is devoted to conclusions and future directions of research.

2   The Bi-level Multiobjective Optimization Problem

Bi-level optimization problems are characterized by the existence of one (lower level) problem embedded in other (upper level) optimization problem. A multi-objective bi-level problem contains, at least, one multiobjective problem in one of its levels. Here we consider both upper and lower level problems as multi-objective optimization problems. A general bi-level multiobjective optimization problem (BLMOP) can be defined as

$$\min F(x) = (F_1(x), F_2(x), \cdots, F_M(x))$$
$$\text{s.t.}\ \ G(x) \geq 0,$$
$$\min f(x_l) = (f_1(x_l), f_2(x_l), \cdots, f_m(x_l)),$$
$$\text{s.t.}\ \ g(x) \geq 0 \tag{1}$$

where $F(x)$ is the upper level objective vector to optimize, $x = (x_u, x_l)$ with $x_u \in \mathbb{R}^{n_1}$ and $x_l \in \mathbb{R}^{n_2}$ the upper and lower level decision vectors, respectively, and $G(x)$ and $g(x)$ are the upper and lower level constraints, respectively.

Definition 1. *1. Let us consider S the constraint region of problem (1), $S = \{(x_u, x_l) \in \mathbb{R}^{n_1+n_2} | G(x_u, x_l) \geq 0, g(x_u, x_l) \geq 0\}$;*
2. The follower's rational reaction set for a given x_u is defined as $P(x_u) = \{x_l \in \mathbb{R}^{n_2} | x_l \in Eff(x_u)\}$, with $Eff(x_u)$ the set of efficient/Pareto optimal solutions of the lower level problem for a given x_u;
3. The feasible set for the leader, which is the feasible set to problem (1), is called induced region. It is given by: $IR = \{(x_u, x_l) | (x_u, x_l) \in S, x_l \in P(x_u)\}$.

It is worth mentioning that the lower level optimization problem is solved only relatively to the x_l variables, keeping the x_u variables fixed.

Definition 2. *If (x_u^*, x_l^*) is a feasible solution to problem (1) and there is no $(x_u, x_l) \in IR$ such that $F_j(x_u, x_l) \leq F_j(x_u^*, x_l^*)$ for all $j=1,\ldots,M$ and $F_j(x_u, x_l) < F_j(x_u^*, x_l^*)$ for at least one $j=1,\ldots,M$, then (x_u^*, x_l^*) is a Pareto optimal solution to problem (1).*

The set of all Pareto optimal solutions to a multiobjective optimization problem is called the Pareto optimal front. Our goal is to approximate the entire Pareto optimal front of the problem. We refer to the solutions found by the algorithm as non-dominated solutions.

3 Bi-level Multiobjective Algorithms

3.1 BLEMO Algorithm

The bi-level algorithm BLEMO [17] uses the evolutionary algorithm NSGA-II [5] operators to solve both the upper and the lower level multiobjective problems composing the bi-level problem (1). The algorithm uses a special structure to represent the population, which is composed by N_u individuals split into n_s sub-populations. All the members of one sub-population have the same x_u vector. Therefore, there are only n_s different upper level vectors of decision variables in the population. The non-dominated sorting mechanism and the crowding distance operator introduced in [5] are used to rank each member of the population. Thus, each individual is assigned a non-dominated rank (ND_u) and a crowding distance (CD_u) relating to upper level objectives and constraints. Considering each individual as a member of one sub-population, the non-dominated rank (ND_l) and the crowding distance (CD_l), relating to lower level objectives and constraints, are also computed. At each upper level iteration of the algorithm, BLEMO starts by using the selection, crossover and mutation genetic operators to create the new population. After that, the non-dominated solutions of the lower level problem for each sub-population are computed. BLEMO uses the NSGA-II algorithm to perform this task. At the end of each iteration, BLEMO stores the non-dominated solutions to problem (1) in an external archive. To prevent dominated individuals of the lower level (i.e. infeasible solutions to problem (1)) from entering the archive, only the individuals that have been in the population for at least r generations can be inserted. This archive is ranked using non-dominating sorting and crowding distance.

3.2 A Bi-level Multiobjective Particle Swarm Optimization (OMOPSO-BL) Algorithm

We propose an algorithm based on PSO which employs a scheme similar to the one used in BLEMO for the interaction between the upper and lower level optimization phases, division of the population into sub-populations with the same x_u vector, and update the external archive with the upper level non-dominated solutions. The OMOPSO-BL algorithm also uses a lower level archive to save the non-dominated solutions found during the lower level iterations. This archive guides the lower level search. The algorithm uses MOPSO operations to solve the bi-level multiobjective optimization problem. In a MOPSO approach it is usual to refer to individuals as particles. Each particle i is assigned a *velocity* vector (v_i) that indicates the direction of the particle movement resulting from the combination of the directions to the best position so far achieved by the particle $(pbest_i)$ and to the best position attained by the whole population $(gbest)$. In multiobjective optimization, $pbest_i$ and $gbest$ are not unique. Each particle i moves itself in iteration t according to the expressions

$$v_i^{t+1} = wv_i^t + c_1 rand()(pbest_i - x_i^t) + c_2 rand()(gbest - x_i^t) \qquad (2)$$

$$x_i^{t+1} = x_i^t + v_i^{t+1}, i = 1, 2, \cdots, n \tag{3}$$

where w is the inertia weight, c_1 and c_2 are the cognitive and social parameters, $rand(\ )$ is a random uniform value in the interval $[0,1]$ and n is the number of particles in the population. The main options and parameters of the OMOPSO [15] algorithm are considered herein, because this algorithm has become one of the most representatives of the state-of-the-art MOPSO approaches [7]. In the case of OMOPSO [15], the parameters are randomly chosen, in each iteration, within a predefined interval: $w \in [0.1, 0.5]$ and $c_1, c_2 \in [1.5, 2]$. As other features of OMOPSO algorithm we refer to the use of a mutation operator, called turbulence, and a crowding distance [5] measure, assigned to each particle of the external archive. This measure is used to select the leader for each particle i of the population $(gbest_i)$ and a particle of the external archive to be replaced when the archive is full. The mutation operator is applied, with a certain probability, after the operations (2) and (3). The population is split in three parts. One third is applied a uniform mutation, another third is applied a non-uniform mutation and the last third of the population is not changed.

Algorithm 1. OMOPSOlowerlevel(P^s, B^s, t_l) - pseudo code

1: **for** each particle $x_i^{t_l} \in P^s$, $i = 1, \cdots, N_l$ **do**
2:      Select the leader $gbest_i$, from the archive B^s, by binary tournament
3:      Update particle velocity $v_i^{t_l+1}$, using (2)
4:      Update particle position $x_i^{t_l+1}$, using (3)
5:      Mutate $x_i^{t_l+1}$, using $turbulence$ operator
6:      Assess the particle $x_i^{t_l+1}$, evaluating $f(x_i^{t_l+1})$
7:      Update $pbest_i$
8: **end for**

The OMOPSO-BL algorithm starts by initializing the population P_t randomly. Then, T iterations of the upper level problem are performed. At each iteration, the OMOPSO-BL algorithm searches for upper level Pareto optimal solutions. Inside the main (upper level) iteration, the OMOPSO-BL algorithm has embedded a lower level routine in order to obtain non-dominated solutions of the lower level problem. The feasible solutions of the upper level problem are among these solutions. In the lower level routine, T_l iterations of the OMOPSO algorithm are performed. At the end of each lower level routine, the archive A_t with upper level non-dominated solutions is updated. The steps of the OMOPSO-BL algorithm are detailed in Algorithm 2, which calls the subroutine *OMOPSOlowerlevel* described in Algorithm 1.

At each new upper level iteration, the algorithm starts by creating the sub-populations (lines 10-22). To create each sub-population, an upper level vector x_u is considered after selecting a particle from the population by binary tournament, using (ND_u) and (CD_u) values lexicographically, and updating it using the OMOPSO operators. Then, N_l individuals are randomly chosen from the current population P_t and the lower level vectors x_l are updated using the

---

Algorithm 2. OMOPSO-BL pseudo code

1: Initialize upper level iteration counter, $t = 1$
2: Randomly initialize each individual of population P_t: $x_{s_j}^t = (x_{u_s}^t, x_{l_j}^t), j = 1, \cdots, N_l,\ s = 1, \cdots, n_s$
3: Assess each individual of the population
4: Rank each sub-population members, evaluating ND_l and CD_l values
5: Rank the population individuals, evaluating ND_u and CD_u values
6: $Pre\_A_t = \emptyset,\ A_t = \emptyset$
7: Insert the individuals of P_t with $ND_u = 1$ and $ND_l = 1$ into Pre_A_t
8: **while** $t \leq T$ **do**
9:      // *Create the new population Q_t as the union of the n_s sub-populations*
10:     $Q_t^s = \emptyset$
11:     **for** $s = 1, \cdots,\ n_s = N_u/N_l$ **do**
12:         Select an individual x_s^t from P_t by binary tournament, using ND_u and CD_u
13:         Update component $x_{u_s}^t$ of individual x_s^t, using (2), (3) and turbulence
14:         Randomly select N_l individuals from the population P_t
15:         **for** $j = 1, \cdots,\ N_l$ **do**
16:             Create $x_{l_j}^t$, using (2), (3) and turbulence
17:             Concatenate $x_j^t = (x_{u_s}^t, x_{l_j}^t)$
18:             $Q_t^s = Q_t^s \cup \{x_j^t\}$
19:         **end for**
20:         Assess Q_t^s sub-population, evaluating F, G, f, g
21:         Assign ND_l and CD_l values to each $x_j^t, j = 1, \cdots, N_l$
22:     **end for**
23:     Set the population $Q_t = \cup_{s=1}^{n_s} Q_t^s$
24:     **for** $s = 1, \cdots,\ n_s$ **do**
25:         Set $t_l = 1$
26:         **if** $t > 1$ **then**
27:             Select $B_t^s = A_t^s$, the sub-archive of A_t nearest Q_t^s
28:         **else**
29:             Select $B_t^s = Pre\_A_t^s$, the sub-archive of Pre_A_t nearest Q_t^s
30:         **end if**
31:         Initialize $pbest_i$ as the individual particle $x_i^{t_l}, i = 1, \cdots, N_l$
32:         **while** $t_l \leq T_l$ **do**
33:             OMOPSOlowerlevel(Q_t^s, B_t^s, t_l)
34:             Rank (Q_t^s) individuals, evaluating ND_l and CD_l values
35:             Update lower level archive B_t^s of non-dominated solutions
36:             Set $t_l = t_l + 1$
37:         **end while**
38:         Update external archive A_t of non-dominated solutions
39:     **end for**
40:     Set the population $Q_t = \cup_{s=1}^{n_s} Q_t^s$
41:     Set $R_t = P_t \cup Q_t$
42:     Assign ND_u and CD_u values to each individual $x_j^t, j = 1, \cdots, 2N_u$ of R_t
43:     Create S_t population, with half of R_t individuals, using ND_u, CD_u and ND_l
44:     **for** each sub-population S_t^i originated from P_t **do**
45:         Set $t_l = 1$

```
46:        if t > 1 then
47:            Select B_t^i = A_t^i, the sub-archive of A_t nearest S_t^i
48:        else
49:            Select B_t^i = Pre_A_t^i, the sub-archive of Pre_A_t nearest S_t^i
50:        end if
51:        Initialize pbest_j as the individual particle x_j^{t_l}, i = 1, ⋯ , N_l
52:        while t_l ≤ T_l do
53:            OMOPSOlowerlevel(S_t^i, B_t^i, t_l)
54:            Rank (S_t^i) individuals, evaluating ND_l and CD_l values
55:            Update lower level archive B_t^i of non-dominated solutions
56:            Set t_l = t_l + 1
57:        end while
58:        Update external archive A_t of non-dominated solutions
59:    end for
60:    Assign ND_u and CD_u values to each individual x_j^t, j = 1, ⋯ , N_u of S_t
61:    Set P_{t+1} = S_t
62:    Set t = t + 1
63: end while
64: Return archive A_t of non-dominated solutions
```

OMOPSO operators. For both upper and lower levels, *gbest* is selected from the non-dominated archive A_t by binary tournament, selecting the least crowded individuals using (CD_u) and (CD_l) values, respectively. Then, the x_l solutions are concatenated with the x_u to form one sub-population. It should be emphasized that all N_l solutions have the same x_u vector. Next, the individuals of the sub-population are evaluated and the ranking (ND_l) and the crowding distance (CD_l) values are assigned to each individual. These operations are repeated until n_s sub-populations have been obtained. Finally all the sub-populations are joined into one population (Q_t) and the ranking (ND_u) and the crowding distance (CD_u) values are assigned to each individual in Q_t.

In order to guide the lower level search for each sub-population (lines 24-39), the algorithm starts by identifying the elements of the archive A_t whose x_u vectors are closer in terms of Euclidean distance to the x_u vector of that sub-population. The selected individuals form a sub-archive, which we call a lower level archive. Considering one sub-population, the OMOPSO algorithm is applied to solve the lower level problem. Initially, for each individual, velocity is null and *pbest* is assigned the individual position. The *gbest* vector is always chosen by binary tournament from the corresponding lower level archive. After updating the individuals' position, the sub-population is evaluated and the *pbest*, the rank and the crowding values of each individual are updated. The $ND_l = 1$ ranked individuals are then tested to enter into the lower level archive. At the end, the lower level archive is sorted by descending order of its members' crowding values. These operations are performed for T_l iterations. This search repeats for all the sub-populations. It may occur that some individuals of the lower level archive are not true Pareto optimal for the problem. Such individuals are not feasible for the upper level problem, although they can dominate some true Pareto optimal individuals of the upper level problem. It should be guaranteed that such solutions do not enter into the archive A_t. To address this problem a parameter r is considered in order to filter particles entering the archive A_t. Only the individuals which have remained in the lower level archive for at least r iterations are tested to enter the archive A_t. In addition, only the solutions that are non-dominated in the upper level with respect to any member of A_t enter the archive A_t. Elements in A_t that become dominated by the new members are then removed from A_t. A higher number of lower level iterations must be performed to approximate the Pareto optimal front of the problem with sufficient accuracy. The n_s updated sub-populations are joined to create the updated population Q_t.

In order to create the new population (lines 41-43), the P_t and the Q_t populations are joined and the R_t population is obtained, consisting of $2N_u$ individuals. The composition of each sub-population from P_t or Q_t is also maintained because it is copied into the new population as an entire block. Aiming at obtaining the fittest individuals, the R_t population is ranked by ND_u values. Starting at $ND_u = 1$ ranked individuals, the ones ranked $ND_l = 1$ are sorted in descending order of their crowding distance values CD_u. Those individuals and the sub-populations they come from are sequentially inserted into the next population S_t. It is worth mentioning that if a new element is tried to enter S_t and its sub-population is already in S_t, it is not copied again in order to prevent the duplication of individuals. After considering every $ND_u = 1$ ranked elements, the individuals ranked $ND_u = 2$ are considered and so on, until N_u members have been copied into the population S_t.

The sub-populations copied to S_t that came from the population P_t are now updated through the OMOPSO algorithm in the same manner as described above (lines 44-59).

The new population P_{t+1} is created joining the n_s sub-populations of S_t. After performing T iterations, the algorithm returns the archive A_t containing the upper level non-dominated solutions as the solutions to the bi-level multiobjective problem.

It should be noted that, at the first iteration of the algorithm, the archive A_t is an empty set and, since elite solutions are needed to perform the search, the non-dominated solutions of the population are considered. Thus, a pre-archive containing the non-dominated solutions existing in the initial population is considered and is used in place of A_t in every step of the iteration $t = 1$.

Because of the structure of the bi-level problem, a certain sequence of upper and lower level operations has to be performed. Thus, performing a sequence of several iterations of the upper level search without running the lower level optimization phase would not be worthwhile, because the solutions obtained may be infeasible to the bi-level problem (if they are dominated in the lower level problem). In our algorithm only particles obtained after a lower level run are accepted to enter the archive A_t. This differs from the algorithm presented in [18],[19], in which a particle enters into the archive directly after an upper level run, without performing the lower level optimization. The authors in [18],[19] do not detail how they deal with the drawbacks of this operation. We have tested this process and we obtained particles entering into the archive A_t which were dominated for the lower level problem. As soon as such particles enter the archive it becomes impossible to insert true Pareto optimal particles dominated by them. These difficulties arise because the particles that are dominated in the lower level may not violate the upper level constraints, although they are not feasible to the problem. Some tests we have performed using that algorithm confirmed this difficulty. The OMOPSO-BL algorithm proposed herein overcomes this problem.

4   Experimental Results

4.1   Test Problems

For the sake of illustration, in this study we use four instances of two multiobjective bi-level problems from [6], considering three different dimensions for the second problem.

Problem 1

$$\min F(x) = (x_{l_1} - x_u, x_{l_2})$$
$$\text{s. t. } G(x_u, x_l) = 1 + x_{l_1} + x_{l_2} \geq 0,$$
$$\min f(x_l) = (x_{l_1}, x_{l_2}), \tag{4}$$
$$\text{s. t. } g(x_u, x_l) = x_u^2 - x_{l_1}^2 - x_{l_2}^2 \geq 0$$
$$-1 \leq x_{l_1}, x_{l_2} \leq 1, \ 0 \leq x_u \leq 1$$

This problem has one upper level variable (x_u) and two lower level variables (x_{l_1}, x_{l_2}). Because of the upper level constraint, some of the lower level Pareto optimal solutions are not feasible to the upper level problem. This increases the difficulty of the problem. The Pareto optimal solutions to the problem was firstly reported in [8]. It is given by:

$$P^* = \{(x_u, x_{l_1}, x_{l_2}) \in \mathbb{R}^3 | x_{l_1} = 1 - x_{l_2}, \ x_{l_2} = -\frac{1}{2} \pm \sqrt{8x_u^2 - 4}, \ x_u \in [\frac{1}{\sqrt{2}}, 1]\}$$

Problem 2

$$\min F(x) = ((x_{l_1} - 1)^2 + \sum_{i=1}^{K} x_{l_{i+1}}^2 + x_u^2, (x_{l_1} - 1)^2 + \sum_{i=1}^{K} x_{l_{i+1}}^2 + (x_u - 1)^2)$$
$$\text{s. t. } \min f(x_l) = (x_{l_1}^2 + \sum_{i=1}^{K} x_{l_{i+1}}^2, (x_{l_1} - x_u)^2 + \sum_{i=1}^{K} x_{l_{i+1}}^2),$$
$$-1 \leq x_{l_1}, \cdots, x_{l_{K+1}}, x_u \leq 2$$

$$\tag{5}$$

This problem has one variable in the upper level and K variables in the lower level. From the entire Pareto optimal set of the lower level problem for a given x_u, only one solution is Pareto optimal solution to the upper level problem. This turns the problem difficult to solve. The set of Pareto optimal solutions of this problem is:

$$P^* = \{x_{l_i} = 0, \ i = 2, \cdots, (K+1), \ x_{l_1} = x_u, \ x_u \in [0.5, 1]\}.$$

For this problem, we consider three instances: $K = 1$, $K = 5$ and $K = 13$.

4.2   Performance Measures

The multiobjective algorithms are assessed in terms of convergence to the Pareto front and with respect to the diversity of the obtained solutions. In order to assess the proposed OMOPSO-BL and compare it with BLEMO, two unary performance measures commonly used in the literature are considered: the Hypervolume indicator (HV), measuring the volume of the space between the non-dominated front obtained and a reference point (usually the nadir point is considered) [10],[20], and the Inverted Generational Distance (IGD), which is the sum of the distances from each point of the true Pareto front to the nearest point of the non-dominated set found by the algorithm. Both indicators measure the convergence and spread of the obtained set of solutions. The lower the IGD value, the better the approximation is. Larger values of HV indicate better approximation sets. A statistical analysis is performed to assess the significance of results. Since the results do not follow a normal distribution, the non-parametric Mann-Whitney test is used to compare the algorithms.

4.3   Parametrization of the Algorithms

The algorithms were implemented in Matlab and the tests were performed on an Intel core 2 Duo 2.4 GHz processor. To assess the IGD measure a true Pareto front sample with 500 solutions was considered for each problem. In order to achieve better results, a good balance should be considered between upper and lower level number of iterations and population size. Thus, after some experiments, with different parametrization, the parameter setting selected to run OMOPSO-BL was: $N_u = 300$, $N_l = 60$, $T_l = 80$ and $T = 100$ for problem 2 and $N_u = 240$, $N_l = 60$, $T_l = 80$ and $T = 100$ for problem 1. The parameters of OMOPSO operations were set as in the original algorithm. For the BLEMO algorithm we used the same parametrization as in OMOPSO-BL. Each algorithm was run 10 times for each problem.

4.4   Results

The solutions obtained by the algorithm OMOPSO-BL for problem 1 are depicted in figure 1. OMOPSO-BL achieved solutions close to the true Pareto optimal front of the problem and the solutions are spread along the entire Pareto optimal front.

Figure 3 represents the solutions obtained by the OMOPSO-BL algorithm for problem 2 ($K = 1$). The results show that the OMOPSO-BL algorithm also obtained a good representation of the true Pareto optimal front of this problem. Figures 2 and 4 contain the results obtained by the BLEMO algorithm to problems 1 and 2 ($K = 1$), respectively. Both algorithms were able to approximate the true Pareto optimal fronts of each problem. In table 1, the hypervolume and IGD values obtained by OMOPSO-BL and BLEMO for all problem instances are presented. The algorithms have similar performance. Observing the IGD values, it may be concluded that both algorithms obtain solutions well spread along the

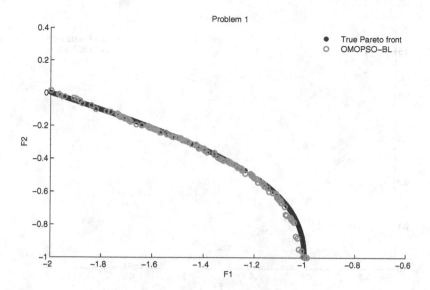

Fig. 1. Problem 1. OMOPSO-BL algorithm

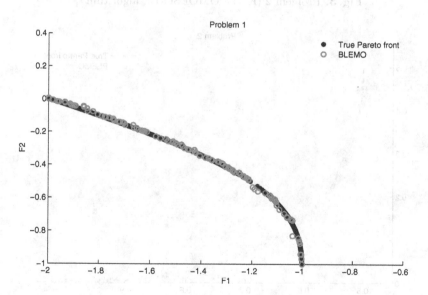

Fig. 2. Problem 1. BLEMO algorithm

entire Pareto optimal front of the instances tested. OMOPSO-BL has achieved slightly better hypervolume values in three over the four instances and BLEMO obtained better IGD values in all instances. In both performance measures the values of OMOPSO-BL and BLEMO are very close.

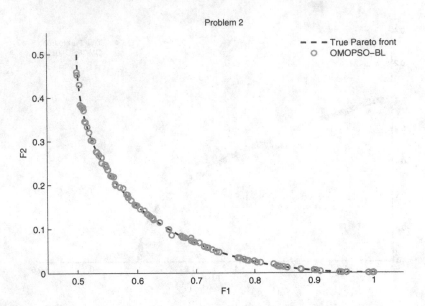

Fig. 3. Problem 2 (K=1). OMOPSO-BL algorithm

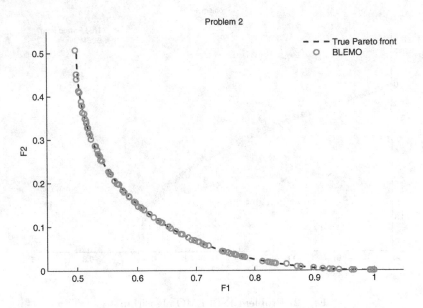

Fig. 4. Problem 2 (K=1) BLEMO algorithm

The Mann-Whitney test confirms the significance of the results at $\alpha = 0.05$ level in all instances except in the problem 2 ($K = 1$ and $K = 5$), for the hypervolume measure.

Table 1. HV and IGD median values obtained by OMOPSO-BL and BLEMO algorithms

Problem	OMOPSO-BL		BLEMO	
	Hypervolume	IGD	Hypervolume	IGD
Problem 1	**0.3068**	0.01562	0.3024	**0.01138**
Problem 2 (K=1)	**0.2074**	0.01020	0.2067	**0.00631**
Problem 2 (K=5)	**0.2064**	0.01289	0.2052	**0.00921**
Problem 2 (K=13)	0.2018	0.03124	**0.2059**	0.01053

5   Conclusions and Future Research

In this paper a multiobjective particle swarm optimization algorithm (OMOPSO-BL) has been proposed to solve the bi-level optimization problem with multiple objective functions in both levels (BLMOP). The main difficulty of these problems is that dominated solutions in the lower level may dominate true Pareto optimal solutions to the bi-level problem and may be accepted to enter the upper level non-dominated archive, although they are not feasible to the problem. In order to avoid this situation, accurate solutions to the lower level problem should be computed. The OMOPSO-BL algorithm has shown its ability to overcome this difficulty in solving the test problems. The algorithm is competitive in comparison with BLEMO, an evolutionary algorithm developed to solve the generic BLMOP. The hypervolume and IGD values obtained by both algorithms are similar.

Further research should be conducted namely to assess the robustness of the OMOPSO-BL algorithm in a larger and more demanding set of problem instances as well as reduce the computational effort. For this purpose, hybridization with other techniques and adaptive parameters will be attempted.

Acknowledgements.. This R&D work has been partially supported by the Portuguese Foundation for Science and Technology (FCT) under projects grant PEst-OE/EEI/UI0308/2014, and QREN Mais Centro Program Projects EMSURE (CENTRO 07 0224 FEDER 002004) and iCIS (CENTRO-07-ST24-FEDER-002003).

References

1. Alves, M.J.: Using MOPSO to Solve Multiobjective Bilevel Linear Problems. In: Dorigo, M., Birattari, M., Blum, C., Christensen, A.L., Engelbrecht, A.P., Groß, R., Stützle, T. (eds.) ANTS 2012. LNCS, vol. 7461, pp. 332–339. Springer, Heidelberg (2012)
2. Alves, M.J., Costa, J.P.: An algorithm based on particle swarm optimization for multiobjective bilevel linear problems. Applied Mathematics and Computation **247**, 547–561 (2014)
3. Abo-Sinna, M.A., Baky, A.I.: Interactive Balance Space Approach for Solving Multi-level Multi-objective Programming Problems. Information Sciences **177**, 3397–3410 (2007)
4. Bard, J.F.: Practical Bilevel Optimization: Algorithms and Applications. Kluwer Academic Publisher, Dordrech (1998)

5. Deb, K., Pratap, A., Agarwal, S., Meyarivan, T.: A Fast and Elitist Multiobjective Genetic Algorithm: NSGA-II. IEEE Transactions on Evolutionary Computation 6(2), 182–197 (2002)
6. Deb, K., Sinha, A.: Solving Bilevel Multi-Objective Optimization Problems Using Evolutionary Algorithms. In: Ehrgott, M., Fonseca, C.M., Gandibleux, X., Hao, J.-K., Sevaux, M. (eds.) EMO 2009. LNCS, vol. 5467, pp. 110–124. Springer, Heidelberg (2009)
7. Durillo, J.J., García-Nieto, J., Nebro, A.J., Coello, C.A.C., Luna, F., Alba, E.: Multi-Objective Particle Swarm Optimizers: An Experimental Comparison. In: Ehrgott, M., Fonseca, C.M., Gandibleux, X., Hao, J.-K., Sevaux, M. (eds.) EMO 2009. LNCS, vol. 5467, pp. 495–509. Springer, Heidelberg (2009)
8. Eichfelder, C.: Solving nonlinear multiobjective bilevel optimization problems with coupled upper level constraints. Technical Report Preprint No. 320, Preprint Series of the Institute of Applied Mathematics, Univ. Erlangen-Nrnberg, Germany (2007)
9. Eichfelder, G.: Multiobjective Bilevel Optimization. Mathematical Programming 123(2), 419–449 (2010)
10. Fonseca, C.M., Paquete, L., López-Ibáñez, M.: An Improved Dimension-Sweep Algorithm for the Hypervolume. In: Proceedings of 2006 IEEE Congress on Evolutionary Computation, pp. 1157–1163 (2006)
11. Halter, W., Mostaghim, S.: Bilevel optimization of multi-component chemical systems using particle swarm optimization. In: Proc. of the 2006 Congress on Evolutionary Computation (CEC 2006), pp. 1240–1247. IEEE Press (2006)
12. Kennedy, J., Eberhart, R.C.: Particle swarm optimization. In: IEEE International Conference on Neural Network, pp. 1942–1948 (1995)
13. Mousa, A.A., El-Shorbagy, M.A., Abd-El-Wahed, W.F.: Local search based hybrid particle swarm optimization algorithm for multiobjective optimization. Swarm and Evolutionary Computation 3, 1–14 (2012)
14. Nishizaki, I., Sakawa, M.: Stakelberg Solutions to Multiobjective Two-Level Linear Programming Problems. Journal of Optimization Theory and Applications 103, 161–182 (1999)
15. Sierra, M.R., Coello, C.A.C.: Improving PSO-Based Multi-objective Optimization Using Crowding, Mutation and ε-Dominance. In: Coello Coello, C.A., Hernández Aguirre, A., Zitzler, E. (eds.) EMO 2005. LNCS, vol. 3410, pp. 505–519. Springer, Heidelberg (2005)
16. Shi, X., Xia, H.: Model and interactive algorithm of bi-level multi-objective decision-making with multiple interconnecting decision makers. J. Multi-Criteria Decision Analysis 10, 27–34 (2001)
17. Sinha, A., Deb, K.: Towards understanding evolutionary bilevel multiobjective optimization algorithm. Technical Report Proceedings of the IFAC Workshop on Control Applications of Optimization (6–8 May, 2009, Jyvskyl, Finland), Kanpur, Indian Institute of Technology, India. (Also KanGAL Report No. 2008006) (2009)
18. Zhang, T., Hu, T., Zheng, Y., Guo, X.: An Improved Particle Swarm Optimization for Solving Bilevel Multiobjective Programming Problem. Journal of Applied Mathematics, Article ID 626717, 13 pages (2012)
19. Zhang, T., Hu, T., Guo, X., Chen, Z., Zheng, Y.: Solving high dimensional bilevel multiobjective programming problem using a hybrid particle swarm optimization algorithm with crossover operator. Knowledge-Based Systems 53, 13–19 (2013)
20. Zitzler, E., Thiele, L.: Multiobjective Evolutionary Algorithms: A Comparative Case Study and the Strength Pareto Approach. IEEE Transactions on Evolutionary Computation 3(4), 257–271 (1999)

An Interactive Simple Indicator-Based Evolutionary Algorithm (I-SIBEA) for Multiobjective Optimization Problems

Tinkle Chugh[(⊠)], Karthik Sindhya, Jussi Hakanen, and Kaisa Miettinen

Department of Mathematical Information Technology, University of Jyvaskyla,
P.O. Box 35, FI-40014 Jyvaskylan yliopisto, Finland
tinkle.t.chugh@student.jyu.fi,
{karthik.sindhya,jussi.hakanen,kaisa.miettinen}@jyu.fi

Abstract. This paper presents a new preference based interactive evolutionary algorithm (I-SIBEA) for solving multiobjective optimization problems using weighted hypervolume. Here the decision maker iteratively provides her/his preference information in the form of identifying preferred and/or non-preferred solutions from a set of nondominated solutions. This preference information provided by the decision maker is used to assign weights of the weighted hypervolume calculation to solutions in subsequent generations. In any generation, the weighted hypervolume is calculated and solutions are selected to the next generation based on their contribution to the weighted hypervolume. The algorithm is compared with a recently developed interactive evolutionary algorithm, W-Hype on some benchmark multiobjective optimization problems. The results show significant promise in the use of the I-SIBEA algorithm. In addition, the performance of the algorithm is demonstrated using a human decision maker to show its flexibility towards changes in the preference information. The I-SIBEA algorithm is found to flexibly exploit the preference information from the decision maker and generate solutions in the regions preferable to her/him.

1 Introduction

Industrial optimization problems often involve multiple conflicting objectives, which usually have multiple Pareto optimal solutions with different trade-offs. Different methods have been proposed in the literature (see e.g. [13]) and evolutionary multiobjective optimization (EMO) algorithms [5,6] have often been applied to solve multiobjective optimization problems and find an approximation of the Pareto front consisting of all the Pareto optimal solutions. However, finding an approximation of the Pareto front is not easy, especially when objective and constraint functions are computationally expensive.

When EMO algorithms are used to find an approximation of the Pareto front, a human decision maker (DM) who is an expert in the domain of the problem is supposed to choose one among several nondominated solutions for implementation or further evaluation. Such an approach is often termed as a posteriori

© Springer International Publishing Switzerland 2015
A. Gaspar-Cunha et al. (Eds.): EMO 2015, Part I, LNCS 9018, pp. 277–291, 2015.
DOI: 10.1007/978-3-319-15934-8_19

approach in multiobjective optimization [13]. Since finding a good approximation of the Pareto front is often difficult, especially when more than two computationally expensive objectives are involved, it is practical to approximate a region of the Pareto front that is of interest to the DM. At least two different approaches involving preference information have been considered in the literature:

1. *a priori approaches* where the DM initially expresses her/his preference information, which is subsequently used to find a set of solutions reflecting her/his preferences [7,9], and
2. *interactive methods* where the DM iteratively provides her/his preference information and drives the algorithm towards her/his preferred region(s) of the Pareto front [8,12,15].

We can easily incorporate DM's preference information in indicator based evolutionary algorithms. These algorithms have been proposed in the literature [3,4] to handle a large number of objectives and in this article we focus our attention on these algorithms. In them, a hypervolume of the dominated region of the objective space is used as the indicator of the quality of the approximation of the Pareto front due to the Pareto compliance of the indicator [16]. However, as the number of objectives increases, the calculation of the hypervolume gets extremely time consuming. Recently, a Monte-Carlo simulation based approach to calculate hypervolume has been proposed to speed up the calculation [2].

In this paper, we propose a new interactive preference based EMO algorithm called *interactive simple indicator-based evolutionary algorithm (I-SIBEA)* where different weights are associated with different regions of the Pareto front such that the importance given by the DM for different regions of the Pareto front can be altered. In the proposed algorithm, we extend the simple indicator-based evolutionary algorithm (SIBEA) [16] to take into account the preference information of the DM iteratively and direct the search towards the preferred regions of the DM. Specifically, the DM is iteratively shown a set of nondominated solutions and asked to provide her/his preferences by classifying this set into preferred and/or non-preferred solutions. The weights of the preferred solutions in the weighted hypervolume calculation are subsequently altered such that their selection pressure is increased. Using preference information for both the preferred and non-preferred solutions simultaneously is a novel approach in preference based EMO algorithms and provides more flexibility to the DM in guiding the search.

The rest of the paper is organized as follows. In Section 2, we introduce the main concepts and discuss how the preference information is incorporated into the method. Then I-SIBEA algorithm is presented in Section 3 with detailed description. In Section 4, we present preliminary numerical experiments used to test the method. Finally, the conclusions are drawn in Section 5.

2 Main Concepts and Principles of Utilizing Preference Information from the Decision Maker in I-SIBEA

2.1 Concepts and Notations

We consider multiobjective optimization problems of the form [13]:

$$\text{minimize } \{f_1(x), \ldots, f_k(x)\}$$
$$\text{subject to } x \in S \tag{1}$$

with $k(\geq 2)$ objective functions $f_i(x) : S \to \Re$. The vector of objective function values is denoted by $f(x) = (f_1(x), \ldots, f_k(x))^T$. For the simplicity of presentation, we assume that all the objective functions are to be minimized. If some objective function f_i is to be maximized, it is equivalent to minimize $-f_i$. The (nonempty) feasible region (set) S is a subset of the decision variable region $\Re^n$ and consists of decision variable vectors $x = (x_1, \ldots, x_n)^T$ that satisfy all the constraints. The image of the feasible region S in the objective region $\Re^k$ is called the feasible objective region (set) denoted by Z. The elements of Z are called feasible objective vectors denoted by $f(x)$ or $z = (z_1, \ldots, z_k)^T$, where $z_i = f_i(x), i = 1, \ldots, k$, are the objective function values. An ideal objective vector $z^* \in \Re^k$ is determined by minimizing each objective function individually, that is $z_i^* = \underset{x \in S}{\text{minimize}} f_i(x)$. We say that a vector $z^1 \in \Re^k$ is said to weakly dominate a vector $z^2 \in \Re^k$ and denoted by $z^1 \preceq z^2$ if and only if for all $1 \leq i \leq k$: $f_i(x^1) \leq f_i(x^2)$.

In this paper, we consider an interactive preference based EMO algorithm, wherein a DM iteratively provides her/his preference information as a set of preferred and/or non-preferred solutions. To emphasize solutions in the preferred region, the weighted hypervolume, $I_H^w(A)$ is used, where A is the set of nondominated solutions in the objective space. The weighted hypervolume is defined as the integral over the product of the weight distribution function $w(z)$ and the attainment function $\alpha(z)$ [16], that is,

$$I_H^w(A) = \int \int_Z w(z) \alpha_A(z) dz$$

where

$$\alpha_A(z) = \begin{cases} 1 & \text{if } A \preceq z \\ 0 & \text{else} \end{cases}$$

and $A \preceq z$ represent that at least one element of A weakly dominates $z \in Z$.

2.2 Incorporating Preference Information into the Algorithm

There are different ways to obtain preference information from the DM. In the preference based EMO algorithms literature where the hypervolume based selection criterion is used [3,4,16], as far as we know, only the preferred solutions are considered as the preference information from the DM. In the proposed I-SIBEA algorithm, we provide the flexibility to the DM to give her/his preferences by selecting preferred and/or non-preferred solutions among a set of nondominated solutions shown to her/him. For example, if the DM selects only preferred solutions, the rest of the solutions can be regarded as either non-preferred solutions or solutions with no preference information. However, in this study we consider them as non-preferred solutions. On the other hand, if the DM selects both preferred and non-preferred solutions, the rest of the solutions are regarded as solutions with no preference information.

It is often assumed that the DM has prior information about the preferred solutions before starting the solution process [3,16]. In the I-SIBEA algorithm, it is not assumed that the DM has some prior information about preferred and/or non-preferred solutions and that the DM is consistent during interaction. The DM iteratively gives her/his preference information, which is used by the I-SIBEA algorithm to focus its search towards solutions that lie in the preferred region. In what follows, we discuss how DM's preferences are incorporated in the I-SIBEA algorithm.

As mentioned in the introduction, the proposed algorithm extends SIBEA to consider the preference information of the DM. After a fixed number of generations of SIBEA, in the first interaction with the DM, a set $A \subset Z$ of nondominated solutions (in the objective space) is shown to the DM. The number of solutions shown is a parameter that the DM can set. Next, we suppose that the DM selects preferred and non-preferred solutions from the set A. Therefore, the obtained preference information creates a partition of A into three non-overlapping subsets:

$$AA = \{z \in A \mid z \text{ is preferred by the DM}\}$$
$$RA = \{z \in A \mid z \text{ is non-preferred by the DM}\}$$
$$IA = \{z \in A \mid \text{no preference information is available from the DM for } z\}$$

and $A = AA \cup RA \cup IA$.

After the partitioning of A into the three subsets, Z is partitioned into three regions based on the preferences from the DM. The regions are called dominated (Do), preferred (Pr) and no preference information (In) and an example illustrating them is shown in Fig. 1 for a biobjective optimization problem. The shaded region in the Fig. 1 represents the infeasible region.

In what follows, three regions, Do, Pr and In are defined based on preference information from the DM. The weight distribution function is then derived using hypervolume of these three regions which is incorporated into the algorithm to calculate the weighted hypervolume.

The region Do is the part of Z which is weakly dominated by at least one element of RA:

$$Do = \{z \in Z \mid \text{there exists } \overline{z} \in RA, \overline{z} \preceq z\}.$$

The hypervolume $\mu(Do)$ and the weighted hypervolume $w(Do)$ for region Do are calculated as

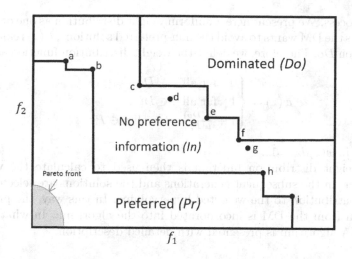

Fig. 1. $AA = \{a, b, h, i\}$, $RA = \{c, e, f\}$, $IA = \{d, g\}$. Regions: dominated, no preference information and preferred

$$\mu(Do) = \int \int_z \alpha_{RA}(z)dz$$

$$w(Do) = I_H^{(w)}(RA).$$

The region Pr is the part of Z which weakly dominates at least one element of AA:

$$Pr = \{z \in Z \mid \text{there exists } \bar{z} \in AA, z \preceq \bar{z}\}.$$

The hypervolume $\mu(Pr)$ and the weighted hypervolume $w(Pr)$ for region Pr are calculated as

$$\mu(Pr) = \int \int_z \alpha_{AA}(z)dz$$

$$w(Pr) = I_H^{(w)}(AA).$$

The region In is the remaining part of Z (Fig. 1):

$$In = Z \setminus \{Do \cup Pr\}$$

with the hypervolume $\mu(In) = 1 - \mu(Do) - \mu(Pr)$.

The reason to partition Z into these three regions is to emphasize the solutions that lie in the preferred region (Pr). There can exist several ways to implement this principle. In the literature [3, 12, 16], several weight distribution functions (e.g. stressing objectives with exponential weights, guiding single solutions with dirac-type weights etc.) are used to incorporate the DM's preference information in the

solution process. We present here a uniform weight distribution as one of the possibilities. As the DM wants to avoid the non-preferred solutions, $w(z)$ remains zero
for the region Do. Therefore, we define the weight distribution function as:

$$w(z) = \begin{cases} 0 & \text{for all } z \in Do \\ 1 & \text{for all } z \in In \\ 1 + \frac{\mu(Do)}{\mu(Pr)} & \text{for all } z \in Pr \end{cases}$$

(so that $\int \int_Z w(z)dz = 1$).

This weight distribution function is then used to calculate the weighted
hypervolume in the subsequent generations and the solutions are selected based
on their contribution to the weighted hypervolume. In this way, the preference
information from the DM is incorporated into the algorithm. In what follows,
the I-SIBEA algorithm is presented with detailed description.

3   Interactive Simple Indicator-Based Evolutionary Algorithm (I-SIBEA)

The main motivation of the proposed algorithm is to direct its search process
towards solutions that lie in the preferred region defined by the DM's preferences.
To do this, solutions having a large contribution to the weighted hypervolume are
selected and solutions having the smallest contribution to the weighted hypervolume are removed from the population after every generation. This criterion
of selecting solutions is common among hypervolume based search algorithms
[16,17]. In the proposed I-SIBEA, in addition to hypervolume based selection
criterion, different preference information from the DM is incorporated into the
algorithm. In this algorithm, the DM gives her/his preference information by
selecting preferred and/or non-preferred solutions. This preference information
guides the algorithm to focus its search direction for solutions that lie in the
preferred region. The algorithm is presented in the I-SIBEA algorithm and we
discuss now the step by step procedure of the algorithm.

Initially, a population P of individuals of size NP is created randomly in step
1. Next in step 2, crossover and mutation operators are used to create an offspring
population Q of the same size (NP). The parent and the offspring populations
are combined $P := P + Q$ and then environmental selection is used to select
individuals as mentioned in step 3 of the I-SIBEA algorithm. Nondominated
sorting [14] is used to rank the individuals of the combined population and
different fronts $F_i, i = 1, 2, \ldots$ are identified. These fronts are added to an empty
set $P1$ as long as the size of the population of $P1$ becomes equal to or exceeds
NP. If the size of $P1$ is NP, the population for next generation is set as $P := P1$.
Otherwise, the set of individuals in the worst rank front in $P1$ is identified
and denoted by P'. To remove solutions from the worst rank front so that the
population size of $P1$ does not exceed NP, the usual hypervolume based selection
is used. For each solution $z \in P'$, the loss in the hypervolume $d(z) = I(P') -
I(P' \setminus z)$ is determined, where I is the hypervolume indicator and represented

Algorithm: An interactive simple indicator-based evolutionary algorithm (I-SIBEA)

Input to algorithm: NP = population size; NG = maximum number of generations

Input from DM: DA = maximum number of solutions to be shown to the DM (default is maximum 5); AA = preferred and RA = non-preferred solutions after each interaction; H = maximum number of interactions

Output: f^* = Pareto optimal solution obtained by projecting the most preferred solution to the Pareto front, where $f^* \subseteq A$ and A is the set of nondominated solutions in the last population

Step 1 (Initialization): Generate an initial set P of decision vectors of size NP; set the generation counter $m := 1$; set the interaction step $intr := 0$; $NA :=$ number of points in A; set number of generation before first interaction $NI := round(NG/H)$; set $N := NI$; set the hypervolume indicator $I := \mu(\cdot)$.

Step 2 (Mating): Create an offspring population Q using crossover and mutation operators. Set $P := P + Q$ (multi-set union).

Step 3 (Environmental Selection): Rank the population P using nondominated sorting and identify different fronts $F_i, i = 1, 2, \ldots$ and do the following four steps $(a-d)$.

   a. Set a new population $P1 = \phi$. Set a count $i = 1$ and perform $P1 = P1 + F_i$ and as long as $|P1| \geq NP$ and set $i = i + 1$. Here, $|P1|$ denotes the cardinality of $P1$.

   b. If $|P1| = NP$, set $P := P1$ and go to step 4 otherwise determine the set of individuals $P' \subseteq P1$ with the worst rank.

   c. if $m \leq N$ and $N = NI$

     For each solution $z \in P'$ determine the loss $d(z)$ in the hypervolume I if it is removed from P', i.e., $d(z) := I(P') - I(P' \setminus z)$.

     else

     Identify P'_{Pr} i.e. the solutions $x \in P'$ belonging to region Pr and perform $P1 = P1 \setminus P' + P'_{Pr}$

       1. If $|P1| \geq NP$, for each solution in $z \in P'$ belonging to Pr determine the loss $d(z)$ in the hypervolume I if it is removed from P', i.e., $d(z) := I(P') - I(P' \setminus z)$.

       2. Else determine the loss in weighted hypervolume $d(z) := I(P') - I(P' \setminus z)$ for each solution $z \in P'$ belonging to the regions Do and In.

   d. Remove the $|P1| - NP$ solutions from P' with the smallest loss $d(z)$ (ties are broken randomly) and include the remaining solutions of P' into $P1$. Set $P := P1$.

Step 4: If $m \geq NG$ or $m > N$ then go to step 5. Otherwise set $m := m + 1$ and go to step 2.

Step 5 (Identify A): Set A as the set of nondominated solutions in P. If $NA > DA$, remove additional solutions by using e.g. clustering.

Step 6 (Interaction with DM): Show DA solutions of A to the DM and set $intr := intr + 1$. If the DM wants to stop or $m \geq NG$, go to step 7 otherwise go to step 8.

Step 7 (Termination): Ask the DM to select the most preferred solution (f^*) from DA. Obtain the final solution by projecting f^* to the Pareto front and terminate the algorithm.

Step 8: Ask the DM to classify DA into AA and/or RA and derive the sets Do, In and Pr to get the updated weighted hypervolume $w(\cdot)$. Set $I := w(\cdot)$; $NI = \frac{NG-NI}{H-intr}$; $N := m + NI$ and $m := m + 1$. Go to Step 2.

as the hypervolume or the weighted hypervolume for a given set. The solution with the smallest loss is removed until the size of the population does no longer exceed NP and the population is set as $P := P1$ for the next generation. After a fixed number of generations, NI in step 4, the DM interacts with the algorithm. In step 5, a fixed number of nondominated solutions $DA \subseteq A$ (input from the DM) is identified and then shown to her/him in step 6, where A is the set of nondominated solutions. There exist different ways to select the fixed number of solutions from A and we use k-means clustering [11] in this study. Here, a solution is selected randomly from each cluster and shown to the DM. In step 7, if the DM wants to quit, s(he) selects the most preferred solution f^* from DA. The final solution is obtained by projecting f^* to the actual Pareto front by optimizing an achievement scalarizing function (ASF) [13], that is by solving the problem

$$\underset{i=1,\ldots,k}{\text{minimize max}} \left[w_i(f_i(x) - f_i^*) \right] + \rho \sum_{i=1}^{k} w_i(f_i(x) - f_i^*) \tag{2}$$

$$\text{subject to } x \in S.$$

where $\rho > 0$ is the augmentation coefficient which takes a small positive value e.g. 10^{-6}. The weight vector $w_i = \frac{1}{z_i^{max} - z_i^{min}}$ is assigned to each objective function. The maximum and minimum values of each objective function in the set A are represented by z_i^{max} and z_i^{min}, respectively. One of the advantages for using an ASF is that the optimal solution of an ASF is always Pareto optimal [13]. Therefore, optimizing an ASF ensures that final solution is locally Pareto optimal. Since we assume that less is preferred to more for the DM, the projected solution is at least as preferred to the DM as the solution s(he) selected. We can utilize an equivalent differentiable formulation of ASF when all the objective functions are differentiable by adding extra real valued variable, δ and k new constraints [13]

$$\text{minimize } \delta + \rho \sum_{i=1}^{k} w_i(f_i(x) - f_i^*)$$

$$\text{subject to } w_i(f_i(x) - f_i^*) \leq \delta \text{ for all } i = 1, \ldots, k \tag{3}$$

$$x \in S \ \delta \in \Re.$$

In addition to termination by the DM, the solution process is ended if the maximum number of generations (NG) is reached. In that case, solutions $DA \subseteq A$ are shown to the DM and (s)he is asked to select the the most preferred solution (f^*). This solution is then projected to the Pareto front and the final solution is obtained by solving problem (2) or (3) with a single objective optimization method appropriate to the characteristics of the problem in question. A local search method can be used since the evolutionary algorithm is supposed to take care of the global search. If the termination criterion is not met, the DM is then asked in step 8 to select preferred (AA) and non-preferred solutions (RA) from DA to get the three non-overlapping subsets AA, RA and IA. The three regions

Do, Pr and In are then derived using this preference information to get the weight distribution function. This completes one interaction with the DM. The weight distribution function is then used to calculate the weighted hypervolume as the selection criterion in the subsequent generations.

In the next generation (after the first interaction), the offspring are created again in step 2 and other steps are then followed. If the population size exceeds NP in $P1$, solutions $z \in P'$ belonging to region Pr are identified and denoted by P'_{Pr}. The set $P1$ is then updated as $P1 := P1 \setminus P' + P'_{Pr}$. If the size of the population of $P1$ exceeds NP, the usual hypervolume based selection is used to remove the solutions from P' belonging to region Pr. Otherwise, the solutions $z \in P'$ belonging to regions Do and In are added to $P1$. If the population size exceeds NP, the weighted hypervolume based selection is used to remove solutions $z \in P'$ belonging to regions Do and In. This principle of selecting individuals emphasizes solutions in the preferred region. The regions Do, Pr and In are updated after every interaction after the DM has classified DA into AA and RA. In this way, the DM gives her/his preference and the weights of the solutions in the weighted hypervolume calculation are altered in such a way that solutions in Pr are emphasized and solutions in Do are avoided.

In the proposed algorithm, the DM has the freedom to choose the number of times (s)he wishes to interact with the algorithm. From this input, the maximum number of generations (NG) is uniformly divided by H to get the number of generations before each interaction. For example, the first interaction will take place after $NI = NG/H$ generations and the second interaction will take place after $N = NI + (NG - NI)/(H - 1)$ generations. Even though, the maximum number of interactions H is given by the DM in the beginning, the DM is free to change it during any interaction. However, this is not presented in the algorithm but if the DM gives her/his updated number of interactions, the remaining generations can be uniformly divided accordingly. In the next section, the algorithm is tested using some benchmark problems.

4   Numerical Experiments

The I-SIBEA algorithm was tested on standard benchmark problems [5] with 2-3 objectives and 7-11 decision variables. Firstly, we compare I-SIBEA against a recently proposed interactive weighted hypervolume based algorithm called W-Hype [4]. One of the main differences between W-Hype and I-SIBEA is that W-Hype considers information for only preferred solutions while I-SIBEA considers information for both preferred and non-preferred solutions. To enable easy comparison, we use the same set of test problems i.e. DTLZ2, ZDT4 and DTLZ1 and the same criterion of [12] to get the number of generations before each interaction as used in W-Hype. The parameter values used in I-SIBEA are provided in Table 1. In all three problems, polynomial mutation (distribution index is 20 and probability of mutation is 1/number of decision variables) and simulated binary crossover (distribution index is 20 and probability of crossover is 0.9) were used. While testing the algorithm for these test problems, we replaced the

DM by a weighted Chebyshev function $\max_{i=1,\ldots,k}[w_i(f_i(x) - z_i^*)]$ at each interaction step with z_i^* as the ideal objective vector. The weight vector $(w_1, \ldots, w_k)^T$ is assigned to each objective function and used to describe the DM's preferences. After each interaction step, the solution that minimized the weighted Chebyshev function was considered as the preferred solution (AA) and the rest of the solutions were considered as non-preferred solutions (RA) i.e. it was assumed that there were no solutions with no preference information (IA). This setting was used to be able to compare I-SIBEA with W-Hype (where this setting had been used).

Table 1. Parameters used in this study

	DTLZ2	ZDT4	DTLZ1
Number of decision variables/objectives	11/2	10/2	7/3
Ideal vector	(0,0)	(0,0)	(0,0,0)
Population size	50	100	200
Number of interactions	2,4,6,8	4,6	4,6
Weight vector	(0.2,0.8)	(0.5,0.5)	(0.7,0.2,0.1)
Number of independent runs	30	50	10
Total number of function evaluations	20,000	40,000	120,000

To measure the performance of the proposed algorithm, mean, standard deviation, absolute deviation and optimal Chebyshev function value were calculated after a maximum number of function evaluations. In the three tables reporting the results of I-SIBEA and W-Hype, the values of performance criteria are written in bold face if the difference was greater than 0.001. The algorithm was also tested by varying the maximum number of interactions. The comparison of the present algorithm with W-Hype for DTLZ2 is shown in Table 2.

Table 2. Results for DTLZ2: algorithm, number of interactions (H), mean, standard deviation (Std.), absolute deviation (Abs.), optimal Chebyshev function value (C^*) and number of function evaluations (nfun)

Algorithm	H	Mean	Std.	Abs.	C^*	nfun
I-SIBEA	2	0.21500	0.052800	0.03230	0.19400	**20,000**
	4	0.19430	0.000571	0.00030	0.19400	**20,000**
	6	0.19410	0.000121	0.00006	0.19400	**20,000**
	8	0.19410	0.000042	0.00003	0.19400	**20,000**
W-Hype	2	**0.19418**	**0.000114**	**0.00016**	0.19403	25,000
	4	0.19413	0.000064	0.00010	0.19403	25,000
	6	0.19411	0.000053	0.00009	0.19403	25,000
	8	0.19410	0.000049	0.00007	0.19403	25,000

The results show that W-Hype performed better than I-SIBEA for $H = 2$. Otherwise, equivalent results were obtained for $H = 4, 6, 8$ and I-SIBEA needed fewer function evaluations when compared with W-Hype. The total number of function evaluations used for this problem by W-Hype and I-SIBEA were $25,000$ and $20,000$, respectively. In addition, better results were obtained by both algorithms with increase in H. We also observed that, after a certain number of generations, the mean of the weighted Chebyshev function did not change for $H = 4, 6$ and 8 which indicates the convergence of the algorithm. Moreover, there was no considerable difference in the results for $H = 6$ and $H = 8$ in I-SIBEA and, therefore, we restricted ourselves to $H = 4$ and 6 when solving the following two problems.

In case of the ZDT4 problem, the weight vector $w = (0.5, 0.5)^T$ was used in the weighted Chebyshev function to identify the preferred solution (AA). The comparison of I-SIBEA with W-Hype is shown in Table 3. I-SIBEA performed better than W-Hype both in terms of results obtained and the number of function evaluations used. I-SIBEA used 40,000 function evaluations and converged in 50% fewer function evaluations as comparison with W-Hype.

Table 3. Results for ZDT4: algorithm, number of interactions (H), mean, standard deviation (Std.), absolute deviation (Abs.) , optimal Chebyshev function value (C^*) and number of function evaluations (nfun)

Algorithm	H	Mean	Std.	Abs.	C^*	nfun
I-SIBEA	4	**0.19180**	**0.001600**	**0.00100**	0.19100	**40,000**
	6	**0.19110**	**0.000262**	**0.00016**	0.19100	**40,000**
W-Hype	4	0.35591	0.203362	0.16493	0.19098	80,000
	6	0.36171	0.230273	0.17073	0.19098	80,000

Next, we tested I-SIBEA on a 3-objective DTLZ1 problem. The weight vector $w = (0.7, 0.2, 0.1)^T$ was used in weighted Chebyshev function to identify the preferred solution (AA). Table 4 shows the results of this study. For this problem as well, equivalent results were obtained in 62.5% fewer function evaluations. I-SIBEA used 120,000 function evaluations in contrast to W-Hype which used 320,000 function evaluations. In all three problems, better or equivalent results were obtained but I-SIBEA always consumed fewer function evaluations. The reason for fewer function evaluations using I-SIBEA can be attributed to the use of preference information of both preferred and non-preferred solutions when compared to W-Hype, where only preferred solutions were considered as the preference information from the DM. This extra information on non-preferred solutions can help the algorithm to avoid solutions in the corresponding regions in subsequent generations and converge faster to solutions in the preferred region. In addition, the DM has more options of how to express one's preferences.

To show the flexibility of the proposed algorithm, a fourth case study was performed on the ZDT4 problem (as it is easy to visualize a biobjective optimization problem), where a DM was involved. In the beginning of the solution process, the DM was asked to provide the maximum number of interactions i.e.

Table 4. Results for DTLZ1: algorithm, Number of interactions (H), mean, standard deviation (std.), absolute deviation (abs.), optimal Chebyshev function value (C^*) and number of function evaluations (nfun)

Algorithm	H	Mean	Std.	Abs.	C^*	nfun
I-SIBEA	4	0.03090	0.000397	0.00035	0.03050	**120,000**
	6	0.03080	0.000167	0.00014	0.03050	**120,000**
W-Hype	4	0.03048	0.000069	0.00005	0.03043	320,000
	6	0.03045	0.000026	0.00002	0.03043	320,000

how many times he wanted to interact with the algorithm. In addition, the flexibility is given to the DM to change the number of nondominated solutions (DA, default is maximum 5) he wanted to see during interaction. In this study, the maximum number of generations (NG) was uniformly divided into 6 times for the interaction with the DM as mentioned in the I-SIBEA algorithm. The other parameters used for this problem are shown in Table 5.

Table 5. Parameters used in the fourth case

	ZDT4
Number of decision variables/objectives	10/2
Population size	50
Number of interactions	6 (Input from the DM)
Maximum number of generation	400

The results of this study are shown in Fig. 2. The first scatter plot shows the nondominated solutions (A) before the first interaction. For the first interaction, the DM wanted to see 5 nondominated solutions and k-means clustering was used to get them. The DM then selected the preferred (AA) and non-preferred solutions (RA) which are shown in the second scatter plot. The solutions obtained before the second interaction are also plotted in the same plot to show the search direction of I-SIBEA. The solution process was then continued until the second interaction. In Fig. 2, the preferred and non-preferred solutions are shown for five interactions. In this study, the DM changed his preferences in the subsequent interactions or in other words, the DM was not consistent with his preferences as shown in Fig. 2. The I-SIBEA algorithm was found to exploit the preference information provided by the DM and generate solutions in the regions preferable to the DM. This shows that the algorithm is flexible to changes in the preferences and can find solutions in the preferred region. After completion of the maximum number of generations, the DM interacted again (6th time) and selected the most preferred solution. This solution was then projected to the Pareto front by solving problem (3). We used here fmincon from MATLAB optimization toolbox to solve problem (3). In this study, $f_1 = 0.096327$ and $f_2 = 0.7074$ were the most preferred objective function values after the final interaction and $f_1 = 0.094345$ and $f_2 = 0.69284$ were the final objective function values.

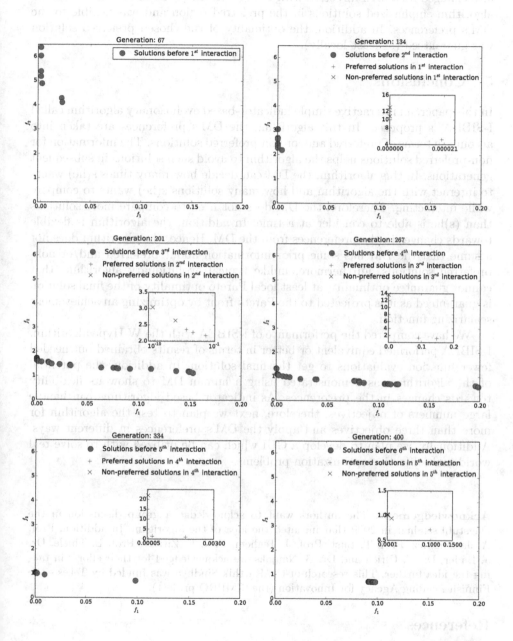

Fig. 2. Decision making process using I-SIBEA algorithm for the ZDT4 problem

The proposed algorithm directed its search towards the DM's preferences and also changed its search direction with changes in the preferences. Therefore, the algorithm emphasized solutions in the preferred region and was flexible to the DM's preferences. In addition, the optimality of the chosen preferred solution was guaranteed (at least locally).

5   Conclusions

In this paper, an interactive simple indicator-based evolutionary algorithm called I-SIBEA is proposed. In this algorithm, the DM's preferences are taken into account in terms of preferred and/or non-preferred solutions. The information for non-preferred solutions helps the algorithm to avoid such solutions in subsequent generations. In this algorithm, the DM can decide how many times s(he) wants to interact with the algorithm and how many solutions s(he) wants to compare while interacting. Therefore, the DM does not need to compare more solutions than (s)he is able to consider at a time. In addition, the algorithm is flexible towards changes in the preferences from the DM. Hence, the algorithm does not assume that the DM has some prior information about preferred and/or non-preferred solutions. Furthermore, unlike typical evolutionary algorithms that cannot guarantee optimality, at least local Pareto optimality of the final solution is guaranteed as it is projected to the Pareto front by optimizing an achievement scalarizing function.

We have compared the performance of I-SIBEA with the W-Hype algorithm. I-SIBEA performed equivalent or better in terms of results obtained but needed fewer function evaluations to get the final solution. In addition, the potential of the algorithm was demonstrated using a human DM to show its flexibility towards changes in the preferences. As indicator based algorithms can handle large numbers of objectives, therefore, next we plan to test the algorithm for more than three objectives and apply the DM's preferences in different ways. Additionally, we plan to develop a GUI which can be utilized with to solve real world multiobjective optimization problems.

Acknowledgement. The authors want to acknowledge a group discussion in the Dagstuhl seminar in 2009 that initiated the idea of the algorithm. In addition, Prof. A. Jaszkiewicz, Prof. T. Lust, Prof. J. Teghem, Prof. E. Zitzler, Prof. L. Thiele, Dr. J. Bader, Dr. T. Ulrich and Dr. B. Naujoks are acknowledged for their efforts in taking the idea further. This research of Dr. Karthik Sindhya was funded by Tekes - the Finnish Funding Agency for Innovation (the SIMPRO project).

References

1. Auger, A., Bader, J., Brockhoff D., Zitzler, E.: Theory of the hypervolume indicator: Optimal μ distributions and the choice of the reference point. In: 10th ACM SIGEVO Workshop on Foundations of Genetic Algorithms, pp. 87–102. ACM (2009)

2. Bader, J., Zitzler, E.: HypE: An algorithm for fast hypervolume-based many-objective optimization. Evolutionary Computation **19**, 45–76 (2011)
3. Brockhoff, D., Bader, J., Thiele, L., Zitzler, E.: Directed multiobjective optimization based on the weighted hypervolume indicator. Journal of Multi-Criteria Decision Analysis **20**, 291–317 (2013)
4. Brockhoff, D., Hamadi, Y., Kaci, S.: Using comparative preference statements in hypervolume-based interactive multiobjective optimization. In: Pardalos, P.M., Resende, M.G.C., Vogiatzis, C., Walteros, J.L. (eds.) LION 2014. LNCS, vol. 8426, pp. 121–136. Springer, Heidelberg (2014)
5. Coello, C.A.C., Lamont, G.B., Veldhuizen, D.A.V.: Evolutionary Algorithms for Solving Multi-Objective Problems, 2nd edn. Springer, New York (2007)
6. Deb, K.: Multi-Objective Optimization using Evolutionary Algorithms. Wiley, Chichester (2001)
7. Deb, K., Kumar, A.: Light beam search based multi-objective optimization using evolutionary algorithms. In: IEEE Congress on Evolutionary Computation (CEC 2007), pp. 2125–2132 (2007)
8. Deb, K., Sinha, A., Korhonen, P.J., Wallenius, J.: An interactive evolutionary multi-objective optimization method based on progressively approximated value functions. IEEE Transactions on Evolutionary Computation **14**, 723–739 (2010)
9. Deb, K., Sundar, J., Rao, U.B.N., Chaudhuri, S.: Reference point based multi-objective optimization using evolutionary algorithms. International Journal of Computational Intelligence Research **2**, 273–286 (2006)
10. Grunert da Fonseca, V., Fonseca, C.M., Hall, A.O.: Inferential Performance Assessment of Stochastic Optimisers and the Attainment Function. In: Zitzler, E., Deb, K., Thiele, L., Coello Coello, C.A., Corne, D.W. (eds.) EMO 2001. LNCS, vol. 1993, pp. 213–225. Springer, Heidelberg (2001)
11. Jain, A.K., Murty, M.N., Flynn, P.J.: Data clustering: A review. ACM Computing Surveys **31**, 264–323 (1999)
12. Köksalan, M., Karahan, I.: An interactive territory defining evolutionary algorithm: iTDEA. IEEE Transactions on Evolutionary Computation **14**, 702–722 (2010)
13. Miettinen, K.: Nonlinear Multiobjective Optimization. Kluwer, Boston (1999)
14. Srinivas, N., Deb, K.: Multiobjective optimization using nondominated sorting in genetic algorithms. Evolutionary Computation **2**, 221–248 (1994)
15. Thiele, L., Miettinen, K., Korhonen, P., Molina, J.: A preference-based evolutionary algorithm for multiobjective optimization. Evolutionary Computation **17**, 411–436 (2009)
16. Zitzler, E., Brockhoff, D., Thiele, L.: The Hypervolume Indicator Revisited: On the Design of Pareto-compliant Indicators Via Weighted Integration. In: Obayashi, S., Deb, K., Poloni, C., Hiroyasu, T., Murata, T. (eds.) EMO 2007. LNCS, vol. 4403, pp. 862–876. Springer, Heidelberg (2007)
17. Zitzler, E., Thiele, L., Bader, J.: On set-based multiobjective optimization. IEEE Transactions on Evolutionary Computation **14**, 58–79 (2010)

Combining Non-dominance, Objective-order and Spread Metric to Extend Firefly Algorithm to Multi-objective Optimization

M. Fernanda P. Costa[1,3]([✉]), Ana Maria A.C. Rocha[2,4], and Edite M.G.P. Fernandes[4]

[1] Department of Mathematics and Applications, University of Minho, 4800-058 Guimaraes, Portugal
mfc@math.uminho.pt
[2] Department of Production and Systems, University of Minho, 4710-057 Braga, Portugal
arocha@dps.uminho.pt
[3] Centre of Mathematics, University of Minho, 4710-057 Braga, Portugal
[4] Algoritmi Research Centre, University of Minho, 4710-057 Braga, Portugal
emgpf@dps.uminho.pt

Abstract. In this paper, we propose an extension of the firefly algorithm (FA) to multi-objective optimization. FA is a swarm intelligence optimization algorithm inspired by the flashing behavior of fireflies at night that is capable of computing global solutions to continuous optimization problems. Our proposal relies on a fitness assignment scheme that gives lower fitness values to the positions of fireflies that correspond to non-dominated points with smaller aggregation of objective function distances to the minimum values. Furthermore, FA randomness is based on the spread metric to reduce the gaps between consecutive non-dominated solutions. The obtained results from the preliminary computational experiments show that our proposal gives a dense and well distributed approximated Pareto front with a large number of points.

Keywords: Multi-objective · Firefly algorithm · Fitness assignment · Spread metric

1 Introduction

This paper aims to extend a global optimization framework, known as firefly algorithm (FA), to tackle nonlinear multi-objective (MO) optimization problems. This is one of the most challenging problems since the goal is to optimize more than one objective. FA is a population-based algorithm and therefore suitable to solve MO problems. It is capable of finding multiple Pareto-optimal solutions in a single run. Here, we consider solving nonlinear bound constrained MO optimization problems with $no > 1$ objectives and $n \geq 1$ decision variables:

$$\min \ (f_1(x), f_2(x), \ldots, f_{no}(x))$$
$$\text{subject to} \ l_i \leq x_i \leq u_i, \ i = 1, \ldots, n \tag{1}$$

© Springer International Publishing Switzerland 2015
A. Gaspar-Cunha et al. (Eds.): EMO 2015, Part I, LNCS 9018, pp. 292–306, 2015.
DOI: 10.1007/978-3-319-15934-8_20

where the conflicting objective functions $f_j : \mathbb{R}^n \to \mathbb{R}$, $j = 1, 2, \ldots, no$, are continuous and possibly nonlinear functions and $l \in \mathbb{R}^n$ and $u \in \mathbb{R}^n$ are the vectors of lower and upper bounds for the decision variables, respectively. We note that the feasible region $\Omega = [l, u]$ is a nonempty compact set and differentiability and convexity of the objectives are not assumed, although the search space of problem (1) is convex.

MO optimization is an important research area mainly for two reasons. First, a large number of real-world applications are formulated as MO problems; second, many issues, such as the statistical interpretation associated with performance comparison, still need to be addressed. For MO no single solution optimizes simultaneously all objectives. In practice, several conflicting objectives arise and the goal is to identify the best compromise solution among a set of Pareto-optimal solutions. The set of optimal solutions in the decision space is in general denoted as the Pareto-optimal set and its image in the objective space is denoted as Pareto-optimal front. The main task of MO algorithms is to support a decision maker to formulate his/her preferences and to identify the best of the Pareto-optimal solutions.

In a MO minimization problem, a solution $\bar{x} \in \mathbb{R}^n$ is said to dominate $\hat{x} \in \mathbb{R}^n$ if and only if $f_j(\bar{x}) \leq f_j(\hat{x})$ for all $j \in \{1, 2, \ldots, no\}$ where $f_j(\bar{x}) < f_j(\hat{x})$ for at least one j. Further, a solution $\bar{x} \in \mathbb{R}^n$ is said to be Pareto-optimal if and only if there is no solution $\hat{x} \in \mathbb{R}^n$ that dominates $\bar{x}$. Thus, the goal with a MO algorithm is to find a good and balanced approximation to the set of Pareto-optimal solutions.

The most popular methods to tackle a MO problem are based on the aggregation of the objectives, on the ϵ–constraint, and on producing an approximation to the Pareto-optimal front directly. The aggregation method transforms the MO formulation into a uni-objective formulation problem by assigning to each objective function f_j a non-negative weight w_j such that $\sum_{j=1}^{no} w_i = 1$, and minimizing an aggregate function that is the weighted sum of the objectives. The approximate Pareto-optimal front is obtained by running as many times as the desired number of points using different weight values [22]. In the ϵ–constraint method one objective is selected to be minimized and all the other objective functions are converted into inequality constraints by setting an upper value to each one [13]. Methods to compute an approximation to the Pareto front in a single run are in general stochastic population-based search techniques. Fitness assignment is a crucial issue in MO algorithms and depends on the entire set of points in the population. Two categories of common strategies to assign fitness are aggregation-based and Pareto/dominance-based. The latter may use more than one dominance order (for example, dominance rank, dominance count or dominance depth) [35]. Fitness assignment strategies may also depend on MO performance metrics, for instance, the hyper-volume, the purity metric or the spread metric [13, 16, 33, 34].

Evolutionary algorithms are widely used when solving MO optimization problems. They are designated as MO evolutionary algorithms (MOEAs) and largely dominate the research area of approximate metaheuristics for MO [16].

From the most classical procedure VEGA to other more recent MOEA variants, like MOGA, MOMGA, NPGA, NSGA, PESA, PAES, SPEA, NSGA-II, SPEA2, RPSGAe and MEGA [8,12,17,33,36], all of them have been used in a variety of real-world applications. In [23], a hybrid multi-objective evolutionary algorithm combining a genetic algorithm and a particle swarm optimization is presented; in [4,11], different robust MO optimization procedures are presented; and in [14], robustness assessment during multi-objective optimization using a MOEA is discussed. Besides MOEAs other metaheuristics have been used in MO optimization [1,6,7,22]. Deterministic-type approaches are also available [5,10].

The contribution of this paper is the extension of the FA paradigm to the MO optimization. FA is a recently developed bio-inspired metaheuristic algorithm that is capable of computing global solutions to optimization problems [15,27,28]. It is a swarm intelligence optimization algorithm inspired by the flashing behavior of fireflies at night, and it competes with the most well-known swarm intelligence algorithms, like ant colony optimization, particle swarm optimization, artificial bee colony, artificial fish swarm, bat algorithm and cuckoo-search.

FA has already been adapted to the MO optimization area [2,20,29]. Recently proposed FA extensions to MO are related with applications in operations research, like fleet planning problems, circuit design problems, production scheduling system, economic emission dispatch problem [31], energy optimization in grid environments [3], hybrid flowshop scheduling problem [21], job shop scheduling problems [18], geometric design of clamped-free beams [19] and optimal hydrocyclone design [25]. Most of these studies transform the MO formulation into a uni-objective one, although others produce approximations to the Pareto front in a single run using an aggregation-based strategy to assign fitness to points.

Our proposal for the MO optimization area uses a non-dominance/dominance ranking combined with an objective-order process based on scaled distances to the minimum values for the fitness assignment procedure. It also incorporates a spread metric-based randomness term into the FA paradigm to generate candidate points from the current ones. This randomness term aims to diversify the search as well as to reduce the gaps between consecutive non-dominated solutions in the approximated Pareto front. The herein proposed non-dominance/dominance ranking aims to favor non-dominated points of the populations giving them ranks that are always lower than those assigned to any of the other dominated points. This way, non-dominated points correspond to the positions of the brightest fireflies. Our algorithm computes candidate points to all current ones, except to the best point of the population, representing the position of the brightest firefly of all. Assuming that all non-dominated and dominated positions in the search space are ordered, the algorithm simulates movements to all fireflies, except the brightest, in direction to the more brighter ones. Then each computed candidate/trial position is accepted just after the movement except when a current non-dominated position generates a dominated trial position. Furthermore, at the end of each iteration, the set of non-dominated solutions found thus far is updated with the accepted non-dominated points, being the dominated solutions removed from the set. Our proposal is designated by Multi-Objective-order Firefly Algorithm (MOoFA).

The remaining part of the paper is organized as follows. In Section 2 the FA paradigm is described and in Section 3 the proposed MOoFA is presented and discussed. Section 4 reports on the preliminary computational experiments carried out using a benchmark set of MO problems and we conclude the paper with Section 5.

2   The FA Paradigm

Throughout the paper, $\| \cdot \|$ represents the Euclidean norm of a vector and the vector $x = (x_1, x_2, \ldots, x_n)^T$ represents the position of a firefly in the search space. The position of the firefly j will be represented by $x^j \in \mathbb{R}^n$. We assume that the size of the population of fireflies is $1 < m < \infty$. In the context of an uni-objective optimization problem, firefly j is brighter than firefly i if the objective function value at x^j is lower than the objective value at x^i.

FA is a bio-inspired metaheuristic algorithm inspired by the flashing behavior of fireflies at night. According to [9, 26–28, 30, 32], the three main rules used to construct the standard algorithm are the following: (i) all fireflies are unisex, meaning that any firefly can be attracted to any other brighter one; (ii) the brightness of a firefly is determined from the encoded objective function; (iii) attractiveness is directly proportional to brightness but decreases with distance. In the FA paradigm, the movement of a firefly i is attracted to another more attractive/brighter firefly j and the new candidate position, also designated by trial position, for firefly i is given by:

$$t^i = x^i + \beta(x^j - x^i) + \alpha \left(z + L(0,1)\sigma^i \right), \tag{2}$$

where x^i represents its current position, $\alpha \in [0, 1]$ and

$$\beta = \beta_0 \exp \left(-\gamma \|x^i - x^j\|^2 \right) \tag{3}$$

is the attractiveness of a firefly which varies with the light intensity seen by adjacent fireflies and the distance between themselves. The parameter β_0 is the attraction parameter when the distance is zero. $L(0,1)$ is a random number from the standard Lévy distribution centered at zero with an unitary standard deviation. The vector $z = z(k)$ is a reference point from the set of best solutions found so far and the vector $\sigma^i = \left(|x_1^i - z_1|, \ldots, |x_n^i - z_n| \right)^T$ gives the variation around z. The notation $z(k)$ means that it varies with the iteration counter, k, of the algorithm. The second term on the right hand side of (2) is due to the attraction while the third term gives randomness, with α being a scale parameter that controls the randomness and aims to maintain the diversity of solutions. The parameter γ characterizes the variation of the attractiveness, and is crucial to speed the convergence of the algorithm. As in [9], we allow α to decrease linearly with k, from $\alpha_{\max}$ to $\alpha_{\min}$, and we use a dynamic update of γ that increases the attractiveness with k from a lower value $\gamma_{\min}$ to an upper value $\gamma_{\max}$. Contrary to the evolutionary strategies and genetic algorithm, in FA all fireflies simulate movement in order to find a better position. Although in the oldest versions of FA, the brightest firefly was not moved, some recent versions move it, either randomly or in a direction in which the brightness increases [26, 30]. Furthermore, the new positions of each firefly are only accepted if they improve over the old ones. This is particularly promising since the best position is never lost.

3   Strategies in MOoFA

Since the proposed MOoFA is of a stochastic nature, the goal is to search for the best approximation to the Pareto-optimal front. MOoFA performs the search in the objective space, i.e., the algorithm selects the positions to be varied (corresponding to fireflies that simulate movement) based on the fitness assigned to the fireflies in the population. This fitness assignment is a crucial issue in FA since a firefly movement depends on brighter fireflies and the brightness is inversely proportional to the fitness value. In this extended FA for MO, the lower the fitness value the brighter is the firefly (and the lower is the order of the position). The simplest way to implement FA in a MO paradigm is to order the positions of fireflies from lowest to highest fitness value. In this paper, we propose two fitness assignment schemes that are based on an ordering strategy of the objective values. To order the positions of the fireflies, the following ranking steps are required.

1. Assign 'non-dominance rank', r_{n-d}, that aims to favor non-dominated points giving them the rank value $r_{n-d} = 1$, and giving to the remaining (the dominated ones) points $r_{n-d} = 2$;
2. Assign 'f–values order', o_f, that aims to give lower order to points with lower function values. Two schemes are proposed. One depends on assigning ranks to the objective function values; the other relies on the difference from the function values themselves to the minimum value. The 'f–values order' aggregates quantities using weights that satisfy $0 \leq w_j \leq 1$ and $\sum_{j=1}^{no} w_j = 1$. Thus,
    (a) Using ranks (integer values ranging from 1 to m), r_j, assigned to the objective values $f_j(x^i)$, $j = 1, \ldots, no$, the 'f–values order' of a point x^i is calculated by

    $$o_f(x^i) = \frac{1}{m} \left(w_1 \mathrm{r}_1 + w_2 \mathrm{r}_2 + \ldots + w_{no} \mathrm{r}_{no} \right). \tag{4}$$

    (b) Using the objective function values, a factor that is a scaled distance to the minimum value of objective f_j is computed,

    $$\mathrm{s}_j = \frac{f_j(x^i) - f_{j,\min}}{f_{j,\max} - f_{j,\min}} \tag{5}$$

where $0 \leq s_j \leq 1$, $f_{j,\max}$ and $f_{j,\min}$ are the maximum and minimum values of f_j attained by the population, respectively. Then, the 'f–values order' is computed by

$$o_f(x^i) = w_1 s_1 + w_2 s_2 + \ldots + w_{no} s_{no}. \qquad (6)$$

3. Finally, for either case (2a) or (2b), the fitness value, $Fit(x^i)$, assigned to each point x^i is defined by

$$Fit(x^i) = r_{n-d} + o_f(x^i). \qquad (7)$$

This way non-dominated points have fitness values in the range $[1, 2]$ and dominated points have fitness in the range $[2, 3]$.

Table 1 shows the fitness assignment scheme (4), for a small example with two objectives, ten points in the population, and $w_1 = w_2 = 0.5$. The last column in the table shows the ordering of the points based on the Fit values. (We note that any occurring tie is broken arbitrarily.) Table 2 depicts the fitness assignment scheme based on (5) and (6). We note that this ordering is not the same as that of previous table. In Table 1 two sets of ties occur in Fit: one originates x^5 and x^6, the other x^8 and x^9. With the factor s_j, the likelihood that ties will occur is much lower than with the scheme (4).

Table 1. Fitness assignment based on ranking the objectives (4), for ten points

i	$f_1(x^i)$	$f_2(x^i)$	r_1	r_2	$o_f(x^i)$	r_{n-d}	$Fit(x^i)$	ordering
1	6.75	**3**	6	7	0.65	2	2.65	x^5
2	**4**	**1**	1	3	0.20	1	1.20	x^1
3	**7**	**0.5**	7	1	0.40	1	1.40	x^3
4	10	2.5	10	5	0.75	2	2.75	x^9
5	5	4	3	10	0.65	2	2.65	x^6
6	4.5	2	2	4	0.30	2	2.30	x^4
7	**6**	**0.75**	4	2	0.30	1	1.30	x^2
8	6.5	3.5	5	9	0.70	2	2.70	x^7
9	9	2.7	9	6	0.75	2	2.75	x^8
10	8	3.25	8	8	0.80	2	2.80	x^{10}

Non-dominated points are in bold style

We now briefly describe some technical issues of MOoFA in Algorithm 1. MOoFA starts by randomly generating m points – positions of the population of fireflies – in the search space Ω. The objective functions are evaluated at all points and the non-dominated points are identified. The set, denoted by ND, of all produced non dominated points (the corresponding no–tuple $(f_1, f_2, \ldots, f_{no})$) is initialized. The fitness assignment strategy described in (7) is applied and the points are ordered according to their fitness value Fit, from lowest to largest, i.e., x^1 is the point with lowest Fit value, x^2 is the point with the second lowest value of Fit, and so forth, x^m is the point with largest Fit value. Now, new

Table 2. Fitness assignment based on scaled distance of objectives to minimum (5) and (6), for ten points

i	$f_1(x^i)$	$f_2(x^i)$	s_1	s_2	$o_f(x^i)$	r_{n-d}	$Fit(x^i)$	ordering
1	6.75	3	0.4583	0.7143	0.5863	2	2.5863	x^6
2	4	1	0	0.1429	0.0714	1	1.0714	x^1
3	7	**0.5**	0.5	0	0.2500	1	1.2500	x^3
4	10	2.5	1	0.5714	0.7857	2	2.7857	x^{10}
5	5	4	0.1667	1	0.5833	2	2.5833	x^5
6	4.5	2	0.0833	0.4286	0.2560	2	2.2560	x^4
7	**6**	**0.75**	0.3333	0.0714	0.2024	1	1.2024	x^2
8	6.5	3.5	0.4167	0.8571	0.6369	2	2.6369	x^7
9	9	2.7	0.8333	0.6286	0.7310	2	2.7310	x^9
10	8	3.25	0.6667	0.7857	0.7262	2	2.7262	x^8

Non-dominated points are in bold style

candidate positions are computed for the current position x^2 and all the others that follow, i.e., x^2 may be moved towards x^1 (meaning that firefly 2 is attracted to firefly 1), x^3 may be moved towards x^1 (in first place) and then x^2, and so on. We use the term 'candidate' because, in the proposed FA extension to MO, the new point may not be a promising position, when compared with the current one, and will not be accepted. This is a crucial issue and arises when a non-dominated current point generates a dominated candidate. In all the other cases, the candidate position is accepted. Furthermore, whenever a position is declared non-dominated, via flag='true' in the algorithm, any subsequent candidate position will be accepted only if it is non-dominated.

When extending FA to MO, the choice of the point z to center the randomization contribution to the firefly movement (see equation (2)) is based on a MO performance measure, a spread metric. Thus, z is one of the arguments of two consecutive non-dominated points with a maximum distance (based on infinity norm) in objective function values, i.e.,

$$z = \arg \max_{j \in \{1,\ldots,no\}} \left\{ \max_{i \in \{1,\ldots,|ND|-1\}} \left\{ f_j^{i+1} - f_j^i \right\} \right\} \quad (8)$$

where $|ND|$ is the cardinal of ND. Our choice falls on the first of the two points. We recall that the f_j values are sorted (from lowest to largest). This choice for the point z aims to force the movement of the firefly i towards the set of non-dominated points, as well as to the region where the distance between consecutive points is largest. This way the algorithm will generate an approximated Pareto front with evenly spread points. Only after all points (except x^1) have potentially moved towards other points, is the set ND updated with the new accepted non-dominated points, being removed the dominated solutions. Finally, at the end of each iteration, all accepted points are ordered based on their fitness values. The output of the algorithm is the set ND that contains an approximation to the Pareto front.

Data: $k_{\max}$, m
Set $k = 1$;
Randomly generate $x^i \in \Omega$, $i = 1, \ldots, m$ and evaluate $f_j(x^i), i = 1, \ldots, m$,
$j = 1, \ldots, m$;
Define the set ND with the non-dominated points;
Assign flag='true' to all non-dominated points of the population;
Assign fitness to all m points, using (7), and order them;
while $k \leq k_{\max}$ **do**
    forall the x^i *such that* $i = 2, \ldots, m$ **do**
        forall the x^j *such that* $j = 1, \ldots, i - 1$ **do**
            Compute randomization term and attractiveness β;
            Move firefly i towards j using (2) and evaluate
            $f_j(t^i), j = 1, \ldots, no$;
            if x^i *has flag='true'* **then**
                if t^i *is a non-dominated point* **then**
                  | Set $x^i = t^i$ and assign flag='true' to x^i;
                end
            else
                Set $x^i = t^i$;
                if x^i *is a non-dominated point* **then**
                  | Assign flag='true' to x^i;
                end
            end
        end
    end
    Set $k = k + 1$;
    Update the set ND with the accepted non-dominated points (remove
    the dominated ones);
    Assign flag='true' to all non-dominated points of the population;
    Assign fitness to all m points, using (7), and order them;
end
Output: set ND

Algorithm 1. MOoFA

The algorithm MOoFA stops when a target number of iterations, $k_{\max}$, is exceeded, although other criteria may be used. We may require that the number of function evaluations reaches a target value, or the largest gap between two consecutive points of the approximated Pareto front falls below a tolerance.

4   Numerical Comparisons

MOoFA is coded in MATLAB programming language (Matlab Version 8.1.0.604 (R2013a)) and the numerical experiments were carried out on a PC Intel Core 2 Duo Processor E7500 with 2.9GHz and 4Gb of memory. To analyze the performance of two variants of MOoFA, a set of nine benchmark problems with different properties in terms of Pareto-optimal front is used (see [12, 29, 34]). The known acronyms are: FON with non-convex Pareto front, $n = 3$ and $no = 2$; KUR with discontinuous Pareto front, $n = 3$ and $no = 2$; POL with discontinuous Pareto front, $n = 2$ and $no = 2$; SCH with convex Pareto front, $n = 1$ and $no = 2$; ZDT1 with convex Pareto front, $n = 30$ and $no = 2$; ZDT2 with non-convex Pareto front, $n = 30$ and $no = 2$; ZDT3 with discontinuous Pareto front, $n = 30$ and $no = 2$; ZDT4 with convex Pareto front, $n = 10$ and $no = 2$; ZDT6 with non-convex Pareto front and $n = 10$ and $no = 2$. We use the following acronyms to identify the two variants of MOoFA: (i) 'MOoFA-rank', for Algorithm 1 based on the objective ranking (4), with fitness (7); (ii) 'MOoFA', for Algorithm 1 based on the scaled objective distance to the minimum (5) and (6), with fitness (7). Each tested variant was run 10 times with each problem. In Algorithm 1, we set $m = 50$, as suggested in [29], and $k_{max} = 100$ when solving FON, KUR, POL, SCH, ZDT1, ZDT2, ZDT3 and ZDT6, and $m = 100$ and $k_{max} = 500$ when solving ZDT4. Some preliminary experiments were carried out to analyze the performance of the algorithms using previously proposed parameter values [9, 15]. The results showed that higher quality solutions are obtained with $\beta_0 = 1$, $\alpha_{min} = 0.01$, $\alpha_{max} = 0.5$, $\gamma_{min} = 0.1$ and $\gamma_{max} = 10$ as presented in [9].

4.1 MO Performance Measures

Three aspects could be considered when comparing the performance of multi-objective optimization algorithms: (i) the closeness to the true Pareto front; (If the true Pareto front for a given problem is known then the closeness can be measured using, for instance, the distance between the true Pareto front and the produced approximation to the Pareto front.) (ii) the spread along the Pareto front; (iii) the number of solutions in the non-dominated set. Here, we aim to compare closeness to the true Pareto front and select two performance metrics known as generational distance, GD_p, and inverted generational distance, IGD, which are defined by

$$GD_p = \frac{1}{|ND|} \left(\sum_{j=1}^{|ND|} d_j^p \right)^{1/p} \quad \text{and} \quad IGD = \frac{1}{N} \left(\sum_{j=1}^{N} D_j \right) \tag{9}$$

respectively, where $p \geq 1$, d_j is the Euclidean distance from the j-th point of the approximated front ND to its nearest point of the true Pareto front [12, 13, 22, 29], D_j is the minimum Euclidean distance between the point j in the true Pareto front and the points in ND and N is the number of uniformly distributed points along the true Pareto front. Smaller values of GD_p and IGD indicate better approximations to the Pareto-optimal front.

4.2 Experimental Results

First, using a visual presentation of our results, we show the approximated Pareto front produced by 'MOoFA-rank' and 'MOoFA'. We plot the ND set that corresponds to the run that gave the lowest GD_2 (corresponding to $p = 2$) value. Figure 1 contains the six plots that are produced by Algorithm 1 and objectives ranking (4), when solving SCH, ZDT1, ZDT2, ZDT3, ZDT4 and ZDT6.

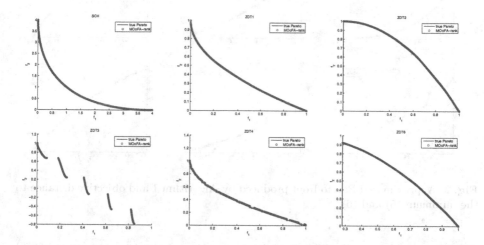

Fig. 1. Approximated Pareto front produced by Algorithm 1 and objective ranking (4)

Figure 2 contains the plots for the six previously referred problems using Algorithm 1 and objective distances to the minimum values (5) and (6). We may conclude that the produced approximated Pareto fronts are dense and have a sufficient large number of uniformly distributed points. The differences between the two variants are not significant, although we observe a slight improvement on closeness and density of MOoFA front, for the problems ZDT4 and ZDT6.

The large number of non-dominated solutions produced by Algorithm 1 requires a moderate computational effort specially when $m = 100$ and the algorithm runs for 500 iterations. We then decided to test another variant that computes candidate solutions only to fireflies that correspond to dominated positions. This means that only the dominated fireflies are attracted to non-dominated and dominated brighter fireflies. Hence, if $m_{nd} \leq m$ represents the number of non-dominated positions in the current population, the outer 'for' loop in Algorithm 1 starts with $x^{m_{nd}+1}$ and finishes with x^m. This variant is denoted by 'MOoFA-dom'. We observed that this variant produced a very small number of non-dominated solutions. However, increasing the size of the population and the maximum number of iterations allow the variant to find a larger number of points while improving GD_p and IGD. Thus, we have used $m = 100$ and $k_{max} = 500$ for all tested problems. Figure 3 displays the plots that correspond to the previously referred six problems. Nevertheless, these results are not as good as those produced by the variants 'MOoFA-rank' and 'MOoFA' of Algorithm 1.

Now, we report on Tables 3 and 4 the numerical results produced by 'MOoFA-rank' and 'MOoFA'. For these comparisons we use both the generational distance

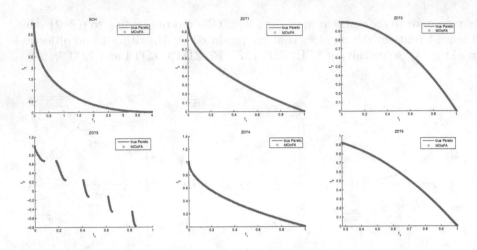

Fig. 2. Approximated Pareto front produced by Algorithm 1 and objective distance to the minimum (5) and (6)

GD_2 (based on $p = 2$), GD_1 (based on $p = 1$) and the inverted generational distance IGD (see (9)).

Table 3 contains the corresponding averaged GD_2 values over the runs. In parentheses, we show the average number of non-dominated solutions $|ND|$. The other results for comparison are from MOFA and three popular MOEAs known as SPEA, NSGA-II and DEMO, that are available from [29]. The author in [29] reports the use of $m = 50$, $k_{max} = 500$, and in FA several values for α_0 (ranging from 0.1 to 0.5) and β_0 (ranging from 0.7 to 1) were tested, with $\alpha = \alpha_0(0.9)^k$. Our results (based on $N = 500$) show that the variant 'MOoFA' gives slightly better values of GD_2 on problems ZDT1, ZDT2, ZDT3 and ZDT6 and variant 'MOoFA-rank' is better on SCH and ZDT4. Furthermore, when compared with MOFA [29], SPEA, NSGA-II and DEMO, our proposed variants of MOoFA give lower averaged GD_2 values when solving problems ZDT1, ZDT2 and ZDT3, but larger values when solving SCH and compared with MOFA and DEMO.

Table 4 contains average values of GD_1 and IGD computed from our results. We now compare with the GD_1 results reported in [12] for SPEA, PAES and NSGA-II, where $m = 100$, $k_{max} = 250$ are used. The results obtained with the problems FON, KUR and POL are also shown for comparison.

We remark that the reference Pareto fronts of problems FON, KUR and POL were obtained from the literature and they are not uniformly distributed. We also note that the IGD values produced by our variants of the Algorithm 1, when solving POL, are large since the set ND has just a few points with $f_1 > 15$ and $f_2 < 0.1$. When comparing GD_1, NSGA-II has slightly lower values on problems FON, KUR and ZDT4, PAES has a lower value on SCH, while the variant 'MOoFA' produces lower values than any of the other four in comparison, when solving problems POL, ZDT1, ZDT2, ZDT3 and ZDT6.

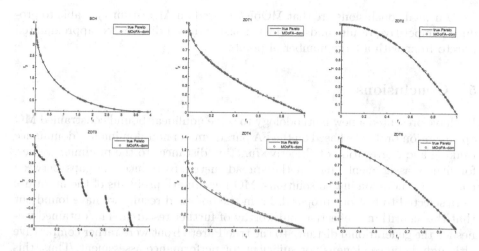

Fig. 3. Approximated Pareto front produced by 'MOoFA-dom', and objective distance to the minimum (5) and (6)

Table 3. Comparison based on GD$_2$ with $|ND|$ in parentheses

Prob.	'MOoFA-rank' GD$_2$	'MOoFA' GD$_2$	MOFA[†] GD$_2$	SPEA[†] GD$_2$	NSGA-II[†] GD$_2$	DEMO[†] GD$_2$
SCH	2.37e-04 (3314)	2.57e-04 (2977)	4.55e-06	5.17e-03	5.73e-03	1.79e-04
ZDT1	3.35e-05 (2644)	2.09e-05 (3325)	1.90e-04	1.78e-03	3.33e-02	1.08e-03
ZDT2	1.96e-05 (3360)	1.35e-05 (3517)	1.52e-04	1.34e-03	7.24e-02	7.55e-04
ZDT3	2.12e-05 (3110)	1.98e-05 (2639)	1.97e-04	4.75e-02	1.14e-01	1.18e-03
ZDT4	3.63e-01 (1201)	6.59e-01 (1033)	–	–	–	–
ZDT6	4.68e-03 (2033)	1.59e-04 (4402)	–	–	–	–

[†] results available in [29] with $m = 50$ and $k_{max} = 500$; – not available

Table 4. Comparison based on GD$_1$ and IGD

Prob.	'MOoFA-rank' GD$_1$	IGD	'MOoFA' GD$_1$	IGD	SPEA[‡] GD$_1$	PAES[‡] GD$_1$	NSGA-II[‡] GD$_1$
FON	9.50e-03	3.57e-03	8.57e-03	3.47e-03	1.26e-01	1.51e-01	1.93e-03
KUR	3.37e-02	4.15e-02	3.51e-02	3.78e-02	4.56e-02	5.73e-02	2.90e-02
POL	1.13e-02	1.57e+04	1.12e-02	1.57e+04	3.78e-02	3.09e-02	1.56e-02
SCH	5.05e-03	1.10e-03	5.40e-03	1.24e-03	3.40e-03	1.31e-03	3.39e-03
ZDT1	1.11e-03	5.29e-04	8.42e-04	3.57e-04	1.80e-03	8.21e-02	3.35e-02
ZDT2	9.63e-04	4.09e-04	7.11e-04	6.12e-02	1.34e-03	1.26e-01	7.24e-02
ZDT3	8.95e-04	3.13e-04	8.22e-04	1.06e-01	4.75e-02	2.39e-02	1.15e-01
ZDT4	3.91e+00	2.27e+00	1.02e+01	8.23e+00	7.34e+00	8.55e-01	5.13e-01
ZDT6	1.13e-02	5.16e-03	9.03e-04	5.81e-04	2.21e-01	8.55e-02	2.97e-01

[‡] results available in [12] with $m = 100$ and $k_{max} = 250$

Our final conclusions are that MOoFA (based on Algorithm 1) is able to produce competitive results and provides dense and well distributed approximated Pareto front with a large number of points.

5   Conclusions

We have presented a new methodology to solve nonlinear bound constrained MO optimization problems based on the FA paradigm, on non-dominance/dominance ranking and aggregation of objective function distances to the minimum values, for fitness assignment, and on the spread metric to reduce the gaps between consecutive non-dominated solutions. MO benchmark problems of the literature were selected to test our proposal. From the obtained results we have found out that the algorithm is effective and worthy of further research. The obtained values for the generational distance to the true Pareto front were rather competitive although distance alone is not sufficient for performance assessment. Thus, this study will be complemented with other performance guided metrics.

Future work will focus on incorporating a clustering technique into MOoFA to reduce the number of archived non-dominated solutions while maintaining the good density-based characteristics, so that computational time can be reduced. Furthermore, experimental tests will be extended to MO problems with three and more objectives and larger number of variables. The effect of increasing the number of objectives on the convergence of the algorithm will be investigated. Results available in the literature from other MOEAs will be used for comparison purposes.

Acknowledgments. The authors thank the anonymous referees for the valuable suggestions. This work has been supported by FCT (Fundação para a Ciência e Tecnologia, Portugal) in the scope of the projects: PEst-OE/MAT/UI0013/2014 and PEst-UID/CEC/00319/2013.

References

1. Akbari, R.B., Hedayatzadeh, R., Ziarati, K., Hassanizadeh, B.: A multi-objective artificial bee colony algorithm. Swarm and Evolutionary Computation **2**, 39–52 (2012)
2. Amiri, B., Hossain, L., Crawford, J.W., Wigand, R.T.: Community detection in complex networks: multi-objective enhanced firefly algorithm. Knowl.-Based Syst. 46, 1–11 (2013)
3. Arsuaga-Ríos, M., Vega-Rodríguez, M.A.: Multi-objective Firefly Algorithm for Energy Optimization in Grid Environments. In: Dorigo, M., Birattari, M., Blum, C., Christensen, A.L., Engelbrecht, A.P., Groß, R., Stützle, T. (eds.) ANTS 2012. LNCS, vol. 7461, pp. 350–351. Springer, Heidelberg (2012)
4. Barrico, C., Antunes, C.H.: Robustness analysis in multi-objective optimization using a degree of robustness concept. In: Proceedings of the 2006 IEEE World Congress on Computational Intelligence (WCCI 2006), pp. 6778–6783 (2006)

5. Biondi, T., Ciccazzo, A., Cutello, V., D'Antona, S., Nicosia, G., Spinella, S.: Multi-objective evolutionary algorithms and pattern search methods for circuit design problems. J. Univers. Comput. Sci. **12**(4), 432–449 (2006)
6. Chica, M., Cordón, O., Damas, S., Bautista, J.: A new diversity induction mechanism for a multi-objective ant colony algorithm to solve a real-world time and space assembly line balancing problem. Memetic Comp. **3**, 15–24 (2011)
7. Coello, C.A.C., Pulido, G.T., Lechuga, M.S.: Handling multiple objectives with particle swarm optimization. IEEE T. Evolut. Comput. **8**(3), 256–279 (2004)
8. Costa, L., Oliveira, P.: An elitist genetic algorithm for multiobjective optimization. In: Resende, M.G.C., Pinho de Sousa, J. (eds.) Metaheuristics, pp. 217–236. Kluwer Academic Publishers, USA (2004)
9. Costa, M.F.P., Rocha, A.M.A.C., Francisco, R.B., Fernandes, E.M.G.P.: Heuristic-based firefly algorithm for bound constrained nonlinear binary optimization. Advances in Operations Research, Article ID 215182, 12 pages (2014)
10. Custódio, A.L., Madeira, J.F.A., Vaz, A.I.F., Vicente, L.N.: Direct multisearch for multiobjective optimization. SIAM J. Optim. **21**, 1109–1140 (2011)
11. Deb, K., Gupta, H.: Introducing robustness in multi-objective optimization. Evolut. Comput. **14**(4), 463–494 (2006)
12. Deb, K., Pratap, A., Agrawal, S., Meyarivan, T.: A fast and elitist multiobjective genetic algorithm: NSGA-II. IEEE T. Evolut. Comput. **6**(2), 182–198 (2002)
13. Denysiuk, R.: Multiobjective Optimization: Review, Algorithms, and Applications. Ph.D. Thesis, University of Minho, Braga, Portugal (2014)
14. Ferreira, J., Fonseca, C.M., Covas, J.A., Gaspar-Cunha, A.: Evolutionary multi-objective robust optimization. In: Kosiński, W. (ed.) Advances in Evolutionary Algorithms, pp. 261–278. I-Tech Education and Publishing (2008)
15. Fister, I., Fister Jr, I., Yang, X.-S., Brest, J.: A comprehensive review of firefly algorithms. Swarm and Evolutionary Computation **13**, 34–46 (2013)
16. Fontes, D.B.M.M., Gaspar-Cunha, A.: On multi-objective evolutionary algorithms. In: Zopounidis, C., Pardalos, P.M. (eds.) Handbook of Multicriteria Analysis, Appl. Optimizat., vol. 103, pp. 287–310. Springer (2010)
17. Gaspar-Cunha, A, Covas, J.A.: RPSGAe - reduced Pareto set genetic algorithm: application to polymer extrusion. In: Gandibleux, X., Sevaux, M., Sörensen, K., T'kindt, V. (eds.) Metaheuristics for Multiobjective Optimisation, Lect. Notes Econ. Math. vol. 535, pp. 221–249 (2004)
18. Karthikeyan, S., Asokan, P., Nickolas, S., Page, T.: A hybrid discrete firefly algorithm for solving multi-objective flexible job shop scheduling problems. International Journal of Bio-Inspired Computation (2014) (in press)
19. Lobato, F.S., Arruda, E.B., Cavalini Jr., A.A., Steffen Jr., V.: Engineering system design using firefly algorithm and multi-objective optimization. In: ASME 2011, 31th Computers and Information in Engineering Conference (Parts A and B), vol. 2, pp. 577–585 (2011)
20. Li, H., Ye, C.: Firefly algorithm on multi-objective optimization of production scheduling system. Advances in Mechanical Engineering and its Applications **3**(1), 258–262 (2012)
21. Marichelvam, M.K., Prabaharan, T., Yang, X.-S.: A discrete firefly algorithm for the multi-objective hybrid flowshop scheduling problems. IEEE T. Evolut. Comput. **18**(2), 301–305 (2014)
22. Molina, J., Laguna, M., Martí, R., Caballero, R.: SSPMO: a scatter tabu search procedure for non-linear multiobjective optimization. INFORMS J. Comput. **19**(1), 91–100 (2007)

23. Mousa, A.A., El-Shorbagy, M.A., Abd-El-Wahed, W.F.: Local search based hybrid particle swarm optimization algorithm for multiobjective optimization. Swarm and Evolutionary Computation **3**, 1–14 (2012)
24. Schutze, O., Esquivel, X., Lara, A., Coello Coello, C.A.: Using the averaged Hausdorff distance as a performance measure in evolutionary multiobjective optimization. IEEE T. Evolut. Comput. **16**(4), 504–522 (2012)
25. Silva, D.O., Vieira, L.G.M., Lobato, F.S., Barrozo, M.A.S.: Optimization of hydrocyclone performance using multi-objective firefly colony algorithm. Separ. Sci. Technol. **48**(12), 1891–1899 (2013)
26. Tilahun, S.L., Ong, H.C.: Modified firefly algorithm. Journal of Applied Mathematics, Article ID 467631, 12 pages (2012)
27. Yang, X.-S.: Firefly Algorithms for Multimodal Optimization. In: Watanabe, O., Zeugmann, T. (eds.) SAGA 2009. LNCS, vol. 5792, pp. 169–178. Springer, Heidelberg (2009)
28. Yang X.-S.: Firefly algorithm. In: Nature-Inspired Metaheuristic Algorithms, 2nd edn., pp. 81–96. Luniver Press, University of Cambridge (2010)
29. Yang, X.-S.: Multiobjective firefly algorithm for continuous optimization. Eng. Comput. **29**(2), 175–184 (2013)
30. Yang, X.-S., He, X.: Firefly algorithm: recent advances and applications. Int. J. Swarm Intelligence **1**(1), 36–50 (2013)
31. Younes, M., Khodja, F., Kherfane, R.L.: Multi-objective economic emission dispatch solution using hybrid FFA (firefly algorithm) and considering wind power penetration. Energy **67**, 595–606 (2014)
32. Yu, S., Yang, S., Su, S.: Self-adaptive step firefly algorithm. Journal of Applied Mathematics, Article ID 832718, 8 pages (2013)
33. Zhou, A., Qu, B.-Y., Li, H., Zhao, S.-Z., Suganthan, P.N., Zhangd, Q.: Multiobjective evolutionary algorithms: A survey of the state of the art. Swarm and Evolutionary Computation **1**(1), 32–49 (2011)
34. Zitzler, E., Deb, K., Thiele, L.: Comparison of multiobjective evolutionary algorithms: empirical results. Evolut. Comput. **8**(2), 173–195 (2000)
35. Zitzler, E., Laumanns, M., Bleuler, S.: A tutorial on evolutionary multiobjective optimization. In: Gandibleux, X., Sevaux, M., Sörensen, K., T'kindt, V. (eds.) Metaheuristics for Multiobjective Optimisation. Lect. Notes Econ. Math., vol. 535, pp. 3–37 (2004)
36. Zitzler, E., Laumanns, M., Thiele, L.: SPEA2: Improving the strength Pareto evolutionary algorithm for multiobjective optimization. In: Evolutionary Methods for Design, Optimisation and Control with Applications to Industrial problems, ICNME, pp. 95–100 (2002)

GACO: A Parallel Evolutionary Approach to Multi-objective Scheduling

Jarosław Rudy and Dominik Żelazny[⊠]

Department of Automatic Control and Mechatronics,
Wrocław University of Technology,
Janiszewskiego 11-17, 50-372 Wrocław, Poland
{jaroslaw.rudy,dominik.zelazny}@pwr.wroc.pl

Abstract. In this paper the job shop scheduling problem with two criteria of minimizing makespan and the sum of tardiness of jobs is considered. This multi-objective problem is strongly NP-hard, as single criterion version is strongly NP-hard as well. A permutation-based representation for the job shop problem is used and a new hybrid parallel multi-agent method, called GACO (Genetic Algorithm Ant Colony Optimization), is proposed. The computation is done in parallel and additional threads concurrently compute certain parts of both algorithms. The researched speed-up is considerable, albeit limited by the need to combine solutions. Approximation of the Pareto front obtained by GACO is superior to the approximations obtained by GA and ACO separately.

Keywords: Multi-criteria optimization · Job shop problem · MCDA · Hybrid algorithm · Nature-based

1 Introduction

Maintaining competitive position in fast changing market requires companies to use new methods of optimization and drives scientists to develop more efficient algorithms. Due to that competitiveness, developing effective, advanced methods is extremely important. The so-called job shop scheduling problem (JSP) represents a class of widely studied cases based on ideas derived from production engineering. Most of the currently researched problems consider single criterion objective value function and are easily adaptable to real world applications, but modern scheduling problems need more advanced models. This applies not only to scheduling problems in manufacturing processes, but also for network scheduling [16] or vehicle routing problems [7].

Multi-objective JSP (MOJSP) is the result of natural evolution of models and optimization methods that put more emphasis on practical applications of JSP. This is because decision making in scheduling usually have to take several economic indexes simultaneously, which naturally take the form of several optimization criteria. Thus, real world applications require adjusting existing models of discrete optimization and objective functions to multi-criteria approach to

© Springer International Publishing Switzerland 2015
A. Gaspar-Cunha et al. (Eds.): EMO 2015, Part I, LNCS 9018, pp. 307–320, 2015.
DOI: 10.1007/978-3-319-15934-8_21

solve modern optimization problems. However, frequently studied cases apply optimization algorithms with only one criterion and few researchers use multi-criteria approach to JSP. One of the biggest concerns of multi-criteria optimization is computational complexity, which generally grows with the number of defined criteria, making solving NP-hard problems even more difficult. Recently researchers try to bypass this issue by harnessing the possibilities of parallel programming, CUDA architecture and distributed computing being the prime examples. These are used to significantly speed up the computation process.

1.1 Multi-objective Job Shop Scheduling Problem

Most common multi-objective algorithms are variances of evolutionary and population based methods. This is caused by the fact that multi-agent algorithms perform multiple searches of the solution space at the same time. One of their qualities is an ability to find an approximation of the Pareto front in short time, which made them useful and efficient in multi-criteria optimization. Below we present a brief review of some of such methods proposed in the literature.

In paper [8], a Two-Stage Genetic Algorithm (2S-GA) was proposed by Kachitvichyanukul *et al.* Its goal was to minimize weighted sum of the criteria, including makespan, total weighted earliness and tardiness. Proposed algorithm was compared with single criterion methods and one multi-objective algorithm. Authors compared their new dispatching rule based representation with others known from literature. Algorithms using Model-based Hybrid Representation (MHR) obtained higher values of Hyper-Volume Indicator (I_H) and were more robust to the representation size. An evolutionary algorithm (EA) was proposed by Lei and Wu [11]. External population was adjusted using crowding measure and assigned different fitness values for individuals. Proposed algorithm performed well in optimizing bi-criteria objective function, consisting of makespan and total tardiness.

A hybrid algorithm called Jumping Genes Genetic Algorithm (JGGA) [15], proposed by Ripon, was capable of searching for near-optimal Pareto solutions, while maintaining convergence. Performed tests have shown better results compared to the other existing evolutionary approaches.

Another multi-agent method, called Particle Swarm Optimization (PSO), was proposed by Lei [10]. This method used global best position selection combined with maintaining crowding measure-based archive. Tests have shown, that the algorithm produced high quality Pareto fronts. Sha and Lin proposed Multi-Objective PSO (MOPSO) [17] with the optimization criteria of the makespan, total tardiness and total idle times. Decoding was performed using the Giffler and Thompson heuristic, which provides active solutions from schedules. Small-sized benchmarks were used in evaluation and MOPSO yielded good quality results. Representative of another multi-agent method was proposed by Udomsakdigool and Khachitvichyanukul [20]. An Ant Colony Optimization (ACO) solved MOJSP with competitive results. Authors applied Local Search (LS) method in order to intensify the search.

Apart from multi-agent algorithms, some methods based on local search (LS) methods were proposed for solving MOJSP. Suresh and Mohanasndaram developed Pareto-archived simulated annealing (PASA) [19] for solving bi-criteria MOJSP. It made use of Pareto dominance and a criteria aggregating to accept the candidate solution from among the solution set generated by the segment random insertion (SRI neighborhood structure. PASA outperformed other tested algorithms in benchmarks. Fattahi *et al.* tackled bi-criteria MOJSP with simulated annealing (SA) approach [4]. Multi-objective problem with makespan and total weighted tardiness was converted to scalar optimization function. Unfortunately, no comparative tests were performed.

1.2 Parallel Job Shop Scheduling Problem

Recently parallelization of algorithms became common in the field of computer-aided optimization. This is caused by the fact that modern computer systems increase their computational power by developing methods of parallel processing (*e.g.* multiple processor cores), instead of increasing the clock rate of processors (and other subsystems). As a result, sequential algorithms, which use only a single core, fail to employ almost all of the available computational power offered by state-of-the-art computers. Along with the further development of multi-core devices the advantage of parallel algorithms over sequential algorithms will increase even further. Below we briefly present some recent developments and approaches to parallel JSP.

Bożejko *et al.* proposed a parallel SA algorithm [1] and significantly reduced the computation time through the parallelization of computing the fitness procedure. Super linear speedup was obtained when using representative-based neighborhood. Gu *et al.* [5] proposed a parallel genetic algorithm (GA). The solution space was divided into so-called islands. Each of the parallel algorithms performed operations on one of those islands. Algorithm instances exchanged information with each other to provide the best individuals. Tests have shown that the proposed algorithm has high convergence speed and provides near-optimal solutions efficiently. Another population based algorithm was proposed by Yusof *et al.* in [21]. It was micro GA, which worked on small populations. When populations reached similar chromosomes, the re-initialization started. The best individual was kept and all others were replaced by randomly-generated population. Parallel approach provided better solutions in less time than sequential GA and micro GA. Parallel tabu search algorithm was proposed by Bożejko *et al.* [2]. Authors solved flexible JSP. The problem has been divided into two subproblems: assigning operations to the machines and determining the order of operations on each machine. The second phase was performed on parallel GPU machines, and the results were collected using the MPI protocol. Computations time for instances of big size was significantly improved.

1.3 Applications

Pfund *et al.* considered complex job shop problem in semiconductor wafer fabrication process. In paper [14], a Modified Shifting Bottleneck Heuristic was proposed. Semiconductor wafer usually need to be processed many times on different machines, sometimes repeatedly on each machine in different stages of the processing. Since each machine can cost from tens to hundreds of thousands of dollars, it is necessary to reduce capital spending by scheduling of jobs.

In further practical use for multi-criteria scheduling, one should concentrate on decision support systems (DSS), which will aid decision-maker (DM) and allow to improve results accordingly. When considering more than two criteria, it is crucial to decide on a method of visualization. Miettinen prepared a survey of visualization techniques in paper [12]. Depending on problem and DMs preferences, there is a number of methods to implement in DSS.

2 Problem Description

We consider a manufacturing system with the set $M = \{1, \ldots, m\}$ of m machines with unit capacity. Moreover, $J = \{1, 2, \ldots, n\}$ is the set of n jobs to be processed. Job j-th, consists of the sequence of n_j operations indexed consecutively $(l_{j-1} + 1, \ldots, l_{j-1} + n_j)$, where $l_j = \sum_{i=1}^{j} n_i$ is the total number of operations of the first j jobs, $j = 1, 2, \ldots, n$, $(l_0 = 0)$, $o = \sum_{i=1}^{n} o_i$ is the total number of all operations and o_i is the number of operations required to complete job i. Operation x is to be processed on machine $\mu_x \in M$ during an uninterrupted processing time $p_x > 0$, $x \in O$. Our aim is to find the schedule under the following constraints: (1) each machine can process at most one product at a time, (2) each product can be process by at most one machine at a time, (3) operations cannot be preempted.

The set of operations O can be decomposed into subsets $O_k = \{x \in O | \mu_x = k\}$, each of them containing the operations to be processed on machine $k \in M$. Let permutation π_k define the processing order of operations from the set O_k on machine k, and let Π_k be the set of all permutations on O_k. The processing order of all operations on machines is determined by m-tuple $\pi = (\pi_1, \pi_2, \ldots, \pi_m)$, where $\pi \in \Pi_1 \times \Pi_2 \times \ldots \times \Pi_m$.

A given schedule π can be described by a pair of vectors $S = (S_1, \ldots, S_o)$ and $C = (C_1, \ldots, C_o)$, where S_j and C_j denote starting and completion time of operation j. The schedule has to satisfy the following constraints:

$$C_{\underline{t}_j} \leq S_j \qquad \underline{t}_j \neq 0, \ j \in O, \tag{1}$$

$$C_{\underline{s}_j} \leq S_j \qquad \underline{s}_j \neq 0, \ j \in O, \tag{2}$$

$$C_j = S_j + p_j \qquad j \in O, \tag{3}$$

A schedule that satisfies inequalities (1–3) is feasible. Constraint (1) follows from technological processing order of operations inside job, whereas (2) follows from the unit capacity of machines.

Our aim is to find feasible processing order $\pi^* \in \Pi$, such that:

$$C_{sum}(\pi^*) = \min_{\pi \in \Pi} C_{sum}(\pi), \tag{4}$$

$$C_{max}(\pi^*) = \min_{\pi \in \Pi} C_{max}(\pi), \tag{5}$$

where $C_{sum}(\pi) = \sum_{i \in O^L}^{n} C_i$ is the sum of jobs completion times and $O^L = \{i : i = l_{i-1} + n_i, \ i \in O\}$ is the set of the last operations of jobs, while $C_{max}(\pi)$ is the time required to complete all jobs on the machines in the processing order.

3   Multi-Criteria Solutions Evaluation

Evaluation of multi-criteria solutions is not as straight-forward as in the case of single-criterion problems. Comparing two solutions requires different approaches. Aggregation of (weighted) objectives is one of the most commonly used techniques, unfortunately this method requires either fine tuning of the weights or running the algorithm using the multi-start method. Thus, we considered techniques from multi-criteria decision analysis (MCDA) to evaluate solutions in those multi-agent algorithms.

Technique for Order of Preference by Similarity to Ideal Solution. Hwang and Yoon proposed TOPSIS, a MCDA method in [6]. The concept is to choose the solution with the shortest geometric distance from the best (ideal) solution and the longest distance from the worst (negative-ideal) solution. The extreme criteria values of given solutions are used to determine this ideal and negative-deal solution. The method also uses weights for each criterion and normalizes all solutions before calculating the geometric distances. The higher the value of the relative closeness the better the solution. This method allows to choose one solution from the Pareto front, without involving decision-maker in the process.

Pareto Efficiency. The solution to a multi-objective problem is the set of non-dominated solutions called the Pareto front, where dominance is defined as follows. In a minimization problem a solution $y = (y_1, y_2, \ldots, y_f)$ dominates $z = (z_1, z_2, \ldots, z_f)$ (denoted $y \prec z$) if and only if:

$$\underset{i \in F}{\forall} \ y_i \leq z_i, \tag{6}$$

$$\underset{i \in F}{\exists} \ y_i < z_i, \tag{7}$$

where $F = \{1, \ldots, f\}$ is the set of f criteria (objective functions).

4   Representation and Decoding

The representation of solutions is an important aspect in algorithm development, as it determines the solution space and some properties of solutions. When solving JSP, it is also important to decide on a method of decoding such representation. For the purpose of this work we decided to use a job-based representation and decoding scheme based on Giffler and Thompson heuristic. In result, solutions decoded by our algorithm are active. Moreover, simplified representation results in decreased computational time of the algorithm.

5   Proposed Method

For the purpose of this article a hybrid parallel algorithm was implemented. Said algorithm used two multi-agent methods – Genetic Algorithm (GA) and Ant Colony Optimization (ACO) – which allow fast approximation of Pareto front and, due to their characteristics, search different areas of solutions space. Parameters of all algorithms were automatically adjusted (self-set parameters). Moreover, a second stop rule was implemented. When execution reaches certain (predetermined by tests) run time, the algorithm stops at current iteration and returns gathered solutions.

5.1   Ant Colony Optimization

The Ant Colony Optimization (ACO) is a probabilistic meta-heuristic technique used to create approximate algorithms for optimization problems. The technique itself is used in order to find good (short) paths in a given graph. It is most commonly used in the Traveling Salesman Problem (TSP). Currently it has found usage in a wide range of discrete optimization problems and is applicable to any problem that can be reduced to short path search in a graph, including the MOJSP.

ACO is a population-based algorithm simulating the foraging behavior of ant colonies, where ants create candidate solutions in each iteration. Solutions are created in steps, each step extends an existing partial solution, by choosing next node in a graph. The probability of selecting a given node is dependent on the visibility of the node and the pheromone trail on the edge that leads to that node, the higher the pheromone value, the more attractive the edge is. All constructed solutions are then evaluated and the pheromone trails are updated depending on the quality of solutions found. Thus, the search can intensify on promising parts of the solution space. The pheromone evaporates over time, meaning the colony can forget unused or bad trails and thus diversify the search process.

A number of variants of ACO have been proposed. In this paper we employ Max-Min Ant System (or MMAS) by Stützle and Hoos [18], which introduces a few new elements. The pheromone trail on a given edge is restricted between values of τ_{min} and τ_{max}. Contrary to the basic ACO variant, MMAS uses an elitist approach, where only one or two ants have the right to update the pheromone matrix per iteration. Both iteration- and global-best ants are used as candidates for this role. The MMAS is easily adapted to MOJSP, since our solution representation is a permutation (the same as in TSP). The main problem is the definition of the visibility of the edge (job) j. We decided on a heuristic approach where the visibility is equal to the sum of processing times of all operations of the job j. The remaining implementation of our MMAS closely follows the original paper by Stützle and Hoos.

5.2   Genetic Algorithm

Genetic Algorithm (GA) is a multi-agent method based on the evolution process found in the nature to find better solutions. Evolutionary algorithms use techniques inspired by natural occurring factors, such as inheritance, mutation, selection and crossover to generate solutions to optimization problems. Usually, the evolution starts from random initial population, which consists of individuals. In each iteration, called generation, those individuals are modified (by means of mutation and/or crossover) and their fitness is evaluated in order to select best solutions for next generation. Over the years, different approaches to the GA were proposed and tested for a variety of optimization problems. Our GA uses external Pareto archive in order to maintain non-dominated solutions through successive iterations. The individuals in population are represented by the following: jobs permutation, values of criteria functions and relative closeness indicator calculated by the TOPSIS method. Initial population includes solutions obtained from certain constructive algorithms, prepared to optimize one of the criteria. Such initialization allows faster designation of the approximation of Pareto front. Mutation is performed by interchanging two random jobs in schedule, while crossover uses a partially matched crossover (PMX) scheme. Fitness values are evaluated using the TOPSIS technique and are then used in tournament selection. After the selection, half of parent and child population is combined into new parent population. Moreover, when relative closeness values converge to zero, an anti-stagnation function is employed.

5.3   Parallel Algorithm – GACO

Both component algorithms, GA and ACO, are population-based meta-heuristics. GA uses its current solution population (let us denote it P_{GA}) in order to perform the selection process: select(P_{GA}). Similarly, ACO employs its constructed solutions (P_{ACO}) to perform the pheromone update: update(P_{GA}). Our idea is to allow both algoritms to use each population of the other algorithm in addition to their own. Let $P_{GACO} = P_{\text{GA}} \cup P_{\text{ACO}}$. Then we define new selection and pheromone update operations: select(P_{GACO}) and update(P_{GACO}). This allows to combine solutions gained through the capabilities and unique features of both algorithms. The resulting hybrid algorithm is named GACO.

In order to properly perform the selection and update process, the GACO algorithm needs both populations. Therefore, a synchronization mechanism is implemented (using condition variables) to ensure that those processes will not

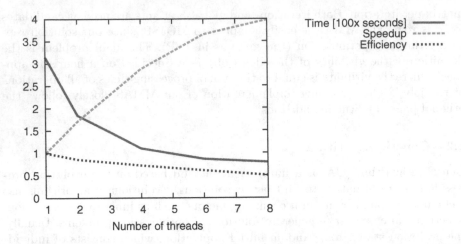

Fig. 1. Processing time, speed-up and speed-up efficiency for the GACO algorithm

start until both P_{GA} and P_{ACO} are prepared for the current iteration. The resulting sequential algorithm has the processing time a little over a sum of the sequential GA and ACO processing times. In case of the ta01 instance this time equals 314 seconds (10 000 iterations, 200 ants and 200 GA specimens). However, we employ two parallelization mechanisms. First, the GA and ACO are computed in parallel, only stopping to wait for the other algorithm before combining their current populations. This uses two parallel threads and reduces the processing time to 183.6 seconds for ta01. Next, a number of subthreads is created for each algorithm. Those threads are created only once and are used to perform time-consuming operations (crossover, mutation, ants solution construction) in each iteration. For 8 concurrent threads this results in and processing time of 79 seconds for ta01 and a total speed-up of roughly 4. This value stays fairly constant for other Taillard instances, meaning (from Amdahl's law) that the parallel fraction of the GACO algorithm is roughly equal to 85%, including the processor overhead. The speed-up and speed-up efficiency for the GACO algorithm are shown in Fig. 1. The values were measured for even number of threads and the computing times in the figure were divided by 100.

6   Computer Experiment

All algorithms were implemented in C++ and compiled with g++ 4.6.3. The programs were tested on i7-3610QM 2.30 GHz machine (8 concurrent threads) with 6 GB of RAM under the Linux operating system. Benchmarks were taken from literature and contain 8 groups. Each group consists of 10 instances of the same size.

6.1   Multi-Criteria Quality Indicators

Our approach is based on a dominance relation, so the result for each instance of GACO is a set of non-dominated solutions – an approximation of the Pareto front. Solutions from all the algorithms were flagged and aggregated into a single set, which was then purged of dominated solutions. A number of solutions in this global Pareto-efficient set was computed for each algorithm and used to compare solution sets [13].

Zitzler *et al.* [9] provided a necessary tool for a better evaluation and comparison of multi-objective algorithms. Hyper-volume indicator (or HVI) I_H measures quality of the Pareto front approximations. It measures the area covered by the approximated Pareto fronts bound by reference point, described as 120% of the worst values of each criterion.

6.2   Results

There are 80 instances divided into 8 instance sizes, thus computation results were combined into groups of fixed instance size.

Table 1. Hyper-Volume indicators and number of Pareto solutions

Group	$|P|$	$|P_{ACO}|$	$|P_{GA}|$	$|P_{GACO}|$	$I_H(ACO)$	$I_H(GA)$	$I_H(GACO)$
15×15	124	10	36	78	0,077	0,064	0,100
20×15	135	22	39	74	0,088	0,069	0,118
20×20	120	20	35	65	0,078	0,067	0,090
30×15	133	11	38	84	0,093	0,068	0,100
30×20	145	9	50	86	0,083	0,068	0,094
50×15	195	12	71	112	0,104	0,059	0,120
50×20	167	18	49	100	0,094	0,066	0,105
100×20	208	25	65	119	0,114	0,061	0,130
Sum	1227	127	383	718			
Average					0,091	0,065	0,107

For each test instance and for each run of the algorithms, we collected the following values:

- $|P|$ – number of non-dominated solutions in aggregated approximations of Pareto fronts from all algorithms,
- $|P_x|$ – number of non-dominated solutions found using algorithm x,
- $I_H(x)$ – HVI value of Pareto front found by algorithm x,

where $x \in \{ACO, GA, GACO\}$.

Tab. 1 shows summed up numbers of Pareto solutions for each instance size. As can be seen, the number of Pareto solutions found by GACO algorithm exceeds the number of Pareto solutions found by ACO ang GA algorithms, both of them in total only found around 41% of all of the non-dominated solutions. Moreover, I_H values of ACO and GA were from 15 to 39% lower than GACO.

Furthermore, a series of statistical hypothesis testing, using Octave software, was conducted in order to assess the mean value of the HVI of GACO algorithm compared to ACO and GA. For all test we used Student's t-test with the significance level of 0.05, *i.e.* the test yeilds the p-value and we compare it with 0.05 in order to reject or accept the null hypothesis. First, we tested the GACO algorithm compared to ACO. For this we defined a new I_H parameter as follows:

$$Q_{\mathrm{ACO}} = \frac{I_{H_i}(\mathrm{GACO})}{I_{H_i}(\mathrm{ACO})}, \tag{8}$$

where i is the tested instance (*i.e.* from TA01 to TA80). Thus, we obtained sample of 80 values of Q_{ACO}. We have concluded that mean value of Q_{ACO} is equal to 1.25 (GACO has 25% better HVI value compared to ACO) with test yielding the p-value of 0.075. We also determined that the mean Q_{ACO} is greater than 1.15 using one-sided t-test (yielded p-value of 0.4, null hypothesis $\mu = 1.15$ is rejected and alternative hypothesis $\mu > 1.15$ is accepted).

Similar test was done for the GACO vs. GA case. The resulting value Q_{GA} indicates that GACO is 70% (test yielded p-value of 0.272) or even 74% (p-value of 0.052) better than pure GA in terms of I_H. Using one-sided testing we also observed that mean I_H of GACO vs. GA is greater than 1.57 (p-value of 0.045, null hypothesis rejected, alternative $\mu > 1.57$ accepted).

As a last test we have decided to put GACO against the combined forces of GA and ACO algorithms:

$$Q_{\mathrm{MAX}} = \frac{I_{H_i}(\mathrm{GACO})}{\max(I_{H_i}(\mathrm{ACO}), I_{H_i}(\mathrm{GA}))}. \tag{9}$$

We observe that the mean for such defined Q_{MAX} is equal to 1.2 (p-value 0.75) or even 1.24 (p-value 0.077). For the one sided test with $\mu > 1.14$ as the alternative hypothesis, we obtained the p-value of 0.032, thus rejecting the null hypothesis in favor of the alternative hypothesis.

All this allows us to conclude that I_H indicator values for GACO are much greater than that for GA and considerably greater than that for ACO and both algorithms combined. This result is also statistically significant.

In order to assist the visual evaluation of the approximations of Pareto fronts obtained by proposed algorithms, we normalized all solutions from all instances. First, for each instance we found minimal and maximal values of all criteria. Second, we computed a normalized value of the solutions using the following formula:

$$norm(x_{i,j}) = \frac{x_{i,j} - \min x_i}{\max x_i - \min x_i}, \tag{10}$$

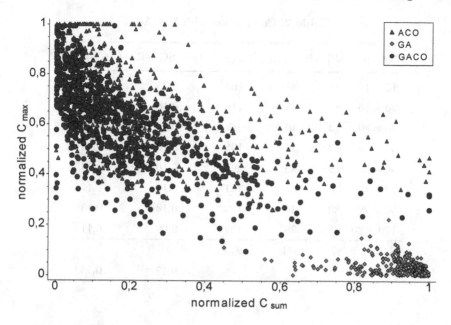

Fig. 2. Normalized solutions from all instances

where:

- $x_{i,j}$ – value of the i-th objective function in the j-th solution,
- $\min x_i$ – minimal value of the i-th objective function in the current instance,
- $\max x_i$ – maximal value the of i-th objective function in the current instance.

The sets of normalized non-dominated solutions, henceforth called clouds, which were obtained using all tested algorithms are shown in Fig. 2. The cloud obtained from the proposed GACO algorithm is similar in shape to the cloud obtained from the results provided by the ACO algorithm, however the GACO cloud has better (smaller) values for both criteria than ACO, showing its superiority over it. Moreover, the GACO cloud also comes closer to the GA cloud and while it does not reaches it, the spread of Pareto-efficient solutions is improved compared both to GA and (to lesser extent) to ACO.

Furthermore, we have compared proposed algorithm with NSGA-II proposed by Deb *et al.* in [3]. The results of the comparison are shown in Tab. 2. The total and individual values of $|P|$ are lower, than those of $|P_{\text{NSGA-II}}|$ and $|P_{\text{GACO}}|$, because a number of locally Pareto-optimal solutions from one algorithm was dominated by the other algorithm and *vice versa*. Removing the dominated and repeated solutions yielded lower value of the total number of Pareto-optimal solutions. In the evaluation the NSGA-II algorithm had overall better values of HVI (especially in case of small instances), but proposed GACO algorithm achieved its assumed purpose of finding an approximation of Pareto front, which is more evenly spread and contains, overall, three times greater number of non-dominated solutions.

Table 2. Comparison with NSGA-II

Group	$|P|$	$|P_{\text{NSGA-II}}|$	$|P_{\text{GACO}}|$	$I_H(\text{NSGA-II})$	$I_H(GACO)$
15×15	81	38	105	0,25	0,19
20×15	93	45	116	0,18	0,10
20×20	72	40	95	0,28	0,21
30×15	92	44	105	0,10	0,08
30×20	78	45	99	0,16	0,18
50×15	110	51	131	0,15	0,17
50×20	80	40	123	0,14	0,13
100×20	68	38	139	0,09	0,11
Sum	674	341	913		
Average				0,17	0,15

7   Conclusions and Further Research

Main idea behind the proposed algorithm was derived from our previous work, where we observed certain properties of Pareto front approximations produced by ACO and GA algorithms. Proposed hybrid algorithm met our expectations and provided Pareto fronts which included the minimization of both objectives. In our previous work we concluded that ACO with its pheromone matrix tends to minimize C_{sum} while GA steers itself for lower C_{max} values. Combining those methods and exchanging information between ants and GA individuals allowed a wider exploration of the solution space. Further research involving different measures of the quality of solutions and interchanging of the non-dominated solutions will be performed.

Acknowledgements.. This work is co-financed by the European Union as part of the European Social Fund.

References

1. Bożejko, W., Pempera, J., Smutnicki, C.: Parallel Simulated Annealing for the Job Shop Scheduling Problem. In: Allen, G., Nabrzyski, J., Seidel, E., van Albada, G.D., Dongarra, J., Sloot, P.M.A. (eds.) ICCS 2009, Part I. LNCS, vol. 5544, pp. 631–640. Springer, Heidelberg (2009)
2. Bożejko, W., Uchroński, M., Wodecki, M.: Flexible job shop problem - parallel tabu search algorithm for multi-GPU. Archives of Control Sciences 22, 389–397 (2012)
3. Deb, K., Pratap, A., Agarwal, S., Meyarivan, T.: A fast and elitist multi-objective genetic algorithm: NSGA-II. IEEE Trans. EVolume Comput. 6(2), 182–197 (2002)
4. Fattahi, P., Saidi Mehrabad, M., Arynezhad, M.B.: An algorithm for multi objective job shop scheduling problem. Journal of Industrial. International 2(3), 43–53 (2006)
5. Gu, J., Gu, X., Cao, C.: A novel parallel quantum genetic algorithm for stochastic job shop scheduling. Journal of Mathematical Analysis and Applications 355(1), 63–81 (2009)
6. Hwang, C.L., Yoon, K.: Multiple Attribute Decision Making: Methods and Applications. Springer, New York (1981)
7. Jagiełło, S., Żelazny, D.: Solving Multi-criteria Vehicle Routing Problem by Parallel Tabu Search on GPU. Procedia Computer Science 18, 2529–2532 (2013)
8. Kachitvichyanukul, V., Sitthitham, S.: A two-stage genetic algorithm for multi-objective job shop scheduling problems. Journal of Intelligent Manufacturing 22(3), 355–365 (2011)
9. Knowles J., Thiele L., Zitzler E.: A tutorial on the performance assessment of stochastic multiobjective optimizers. ETH Zurich (2006)
10. Lei, D.: A Pareto archive particle swarm optimization for multi-objective job shop scheduling. Computers and Industrial Engineering 54(4), 960–971 (2008)
11. Lei, D., Wu, Z.: Crowding-measure-based multi-objective evolutionary algorithm for job shop scheduling. International Journal of Advanced Manufacturing Technology 30, 112–117 (2006)
12. Miettinen, K.: Survey of methods to visualize alternatives in multiple criteria decision making problems. OR Spectrum 36(1), 3–37 (2014)
13. Pempera J., Smutnicki C., Żelazny D.: Optimizing bicriteria flow shop scheduling problem by simulated annealing algorithm. Procedia Computer Science 18, 936–945 (2013)
14. Pfund, M.E., Balasubramanian, H., Fowler, J.W., Mason, S.J., Rose, O.: A multi-criteria approach for scheduling semiconductor wafer fabrication facilities. Journal of Scheduling 11(1), 29–47 (2008)
15. Ripon, K.S.N.: Hybrid evolutionary approach for multi-objective job shop scheduling problem. Malaysian Journal of Computer Science 20(2), 183–198 (2007)
16. Rudy J., Żelazny D.: Memetic algorithm approach for multi-criteria network scheduling. In: Proceeding of the International Conference on ICT Management for Global Competitiveness and Economic Growth in Emerging Economies (ICTM 2012), pp. 247–261 (2012)
17. Sha, D.Y., Lin, H.-H.: A multi-objective PSO for job-shop scheduling problems. Expert Systems with Applications 37(2), 1065–1070 (2010)
18. Stützle, T., Hoos, H.H.: MAX-MIN Ant System. Future Generation Computer Systems 16(8), 889–914 (2000)

19. Suresh, R.K., Mohanasndaram, K.M.: Pareto archived simulated annealing for job shop scheduling with multiple objectives. International Journal of Advanced Manufacturing Technology **29**, 184–196 (2006)
20. Udomsakdigoola, A., Khachitvichyanukul, V.: Ant colony algorithm for multi-criteria job shop scheduling to minimize makespan, mean flow time and mean tardiness. International Journal of Management Science and Engineering Management **6**(2), 117–123 (2011)
21. Yusof, R., Khalid, M., Hui, G.T., Yusof, S.M., Othman, M.F.: Solving job shop scheduling problem using a hybrid parallel micro genetic algorithm. Applied Soft Computing **11**, 5782–5792 (2011)

Kriging Surrogate Model Enhanced by Coordinate Transformation of Design Space Based on Eigenvalue Decomposition

Nobuo Namura[✉], Koji Shimoyama, and Shigeru Obayashi

Institute of Fluid Science, Tohoku University, 2-1-1 Katahira, Aoba-ku, Sendai, Miyagi 980-8577, Japan
{namura,shimoyama,obayashi}@edge.ifs.tohoku.ac.jp

Abstract. The Kriging surrogate model, which is frequently employed to apply evolutionary computation to real-world problems, with coordinate transformation of design space is proposed to improve the approximation accuracy of objective functions with correlated design variables. Eigenvalue decomposition is used to extract significant trends in the objective function from its gradients and identify suitable coordinates. Comparing with the ordinary Kriging model, the proposed method shows higher accuracy in the approximation of two-dimensional test functions and reduces the computational cost to achieve the global optimization. In the application to an airfoil design problem with spline curves as correlated design variables, the proposed method achieves better performances not only in the approximation accuracy but also the ability to explore the optimal solution.

Keywords: Kriging model · Efficient global optimization · Eigenvalue decomposition · Airfoil design · Spline curve

1 Introduction

Optimization in real-world problems is usually time consuming and computationally expensive in the evaluation of objective functions [1,2]. Surrogate models are often useful to solve this difficulty. Surrogate models are constructed to promptly estimate the values of the objective functions at any point from sample points where real values of the objective functions are obtained by expensive computations. Therefore, it is important that accurate models can be constructed even with only a small number of sample points.

The most common surrogate model is the polynomial regression (PR) [3]. In construction of the PR model, users give the polynomial order arbitrarily and then compute the coefficients of each term in the polynomial so as to fit the sample points by the least-squares method. The accuracy of the PR model significantly depends on the polynomial order, which corresponds to the number of local maxima and/or minima in an objective function. However, it is not always possible to achieve sufficient accuracy by adjusting the order because the real shape of the objective function is usually

© Springer International Publishing Switzerland 2015
A. Gaspar-Cunha et al. (Eds.): EMO 2015, Part I, LNCS 9018, pp. 321–335, 2015.
DOI: 10.1007/978-3-319-15934-8_22

not known. Generally, quadratic functions are employed to approximate the function locally in the real-world problems [4]. In this case, adequate optimization cannot be performed if the objective function has some local optima and too many sample points are required to obtain the global optimum. Furthermore, a Pareto dominance-based evolutionary multi-objective optimization algorithm explores large design space where diverse Pareto-optimal solutions exist. The surrogate models are also required to approximate large design space as accurately as possible. From this point of view, the local PR model with quadratic functions is not suitable for multi-objective problems.

To approximate complex functions, radial basis function (RBF) networks [5] and the Kriging model [6] are often used. Both of them can adapt well to complex functions because they approximate a function as a weighted superposition of basis functions such as Gaussian function. Thus, the model complexity can be controlled by changing the weight coefficients and variance of each basis function. Gaussian basis functions in the Kriging model have independently different variance values along each design variable direction to fit the complexity and scale while those of RBF have the same values (Fig. 1). This anisotropy enhances the accuracy of the Kriging model. In addition, the Kriging model gives not only estimated function values but also approximation errors, which help users determine the locations of the additional sample points to improve the accuracy of the surrogate model.

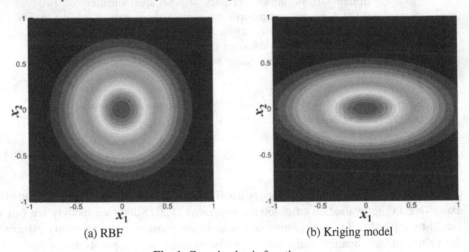

(a) RBF   (b) Kriging model

Fig. 1. Gaussian basis functions

It is desirable that one of these models approximates the function accurately. However, sometimes more complex models are needed. Hybrid methods that combine two surrogate models may be effective if the function consists of complex of macro and micro trends [7,8]. The universal Kriging model (PR+Kriging) [3] and the extended RBF (PR+RBF) [9] are typical hybrid methods. On the other hand, Xiong et al. proposed the non-stationary covariance based Kriging model whose variance values of each basis function vary depending on the location of the basis functions in the design space [10]. This model showed good performance if the complexity of the objective function changes according to the location.

Some design variables can be correlated with each other in the real-world problems, e.g. control points of spline curves and free-form deformation [11]. Optimization with such design variables can express various configurations while the problem tends to become difficult to solve. However, the only PR model takes account of the correlation as the cross-terms among different variables though the PR model does not approximate the complex function accurately for the reason described above.

In this study, we propose modified Kriging models suitable for the problems with correlated design variables by focusing on the anisotropy of Gaussian basis functions. It means finding out the suitable coordinates in the design variable space, which represent significant trends in the objective function from its gradients (Fig. 2(a)) and then defining the variance of basis functions along each coordinate in the transformed system (Fig. 2(b)). The proposed method and the ordinary Kriging (OK) model are applied to test functions and airfoil design problems to investigate the feature of the proposed method.

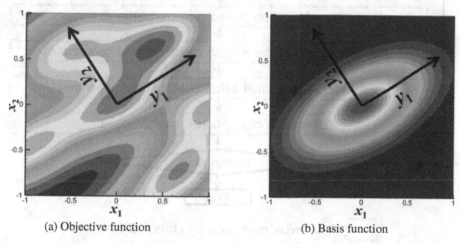

(a) Objective function   (b) Basis function

Fig. 2. Extraction of suitable coordinates

2   Construction of Surrogate Model

The flowchart explaining the construction of the Kriging model with coordinate transformation (KCT) is summarized in Fig. 3. Initial sample points are generated by Latin hypercube sampling (LHS) method [12]. Construction of KCT consists of three parts: construction of the Kriging model in the original coordinate system, coordinate transformation, and reconstruction of the Kriging model in the transformed coordinate system. The latter two parts are skipped if OK is employed for comparison. Optimal solution is explored by the efficient global optimization (EGO) framework [6], which explores the solution where "Expected improvement (EI)" described below becomes maximum by an optimizer such as evolutionary algorithm (the non-dominated sorting genetic algorithm II (NSGA-II) [13] is employed in our EGO system toward the future application to multi-objective optimization in real-world problems) and adds it as an additional sample point. EGO is performed by iterating the procedure illustrated in Fig. 3 until a termination condition is satisfied. Generally, execution time and cost consumed for surrogate-based optimization are dominated by the function evaluation at sample points while those for surrogate model construction and optimal solution exploration on the model are ignorable. Hence, it can be said that better surrogate models should approximate functions more accurately with less sample points. Followings are the details of KCT.

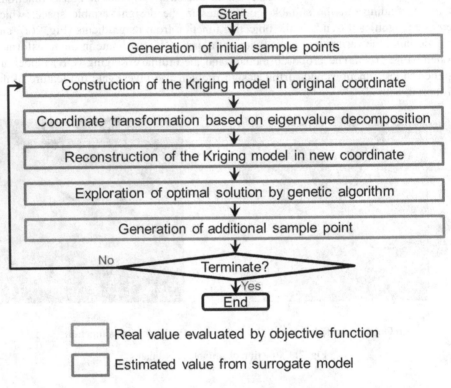

Fig. 3. Flowchart of KCT construction

2.1   Kriging Model

The Kriging model expresses the unknown function $f(\mathbf{x})$ as

$$f(\mathbf{x}) = \mu(\mathbf{x}) + \varepsilon(\mathbf{x}), \tag{1}$$

where $\mathbf{x}$ is an m-dimensional vector (m design variables), $\mu(\mathbf{x})$ is a global model, and $\varepsilon(\mathbf{x})$ represents a local deviation from the global model, which is defined as the Gaussian process following $N(0, \sigma^2)$. The correlation between $\varepsilon(\mathbf{x}_i)$ and $\varepsilon(\mathbf{x}_j)$ is strongly related to the distance between the two corresponding points, $\mathbf{x}_i$ and $\mathbf{x}_j$. In the Kriging model, a specially weighted distance is used instead of the Euclidean distance because the latter weighs all design variables equally. The distance function between the points at $\mathbf{x}_i$ and $\mathbf{x}_j$ is expressed as

$$d(\mathbf{x}^i, \mathbf{x}^j) = \sum_{k=1}^{m} \theta_k \left(x_k^i - x_k^j \right)^2,\tag{2}$$

where θ_k $(0 \le \theta_k < \infty)$ is the weight coefficient and the k-th element of an m-dimensional weight vector $\boldsymbol{\theta}$. These weights give the Kriging model anisotropy and enhance its accuracy. The correlation between the points $\mathbf{x}_i$ and $\mathbf{x}_j$ is defined as

$$Corr(\varepsilon(\mathbf{x}^i), \varepsilon(\mathbf{x}^j)) = \exp(-d(\mathbf{x}^i, \mathbf{x}^j)).\tag{3}$$

The Kriging predictor is

$$\hat{f}(\mathbf{x}) = \hat{\mu}(\mathbf{x}) + \mathbf{r}^{\mathrm{T}} \mathbf{R}^{-1}(\mathbf{f} - \hat{\boldsymbol{\mu}}),\tag{4}$$

where $\hat{\mu}(\mathbf{x})$ is the estimated value of $\mu(\mathbf{x})$, $\mathbf{R}$ denotes the $n \times n$ matrix whose (i, j) entry is $Corr(\varepsilon(\mathbf{x}^i), \varepsilon(\mathbf{x}^j))$, $\mathbf{r}$ is an n-dimensional vector whose i-th element is $Corr(\varepsilon(\mathbf{x}), \varepsilon(\mathbf{x}^i))$, and $\mathbf{f}$ and $\hat{\boldsymbol{\mu}}$ denote as follows (n sample points):

$$\mathbf{f} = (f(\mathbf{x}^1) \quad \cdots \quad f(\mathbf{x}^n))^{\mathrm{T}},\tag{5}$$

$$\hat{\boldsymbol{\mu}} = (\hat{\mu}(\mathbf{x}^1) \quad \cdots \quad \hat{\mu}(\mathbf{x}^n))^{\mathrm{T}}.\tag{6}$$

Thus, the unknown parameters in the Kriging model are $\hat{\sigma}^2$ (estimated σ^2), $\hat{\mu}(\mathbf{x})$, and $\boldsymbol{\theta}$, which are obtained by maximizing the following log-likelihood function:

$$Ln(\hat{\mu}, \hat{\sigma}^2, \boldsymbol{\theta}) = -\frac{n}{2}\ln(2\pi) - \frac{n}{2}\ln(\hat{\sigma}^2) - \frac{1}{2}\ln(|\mathbf{R}|) - \frac{1}{2\hat{\sigma}^2}(\mathbf{y} - \hat{\boldsymbol{\mu}})^{\mathrm{T}} \mathbf{R}^{-1}(\mathbf{y} - \hat{\boldsymbol{\mu}}).\tag{7}$$

$\hat{\sigma}^2$ is analytically determined through partial differentiation as

$$\hat{\sigma}^2 = \frac{(\mathbf{f} - \hat{\boldsymbol{\mu}})^{\mathrm{T}} \mathbf{R}^{-1}(\mathbf{f} - \hat{\boldsymbol{\mu}})}{n}.\tag{8}$$

The definition of $\hat{\mu}(\mathbf{x})$ has some variations. The OK model, which is the most wide-ly used Kriging model, assumes the global model to be a constant value as $\hat{\mu}(\mathbf{x}) = \hat{\mu}$. In this case, $\hat{\mu}$ is also analytically determined as

$$\hat{\mu}(\mathbf{x}) = \hat{\mu} = \frac{\mathbf{1}^{\mathrm{T}} \mathbf{R}^{-1} \mathbf{f}}{\mathbf{1}^{\mathrm{T}} \mathbf{R}^{-1} \mathbf{1}},\tag{9}$$

where $\mathbf{1}$ denotes an n-dimensional unit vector. Plugging in Eq. (8) for Eq. (7), the log-likelihood function becomes

$$Ln(\hat{\mu},\hat{\sigma}^2,\boldsymbol{\theta}) = -\frac{n}{2}(\ln(2\pi)+1) - \frac{n}{2}\ln(\hat{\sigma}^2) - \frac{1}{2}\ln(|\mathbf{R}|) \cdot \tag{10}$$

The first term can be ignored in the maximization because it has a constant value. Therefore, the log-likelihood maximization becomes an m-dimensional unconstrained non-linear optimization problem. In this study, a simple genetic algorithm is adopted to solve this problem.

2.2   Expected Improvement

The accuracy of the function value predicted by the Kriging model depends largely on the distance from sample points. The closer point $\mathbf{x}$ is to the sample points, the more accurate the prediction, $\hat{f}(\mathbf{x})$, becomes. This is expressed in the following equation:

$$s^2(\mathbf{x}) = \hat{\sigma}^2\left(1 - \mathbf{r}^T\mathbf{R}^{-1}\mathbf{r} + \frac{(1-\mathbf{1}^T\mathbf{R}^{-1}\mathbf{r})^2}{\mathbf{1}^T\mathbf{R}^{-1}\mathbf{1}}\right), \tag{11}$$

where $s^2(\mathbf{x})$ is the mean square error at point $\mathbf{x}$, which indicates the uncertainty of the estimated value. Thus, estimated values in the Kriging model do not have deterministic values but follows the Gaussian distribution denoted by $N(\hat{f}(\mathbf{x}),s^2(\mathbf{x}))$, from which the probability that the solution at point $\mathbf{x}$ may achieve a new global optimum can be calculated. The EI value, which corresponds to the expected value of the objective function improvement from the current optimal solution among the sample points, is also derived by using this probability. In $f(\mathbf{x})$ minimization problem, the improvement value $I(\mathbf{x})$ and the EI value, $E(I(\mathbf{x}))$ of $f(\mathbf{x})$ are expressed, respectively, as

$$I(\mathbf{x}) = \max(f_{ref} - f, 0), \tag{12}$$

$$E(I(\mathbf{x})) = \int_{-\infty}^{f_{ref}} (f_{ref} - f)\varphi(f)df, \tag{13}$$

where f_{ref} is the reference value of f and corresponds to the minimum value of f among the sample points in this study. φ is the probability density function denoted by $N(\hat{f}(\mathbf{x}),s^2(\mathbf{x}))$ and represents uncertainty about f.

Special modification for EI has been proposed to enhance the constrained optimization [14]. Modified EI value is expressed by multiplying the probability satisfying the constraint to the conventional EI value. If the constraint function which should be approximated by the Kriging model is expressed as $g(\mathbf{x}) > c$, the modified EI (E_cI) value is calculated as follows:

$$E_c(I(\mathbf{x})) = E(I(\mathbf{x}))\int_c^\infty \varphi(g)dg \cdot \tag{14}$$

2.3   Coordinate Transformation of Design Space

In order to identify the suitable coordinates and improve the approximation accuracy, gradients of objective function to each design variable are employed as is the case in the active subspace method [15]. First, the $m \times m$ covariance matrix $\mathbf{C}$, whose (k, l) entry is

$$C_{kl} = \frac{\partial \hat{f}}{\partial x_k} \frac{\partial \hat{f}}{\partial x_l}, \tag{15}$$

is defined. The objective function estimated by the Kriging model, $\hat{f}$ is used in this study while gradients of real objective function are used in [15]. Estimated gradients are calculated by differentiating Eq. (4) analytically as follows:

$$\frac{\partial \hat{f}(\mathbf{x})}{\partial x_k} = \left(\frac{\partial \mathbf{r}}{\partial x_k}\right)^{\mathrm{T}} \mathbf{R}^{-1}(\mathbf{f} - \hat{\boldsymbol{\mu}}), \tag{16}$$

where i-th element of $\partial \mathbf{r}/\partial x_k$ is

$$\left.\frac{\partial \mathbf{r}}{\partial x_k}\right|_i = -2\theta_k\left(x_k - x_k^i\right)\exp\left(-\sum_{l=1}^{m}\theta_l\left(x_l - x_l^i\right)^2\right). \tag{17}$$

Using estimated gradients, neither finite difference of real objective function nor adjoint computation is needed and function evaluation costs are drastically reduced. Note that we can deal with the objective function as a black box function. This study calculates $\mathbf{C}$ at 10,000 points in the design space, which are randomly sampled by the Monte Carlo method, and averages them as $\overline{\mathbf{C}}$. Second, eigenvalue decomposition is performed to $\overline{\mathbf{C}}$ as

$$\overline{\mathbf{C}} = \mathbf{W}\boldsymbol{\Lambda}\mathbf{W}^{\mathrm{T}}, \tag{18}$$

where $\mathbf{W} = (\mathbf{w}_1 \quad \cdots \quad \mathbf{w}_m)$ are the eigenvectors which represent the suitable coordinates and $\boldsymbol{\Lambda} = \mathrm{diag}(\lambda_1 \quad \cdots \quad \lambda_m)$ is the eigenvalue matrix. Third, the design variable vector in the new coordinate system $\mathbf{y}$ is calculated from the original vector $\mathbf{x}$ as

$$\mathbf{y} = \mathbf{W}^{\mathrm{T}}\mathbf{x}. \tag{19}$$

3   Application to Test Functions

3.1   Test Problem Definition

KCT and OK are applied to a two-dimensional minimization test function defined as

$$f(x_1, x_2) = (1 - y_1)^2 + 100(y_2 - y_1^2)^2, \tag{20}$$

$$\begin{bmatrix} y_1 \\ y_2 \end{bmatrix} = \begin{bmatrix} \cos\phi & -\sin\phi \\ \sin\phi & \cos\phi \end{bmatrix} \begin{bmatrix} x_1 \\ x_2 \end{bmatrix}, \qquad (21)$$

$$-1 \le x_1, x_2 \le 1. \qquad (22)$$

Equation (20) shows the well-known two-dimensional Rosenbrock function [16] if the rotation angle in Eq. (21) is set as $\phi = 0$ [deg]. Two test functions with different rotation angles ($\phi = 0$ and 30[deg]) are considerd to investigate KCT capability to find out the suitable coordinates. The ranges of design variables in Eq. (22) are smaller than those in the original Rosenbrock function because the function in the original ranges has extremely high values in some regions, which may disturb the fair evaluation of model's accuracy. Figure 4 shows the shape of these functions with two angles. 10 initial sample points are generated by LHS and 20 additional sample points are employed one after another at the location where the EI value calculated by Eq. (13) becomes a maximum.

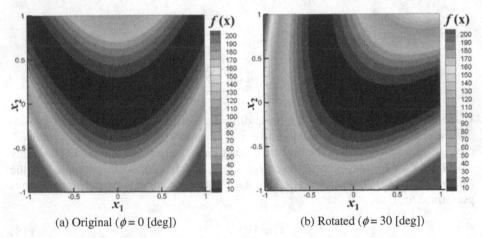

(a) Original ($\phi = 0$ [deg])   (b) Rotated ($\phi = 30$ [deg])

Fig. 4. Shapes of the various Rosenbrock test functions

To compare the accuracy of two models, the following root mean square error (RMSE) between the surrogate model and the real function is calculated at $N = 41 \times 41$ validation points.

$$RMSE = \sqrt{\frac{1}{N} \sum_{i=1}^{N} \left(f(\mathbf{x}^i) - \hat{f}(\mathbf{x}^i) \right)^2} \qquad (23)$$

100 independent trials starting with different initial sample points are performed and their average and standard deviation of RMSE are evaluated for comparison. The numbers of population and generation in NSGA-II are 500 and 100, respectively.

3.2   Results and Discussion for Original Rosenbrock Function ($\phi = 0$ [deg])

Figure 5 shows the histories of average and standard deviation of RMSE obtained by KCT and OK. Averaged RMSEs of both models are converged on zero at 23 sample points, though OK has slightly lower averaged RMSE than KCT when the number of sample points is fewer than 20. The standard deviations of both models are almost the same at any point in the history and drastically decrease from 16 to 23 sample points. Additionally, transformed coordinates have converged toward 0 [deg] because the original Rosenbrock function is roughly symmetric about $x_1 = 0$, which leads the off-diagonal elements in $\overline{C}$ to zero. These results indicate that KCT has comparable accuracy to OK even if the original coordinates are just suitable.

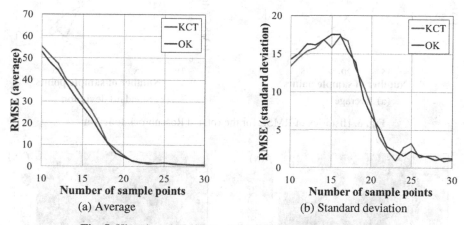

Fig. 5. Histories of RMSE for the original Rosenbrock function

3.3   Results and Discussion for Rotated Rosenbrock Function ($\phi = 30$ [deg])

Figure 6 shows the histories of average and standard deviation of RMSE. KCT has lower RMSE than OK when the number of sample points is over 12, which means KCT can approximate the function more accurately. Carefully comparing Figs. 5(a) and 6(a), the averaged RMSEs of KCT converge on zero with about 24 sample points regardless of the coordinates (i.e. ϕ) although OK yields greatly higher averaged RSME and need more sample points to converge for the rotated Rosenbrock function than the original one. In general, sample points are added until the surrogate model converges. From this point of view, the KCT requires only 24 sample points to achieve the global optimization while OK requires 28 sample points. Moreover, KCT yields lower standard deviation of RMSE than OK with over 23 sample points, which suggests that KCT can obtain the optimal solution robustly.

Typical shapes of KCT and OK with 22 sample points are shown in Fig. 7 in which black dots denote the sample points. RMSEs of these models give the closest agreement with averaged RMSEs of each model with 22 sample points in 100 independent trials. A comparison of Figs. 4(b) and 7 indicates that KCT approximates the entire function shape accurately although OK has a distorted shape. Therefore, it can be said that KCT is useful in the optimization problems with correlated design variables. Transformed coordinates of KCT converges toward 40 [deg] which is slightly different from the exact rotation angle ($\phi = 30$ [deg]) due to the asymmetric property of the rotated Rosenbrock function and its large gradient around $(x_1, x_2) = (1, -1)$.

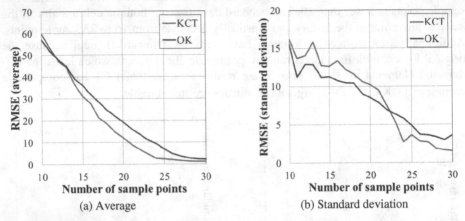

(a) Average   (b) Standard deviation

Fig. 6. Histories of RMSE for the rotated Rosenbrock function

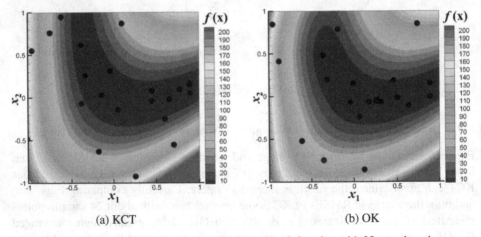

(a) KCT   (b) OK

Fig. 7. Surrogate models for the rotated Rosenbrock function with 22 sample points

4   Application to an Airfoil Design Problem

The results in Section 3 show that KCT approximates the function accurately regardless of the coordinates and has an advantage in the optimization with correlated design variables. However, the test functions in Section 3 are roughly symmetric and have only two design variables. We must evaluate practicality of KCT through a real-world shape design optimization problem, which includes correlated design variables for shape representation such as the control points of spline curves. In Section 4, airfoil design optimization is considered with KCT and OK to investigate the effects of coordinate transformation in real-world problem.

4.1   Design Problem Definition

The objective function and the constraints in the airfoil design problem are defined as follows:

$$\text{Maximize} \quad L/D, \tag{24}$$

$$\text{subject to} \quad C_m \geq -0.1535, \tag{25}$$

$$t_{max}/c \geq 0.11, \tag{26}$$

at the angle of attack $\alpha = 4$ [deg] and the Reynolds number $Re = 5 \times 10^5$. L/D and C_m denote the lift-drag ratio and the pitching moment coefficient, respectively. t_{max} is the maximum thickness of the airfoil and c is the chord length. L/D and C_m at the sample points are evaluated by a subsonic flow solver "XFOIL" [17] which calculates incompressible viscous flow in this study. Hence, KCT and OK are used to estimate these two values while t_{max}/c is calculated directly by representing the airfoil. Generally, computational time of XFOIL is less than one second for one flow condition. This study employs XFOIL to achieve many independent trials of airfoil design optimization and evaluate statistics of the results properly. The constraint values in Eqs. (25) and (26) correspond to those of DAE31 airfoil whose L/D is 138.6.

9 design variables correspond to the locations of 9 control points for two non-uniform rational basis spline (NURBS) curves defining the airfoil's thickness distribution and camber line in Fig. 8. 5 dots filled with red and 4 dots filled with blue are the control points for thickness and camber, respectively. Red and blue bars show the ranges of each design variable. Only x_3 has a relatively small range to help the maximum thickness meet the constraint in Eq. (26). Each range is shown as follows:

$$0.02 \leq x_1 \leq 0.04, \quad 0.05 \leq x_2 \leq 0.09, \quad 0.12 \leq x_3 \leq 0.13,$$

$$0.09 \leq x_4 \leq 0.13, \quad 0.00 \leq x_5 \leq 0.04, \quad 0.02 \leq x_6 \leq 0.06, \tag{27}$$

$$0.02 \leq x_7 \leq 0.06, \quad 0.02 \leq x_8 \leq 0.06, \quad 0.00 \leq x_9 \leq 0.04.$$

50 initial sample points are generated by LHS and 150 additional sample points are employed one after another at the location where the E_cI value calculated by Eq. (14) becomes a maximum. KCT and OK are compared by optimal solutions obtained by EGO with each model and RMSE in Eq. (23) where validation points are generated by LHS and $N = 10,000$. 70 independent trials are performed and their average and standard deviation of L/D and RMSE are evaluated. The numbers of population and generation in NSGA-II are 500 and 100, respectively, as in the case of Section 3.

Flow computation with XFOIL sometimes does not converge depending on the airfoil shape. An alternative initial sample point and validation point are randomly selected from the entire design space if the flow computation does not converge. The same treatment is applied if an initial sample point and a validation point do not meet at least one of the constraints.

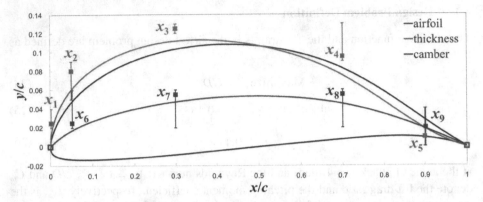

Fig. 8. Design variables and their ranges

4.2   Results and Discussion for Airfoil Design Problem

Figure 9 shows the histories of average and standard deviation of RMSE for L/D and C_m. KCT reduces the averaged RMSEs of L/D compared to OK at any number of sample points in EGO process while both models show the same trend in the standard deviations. Therefore, it is shown that KCT can approximate the function in the real-world problem using spline curves as design variables more accurately than OK. Regarding RMSEs in C_m, OK has lower average and standard deviation than KCT although both models converge on the same trend with the increase of sample points. From aerodynamic theory, C_m is regarded as a function that depends on the camber line and is not affected by the thickness, i.e. the effective number of design variables for C_m is almost 4. OK does not consider the correlation between camber and thickness, which enables OK to easily ignore the variables related with the thickness by decreasing the weight coefficients in Eq. (2). KCT can also ignore these variables though coordinate transformation may disturb it. Thus, KCT is expected to obtain better results by performing coordinate transformation for camber design variables and thickness design variables separately.

Histories of average and standard deviation of maximum L/D among feasible sample points are shown in Fig. 10. KCT obtains better solutions than OK on average and yields lower standard deviations of L/D than OK when the number of sample points is over 67. Therefore, it is suggested that KCT has an advantage over OK not only in approximation accuracy but also in the ability to explore the optimal solution if design variables are correlated with each other.

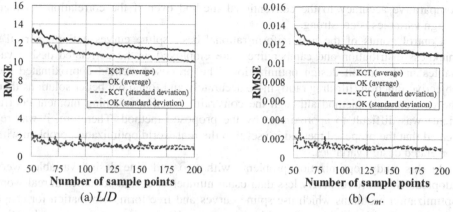

(a) *L/D*   (b) C_m.

Fig. 9. Histories of RMSE for airfoil design problem

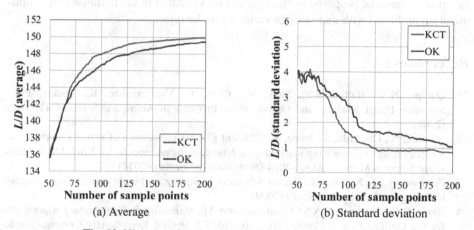

(a) Average   (b) Standard deviation

Fig. 10. Histories of optimal solution for airfoil design problem

5   Conclusions

The Kriging model with coordinate transformation was proposed and validated in two-dimensional test functions and an airfoil design problem with correlated design variables. Eigenvalue decomposition was applied to the covariance matrix of estimated objective function gradients to each design variables to identify suitable coordinates. The statistics of root mean square errors between surrogate models and real objective function and the optimal solutions obtained by efficient global optimization were used to evaluate practicality of the proposed method.

In the application to test functions, the proposed method approximated the entire function shape accurately and reduced the number of function evaluations consumed to obtain the optimal solution comparing with the ordinary Kriging model if design variables were correlated with each other. Additionally, the proposed method showed comparative accuracy to the conventional method even if the correlation between design variables is not strong.

Control points of the non-uniform rational basis spline curves defining airfoil's thickness distribution and camber line were employed as the correlated design variables in the airfoil design optimization. The proposed method approximated the objective function (lift-drag ratio) more accurately and found out better solutions than the conventional method although the constraint function (pitching moment coefficient) was difficult to approximate by the proposed method. Therefore, it was revealed that the proposed method is useful in the real-world optimization problem with correlated design variables.

In this study, optimization problems with two and nine design variables were adopted. These are relatively less than usual number of design variables in real-word optimization problems which use spline curves and free form deformation for shape definition. Besides, these real-world problems tend to have more than two objective functions. Thus, the proposed method should be validated in the multi-objective optimization problems with more design variables in the future.

References

1. Queipo, N.V., Haftka, R.T., Shyy, W., Goel, T., Vaidyanathan, R., Tucker, P.K.: Surrogate-Based Analysis and Optimization. Progress in Aerospace Sciences **41**, 1–28 (2005)
2. Namura, N., Obayashi, S., Jeong, S.: Efficient global optimization of vortex generators on a super critical infinite-wing using kriging-based surrogate models. In: 52nd AIAA Aerospace Sciences Meeting, AIAA-2014-0904. National Harbor (2014)
3. Forrester, A.I.J., Keane, A.J.: Recent Advances in Surrogate-Based Optimization. Progress in Aerospace Sciences **45**, 50–79 (2009)
4. Regis, R.G., Shoemaker, A.S.: Local Function Approximation in Evolutionary Algorithms for the Optimization of Costly Functions. IEEE Trans. on Evolutionary Computation **8**, 490–505 (2004)
5. Broomhead, D.S., Lowe, D.: Multivariate Functional Interpolation and Adaptive Networks. Complex Systems **2**, 321–355 (1988)
6. Jones, D.R., Schonlau, M., Welch, W.J.: Efficient Global Optimization of Expensive Black-Box Function. J. of Global Optimization **13**, 455–492 (1998)
7. Joseph, V.R., Hung, Y., Sudjianto, A.: Blind Kriging: a New Method for Developing Metamodels. ASME J. of Mechanical Design **130**, 031102-1-8 (2008)
8. Namura, N., Shimoyama, K., Jeong, S., Obayashi, S.: Kriging/RBF-Hybrid Response Surface Methodology for Highly Nonlinear Functions. J. of Computational Science and Technology **6**, 81–96 (2012)
9. Mullur, A.A., Messac, A.: Extended radial basis functions: more flexible and effective metamodeling. In: 10th AIAA/ISSMO Multidisciplinary Analysis and Optimization. Conference, AIAA-2004-4573. Albany (2004)
10. Xiong, Y., Chen, W., Apley, D., Ding, X.: A Non-Stationary Covariance-Based Kriging Method for Metamodeling in Engineering Design. International J. for Numerical Methods in Engineering **71**, 733–756 (2007)

11. Samareh, J.A.: Aerodynamic shape optimization based on free-form deformation. In: 10th AIAA/ISSMO Multidisciplinary Analysis and Optimization Conference, AIAA-2004-4630. Albany (2004)
12. McKay, M.D., Beckman, R.J., Conover, W.J.: A Comparison of Three Methods for Selecting Values of Input Variables in the Analysis of Output from a Computer Code. Technometrics **21**, 239–245 (1979)
13. Deb, K., Pratap, A., Agarwal, S., Meyarivan, T.: A Fast and Elitist Multiobjective Genetic Algorithm: NSGA-II. IEEE Trans. on Evolutionary Computation **6**, 182–197 (2002)
14. Jeong, S., Yamamoto, K., Obayashi, S.: Kriging-based probabilistic method for constrained multi-objective optimization problem. In: AIAA 1st Intelligent Systems Technical Conference, AIAA-2004-6437. Chicago (2004)
15. Constantine, P.G., Dow, E., Wang, Q.: Active Subspace Methods in Theory and Practice: Applications to Kriging Surfaces. SIAM J. on Scientific Computing **36**, A1500–A1524 (2014)
16. Rosenbrock, H.H.: An Automatic Method for Finding the Greatest or Least Value of a Function. The Computer J. **3**, 175–184 (1960)
17. Drela, M.: XFOIL: An Analysis and Design System for Low Reynolds Number Airfoils. In: Mueller, T.J. (ed.) Low Reynolds Number Aerodynamics. Lecture Notes in Engineering, pp. 1–12. Springer-Verlag, New York (1989)

A Parallel Multi-Start NSGA II Algorithm for Multiobjective Energy Reduction Vehicle Routing Problem

Iraklis-Dimitrios Psychas, Magdalene Marinaki, and Yannis Marinakis[✉]

School of Production Engineering and Management,
Technical University of Crete, Chania, Greece
ipsychas102@gmail.com, magda@dssl.tuc.gr, marinakis@ergasya.tuc.gr

Abstract. The Multiobjective Energy Reduction Vehicle Routing Problem is a variant of the classic Vehicle Routing Problem where simultaneous optimization of more than one objective functions is required. In this paper, the problem is formulated with three different competitive objective functions. The first objective function corresponds to the optimization of the time needed for the vehicle to travel between two customers or between the customer and the depot, the second objective function is the minimization of the distance and the fuel consumption when a delivery route is planned and the third objective function is the minimization of the distance and the fuel consumption when a pickup route is planned. The problem is solved with a modified version of the NSGA II, with a use of more than one population, a multi start method for the creation of the initial population and a Variable Neighborhood Search algorithm for the improvement of the solution of each individual separately. In order to give the quality of the methodology, experiments are conducted using appropriately modified for the Vehicle Routing Problem instances based on the classic Euclidean Traveling Salesman Problem benchmark instances taken from the TSP library.

Keywords: Multiobjective energy reduction vehicle routing problem ·
NSGA II · VNS · GRASP

1 Introduction

In real world applications, optimization problems with more than one objectives are very common. In these problems, usually, there is no single solution and the optimization of two or more competitive objective functions leads to the calculation of a set of non-dominated solutions, called Pareto Front [3]. The **Vehicle Routing Problem (VRP)** is a Supply Chain Management Problem of designing delivery or collection routes from a depot (or more than one depots) to a number of customers (or cities), taking into account a number of side constraints. The VRP is a variant of the Traveling Salesman Problem which is one of the most famous and extensively studied problem in the field of Combinatorial

© Springer International Publishing Switzerland 2015
A. Gaspar-Cunha et al. (Eds.): EMO 2015, Part I, LNCS 9018, pp. 336–350, 2015.
DOI: 10.1007/978-3-319-15934-8_23

Optimization [7,15] and belongs to the class of NP-hard optimization problems [11]. For an overview of the VRP please see [14,20].

The objective function of the Capacitated Vehicle Routing Problem (CVRP) is the optimization of the distance that the vehicles will travel in order to fulfill the customers demands. However, in real world applications the optimization only of the distance may not be enough to give to the decision maker a safe conclusion about the quality of the routes and if these routes can lead to a decrease of the cost of the routing plan. Thus, in recent years a growth in the publications of Multiobjective Vehicle Routing problems has been noted [13]. The **Multiobjective Vehicle Routing Problem (moVRP)** is the variant of the classic Vehicle Routing problem where simultaneous optimization of distance, time, or other relevant objectives are required. Also, in recent years the optimization of energy or fuel consumption in the Vehicle Routing Problems has been studied [10,21].

NSGA II is an improved version of **NSGA (Non-dominated Sorting Genetic Algorithm)** and was originally proposed by Deb et al. [4,5]. A number of variants of the NSGA II algorithm have been used for solving multiobjective Vehicle Routing Problems, e.g. for solving VRP with route balancing [12], for solving multiobgective VRP problems with Time Windows [9] and for solving a Green Vehicle Routing Problem [10]. Multiobjective Genetic Algorithms for the solution of Multiobjective Vehicle Routing Problems have been used in [1,18].

In this paper, a new variant of the NSGA II algorithm is presented for the solution of a **Multiobjective Energy Reduction Vehicle Routing Problem (MERVRP)**. In this problem, a symmetric case is considered where the distance and the time needed between two customers or a customer and the depot are known and symmetric. Three different objective functions are used. In the one objective function, the time needed between the two customers or a customer and a depot is optimized and in the other two objective functions, the distance is calculated taking into account the fuel consumption in two different cases, in the one where deliveries are realized and in the other where pickups are realized. We solved a number of problems with two or three objective functions. When the second and the third objective functions are used, we considered that all customers can be used as deliver customers or pickup customers and the vehicle will not make the pickups and the delivers simultaneously.

The proposed **Parallel Multi-Start NSGA II (PMS-NSGA II)** algorithm for the solution of the above mentioned problem can be used directly in combinatorial optimization problems and its main characteristics are the following:

1. A Multi-Start method, based on Greedy Randomized Adaptive Search Procedure (GRASP) [6], is used for the creation of the initial population.
2. The algorithm uses more than one populations that are evolved in parallel and a number of Pareto Fronts are used (equal to the number of populations).
3. An external archive is used with the Pareto Front of the whole population based on the crowding distance and the rank of each of the populations

(in the Global Pareto Front, the members of the other Pareto Fronts with rank equal to 1 are used).
4. The combination of the proposed method with a very powerful metaheuristic algorithm, the Variable Neighborhood Search (VNS) [8], is performed.

The structure of the paper is as follows. In Section 2, the optimization models of the MERVRP are described. In Section 3, an analytical description of the proposed algorithm is presented. In Section 4, the other two variants of NSGA II which are used to compare the proposed algorithm are described while in Section 5, the evaluation measures used in the comparisons are presented. In Section 6, the computational results are presented and, finally, concluding remarks and the future research are given in the last Section.

2   Multiobjective Energy Reduction Vehicle Routing Problem

In this paper, a **Multiobjective Energy Reduction Vehicle Routing Problem (MERVRP)** is formulated. We use two or three different objective functions where the first one is the minimization of the time needed for a vehicle to travel between two customers or a customer and the depot, the second one is the minimization of the distance travel and the fuel consumption when the decision maker plans delivery routes where all the customers have only demands and the third objective function is the minimization of the distance travel and the fuel consumption when the decision maker plans pickup routes where all the customers have only pickups. We assume that the customers for the second and the third objective functions are different between them and, thus, they have different coordinates. If we have a two objective functions problem where the first objective function is used and the other objective function is one of the second or the third objective functions, then, the number of customers that corresponds to the first objective function is equal to the number of customers of the second (or third) objective function. However, when we solve a three objective functions problem, where both pickup and delivery customers are included, then, the number of customers that are used in the first objective function (the minimization of the time) is equal to the summation of the number of customers that are used in the second and the third objective functions. The main difference between the second and the third objective functions concerns the load of the vehicles. When a delivery problem is solved, then, the vehicle begins with a full load and returns to the depot when the next customer could not be served. On the other hand, when a pickup problem is solved, then, the vehicle begins with empty load, collects from each one of the customers and returns to the depot when the next customer could not be served.

The first objective function is used for the minimization of the time needed to travel between two customers or a customer and the depot. Thus, if $t_{ij}^{i_1}$ is the time needed to visit customer j immediately after customer i using vehicle i_1

and $s_j^{i_1}$ is the service time of customer j using vehicle i_1, then the first objective function is:

$$\min OF1 = \sum_{i=1}^{n} \sum_{j=1}^{n} \sum_{i_1=1}^{m} (t_{ij}^{i_1} + s_j^{i_1}) x_{ij}^{i_1} \tag{1}$$

where n is the number of customers and m is the number of homogeneous vehicles and the depot is denoted by $i = j = 1$.

The second objective function is used for the minimization of the distance and of the fuel consumption that a vehicle consumes when it travels between two customers or a customer and the depot in the case that the vehicle performs only deliveries in its route. The vehicle should begin with full load and after a visitation of a customer the load is reduced based on the demand of the customer. If we consider that the most loaded is the vehicle the more fuel it consumes, we take the following objective function:

$$\min OF2 = \sum_{j=1}^{n} \sum_{i_1=1}^{m} d_{1j} x_{1j}^{i_1} (1 + \frac{y_{1j}^{i_1}}{Q}) + \sum_{i=2}^{n} \sum_{j=1}^{n} \sum_{i_1=1}^{m} d_{ij} x_{ij}^{i_1} (1 + \frac{y_{i-1,i}^{i_1} - D_i}{Q}) \tag{2}$$

with the maximum capacity of the vehicle denoted by Q, the i customer has demand equal to D_i, $x_{ij}^{i_1}$ denotes that the vehicle i_1 visits customer j immediately after customer i with load $y_{ij}^{i_1}$ and $y_{1j}^{i_1} = \sum_{i=1}^{n} D_i$ for all vehicles as the vehicle begins with load equal to the summation of the demands of all customers assigned in its route and d_{ij} is the distance from node i to node j.

Finally, the third objective function is used for the minimization of the distance and of the fuel consumption that a vehicle consumes when it travels between two customers or a customer and the depot in the case that the vehicle performs only pickups in its route. The vehicle should begin with empty load and after a visitation of a customer the load is increased based on the demand of the customer. If we consider, as previously, that the most loaded is the vehicle the more fuel it consumes we take the following objective function:

$$\min OF3 = \sum_{j=1}^{n} \sum_{i_1=1}^{m} d_{1j} x_{1j}^{i_1} + \sum_{i=2}^{n} \sum_{j=1}^{n} \sum_{i_1=1}^{m} d_{ij} x_{ij}^{i_1} (1 + \frac{y_{i-1,i}^{i_1} + D_i}{Q}) \tag{3}$$

with $y_{1j}^{i_1} = 0$ for all vehicles as the vehicle begins with empty load.

The constraints of the problem are the following:

$$\sum_{j=1}^{n} \sum_{i_1=1}^{m} x_{ij}^{i_1} = 1, i = 1, \cdots, n, \tag{4}$$

$$\sum_{i=1}^{n} \sum_{i_1=1}^{m} x_{ij}^{i_1} = 1, j = 1, \cdots, n, \tag{5}$$

$$\sum_{j=1}^{n} x_{ij}^{i_1} - \sum_{j=1}^{n} x_{ji}^{i_1} = 0, i = 1, \cdots, n, i_1 = 1, \cdots, m \tag{6}$$

$$\sum_{j=0,j\neq i}^{n} y_{ji}^{i_1} - \sum_{j=0,j\neq i}^{n} y_{ij}^{i_1} = D_i, i = 1, \cdots, n, i_1 = 1, \cdots, m \tag{7}$$

$$Qx_{ij}^{i_1} \geq y_{ij}^{i_1}, i, j = 1, \cdots, n, i_1 = 1, \cdots, m \tag{8}$$

$$x_{ij}^{i_1} \in 0, 1 \, i, j = 1, ..., n, i_1 = 1, \cdots, m \tag{9}$$

Constraints (4) and (5) represent that each customer must be visited only by one vehicle; constraints (6) ensure that each vehicle that arrives at a node must leave from that node also. Constraints (7) indicate that the reduced (if it concerns deliveries) or increased (if it concerns pickups) load (cargo) of the vehicle after it visits a node is equal to the demand of that node. Constraints (8) is used to limit the maximal load carried by the vehicle and to force $y_{ij}^{i_1}$ to be equal to zero when $x_{ij}^{i_1} = 0$ while constraints (9) ensure that only one vehicle will visit each customer.

3 Parallel Multi-Start NSGA II Algorithm

3.1 Representation of the Solutions

The first problem that we have to solve is to find a suitable mapping between the Vehicle Routing Problem solutions and individuals in NSGA II. As, for the NSGA II, Equations 10 and 11 (Section 3.3) are needed to be used, we can not use directly a path representation of the route but we should have a transformation of the solutions from continuous to discrete space and vice versa. Thus, a solution is represented by a d-dimensional vector in problem space and its performance is evaluated on the predefined fitness functions (Section 2). Initially, each individual is recorded via the path representation of the tour, that is, via the specific sequence of the nodes. As the calculation of Equations 10 and 11 is performed, the above mentioned representation should be transformed appropriately. Each element of the solution is transformed into a floating point in the interval (0,1], the equations 10 and 11 for all individuals are calculated and, then, a conversion back into the integer domain is performed using relative position indexing [16]. Thus, initially, each element of the solution is divided by the vector's largest element and after the calculation of the equations 10 and 11, the elements of the vectors are transformed back into the integer domain by assigning the smallest floating value to the smallest integer, the next floating value to the next integer and, so on, until the largest floating value is assigned to the largest integer.

3.2   Initialization of the Population

Usually in a NSGA II algorithm, the initial population is calculated at random but as we would like to give more exploration and exploitation abilities in the initial population of the proposed algorithm, the Parallel Multi-Start NSGA II (PMS-NSGA II) algorithm, we use the following strategy. Initially, a selection of X different populations and of W individuals for each population is performed. Let K the number of objective functions, then $w = W/K$.

Afterwards, we have to create the initial solution for each one of the populations. For the first 40% of the populations, the first solution of the first set of w is a solution that is produced by solving a single objective problem with a VNS algorithm (see Section 3.4) using the first objective function, the first solution of the second set of w is a solution that is produced by solving a single objective problem with a VNS algorithm (see Section 3.4) using the second objective function etc. For these first solutions, the value of the other objective functions are calculated without affecting the procedure in this phase of the algorithm. As we would like to start with a good solution, we increase the number of the two main parameters of the VNS (vns_{max}, $local_{max}$). For the next 20% of the populations, the first solution of each set of w is a solution that is produced by solving a single objective problem by using the Nearest Neighborhood method [15] for each corresponding objective function. For the last 40% of the populations, the first solution of each set of w is a solution that is produced by solving a single objective problem by using a variant of GRASP method [6] for each corresponding objective function. In this GRASP algorithm, instead of using the Restricted Candidate List (RCL), the following procedure is used: Always node 1 is used as a starting node. Then, a random number ($Rand$) equal to 0 or 1 is generated. If $Rand = 0$, then, the nearest node to a node i is visited. If $Rand = 1$, the second nearest node to a node i is visited. For the calculation of the rest members of the populations the following procedure is used: For each population and for each set of w, the Swap method [15] is applied at the first solution and is used for the calculation of the second to $w/3$ individuals, while the 2-opt method [17] is applied at the first solution and is used in order to produce the $w/3 + 1$ to $2w/3$ individuals. Finally, the last individuals are produced randomly.

3.3   Parallel Multi-Start NSGA II

In each iteration, for each solution of the different populations the $rank$ and the $crowding\ distance$ are calculated [5] and, initially, the solutions of each population are sorted using the $rank$ and, afterwards, using the $crowding\ distance$. Then, we have to select two parents. For each parent solution, we randomly select two solutions and the parent solution is the one with the best $rank$. If the $rank$ is the same, the parent solution is the one with the best $crowding\ distance$. In the next step, a crossover procedure between two parents is performed in order to produce two offspring. The equations for the two offspring are:

$$offspring_l(t) = (1 - g) * parent_m(t) + g * parent_n(t) \qquad (10)$$

$$offspring_f(t) = g * parent_m(t) + (1 - g) * parent_n(t) \qquad (11)$$

where g is a random number in $(0,1)$, m, n are the indices denoting the two parents $(m, n = 1, \cdots, W)$ and l, f are the indices denoting the two offspring $(l, f = 1, \cdots, W)$. We repeat the two previous steps until $offspring_i$ has W solutions (offspring). After the production of the offspring, in order to improve the solutions, a Variable Neighborhood Search (VNS) algorithm (see Section 3.4) is applied in each solution for a specific number of iterations. In the next step, the parents $(parent_i)$ and offspring $(offspring_i)$ vectors are combined in a new one $(offspring'_i)$ and, then, the members of the $offspring'_i$ are sorted using the *rank* and the *crowding distance* as in the previous step. From these solutions, a number of individuals for each population equal to the initial population is survived in the next iteration. At the end of each iteration, the solutions with rank equal to 1 from all populations are combined into one single population and a new Global Pareto Front, using the *rank* and the *crowding distance*, is calculated. We use these multipopulation version of NSGA II procedure in order to give more exploration and exploitation abilities to the algorithm. Thus, if we have only one population and a number of solutions were dominated from other solutions, these solutions will have a small *rank* number and will be removed from the set of solutions, however, with the use of the multi-population version these solutions have the possibility to have a small *rank* number in their population and to help the algorithm to search for a better solution in unexplored areas of the solution space.

3.4   Variable Neighborhood Search

A Variable Neighborhood Search (VNS) algorithm [8] is applied in each individual in the algorithm. In this research, the following procedure is applied for a certain number of iterations (vns_{max}) for each individual. Initially, the 2-opt local search algorithm [15] is applied for a certain number of iterations $(local_{max})$. For the proposed algorithm, the vns_{max} and the $local_{max}$ were set equal to 10 in order not to increase the computational time of the algorithm. If 2-opt improves the solution (the new solution dominates the old solution), then, 2-opt algorithm is applied for $local_{max}$ number of iterations. On the other hand, if 2-opt is trapped in a local optimum (the new solution is dominated by the old solution or the two solutions are non-dominated between them) when $local_{max}$ number of iterations has been reached, the 3-opt algorithm [15] is applied with the same procedure used when 2-opt was applied and when the 3-opt is trapped in a local optimum, a Swap algorithm [15] is applied. When the swap algorithm is trapped in a local optimum, a 2-2 exchange algorithm [15] is applied as previously. When 2-2 exchange algorithm is trapped in a local optimum, a 1-1 relocate [15] algorithm is applied and, finally, the last algorithm that is used is a 2-2 relocate algorithm [15].

4 Algorithms Used in the Comparisons

In order to see the efficiency of the proposed algorithm two other versions of the NSGA II procedure have been developed for comparing the proposed algorithm with them. Each one of them uses a number of characteristics of the proposed algorithm in order to give a competitive algorithm with the proposed algorithm and to see if any of the new characteristics help the algorithm to improve the solutions. In the results, the proposed algorithm is denoted as **Method 3**.

A single population NSGA II - Method 1

The difference of this variant with the proposed method is that in this variant one population of initial solutions is created. The first member of the initial population of W individuals is calculated by a solution that is created with the nearest neighborhood algorithm and is improved using the proposed variant of VNS (see Section 3.4). The Swap method [15] is used for the calculation of the second to $W/3$ individuals, while the 2-opt method [17] is used in order to produce the $W/3 + 1$ to $2W/3$ individuals. All the other members of the population are created at random. The algorithm continues as the proposed algorithm.

A different multipopulation NSGA II - Method 2

The difference of this variant with the previous methods is at the creation of the initial populations. In this algorithm, a number X of different populations and of W individuals for each population is created. The first member of the first population is calculated by using a solution produced by a random initial solution and is improved using the VNS method (see Section 3.4). The first member of the second population is calculated by using the Nearest Neighborhood method [15] and the first member of the others $X - 2$ populations is calculated by using the GRASP method as it is described previously. For the other members of each population, the swap method, the 2-opt and the random method are applied with the same way they are used for the initial population of the proposed algorithm.

5 Evaluation Measures

The evaluation of a multiobjective optimization problem is a very interesting and complicated procedure as there are many different measures that have been proposed for different problems. In the selected problem, we have one more difficulty. This is the fact that as the problem is NP-hard we do not know the optimum Pareto Front and, thus, it is very difficult to prove if the set of the non-dominated solutions found by the proposed algorithm belongs to the optimum Pareto Front or if we have just calculated a very good and efficient Pareto Front. In general, the main goals of a set of non-dominated solutions are the minimization of the distance in relation with the optimum Pareto Front (if it is known), the finding of a uniform distribution of the solutions in the Pareto Front (spread and distribution), the expanding of the diagram in greater extend in all axes and the finding of as much as possible solutions of the Pareto Front [19, 22].

In this paper, as the optimum Pareto Front is not known, four different measures are used:

- In order to evaluate how well the proposed algorithm has distributed individuals over the non-dominated region, the observations made by Zitzler et al. [22] are used. In [22], the authors mentioned that there are three criteria that an efficient multiobjective algorithm should have. First, the distance of the resulting non-dominated set to the Pareto-optimal front should be minimized. Second, a good (in most cases uniform) distribution of the solutions found is desirable. The assessment of this criterion is based on a certain distance metric. Finally, the extent of the obtained non-dominated front should be maximized. In the proposed algorithm, we use the maximum extent in each dimension to estimate the range to which the front spreads out. This is described by the following equation [22]:

$$M_k = \sqrt{\sum_{i=1}^{K} max\{\| p' - q' \|\}} \tag{12}$$

where K is the number of objectives and p', q' are the values of the objective functions of two solutions that belong to the Pareto front.
- The solutions l of the Pareto front.
- The spread or distribution of solutions [19]. For the calculation of the spread the following equations are used:

$$Spacing = \sqrt{\frac{1}{|L-1|} \sum_{i=1}^{|L|} (dist_i - \overline{dist})^2} \tag{13}$$

$$dist_i = min \sum_{k=1}^{K} |z_k(l_i) - z_k(l_j)|, \quad l_j \in L \ and \ l_j \neq l_i \tag{14}$$

where K is the number of objective functions, L is the number of solutions of the front, z_k is the value of the k objective function, $dist_i$ is the minimum distance of solution i of its nearest solution and $\overline{dist}$ the average value of all distances.
- Coverage [22]: for a pair (A,B) of approximation sets the fraction of solutions in B that are weakly dominated by one or more solutions in A. The coverage measure is calculated by the following equation:

$$C(A,B) = \frac{|\{b \in B; \exists a \in A : a \leq b\}|}{|B|}. \tag{15}$$

6   Computational Results

The whole algorithmic approach was implemented in Visual C++. As it is mentioned previously in the multiobjective (K-objective) VRP, K different objective functions are defined. The first objective function corresponds to the time and the second and third objective functions correspond to the euclidean distance between the nodes when a delivery and a pickup problem with optimizing the energy is solved. As there are no data sets available for the solution of this kind of multiobjective VRPs, we created a number of data sets as follows. Initially, we took five instances with 100 cities from the TSPLIB (kroA100, kroB100, kroC100, kroD100, and kroE100). However, these instances, as they are used for the solution of the Traveling Salesman Problem, include only data for the coordinates of the nodes. All the other data needed for the Multiobjective Energy Reduction VRP (capacity, time limits and demands) were taken from the third instance (par3) of the classic Christofides benchmark instances [2] that is used for the solution of the Capacitated Vehicle Routing Problem (CVRP). Thus, we created a new data set combining the Kro#100 instances (where # corresponds to A or B or C or D or E) with the par3 instance where the coordinates of 100 nodes are taken from the corresponding kro#100 data set and the corresponding demand of each of the 100 nodes was taken from the par3 instance. Also, the maximum tour length, the service time and the capacity of each vehicle were taken from the par3 instance. We created 2- and 3-objective function problems by combining these five instances. For example, in order to create a 3-objective function problem, we used kroA100par3, kroB100par3 and kroC100par3 as the data used for the first, second, and third objective functions, respectively. More precisely, the first objective function is used as described in Section 2 and the necessary data are taken from kroA100par3, the second objective function is used as described in Section 2 and the necessary data are taken from kroB100par3 and the third objective function is used as described in Section 2 and the necessary data are taken from kroC100par3.

A number of different alternative values for the parameters of the algorithm were tested and the ones selected are those that gave the best computational results concerning both the quality of the solution and the computational time needed to achieve this solution and, also, taking into account the fact that we would like to test the algorithms with the same function evaluations. Thus, the selected parameters for all methods are given in the following:

Method 1

- Number of individuals: 1000.
- Number of generations: 500.
- Number of initial populations: 1.

Method 2 and Method 3

- Number of individuals for each initial population: 100.
- Number of generations: 500.
- Number of initial populations: 10.

After the selection of the final parameters, the three versions of the Non-dominated Sorting Genetic Algorithm II (NSGA II) (Methods 1, 2 and 3) were

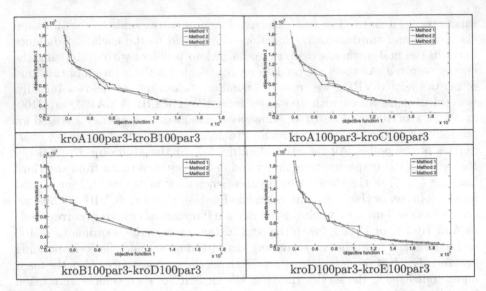

| kroA100par3-kroB100par3 | kroA100par3-kroC100par3 |

| kroB100par3-kroD100par3 | kroD100par3-kroE100par3 |

Fig. 1. Pareto fronts of the three methods for different combinations

tested for four combinations for two objective functions (i.e., kroA100par3-kroB100par3, kroA100par3-kroC100par3, kroB100par3-kroD100par3, kroD100-par3-kroE100par3). In the following tables the comparisons performed based on the five evaluation measures presented previously and the Pareto front are given. In all Tables, kroA100par3 is denoted with A, kroB100par3 is denoted with B, and so on. If we have a combination of two problems, the problem is denoted by the combination of the two letters, for example kroA100par3-kroB100par3 is denoted with A-B in all Tables. More precisely, we use the number of solutions (L) in the non-dominated set, the maximum extend in each dimension (M_k), the minimization of the spread of solutions ($Spacing$), the CPU time ($CPUtime$) (in seconds) and the $Coverage$ for evaluation measures. In Table 1, the results of the first four measures for the three methods and for the four combinations are presented while in Table 2, the results of the $Coverage$ measure are, also, presented. In Figure 1, four Pareto fronts are presented. In general, it is preferred to find as many as possible non-dominated solutions, the expansion of the Pareto front to be as large as possible which shows that better solutions have been found in every dimension and the spacing of solutions to be as smaller as possible which means that the non-dominated solutions are close between them. In Table 1, the spacing seems to have large values, however, this is due to the fact that the values of the objective functions found in a Multiobjective Vehicle Routing Problem are in the interval $(0.2 \times 10^5, 2 \times 10^5)$.

In Table 1, the results of the three Methods using two objective functions are presented. Regarding the combination kroA100par3-kroB100par3, Method 3 performs better than the other two variants of the algorithm (Methods 1 and 2) for all the measures except the M_k measure. The Method 2 performs better

Table 1. Results of the first four measures for the three methods and for the four combinations

Method	A-B				A-C			
	L	M_k	Spacing	CPU time	L	M_k	Spacing	CPU time
1	22	531.74	6126.40	5509.66	19	536.45	6717.00	5237.46
2	26	533.64	6010.00	6982.48	26	545.36	6595.60	6530.45
3	32	528.87	4624.50	4986.46	34	527.61	4914.40	6314.26
Method	B-D				D-E			
	L	M_k	Spacing	CPU time	L	M_k	Spacing	CPU time
1	19	519.20	6471.80	5755.46	21	541.33	7020.60	5877.69
2	31	522.61	7058.70	6740.49	28	533.27	6972.70	5611.38
3	34	506.21	4534.10	6246.43	34	547.65	4912.00	5310.70

than the Method 1 for all the measures. For the combinations kroA100par3-kroC100par3 and kroB100par3-kroD100par3, Method 3 performs better than the other two Methods for the number of Pareto solutions L and the *Spacing*. Method 2 performs better for the M_k measure and Method 1 performs better for the *CPUtime*. Regarding the combination kroD100par3-kroE100par3, Method 3 performs better than the other two Methods for all the measures. The Method 2 performs better than the Method 1 for three measures (L, *Spacing* and *CPUtime*).

Table 2. Results of the Coverage measure

	Coverage						
A-B	1	2	3	A-C	1	2	3
1	0	0.53	0.37	1	0	0.53	0.58
2	0.04	0	0.28	2	0	0	0.23
3	0.13	0.3	0	3	0.05	0.26	0
B-D	1	2	3	D-E	1	2	3
1	0	0.19	0.26	1	0	0.17	0.29
2	0.26	0	0.32	2	0.52	0	0.32
3	0.47	0.32	0	3	0.47	0.35	0

In Table 2, the results of the *Coverage* measure for the three Methods for the four tested combinations are presented. For the first two combinations, Method 1 performs better than the others. Method 3 is ranked second and, then, is the Method 1. Regarding the combination kroB100par3-kroD100par3, Method 3 performs better than the other two. Finally, for the last combination Method 2 performs better than the other two and Method 3 is ranked second. In general, considering the number of the non-dominated solutions (L measure) and the *Spacing* measure, the proposed Method (Method 3) performs better for all the tested combinations and has the same performance with the Method 1 concerning the *CPUtime* measure. Method 2 performs better for the M_k measure and Method 1 performs better for the *Coverage* measure, respectively. Also it is important to mention that in contrast with the other two methods the proposed Method performs better for at least one combination for every measure and performs better at all combinations for the measures L and M_k. In Table 3, the results of all methods in three objective functions are given. As it can be seen, the proposed method (Method 3) performs better than the other two methods in all measures except of the measure of the CPU time. In Table 4, the results of the proposed Method for all the combination for two and three objective functions are presented. The third objective function is the Pickup Energy Vehicle Routing Problem's function.

Table 3. Results of all methods in three objective functions

Method	A-B-C				Coverage			
	L	Mk	Spacing	CPU time	A-B-C	1	2	3
1	87	661.37	5120.40	10298.04	1	0	0.17	0.17
2	86	667.08	8123.70	8846.88	2	0.32	0	0.25
3	100	674.64	4556.50	9093.03	3	0.44	0.27	0

Table 4. Results for all the combinations of the proposed Algorithm

	L	M_k	Spacing	Time		L	M_k	Spacing	Time
A-B	32	528.87	4624.50	4986.46	A-B-C	100	674.64	4556.50	9093.03
A-C	34	527.61	4914.40	6314.26	A-B-D	73	633.50	5592.30	8077.07
A-D	30	514.26	4846.40	6865.33	A-B-E	87	656.52	7054.50	8432.32
A-E	39	522.82	6726.00	6310.04	A-C-D	83	665.51	5796.70	7851.14
B-C	31	519.42	6014.40	5808.93	A-C-E	81	659.16	4925.40	8318.95
B-D	34	506.21	4534.10	6246.43	A-D-E	103	655.14	4488.50	8026.24
B-E	30	533.76	5423.70	6034.07	B-C-D	86	645.19	4921.50	8341.01
C-D	40	531.34	3497.10	5567.38	B-C-E	93	657.49	4508.30	9229.34
C-E	31	524.52	5847.90	5745.89	B-D-E	84	658.30	5327.50	8364.68
D-E	34	547.66	4912.60	5310.70	C-D-E	91	653.86	4325.60	8126.32

7   Conclusions and Future Research

In this paper, an efficient hybridized version of NSGA II algorithm, the Parallel Multi-Start NSGA II (PMS-NSGA II), for the solution of the Multiobjective Energy Reduction Vehicle Routing Problem is presented. The algorithm was hybridized with a Variable Neighborhood Search algorithm. In order to test the efficiency of the proposed algorithm, two other hybridized versions of NSGA II algorithm were developed and a number of different evaluation measures were used. The differences in the results in every evaluation measure indicate the fact that all three methods were efficient for the solution of the problem studied in this paper. However, the proposed NSGA II algorithm (PMS-NSGA II -Method 3) performed slightly better than the other two methods. Our future research will be, mainly, focused on the application of this algorithm in other multiobjective combinatorial optimization problems, especially, problems arising in supply chain management, like Multiobjective Location Routing problem.

References

1. Chand, P., Mohanty, J.R.: Multi objective genetic approach for solving vehicle routing problem. International Journal of Computer Theory and Engineering **5**(6), 846–849 (2013)
2. Christofides, N., Mingozzi, A., Toth, P.: The vehicle routing problem. In: Christofides, N., Mingozzi, A., Toth, P., Sandi, C. (Eds.) Combinatorial Optimization. Wiley, Chichester (1979)
3. Coello Coello, C.A., Van Veldhuizen, D.A., Lamont, G.B.: Evolutionary Algorithms for Solving Multi-Objective Problems. Springer (2007)
4. Deb, K., Agrawal, S., Pratap, A., Meyarivan, T.: A fast elitist non-dominated sorting genetic algorithm for multi-objective optimization: NSGA-II. In: Deb, K., Rudolph, G., Lutton, E., Merelo, J.J., Schoenauer, M., Schwefel, H.-P., Yao, X. (eds.) PPSN 2000. LNCS, vol. 1917, pp. 849–858. Springer, Heidelberg (2000)
5. Deb, K., Pratap, A., Agarwal, S., Meyarivan, T.: A fast and elitist Multiobjective Genetic Algorithm: NSGA-II. IEEE Transactions on Evolutionary Computation **6**(2), 182–197 (2002)
6. Feo, T.A., Resende, M.G.C.: Greedy randomized adaptive search procedure. Journal of Global Optimization **6**, 109–133 (1995)
7. Gutin, G., Punnen, A.P.: The Traveling Salesman Problem and its Variations. Kluwer Academic Publishers, Dordrecht (2002)
8. Hansen, P., Mladenovic, N.: Variable neighborhood search: Principles and applications. European Journal of Operational Research **130**, 449–467 (2001)
9. Huayu, X., Wenhui, F., Tian, W., Lijun, Y.: An Or-opt NSGA-II algorithm for multi-objective vehicle routing problem with time windows. In: 4th IEEE Conference on Automation Science and Engineering, 309–314 (2008)
10. Jemai, J., Zekri, M., Mellouli, K.: An NSGA-II Algorithm for the Green Vehicle Routing Problem. In: Hao, J.-K., Middendorf, M. (eds.) EvoCOP 2012. LNCS, vol. 7245, pp. 37–48. Springer, Heidelberg (2012)
11. Johnson, D.S., Papadimitriou, C.H.: Computational complexity. In: Lawer, E.L., Lenstra, J.K., Rinnoy Kan, A.H.D., Shmoys, D.B., (Eds.) The traveling salesman problem: a guided tour of combinatorial optimization. Wiley and Sons, 37–85 (1985)
12. Jozefowiez, N., Semet, F., Talbi, E.-G.: Enhancements of NSGA II and Its Application to the Vehicle Routing Problem with Route Balancing. In: Talbi, E.-G., Liardet, P., Collet, P., Lutton, E., Schoenauer, M. (eds.) EA 2005. LNCS, vol. 3871, pp. 131–142. Springer, Heidelberg (2006)
13. Jozefowiez, N., Semet, F., Talbi, E.G.: Multi-objective vehicle routing problems. European Journal of Operational Research **189**, 293–309 (2008)
14. Laporte, G.: The vehicle routing problem: An overview of exact and approximate algorithms. European Journal of Operational Research **59**, 345–358 (1992)

15. Lawer, E.L., Lenstra, J.K., Rinnoy Kan, A.H.G.R., Shmoys, D.B.: The Traveling Salesman Problem: a guided tour of combinatorial optimization. Wiley and Sons (1985)
16. Lichtblau, D.: Discrete optimization using Mathematica. In: Callaos, N., Ebisuzaki, T., Starr, B., Abe, J.M., Lichtblau, D. (Eds.) World multi-conference on systemics, cybernetics and informatics (SCI 2002), International Institute of Informatics and Systemics, 16, 169–174 (2002)
17. Lin, S.: Computer solutions of the traveling salesman problem. Bell Systems Technical Journal 44(10), 2245–2269 (1965)
18. Ombuki, B., Ross, B.J., Hanshar, F.: Multi-objective genetic algorithms for vehicle routing problem with time windows. Applied Intelligence 24, 17–30 (2006)
19. Sarker, R., Coello Coello, C.A.: Assessment methodologies for multiobjective evolutionary algorithms. Evolutionary Optimization, International Series in Operations Research and Management Science 48, 177–195 (2002)
20. Toth, P., Vigo, D.: The Vehicle Routing Problem. Monographs on Discrete Mathematics and Applications, Siam (2002)
21. Xiao, Y., Zhao, Q., Kaku, I., Xu, Y.: Development of a fuel consumption optimization model for the capacitated vehicle routing problem. Computers and Operations Research 39(7), 1419–1431 (2012)
22. Zitzler, E., Deb, K., Thiele, L.: Comparison of multiobjective evolutionary algorithms: Empirical results. Evolutionary Computation 8(2), 173–195 (2000)

Evolutionary Inference
of Attribute-Based Access Control Policies

Eric Medvet[1]($\boxtimes$), Alberto Bartoli[1], Barbara Carminati[2], and Elena Ferrari[2]

[1] Dip. di Ingegneria e Architettura, Università degli Studi di Trieste, Trieste, Italy
emedvet@units.it
[2] Dip. di Scienze Teoriche e Applicate,
Università degli Studi dell'Insubria, Como, Italy

Abstract. The interest in attribute-based access control policies is increasingly growing due to their ability to accommodate the complex security requirements of modern computer systems. With this novel paradigm, access control policies consist of attribute expressions which implicitly describe the properties of subjects and protection objects and which must be satisfied for a request to be allowed. Since specifying a policy in this framework may be very complex, approaches for policy mining, i.e., for inferring a specification automatically from examples in the form of logs of authorized and denied requests, have been recently proposed.

In this work, we propose a multi-objective evolutionary approach for solving the policy mining task. We designed and implemented a problem representation suitable for evolutionary computation, along with several search-optimizing features which have proven to be highly useful in this context: a strategy for learning a policy by learning single rules, each one focused on a subset of requests; a custom initialization of the population; a scheme for diversity promotion and for early termination. We show that our approach deals successfully with case studies of realistic complexity.

1 Introduction

Data are today one of the most strategic asset of any company and organization and, as such, their protection from any kind of improper modifications or unauthorized disclosures is a fundamental service to be provided by any Data Management System. In a data management system, accesses are regulated through access control policies [1] that are then encoded into a set of authorizations and checked by the reference monitor, a trusted software module in charge of enforcing access control. Since the 1970s, several access control models for policy specification have been proposed, including Discretionary Access Control (DAC), Mandatory Access Control (MAC), and Role-based Access Control (RBAC). The common characteristic of these models is that they are identity-based, that is, access control is based on the identity of subjects and protection objects. These models are not scalable and flexible and thus they do not fit very well in the current scenario, characterized by open and distributed systems. Due to

A. Gaspar-Cunha et al. (Eds.): EMO 2015, Part I, LNCS 9018, pp. 351–365, 2015.
DOI: 10.1007/978-3-319-15934-8_24

these limitations the new Attribute-based Access Control (ABAC) paradigm has recently emerged [2]. The main advantage of ABAC is that the access control process is not identity-based, rather it exploits attributes of the requestor and resource (e.g., the age of the requestor). Attribute expressions are then used to implicitly denote the sets of users and resources to which a policy applies (e.g., a nurse can add an item in a HR for a patient in the ward in which he/she works). Clearly, the main advantage of ABAC is in terms of flexibility in the specification of protection requirements. In contrast, the drawback is that policy specification becomes more complex and can result in an expensive and time consuming task.

A promising approach to diminish the burden of policy specification is represented by *policy mining*, whose goal is to partially or totally automate the construction of an ABAC policy from available access control information (e.g., access control logs, RBAC policies). Therefore, in this paper, we propose a multi-objective evolutionary approach for learning ABAC policies from sets of authorized and denied access requests. The approach is multi-objective because it aims at learning a policy which, at the same time, is consistent with the input requests, exhibits low complexity and does not use those attributes which uniquely represent user and resource identities, hence exploiting the true potential of the ABAC paradigm.

The evolutionary approach here proposed includes several contributions:(i) a domain-specific phenotypic representation, along with a set of custom genetic operators, which allow individuals to represent valid policy rules in the ABAC paradigm; (ii) an incremental strategy for learning a policy by learning single rules, each one fitting a subset of requests—a form of separate-and-conquer; (iii) a custom initialization of the population; (iv) a diversity promotion scheme; and (v) an early termination criterion.

We experimentally evaluated our proposal on a set of realistic case studies and found that it is always able to obtain a policy which meets the objectives. We also assessed our solution in case of incomplete input—i.e., when the input requests do not fully represent the access control information—and found that it is robust to missing information rates up to 50%.

We would like to remark that the approach presented in this paper can also be a valuable contribution in other security domains, besides policy mining, such as the strategic one of emergency management. Indeed, one of the most widely used approaches to deal with the information needs arising during emergency situations is the Break-the-Glass (BtG) paradigm [3], which allows users to override access control decisions on demand by logging their accesses. The main issue with BtG models is that they could bring the system to an unsafe state due to abuse of BtG policies. An alternative and more secure way to deal with emergency management, which has been recently proposed in [4], is to use a policy-based approach, according to which a set of emergency policies are specified, overriding regular ones during emergency situations. By properly extending the approach presented in this paper, in terms of objectives to be met, emergency policies can be learned from the BtG logs.

2   Related Work

The problem of deriving new ABAC policies from access request logs has been first and only investigated in [5] (extended by [6]). The authors propose an algorithm which incorporates some heuristics aimed at merging and simplifying single rules. In this paper, we use the same ABAC language and the same case studies of the cited papers: our method exhibits the same high effectiveness of the method in [5,6], which is not evolutionary. We think that our proposal could be easier to extend—by incorporating new objectives to be met—in order to fit more specific needs of similar scenarios, such as emergency policy learning from BtG logs.

Other non-evolutionary approaches have been proposed for mining policies from logs for less expressive access control models (e.g., RBAC [7,8]). In some cases, additional information, besides the request logs, is needed as training data [9].

Usage of evolutionary techniques for inferring RBAC rules explaining the observed actions in environments with tree-structured role hierarchies was proposed in [10]. The aim of the proposal was using the inferred rules for identifying mismatches between user roles and actual processes, as a tool for insider threat detection. No actual assessment was provided. An exercise in security policy inference through evolutionary techniques was proposed in [11]. This work considered rules based on boolean expressions constructed in a simple language and applied Genetic Programming for discovering a single expression capturing all the rules provided as examples. The case studies were composed of very few examples and were mainly a proof-of-concept demonstrating the feasibility of policy inference by means of Genetic Programming. We consider instead a full-fledged security policy language capable of expressing attribute-based rules, and demonstrate that our approach can indeed be applied successfully on realistic testbeds.

An evolutionary framework for learning security policies which need to be updated dynamically is proposed in [12]. This work introduces a stochastic risk-based security policy model based on a few numerical or boolean variables and uses this model for generating examples of access control decisions. These examples are then used for driving a Genetic Programming search aimed at inferring a formula leading to the same decisions as those in the examples. The cited work uses the SPEA2 multi objective evolutionary algorithm [13] for minimizing the error rate and the size of each formula. Evolutionary multi-objective optimization techniques have been applied in other security-related problems as a tool for systematically coping with problem-specific constraints, e.g., performance and usability in network reconfiguration strategies [14] and run-time efficiency in deep packet inspection [15].

3   Scenario

3.1   ABAC Policy Language

We consider the ABAC policy language defined in [5], which is stated to be, according to the authors, significantly more complex than policy languages handled in previous work on security policy mining. We briefly describe the language below in order to provide the appropriate context for our work.

Let U be a set of *users* and A_U a set of *user attributes*. The *value* of attribute $a \in A_U$ for user $u \in U$ is represented by a function $d_U(u, a)$. This function can assume a special value $\perp$ to indicate that the value of attribute a for user u is undefined.

The set of user attributes A_U can be partitioned in two sets: $A_{U,1}$, containing *single-valued* attributes, and $A_{U,\infty}$ containing *multi-valued* attributes, i.e., attributes whose values are sets of single values. Set $A_{U,1}$ includes a special attribute uid which has a unique value (different from $\perp$) for each user.

We denote with $V_U(a)$ the set of possible single values assumed by user attribute a, i.e., the range of d_U for $a \in A_{U,1}$ and the union of the range elements for $a \in A_{U,\infty}$.

Similarly, let R be a set of *resources* and A_R a set of *resource attributes*. The value of attribute $a \in A_R$ for resource $r \in R$ is represented by a function $d_R(r, a)$, which can assume a special value $\perp$ to indicate that the value of attribute a for resource r is undefined. The set A_R can be partitioned in two sets $A_{R,1}$ and $A_{R,\infty}$ containing single-valued and multi-valued attributes, respectively. Set $A_{R,1}$ includes a special attribute rid which has a unique value (different from $\perp$) for each resource. We denote with $V_R(a)$ the set of possible single values assumed by resource attribute a.

Subsets of users and resources can be described by means of *attribute expressions*, as follows. We denote by $\text{Set}(S)$ the powerset of set S. A user attribute expression is a function $e_U : A_U \to E$, where $E = \text{Set}(V_U(a)) \cup \top$ when $a \in A_{U,1}$ (see below for the meaning of $\top$) and $E = \text{Set}(\text{Set}(V_U(a))) \cup \top$ when $a \in A_{U,\infty}$. We say that a user u *satisfies* a user attribute expression e_U if and only if, $\forall a \in A_{U,1}, e_U(a) = \top \vee e_U(a) \ni d_U(u, a)$ and $\forall a \in A_{U,\infty}, e_U(a) = \top \vee \exists s \in e_U(a), d_U(u, a) \supseteq s$. In other words, $\top$ is used to indicate that attribute a is irrelevant for determining whether a user satisfies user attribute expression e_U (i.e., $e_U(a) = \top$).

Resource attribute expressions are defined similarly, except that the satisfaction criterion for multi-valued attributes requires equality rather than $\supseteq$ (i.e., $\exists s \in e(a), d_R(r, a) = s$). The reason is because user attributes which are multi-valued represent capabilities.

A *constraint* represents a relationship between users and resources which may or may not be satisfied, as follows. A constraint c is a function $c : A_U \times A_R \to \{\neg\top, \top\}$. A pair composed of a user u and a resource r satisfies a constraint c if and only if $\forall a_U \in A_{U,\infty}, a_R \in A_{R,\infty}, c(a_U, a_R) = \top \vee d_U(u, a_U) \supseteq d_R(r, a_R)$ and $\forall a_U \in A_{U,\infty}, a_R \in A_{R,1}, c(a_U, a_R) = \top \vee d_U(u, a_U) \ni d_R(r, a_R)$ and $\forall a_U \in A_{U,1}, a_R \in A_{R,1}, c(a_U, a_R) = \top \vee d_U(u, a_U) = d_R(r, a_R)$.

A *rule* ρ is a tuple $\langle e_U, e_R, O, c \rangle$, where e_U is an attribute expression, e_R is a resource expression, c is a constraint and $O \subseteq \mathcal{O}$ is a set of *operations*. A *policy* P is a set of rules. Finally, an *access request* is a tuple $\langle u, r, o \rangle$ which means that user u wants to perform the operation $o \in \mathcal{O}$ on the resource r.

An access request $\langle u, r, o \rangle$ is either *accepted* or *denied* by a rule $\rho = \langle e_U, e_R, O, c \rangle$. The former occurs if and only if u satisfies e_U, r satisfies e_R, u, r satisfies c and $o \in O$. An access request is accepted by a policy P if and only if the access request is accepted by at least one rule in P, otherwise the access request is denied by P. We denote with $\langle u, r, o \rangle \models \rho$ and $\langle u, r, o \rangle \models P$ the acceptance of a request $\langle u, r, o \rangle$ by a rule ρ or a policy P, respectively.

We describe attribute expressions and rules by means of the concrete syntax proposed in [5] and outlined in the following example. Let us consider an university domain in which $A_{U,1} = \{\text{uid}, \text{position}, \text{isDean}\}$, $A_{U,\infty} = \{\text{courses}\}$, $A_{R,1} = \{\text{rid}, \text{type}, \text{course}\}$, $A_{R,\infty} = \emptyset$ and $\mathcal{O} = \{\text{writeGrade}, \text{readGrade}, \text{deploy}\}$. A policy P may be composed of the following 4 rules:

$\rho_1 = \langle \text{position} = \text{student}, \text{type} = \text{gradebook}, \{\text{readGrade}\}, \text{courses} \ni \text{course} \rangle$

$\rho_2 = \langle \text{position} = \text{faculty}, \text{type} = \text{gradebook}, \{\text{writeGrade}, \text{readGrade}\}, \text{courses} \ni \text{course} \rangle$

$\rho_3 = \langle \text{position} \in \text{faculty} \wedge \text{isDean} = \text{true}, \text{type} = \{\text{gradebook}\}, \{\text{readGrade}\}, \emptyset \rangle$

$\rho_4 = \langle \text{courses} \supseteq \{\{\text{CS04}\}, \{\text{WD01}\}\}, \text{type} = \{\text{testWebServer}\}, \{\text{deploy}\}, \emptyset \rangle$

Rule ρ_1 says that students can read the grades they got for their courses (i.e., those they attend); ρ_2 says that faculty members can read and write grades for their courses (i.e., those they teach); ρ_3 says that the dean can read all grades; ρ_4 says that users whose courses include one among CS04 and WD01 can deploy on the test web server. Note that ρ_1 and ρ_2 pose a constraint on the relationship between the user and the resource, whereas ρ_3 and ρ_4 do not (i.e., for ρ_3 and ρ_4, $c(a_U, a_R) = \top, \forall a_U \in A_U, a_R \in A_R$).

3.2   Problem Statement

Let us consider two sets A_U and A_R of user and resource attributes along with their possible values V_U and V_R, and the set $\mathcal{O}$ of operations which may be applied to resources. The problem which we aim to solve consists in generating a policy P which accepts all access requests in a specified set S_A and denies all access requests in another specified set S_D. In other words, the problem consists in inferring a policy *consistent* with specified examples of the desired behavior. A *problem instance* is a tuple $\langle S_A, S_D, A_U, A_R, V_U, V_R, d_U, d_R, \mathcal{O} \rangle$.

A trivial solution for every problem instance always exists in the form of an Access Control List (ACL) policy. Such a policy may be constructed by generating, for each request $\langle u, r, o \rangle \in S_A$, a rule $\rho = \langle e_U, e_R, O, c \rangle$ which accepts only a request from user u to perform the operation o to resource r, using only special attributes uid and rid. In other words, $e_U(a) = d_U(u, a)$ if $a = \text{uid}$, $e_U(a) = \top$ otherwise; $e_R(a) = d_R(r, a)$ if $a = \text{rid}$, $e_R(a) = \top$ otherwise; $O = \{o\}$; and $c(a_U, a_R) = \top$.

In order to generate policies which not only are consistent but indeed generalize beyond the provided examples by taking user and resource attributes into account, we add two further requirements:(i) policy rules should use uid and rid special attributes as little as possible and (ii) the complexity of the policy should be minimized. We assess the complexity of a policy P with the weighted structural complexity (WSC) [8]. WSC is a weighted sums of the complexity of rule components (e_U, e_R, O and c)—see [5] for the details. In this paper we used, without loss of generality, equal weights.

4   Our Evolutionary Approach

4.1   Overview

We propose an evolutionary approach for solving the policy generation problem. Each individual represents a rule and we define custom genetic operators which operate on rules and are guaranteed to generate valid rules. In other words, we define a domain-specific phenotypic representation of candidate solutions rather than adopting more general representations which would hardly fit this application domain (e.g., trees as in Genetic Programming or numeric vectors as in Genetic Algorithm).

We construct the required policy *incrementally*, by means of successive itera-tions—a form of separate-and-conquer [16]. At each iteration we execute an evolutionary search which generates one rule $\rho^\star$ and then we drop from the set S_A of requests to be accepted those which are accepted by $\rho^\star$. Each iteration thus operates on a problem instance which differs from the problem instance at the previous iteration—S_A at the $(i + 1)$-th iteration being a subset of S_A at the i-th iteration. The procedure terminates when S_A is empty. An intermedi-ate policy is then constructed as the set of rules generated at each iteration. Finally, the required policy is obtained from a further optimization applied to the intermediate policy.

The evolutionary search includes further key contributions:

- We initialize the population based on the problem instance (in particular using the requests in S_A), rather than generating random individuals.
- We promote population diversity by imposing that no identical individuals can be contained in the population.
- We use an *early termination* criterion based on counting how many times the search would attempt to generate an individual which already exists.

4.2   Evolutionary Search

An evolutionary search takes a problem instance $\langle S_A, S_D, A_U, A_R, V_U, V_R, d_U, d_R, \mathcal{O} \rangle$ as input and produces a single rule $\rho^\star$. Each individual ρ is associated with a counter c_ρ, initially set to 1, and with a *fitness* $f(\rho)$, defined below. First, an initial population of $|S_A|$ individuals (i.e., rules) is built. These individuals are generated from S_A rather than randomly, as follows. For each request $\langle u, r, o \rangle \in S_A$ a rule $\rho = \langle e_U, e_R, O, c \rangle$ is built such that:

$$e_U(a_U) = \begin{cases} d_U(u, a_U) & \text{if } a_U \neq \text{uid} \wedge d_U(u, a_U) \neq \bot \wedge \forall a_R \in A_R, c(a_U, a_R) = \top \\ \top & \text{otherwise} \end{cases}$$

$$e_R(a_R) = \begin{cases} d_R(r, a_R) & \text{if } a_R \neq \text{rid} \wedge d_R(r, a_R) \neq \bot \wedge \forall a_U \in A_U, c(a_U, a_R) = \top \\ \top & \text{otherwise} \end{cases}$$

$$O = \{o\}$$

$$c(a_U, a_R) = \begin{cases} \neg\top & \text{if } d_U(u, a_U) \neq \bot \wedge d_R(r, a_R) \neq \bot \wedge d_U(u, a_U) \supseteq d_R(r, a_R) \\ \neg\top & \text{else if } d_U(u, a_U) \neq \bot \wedge d_R(r, a_R) \neq \bot \wedge d_U(u, a_U) \ni d_R(r, a_R) \\ \neg\top & \text{else if } d_U(u, a_U) \neq \bot \wedge d_R(r, a_R) \neq \bot \wedge d_U(u, a_U) = d_R(r, a_R) \\ \top & \text{otherwise} \end{cases}$$

In practice, in order to build a rule $\rho = \langle e_U, e_R, O, c \rangle$ from the request $\langle u, r, o \rangle$ we first find the user and resource attributes which can be used to define the constraint c; then, we set user and attribute expression e_U and e_R according to the values of the respective u and r attributes; in doing so, we consider only attributes which have not been used for defining c and we do not use either uid or rid.

With reference to the university domain example in Section 3.1, let us consider the following request in S_A:

$$u = \langle \text{uid} = \text{stud111013}, \text{position} = \text{student}, \text{courses} = \{\text{CS01}, \text{CS03}\}\rangle$$
$$r = \langle \text{rid} = \text{gradebook7211}, \text{type} = \text{gradebook}, \text{course} = \text{CS03}\rangle$$
$$o = \text{readGrade}$$

The rule generated from this request will be:

$$\rho = \langle \text{position} = \text{student}, \text{type} = \text{gradebook}, \{\text{readGrade}\}, \text{courses} \ni \text{course}\rangle$$

Note that $c(\text{courses}, \text{course}) = \neg\top$ because $d_U(u, \text{courses}) = \{\text{CS01}, \text{CS03}\} \ni$ $\text{CS03} = d_R(r, \text{course})$.

Having generated the initial population, we execute the following iterative procedure:

1. Choose randomly whether to apply a *mutation* operator or a *crossover* operator; the choice between the two options is made with probability p_{mutation} and $1 - p_{\text{mutation}}$, respectively.
2. Choose randomly the specific operator within the chosen category with uniform probability—we defined 10 mutation operators and 5 crossover operators.
3. If a mutation operator has been chosen, then select one rule in the current population, otherwise (a crossover operation has been chosen) select two rules; each rule selection is made by picking $n_{\text{tournament}}$ rules at random and then selecting the best one.
4. Generate a new rule ρ' with the chosen genetic operator applied to the chosen rule(s); if the current population does not already contain a rule $\rho = \rho'$, then add ρ' to the current population and evaluate its fitness $f(\rho)$; otherwise, increment counter c_ρ by one and discard ρ'.

5. If the current population size is greater than n_{pop}, then iteratively remove the worst rule until the population size is equal to n_{pop}.

The iterative procedure terminates when one of the following holds:(a) a predefined number n_{eval} of fitness evaluations has been performed, or (b) the counter c_{ρ^*} of the best rule ρ^* is larger than a predefined number n_{stop}. At the end, the best rule ρ^* in the current population is the result of the search. Note that the fitness of generated rules are evaluated only when they are different from all existing—and hence already evaluated—rules (step 4).

We defined 10 mutation operators and 5 crossover operators. Their full description is not included in this paper for space constraints but is available separately[1]. For example, we defined a constraint donation crossover operator as follows: let $\rho_1 = \langle e_{U,1}, e_{R,1}, O_1, c_1 \rangle$ and $\rho_2 = \langle e_{U,2}, e_{R,2}, O_2, c_2 \rangle$ be the parent rules. The rule $\rho = \langle e_U, e_R, O, c \rangle$ generated by the operator is initially set to $\rho = \rho_1$; next, a pair $a_U, a_R \in A_U \times A_R$ is randomly chosen such that $c_1(a_U, a_R) = \top \land c_2(a_U, a_R) = \neg\top$; finally, $c(a_U, a_R) := c_2(a_U, a_R)$.

The fitness $f(\rho)$ of a rule ρ is defined as a tuple composed of 4 numbers: $f(\rho) = \langle \text{FAR}(\rho), \text{FRR}(\rho), \text{ID}(\rho), \text{WSC}(\rho) \rangle$, where FAR and FRR are the False Acceptance Rate on the requests in S_D and the False Rejection Rate on the requests in S_A, respectively; $\text{ID}(\rho)$ is a measure of the usage of the special attributes uid and rid; $\text{WSC}(\rho)$ is the WSC index defined in Section 3.2. In detail:

$$\text{FAR}(\rho) = \frac{|\{\langle u, r, o \rangle \in S_D, \langle u, r, o \rangle \models \rho\}|}{|S_D|}$$

$$\text{FRR}(\rho) = \frac{|\{\langle u, r, o \rangle \in S_A, \langle u, r, o \rangle \not\models \rho\}|}{|S_A|}$$

$$\text{ID}(\rho) = \begin{cases} 2 & \text{if } e_U(\text{uid}) \neq \top \land e_R(\text{rid}) \neq \top \\ 1 & \text{if } e_U(\text{uid}) \neq \top \veebar e_R(\text{rid}) \neq \top \\ 0 & \text{otherwise} \end{cases}$$

For all the elements of the fitness tuple, the lower the better.

Rules are ranked basing on lexicographical order of their fitnesses $f(\rho)$: the rule with lower FAR is considered the best; in case two or more have the same lowest FAR, the rule with lowest FRR is considered the best; in case two or more have the same lowest FRR, the rule with lowest ID is considered the best; in case two or more have the same lowest ID, the rule with lowest WSC is considered the best. This method of ranking solutions in a multi-objective problem where objectives are sorted by decreasing importance is also known as *multi-layered fitness* [17]. In our case, in the fitness $f(\rho) = \langle \text{FAR}(\rho), \text{FRR}(\rho), \text{ID}(\rho), \text{WSC}(\rho) \rangle$, the first two components represent the ability of the rule to be consistent with the problem instance, whereas the other two reflect the further problem objectives concerning use of special attributes and complexity (see Section 3.2).

---

[1] http://machinelearning.inginf.units.it/data-and-tools/appendices/
   2014-EMO-EvolutionaryABACInference-Appendix.pdf

4.3 Incremental Strategy

We construct the required policy incrementally, by means of successive evolutionary searches, as follows. Initially, let $P = \emptyset$ and $S'_A = S_A$, then:

1. execute an evolutionary search (Section 4.2) on problem instance $\langle S'_A, S_D,$ $A_U, A_R, V_U, V_R, d_U, d_R, \mathcal{O}\rangle$ and obtain $\rho^\star$;
2. if $\mathrm{FAR}(\rho^\star) = 0$ and $\mathrm{FRR}(\rho^\star) < 1$ then $P := P \cup \{\rho^\star\}$, otherwise terminate;
3. assign $S'_A = S_A \setminus \{\langle u, r, o\rangle \in S_A, \langle u, r, o\rangle \models P\}$;
4. if $S'_A = \emptyset$, terminate.

In other words, at each iteration we obtain a new rule $\rho^\star$ (step 1). As long as this new rule accepts at least one request in S'_A (step 2, $\mathrm{FRR}(\rho^\star) < 1$) while not accepting any request in S_D ($\mathrm{FAR}(\rho^\star) = 0$), the new rule is added to the policy being constructed and the iteration continues. The next iteration will operate on a smaller S'_A, which does not contains any request accepted by the current policy, including $\rho^\star$ (step 3). In case all requests to be accepted are already accepted by the current policy, the iteration terminates (step 4).

Since each $\rho \in P$ has $\mathrm{FAR}(\rho) = 0$—see step 2 above—and since a request is accepted if at least one rule in P accepts it, it follows that $\mathrm{FAR}(P) = 0$ and $\forall \rho \in P, \mathrm{FRR}(P) \leq \mathrm{FRR}(\rho)$, where FRR and FAR are defined for policy P similarly to for requests.

The (intermediate) policy P obtained by the above procedure is optimized further by executing the following procedure for a predefined number of n_{eval} iterations:

1. choose a rule ρ in P at random;
2. generate a new rule ρ' by applying a randomly selected mutation operator on ρ;
3. build a policy P' by replacing ρ with ρ' in P;
4. if P' is better than P, than $P := P'$.

The comparison criterion between two policies P_1, P_2 is based on the same lexicographical order used for rules: the policy with lowest FAR is considered the best; otherwise, the policy with lowest FRR is best; otherwise, the policy with lowest ID is best (where $\mathrm{ID}(P) = \sum_{\rho \in P} \mathrm{ID}(\rho)$); otherwise, the policy with lowest WSC is best.

We chose to use a lexicographical order (both for rules and policies) because it reflects the order of in which the problem objectives are defined (consistency first, then use of special attributes, then complexity). Moreover, concerning consistency, we chose to favor—i.e., minimizing first—FAR instead of FRR because of the way we compose a policy starting from rules: in particular, we aim at obtaining rules with $\mathrm{FAR} = 0$ (see the condition in step 2 above).

5 Experimental Evaluation

We evaluated our proposal experimentally on the same case studies considered in [5]. Each case study consists of a set of users U, a set of resources R and a set of rules P_0. Users and resources are associated with various attributes. Rules were carefully constructed to express non-trivial policies and exercise all the features of the policy language, including use of set membership and superset relations in attribute expressions and constraints. The experimental data consist of 7 case studies: 4 of them were hand-crafted and 3 of them were synthetically generated from the hand-crafted ones. The synthetic case studies include a much larger number of users and resources. Table 1 summarizes the case studies. The set of operations $\mathcal{O}$ is obtained from P_0 as $\mathcal{O} = \bigcup_{\langle e_U, e_R, O, c \rangle \in P_0} O$. The set of requests includes all possible requests, i.e., $S = U \times R \times \mathcal{O}$; this set is then partitioned in S_A and S_D, basing on whether each request in S was accepted or denied by P_0, respectively.

Table 1. Salient information about the hand-crafted (above) and synthetic (below, with a † suffix) case studies

Case study	$|P_0|$	$|U|$	$|R|$	$|\mathcal{O}|$	$|A_U|$	$|A_R|$	$|S_A|$	$|S_D|$	WSC(P_0)
Healthcare	9	21	16	3	6	7	51	957	33
Online video	6	12	13	1	3	3	78	78	20
Project management	11	19	40	7	8	6	189	5131	49
University	10	22	34	9	6	5	168	6564	37
Healthcare†	9	1600	5760	3	6	7	10 097	27 637 903	33
Project management†	11	800	1600	7	8	6	7680	8 952 320	49
University†	10	1320	2520	9	6	5	148 624	29 788 976	37

We executed our approach on each case study for several values of the n_{eval} parameter. We repeated each experiment 3 times for each n_{eval} value, with different random seeds. We set the other parameters as follows: $n_{\text{pop}} = 100$, $n_{\text{tournament}} = 3$, $n_{\text{stop}} = 100$ and $p_{\text{mutation}} = 0.5$—we verified experimentally that reasonable variations in these values do not cause significant variations in the results.

Since the synthetic case studies are associated with several millions of requests to be denied, in these cases we generated the corresponding policies based on a random sample S_D^* of those requests such that $|S_D^*| = 5|S_A|$. The results in terms of FRR and FAR have always been computed on the full set, though.

Table 2 presents the results, averaged across the executions with different random seeds, in terms of FRR(P), FAR(P) and $\frac{\text{WSC}(P_0)}{\text{WSC}(P)}$; the table also shows the actual number $\hat{n}_{\text{eval}}$ of fitness evaluations—recall that the number of iterations of the incremental strategy is not known in advance—and the execution time.

The first crucial finding is that our approach indeed succeeds in generating consistent policies, i.e., policies with FRR$(P) = 0$, FAR$(P) = 0$. Moreover, we verified that our method never produced policies which use special attributes uid and rid, as desired.

Another important result is that our approach definitely tends to generate a policy which is less complex than the baseline: in most cases WSC(P) is not larger than the WSC(P_0) (we remark that P_0 is unknown to our approach).

Table 2. Results

Case study	n_{eval}	FRR(P)	FAR(P)	$\frac{WSC(P_0)}{WSC(P)}$	$\hat{n}_{eval}$	t [s]
Healthcare	500	0	0	1.07	5536	1.2
	2500	0	0	1.18	19 776	4
	5000	0	0	1.18	22 691	5.3
Online video	500	0	0	1	2768	0.6
	2500	0	0	1	5215	0.8
	5000	0	0	1	7715	1.1
Project management	500	0	0	0.96	6646	3.5
	2500	0	0	1.06	24 368	14.7
	5000	0	0	1.06	27 791	22.2
University	500	0	0	0.95	5904	3.1
	2500	0	0	0.98	22 846	14.1
	5000	0	0	1	26 487	21.8
Healthcare[†]	500	0	0	1.18	23 704	228.4
	2500	0	0	1.2	33 864	398.9
	5000	0	0	1.2	36 364	511.9
Project management[†]	500	0	0	0.92	35 037	241.7
	2500	0	0	1.06	45 591	626.9
	5000	0	0	1.06	49 549	790
University[†]	500	0	0	0.89	35 822	1688.8
	2500	0	0	1	52 513	3525.4
	5000	0	0	1	56 718	4784.9

This effect is more apparent for Healthcare and Project management and the corresponding synthetic versions, but can be observed also for University and its synthetic counterpart, for sufficiently large values of n_{eval}. In other words, our approach aims at obtaining the least complex policy which is consistent with the desired behavior in terms of S_A, S_D; as it turns out, the generated policy tends to be less redundant than the baseline policy.

The average execution time for generating a policy is in the order of seconds for the hand-crafted case studies and in the order of minutes or a few tens of minutes for the synthetic ones. It seems reasonable to claim that the computational load is fully practical for this application domain. The experiments have been executed with a single-threaded Java prototype implementation on a quad-core Intel CPU 2.50 GHz with 8 GB RAM. As expected, the execution time is roughly linear with $\hat{n}_{eval}|S_A||S_D|$; with respect to n_{eval}, it can be seen that time is slightly sublinear due to the intervention of the early termination criterion given by n_{stop}.

5.1 Results with Incomplete Input

We wanted to gain insights in our method effectiveness when the input information is incomplete. In particular, we considered the case where the input requests sets S_A and S_D do not contain all the requests.

To this end, we repeated the experimental procedure detailed in the previous section, by removing, before applying our method, a random portion γ of requests in S_A and S_D. The performance in terms of FRR and FAR has obviously been computed on full S_A and S_D. We repeated each experiment 3 times—i.e., with 3 different S_A, S_D and random seeds—for each value of γ.

The corresponding results are in Figure 1 (we executed these experiments only on the hand-crafted case studies). It can be seen that the removal of part of requests has little or no impact on FAR, even for large values of γ. Indeed, FAR is always 0 when $\gamma \leq 0.25$ and remains low ($\leq 1\%$) even when half of the requests in S_D are not available for inferring the policy. The impact of the removal of part of requests is slightly higher on FRR, but always lower than 4%. We believe the reason is because our approach tends to produce a policy which contains only the rules which are needed to accept all the requests contained in the input S_A. This interpretation is supported by the values of $\frac{\text{WSC}(P_0)}{\text{WSC}(P)}$, which become larger with large values of γ: in other words, our approach produces the least complex policy which is consistent with the input.

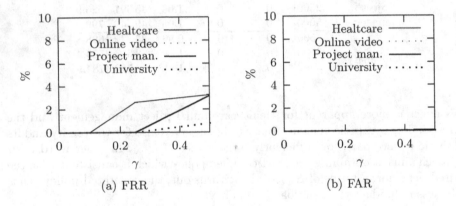

(a) FRR

(b) FAR

Fig. 1. Results with incomplete input

5.2   Assessment of the Contributions

We wanted to assess the specific impact of our key contributions described in Section 4.1: population initialization from requests in S_A, incremental strategy, diversity promotion and early termination. Rather than assessing all combinations, we executed pairwise comparisons between the full approach and the approach with one of these contributions disabled—note that disabling the diversity promotion implies disabling also the early termination. We considered only the hand-crafted case studies.

Concerning the population initialization from the examples in S_A, we experimented with random creation of individuals in the initial population. It was clear from the early experiments that generating a consistent policy from a random initial population is very difficult. For this reason, we investigated the impact of increasing the population size $n_{pop} = 100$ by an order of magnitude $(1000, 2000)$ and disabled early termination. Table 3 shows the results obtained with random population initialization (for ease of comparison we provide the corresponding results of the full method previously shown in Table 2). It is clear that the population initialization from S_A is an essential ingredient for generating consistent policies and greatly improves the method effectiveness. Table 3 shows that, if one favors the exploration ability of the evolutionary search by increasing n_{pop} and n_{eval}, the method can indeed obtain better solutions also with random initialization. These improved results are still far from those with the initialization from S_A, though, despite the increased execution time.

Table 3. Results with and without the population initialization from S_A

Pop. init.	n_{eval}	n_{pop}	Healthcare		Online video		Proj. man.		University	
			FRR(P)	t[s]	FRR(P)	t[s]	FRR(P)	t[s]	FRR(P)	t[s]
From S_A	5000	100	0	5.3	0	1.1	0	22.2	0	21.8
Random	5000	100	37.3	5.5	3.4	7.1	56.6	11.4	73.4	9.1
	25 000	100	32.7	28.2	1.7	37.2	53.1	61.8	69.4	49.1
	25 000	1000	5.2	203.7	0	300.8	38.4	224.7	9.5	276.5
	25 000	2000	2.6	432.1	0	642.0	34.9	394.1	36.1	387.9

Concerning the incremental strategy, we executed a single evolutionary search followed by the final optimization step executed with $n_{eval} = 5000$—results with larger values for n_{eval} are essentially the same. The results in terms of FRR are significantly worse than with the incremental strategy enabled: 82.4%, 53.9%, 83.1% and 74.6% for the Healthcare, Online video, Project management and University case studies respectively, as opposed to FRR = 0. In other words, a single evolutionary search is not able to generate a rule capable of accepting all the requests in S_A and denying all the requests in S_D.

Concerning diversity promotion, we experimented without this contribution and $n_{eval} = 5000$. We found that our approach still generates consistent policies (i.e., with FAR = FRR = 0), but with larger complexity: $\frac{\text{WSC}(P_0)}{\text{WSC}(P)}$ is lower than with the diversity promotion for the Healthcare and Project management case studies: 1.15 vs. 1.18 and 1.01 vs. 1.06, respectively.

Concerning the early termination criterion, we experimented by stopping the evolutionary search after having performed $n_{eval} = 5000$ fitness evaluations irrespective of the number of times the search attempts to generate identical individuals. The quality of the results is unaffected, the only impact being on execution times which are, on the average, 128% longer than with early termination enabled.

Finally, we investigated on the effectiveness of the further optimization of the policy P which we perform at the end of the incremental strategy (see Section 4.3). To this end, we experimented without this step. We found that this procedure impacts on the complexity of the generated policies: $\frac{\text{WSC}(P_0)}{\text{WSC}(P)}$ is equal to 1.18, 0.91, 1.04 and 0.98 for the Healthcare, Online video, Project management and University case studies respectively, as opposed to 1.18, 1, 1.06 and 1, with the optimization. By looking at the raw data, we found that the optimization makes our approach cope with those cases where a rule ρ_1, which has been generated at a given iteration of the incremental strategy, could be made less complex because of a rule ρ_2 generated later, which accepts some requests accepted also by ρ_1.

6   Concluding Remarks and Future Work

We have proposed an evolutionary approach for performing mining of ABAC policies. The approach is based on the design and implementation of a domain-specific phenotypic representation, along with the corresponding genetic operators, which allow attacking ABAC policy mining by means of evolutionary computation. We used a multi-objective optimization framework based on a lexicographic criterion, in which we incorporate requirements on correctness (FAR, FRR) and on expressiveness (WSC, usage of uid and rid). We incorporated in the search several optimizations that have proven to be essential for this task, in particular, we defined a strategy for building a policy incrementally, by learning single rules each one on a different subset of the requests. We showed that our approach deals successfully with case studies of realistic complexity, being highly robust even in scenarios where the access requests available for learning do not fully represent the access control information. We believe that our proposal may indeed form the basis for a practical implementation of ABAC policy mining and we intend to extend the scope of our investigation to emergency management.

Acknowledgement. The research presented in this paper is partially supported by the project *Dynamic Information Management and Exchange for Command and Control Applications*, grant number FA8655-10-1-3080, funded by the European Office of Aerospace Research and Development (EOARD) and the Air Force Office of Scientific Research (AFOSR).

References

1. Ferrari, E.: Access Control in Data Management Systems. Synthesis Lectures on Data Management. Morgan & Claypool Publishers (2010)
2. Hu, V.C., Ferraiolo, D., Kuhn, R., Schnitzer, A., Sandlin, K., Miller, R., Scarfo, K.: Guide to Attribute Based Access Control (ABAC) Definition and Considerations. NIST Special Publication (SP) 800-162, Guide, October 2014
3. Brucker, A.D., Petritsch, H.: Extending access control models with break-glass. In: Proceedings of the 14th ACM Symposium on Access Control Models and Technologies, pp. 197–206. ACM (2009)
4. Carminati, B., Ferrari, E., Guglielmi, M.: A System for Timely and Controlled Information Sharing in Emergency Situations. IEEE Transactions on Dependable and Secure Computing **10**(3), 129–142 (2013)
5. Xu, Z., Stoller, S.D.: Mining attribute-based access control policies. arXiv preprint arXiv:1306.2401 (2013)

6. Xu, Z., Stoller, S.D.: Mining attribute-based access control policies from RBAC policies. In: 2013 10th International Conference and Expo on Emerging Technologies for a Smarter World (CEWIT), pp. 1–6. IEEE (2013)
7. Gal-Oz, N., Gonen, Y., Yahalom, R., Gudes, E., Rozenberg, B., Shmueli, E.: Mining roles from web application usage patterns. In: Furnell, S., Lambrinoudakis, C., Pernul, G. (eds.) TrustBus 2011. LNCS, vol. 6863, pp. 125–137. Springer, Heidelberg (2011)
8. Molloy, I., Chen, H., Li, T., Wang, Q., Li, N., Bertino, E., Calo, S., Lobo, J.: Mining roles with multiple objectives. ACM Trans. Inf. Syst. Secur. **13**(4), 36:1–36:35 (2010)
9. Ni, Q., Lobo, J., Calo, S., Rohatgi, P., Bertino, E.: Automating role-based provisioning by learning from examples. In: Proceedings of the 14th ACM Symposium on Access Control Models and Technologies, pp. 75–84. ACM (2009)
10. Hu, N., Bradford, P.G., Liu, J.: Applying role based access control and genetic algorithms to insider threat detection. In: Proceedings of the 44th Annual Southeast Regional Conference, pp. 790–791. ACM (2006)
11. Lim, Y.T., Cheng, P.C., Rohatgi, P., Clark, J.A.: MLS security policy evolution with genetic programming. In: Proceedings of the 10th Annual Conference on Genetic and Evolutionary Computation, pp. 1571–1578. ACM (2008)
12. Lim, Y.T., Cheng, P.C., Rohatgi, P., Clark, J.A.: Dynamic security policy learning. In: Proceedings of the First ACM Workshop on Information Security Governance, pp. 39–48. ACM (2009)
13. Bleuler, S., Brack, M., Thiele, L., Zitzler, E.: Multiobjective genetic programming: reducing bloat using SPEA2. In: Proceedings of the 2001 Congress on Evolutionary Computation, vol. 1, pp. 536–543. IEEE (2001)
14. Tapiador, J.E., Clark, J.A.: Learning autonomic security reconfiguration policies. In: 2010 IEEE 10th International Conference on Computer and Information Technology (CIT), pp. 902–909. IEEE (2010)
15. Bartoli, A., Cumar, S., De Lorenzo, A., Medvet, E.: Compressing regular expression sets for deep packet inspection. In: Bartz-Beielstein, T., Branke, J., Filipič, B., Smith, J. (eds.) PPSN XIII 2014. LNCS, vol. 8672, pp. 394–403. Springer, Heidelberg (2014)
16. Fürnkranz, J.: Separate-and-conquer rule learning. Artificial Intelligence Review **13**(1), 3–54 (1999)
17. Eggermont, J., Kok, J.N., Kosters, W.A.: Genetic programming for data classification: partitioning the search space. In: Proceedings of the 2004 ACM Symposium on Applied Computing, pp. 1001–1005. ACM (2004)

Hybrid Dynamic Resampling for Guided Evolutionary Multi-Objective Optimization

Florian Siegmund[1]([✉]), Amos H.C. Ng[1], and Kalyanmoy Deb[2]

[1] Virtual Systems Research Center, University of Skövde, Skövde, Sweden
[2] Department of Electrical and Computer Engineering, Michigan State University,
East Lansing, USA

Abstract. In Guided Evolutionary Multi-objective Optimization the goal is to find a diverse, but locally focused non-dominated front in a decision maker's area of interest, as close as possible to the true Pareto-front. The optimization can focus its efforts towards the preferred area and achieve a better result [7,9,13,17]. The modeled and simulated systems are often stochastic and a common method to handle the objective noise is Resampling. The given preference information allows to define better resampling strategies which further improve the optimization result. In this paper, resampling strategies are proposed that base the sampling allocation on multiple factors, and thereby combine multiple resampling strategies proposed by the authors in [15]. These factors are, for example, the Pareto-rank of a solution and its distance to the decision maker's area of interest. The proposed hybrid Dynamic Resampling Strategy DR2 is evaluated on the Reference point-guided NSGA-II optimization algorithm (R-NSGA-II) [9].

Keywords: Evolutionary multi-objective optimization · Guided search · Reference point · Dynamic resampling · Budget allocation

1 Introduction

In Guided Evolutionary Multi-objective Optimization the decision maker is looking for a diverse, but locally focused non-dominated front in a preferred area of the objective space, as close as possible to the true Pareto-front. As solutions found outside of the area of interest are considered less important or even irrelevant, the optimization can focus its efforts towards the preferred area and find the solutions that the decision maker was looking for in a faster way, i.e. with less simulation runs. This is particularly important if the available time for optimization is limited, as for many real-world applications. Focusing the search effort sets time resources free which, if not needed elsewhere, can be used to achieve a better result. Multi-objective evolutionary algorithms that can perform guided search with preference information are, for example, the R-NSGA-II algorithm [9], Visual Steering [17], and interactive EMO based on progressively approximated value functions [7].

© Springer International Publishing Switzerland 2015
A. Gaspar-Cunha et al. (Eds.): EMO 2015, Part I, LNCS 9018, pp. 366–380, 2015.
DOI: 10.1007/978-3-319-15934-8_25

In Simulation-based Optimization the modeled and simulated systems are often stochastic. To obtain an as exact as possible simulation of the system behavior the stochastic characteristics are often built into the simulation models. When running the stochastic simulation this expresses itself in deviating result values. That means that if the simulation is run multiple times for a selected parameter setting the result value is slightly different for each simulation run. In the literature this phenomenon of stochastic evaluation functions is sometimes called Noise, respectively Noisy Optimization [1, 3].

If an evolutionary optimization algorithm is run without countermeasure on an optimization problem with a noisy evaluation function the performance will degrade in comparison with the case if the true mean objective values would be known. The algorithm will have wrong knowledge about the solutions' quality. Two cases of misjudgment will occur. The algorithm will see bad solutions as good and select them into the next generation. Good solutions might be assessed as inferior and might be discarded. The performance can therefore be improved by increasing the knowledge of the algorithm about the solution quality.

Resampling is a way to reduce the uncertainty of the knowledge the algorithm has about the solutions. Resampling algorithms evaluate solutions several times to obtain an approximation of the expected objective values. This allows EMO algorithms to make better selection decisions, but it comes with a cost. As the modeled systems are usually complex they require long simulation times, which limits the number of available solution evaluations. The additional solution evaluations needed to increase objective value knowledge are therefore not available for exploration of the objective space. This exploration vs. exploitation trade-off can be optimized, since the required knowledge about objective values varies between solutions. For example, in a dense, converged population it is important to know the objective values well, whereas an algorithm which is about to explore the objective space is not harmed much by noisy objective values. Therefore, a resampling strategy which samples the solution carefully according to their resampling need, can help an EMO algorithm to achieve better results than a static resampling allocation. Such a strategy is called Dynamic Resampling. This has been done previously for single-objective optimization problems by [11] and [4]. In this paper we study dynamic resampling algorithms that can handle multi-objective evaluation functions, based on a previous study [15].

The paper is structured as follows. In Section 2 background information to Dynamic Resampling and an introduction to the R-NSGA-II algorithm is given. In Section 3 different resampling techniques for EMO and for guided EMO are explained. A new resampling algorithm is proposed combining several resampling techniques in Section 4. In Section 5 numerical experiments on benchmark functions are performed. The test environment is explained and the experiment results are analyzed. In Section 6 conclusions are drawn and possible future work is pointed out.

2   Background

In this section, background information is given regarding Resampling as a noise handling method in Evolutionary Multi-objective Optimization and the preference-based multi-objective optimization algorithm R-NSGA-II [9], which are the basis for the proposed algorithms in this paper.

2.1   Noise Compensation via Sequential Dynamic Resampling

To be able to assess the quality of a solution according to a stochastic evaluation function statistical measures like sample mean and sample standard deviation can be used. By executing the simulation model multiple times a more accurate value of the solution quality can be obtained. This process is called Resampling. We denote the sample mean value of objective function F_i for solution s as follows: $\mu_n(F_i(s)) = \frac{1}{n}\sum_{j=1}^{n} F_i^j(s)$, where $F_i^j(s)$ is the j-th sample of s, and the sample variance of objective function i: $\sigma_n^2(F_i(s)) = \frac{1}{n-1}\sum_{j=1}^{n}(F_i^j(s) - \mu_n(F_i(s)))^2$.

The general goal of resampling a stochastic objective function is to reduce the standard deviation of the mean of an objective value $\sigma(\mu(F_i(s)))$ which increases the knowledge about the objective value. With only a limited number of samples available the standard deviation of the mean can be estimated by the sample standard deviation of the mean which usually is called standard error of the mean. It is calculated as follows:

$$se_n(\mu_n(F_i(s))) = \frac{\sigma_n(F_i(s))}{\sqrt{n}}$$

By increasing the number of samples n of $F_i(s)$ the standard deviation of the mean and its estimate $se_n(\mu_n(F_i(s)))$ is reduced.

Dynamic resampling allocates a different sampling budget to each solution based on the evaluation characteristics of the solution. A basic dynamic resampling procedure would be to reduce the standard error until it gets below a certain threshold $se_n(\mu_n(F(s))) < se_{thr}$. The required sampling budget for the reduction can be calculated as $n > \left(\frac{\sigma_n(F(s))}{se_{thr}}\right)^2$. However, since the sample mean changes as new samples are added this one-shot sampling allocation might not be optimal. The number of fitness samples drawn might be too small for reaching the error threshold, in case the sample mean has shown to be larger than the initial estimate. On the other hand, a one-shot strategy might add too many samples if the initial estimate of the sample mean was too big. Therefore dynamic resampling is often done sequentially. Through this sequential approach the number of required samples can be determined more accurately. For Sequential Dynamic Resampling often the shorter term Sequential Sampling is used. Dynamic resampling techniques can therefore be classified in one-shot sampling strategies and sequential sampling strategies.

2.2 Reference Point-Guided NSGA-II

The resampling techniques described in this paper will be tested how well they can support the Reference point-Guided NSGA-II algorithm (R-NSGA-II) [9] as an example for a guided Evolutionary Multi-objective Optimization (EMO) Algorithm. It is particularly suitable for evaluation in this paper since it uses fitness functions which are used as resampling criteria in resampling algorithms. Therefore, the resampling algorithms can support R-NSGA-II particularly well.

R-NSGA-II is based on the Non-dominated Sorting Genetic Algorithm II [6] which is a widely-used and representative multi-objective evolutionary algorithm. NSGA-II sorts the solutions in population and offspring into different non-dominated fronts. Selected are all solutions in all fronts that fit into the next population. From the front that only fits partially those solutions are selected into the next population that have big distances to their neighbors and thereby guarantee that the result population will be diverse. As diversity measure the crowding distance is used. After selection is completed offspring solutions are generated by tournament selection, crossover, and mutation. The offspring are evaluated and the selection step is performed again. The R-NSGA-II algorithm replaces the crowding distance operator by the distance to reference points. Solutions that are closer to a reference point get a higher selection priority. The reference points are defined by the user in areas that are interesting and where solutions shall be found. As a diversity preservation mechanism R-NSGA-II uses clustering. The reference points can be created, adapted or deleted interactively during the optimization run. Since R-NSGA-II uses non-domination sorting it has a tendency to prioritize population diversity before convergence to the reference points. Therefore, extensions have been proposed which limit the influence of the Pareto-dominance and allow the algorithm to focus faster towards the reference points [14].

3 Resampling Algorithms

In this chapter several resampling algorithms are described which are used in this study to support the R-NSGA-II algorithm at stochastic simulation optimization problems. We denote: Sampling budget for solution s: b_s, minimum and maximum number of samples for an individual solution: b_{min} and b_{max}, acceleration parameter for the increase of the sampling budget: $a > 0$. Increasing a decreases the acceleration of the sampling budget, decreasing a increases the acceleration. The calculated normalized sampling need x_s is discretised in the following way which guarantees that b_{max} is assigned already for $x_s < 1$:
$$b_s = \min\{b_{max}, \lfloor x_s(b_{max} - b_{min} + 1)\rfloor + b_{min}\}$$

3.1 Generally Applicable Resampling Techniques

This section contains resampling strategies that can be used in all multi-objective optimization problems regardless of preference information is given by a decision maker or not. They can support the goal of finding the whole Pareto-front as well as the goal of exploring a limited, preferred area in the objective space.

Static Resampling samples each solution the same amount of times. The sampling budget is constant, $b_s = b_{min} = b_{max}$. If the objective noise is high enough this technique can lead to an improvement of the optimization result for $b_s > 1$. However, since many samples are wasted on less important solutions, this technique is inferior to more advanced resampling techniques. The reason for its popularity is the low implementation effort it requires.

Time-Based Dynamic Resampling allocates a small sampling budget in the beginning of the optimization and a high sampling budget towards the end of the optimization [15]. The strategy of this resampling technique is to support the algorithm when the solutions in the population are close to the Pareto-front and to save sampling budget in the beginning of the optimization when the solutions are still far away from the Pareto-front.

Time-based Resampling is a dynamic resampling technique that is not considering variance and a one-shot allocation. We denote: B = maximum overall number of simulation runs, B_t = current overall number of simulation runs. The normalized time-based resampling need x_s^T is calculated as in Equation 1.

$$x_s^T = \left(\frac{B_t}{B}\right)^a \tag{1}$$

Rank-Based Dynamic Resampling assigns more samples to solutions in the first few fronts and less samples to solutions in the last fronts, to save evaluations [15]. In a well-converged population most solutions will have Pareto-rank 1 and get the maximum number of samples. This technique is not effective in many-objective optimization, since there most solutions are non-dominated. Rank-based Resampling performs sequential sampling and is a comparative resampling technique. We denote: S = solution set of current population and offspring, R_s = Pareto-rank of solution s in S, $R_{max} = \max_{s \in S} R_s$. The normalized rank-based resampling need x_s^R is calculated as in Equation 2.

$$x_s^R = 1 - \left(\frac{R_s - 1}{R_{max} - 1}\right)^a \tag{2}$$

We propose a modification of Rank-based Resampling that allocates additional samples only to the first n fronts. This allows to concentrate the additional samples on the first fronts where they can be more beneficial. The allocation function of MaxN-Rank-based Resampling is $x_s^{Rn} = 1 - \left(\frac{\min\{n, R_s\} - 1}{\min\{n, R_{max}\} - 1}\right)^a$.

We propose a Hybrid Dynamic Resampling algorithm: Rank-based Resampling can be combined with Time-based Resampling to avoid allocating samples in the beginning of the optimization where the optimization algorithm has only little gain of knowing the accurate objective values. Similar to a logical conjunction x_s^R and x_s^T can be combined to form the Rank-Time-based Resampling allocation x_s^{RT} for solution s as in Equation 3.

$$x_s^{RT} = \min\{x_s^T, x_s^R\} \tag{3}$$

3.2   Preference-Based Resampling Techniques

In this section two resampling algorithms are described that use the preference information given by a decision maker in form of a reference point r in the objective space for the R-NSGA-II algorithm [9].

Progress-Based Dynamic Resampling allocates samples to solutions depending on the progress of the population towards a reference point r. If the population is more converged better knowledge of the objective values is required. The progress is defined as the average distance from the population members to r. Since the progress in Evolutionary Multi-objective optimization can be fluctuating the average progress $\overline{P}$ from the last n populations is used. Progress-based Resampling is a sequential sampling algorithm. We denote: P_{max} = the maximum progress threshold. If $\overline{P} > P_{max}$ then b_{min} is allocated. If a progress measurement is negative the absolute value of the progress will be used, multiplied by a penalty factor. This method also works for the case of a feasible reference point, where a population moving away from r is a wanted behavior. At convergence the absolute progress value becomes smaller and smaller, leading to higher sampling allocations. The normalized progress-based resampling need x_s^P is calculated as in Equation 4.

$$x_s^P = 1 - \left(\frac{\min\{\overline{P}, P_{max}\}}{P_{max}} \right)^a \tag{4}$$

The disadvantage of Progress-based resampling is that all solutions in the population are assigned the same budget. This means that solutions, population or offspring, which are dominated by many solutions or distant to r and thereby less relevant, will be assigned an unnecessary high number of samples. They might be discarded in the next selection step of the evolutionary algorithm which reduces the benefit of the assigned samples even more. Progress-based resampling without using the distance information to the reference point is often of little use. As soon as the optimization gets stuck in a local optimum far from r, too many samples are wasted, in a situation where less samples and more uncertainty would actually help to escape the local optimum. Progress-based Resampling has been evaluated in [16].

Distance-Based Dynamic Resampling (DDR). The Hybrid Dynamic Resampling strategy Distance-based Dynamic Resampling DDR was proposed in [15]. A summary is given in the following. This paper focuses on the more important case of infeasible reference points, and only the description for this case is given. Infeasible reference points are outside the feasible objective space and cannot be attained by the optimization. Distance-based Resampling requires the use of the factors of progress and time. In the case of an infeasible reference point r the solution at the minimum possible distance to r should be assigned b_{max}, otherwise the decision maker would not be guaranteed full control over the sampling allocation. The time factor is required because in case the optimization gets stuck at an early point in time in a local optimum with a low progress, then no samples should be wasted during this stage. DDR is a sequential sampling algorithm.

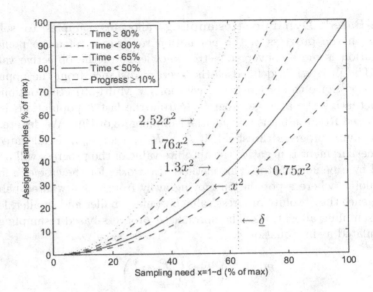

Fig. 1. Sampling allocation of the Distance-based Dynamic Resampling (DDR) algorithm [15]. Acceleration parameter $a = 2$.

If d_s is the distance of solution s to r then the sampling budget is assigned according to $x_s^{DDR} = (1 - d_s)^a$. Distance-based Dynamic Resampling guarantees b_{max} samples for a certain percentage of the solutions in the current population, depending on the average progress $\overline{P}$ in the population. If $10\% > \overline{P} \geq 5\%$ then only the best (hypothetical) solution that is closest to r is allocated b_{max}. If $5\% > \overline{P} \geq 2.5\%$ then the 10% best solutions are allocated b_{max}. $2.5\% > \overline{P} \geq 1\%$ corresponds to 20% of the solutions and $\overline{P} < 1\%$ to 40%. For this purpose the maximum distance $\underline{\delta}$ to r of 40% of the population is calculated and the allocation function is increased by the factor $1/(1 - \underline{\delta})$ (cf. Equation 5). If $\overline{P} > 10\%$ then the closest (hypothetical) solutions are allocated less than b_{max} to be prepared in case r is feasible and b_{max} should only be allocated to solutions dominating r. As mentioned above, the time criterion is used to slow down the allocation. In several steps during the optimization runtime the slowing effect is reduced. This is shown in Figure 1.

$$x_s^{DDR} = \min\left\{1, \left(\frac{1 - d_s}{1 - \underline{\delta}}\right)^a\right\} \tag{5}$$

4   Distance-Rank Dynamic Resampling (DR2)

As an attempt to use several different resampling criteria in a resampling algorithm and to create a hybrid resampling strategy for guided EMO, we propose to combine the Rank-based Dynamic Resampling and the Distance-based Dynamic Resampling (DDR) strategies described previously in Section 3. We call it Distance-Rank Dynamic Resampling (DR2). It uses four different factors to determine the resampling allocation for individual solutions: Pareto-rank, time, reference point distance, and progress. Whereas the elapsed optimization time can be combined with any other resampling criterion with little effort, as seen at Rank-Time-based Dynamic Resampling mentioned above, the Pareto-rank and the reference point distance are two truly different resampling criteria. Therefore we emphasize the term Hybrid for the DR2 algorithm in this paper.

Equation 6 describes the sampling allocation of DR2. Similarly to Rank-Time-based Resampling, the minimum of both the normalized sampling need of Distance-based Resampling and Rank-based Resampling is used to create a combined sampling allocation. However, the Distance-based sampling need is not calculated individually for each solution. Instead, DR2 identifies the solution s_m closest to r and the normalized sampling need for x_m^{DDR} is used in the formula as fixed value for all solutions in the current generation. This way of combining rank and distance information has shown the best optimization results.

$$x_s^{DR2} = \min\left\{x_m^{DDR}, x_s^R\right\} \tag{6}$$

5   Numerical Experiments

In this section the described resampling techniques in combination with R-NSGA-II are evaluated on two benchmark functions.Stepwise, more and more advanced techniques using multiple resampling criteria are compared in different configurations, showing superior results. To facilitate comparison, the experiments are grouped in experiments where no preference information is used for resampling and experiments where the resampling algorithm uses information about the distance to a reference point r defined for R-NSGA-II. For reasons of simplicity only experiments with one reference point are run, even though R-NSGA-II is capable of guiding multiple sub-populations to different reference points in the objective space. Also, the reference point is chosen to be infeasible in order to keep R-NSGA-II and the preference-based resampling algorithms as simple as possible. The combinations of R-NSGA-II with different resampling strategies are tested on two bi-objective benchmark functions ZDT1 and ZDT4 [18]. The ZDT1 function is used for evaluation due to its popularity in the literature. ZDT4 is more difficult to solve and features many local Pareto fronts. This allows a more detailed analysis of the algorithm behavior. The ZDT benchmark functions are deterministic in their original version. In order to create noisy problems, a zero-mean normal distribution is added on both objective functions.

5.1   Problem Settings

The used benchmark functions are deterministic in their original version. Therefore zero-mean normal noise has been added to create noisy optimization problems. The ZDT1 objective functions are for ex. defined as $f_1(x) = x_1 + \mathcal{N}(0, \sigma_1)$ and $f_2(x) = g(x)\left(1 - \sqrt{x_1/g(x)}\right) + \mathcal{N}(0, \sigma_2)$, where $g(x) = 1 + 9\sum_{i=2}^{30} x_i/29$. For the ZDT1 and ZDT4 functions the two objectives have different scales. Therefore the question arises if the added noise should be normalized according to the objective scales. We consider the case of noise strength relative to the objective scale as realistic which can occur in real-world problems, and therefore this type of noise is evaluated in this paper. For the ZDT1 function the relative added noise (5%) is $(\mathcal{N}(0, 0.05), \mathcal{N}(0, 0.5))$ (considering the relevant objective ranges of $[0, 1] \times [0, 10]$), and for ZDT4 it is $(\mathcal{N}(0, 0.05), \mathcal{N}(0, 5))$ (relevant objective ranges $[0, 1] \times [0, 100]$). In the following these problems are called ZDT1-(0.05,0.5) and ZDT4-(0.05, 5).

5.2   Algorithm Parameters

The limited simulation budget is chosen as 2000 solution replications for ZDT1 and 5000 replications for ZDT4. This corresponds to a 1 day optimization runtime on a cluster with 50 computers and a 15 minutes function evaluation time, which could be a realistic real-world optimization scenario. R-NSGA-II is run with a crossover rate $p_c = 0.8$, SBX crossover operator with $\eta_c = 2$, Mutation probability $p_m = 0.07$ and Polynomial Mutation operator with $\eta_m = 5$. The Epsilon clustering parameter is chosen as $\epsilon = 0.001$. This corresponds to a 1 day optimization runtime on a cluster with 50 computers and a 15 minutes function evaluation time, which could be a realistic real-world optimization scenario. R-NSGA-II is run with a crossover rate $p_c = 0.8$, SBX crossover operator with $\eta_c = 2$, Mutation probability $p_m = 0.07$ and Polynomial Mutation operator with $\eta_m = 5$. The Epsilon clustering parameter is chosen as $\epsilon = 0.001$. For ZDT1 and ZDT4 the reference point $r = (0.05, 0.5)$ is used which is close to the Ideal point $(0, 0)$.

Since there is no perfect parameter configuration for Dynamic Resampling algorithms that works well on all optimization problems, we chose one configuration and did not do any parameter optimization. We chose the parameter values that seemed most intuitive to us and used them for all experiments. For all resampling algorithms the minimum budget to be allocated is $b_{min} = 1$ and the maximum budget is $b_{max} = 5$. Static Resampling is run in two configurations with $b_s = 1$ and 2. Time-based Resampling uses uses a linear allocation, $a = 1$. Rank-based Resampling and Rank-Time-based Resampling are run as Max5 Rank-based Resampling and use linear allocation ($a = 1$) for both the rank-based and time-based criteria. Progress-based Dynamic Resampling is not evaluated due to the described disadvantages. Distance-Progress-Time-based Dynamic Resampling DDR uses delayed ($a = 2$) distance-based allocation. Distance-Rank-based Dynamic Resampling DR2 uses the same parameters as the constituting resampling techniques.

5.3 Evaluation, Replication, and Interpolation

In order to obtain a reliable performance measurement the accurate objective values for each evaluated solution are used. For the benchmark problems, either the added noise landscape can be removed and the solution is evaluated deterministically, or the solution can be sampled a high number of times to obtain accurate objective values, which was done in this study. 2500 samples on a benchmark problem solution reduce the uncertainty of the objective values by a factor of 50.

All experiments performed in this study are replicated 10 times and mean performance metric values are calculated. To be able to see the performance development over time a performance metric is evaluated after every generation of the optimization algorithm. This is shown in Figure 2. However, due to the resampling, each generation uses a different number of solution evaluations. Since it is assumed that solution evaluations are equally long and that the runtime of the optimization algorithm and resampling algorithms is negligible compared to the evaluation time, the number of solution evaluations corresponds to the optimization runtime. Therefore, an interpolation is required which calculates the performance metric values at equidistant evaluation number intervals where the mean performance measure values for all experiment replications can be calculated. Not only differ the number of solution evaluations per generation (measurement points) between experiment replications, but also between different experiments with different resampling algorithms of the same optimization problem, which shall be compared.

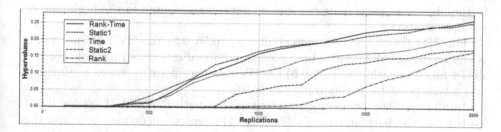

Fig. 2. Focused Hypervolume chart showing the R-NSGA-II progress development over time on ZDT1-(0.05,0.5) for resampling methods that do not rely on preference information. Reference point (0.05, 0.5).

5.4 Focused Hypervolume

To measure and compare the results of the different resampling algorithms together with R-NSGA-II the Focused Hypervolume performance metric for ϵ-dominance based EMOs (F-HV) is used [16]. The F-HV allows to measure the convergence and diversity of a population limited to a preferred area in the objective space. The limits are defined based on the intended diversity and the population size of the optimization algorithm. This allows to measure to which degree the optimization algorithm can achieve the intended diversity. For the R-NSGA-II algorithm the intended diversity is controlled by the user parameter epsilon. F-HV is based on the Hypervolume metric (HV) [19]. In F-HV the population is filtered before it is judged by the HV. The filter is a cylindrical subspace of the objective space retaining only solutions close to the reference point r dominating the HV-reference point. The cylinder axis is defined by r and a second point determining the direction, approximately orthogonal to the potentially not completely known true Pareto-front. For R-NSGA-II and a bi-objective problem, solutions within the distance $d = \epsilon \frac{N}{2}$ from the cylinder axis, with N being the population size, are passed on to the standard HV, the rest is discarded. In this way, within the cylinder there is enough space for N non-dominated solutions, each with the distance ϵ to its neighbors. The cylinder filter must be applied before the non-domination sorting is performed. Otherwise, dominated solutions are filtered out during non-domination sorting which would be non-dominated after the application of the cylinder filter. The F-HV is similar to the R-Metric (R-HV) [8], which is a metric to assess the quality of converged, focused populations on the Pareto-front. It filters the solutions with a box around a representative solution of the population and then projects the remaining solutions on an axis defined by r and HV reference point. Due to the limited optimization time in our experiments, the population will never fully converge towards the Pareto-front, or r. Therefore, the R-HV representative point filter and the shifting operation cannot be applied. Instead, the F-HV is used which filters the solutions close to r, regardless of the population position in the objective space. This can lead to zero metric values in the beginning of the optimization.

For ZDT1 with the reference point $r = (0.05, 0.5)$ the HV reference point is chosen as $(0.1, 1.1)$ and a base point for normalization as $(0, 0.68)$. For the ZDT4 function with $r = (0.05, 0.5)$ the HV reference point is chosen as $(0.1, 30)$ and a base point for normalization as $(0, 0)$.

In all cases, the population size 50 together with R-NSGA-II epsilon 0.001 leads to a cylinder diameter of 0.05. The cylinder axis is defined by r and the direction point. For ZDT1 and this direction point is defined as $(0.06, 1.1)$ and for ZDT4 as $(0.06, 30)$.

5.5   Resampling Without Preference Information

In this section the resampling algorithms from Section 3.1 are evaluated and compared: Static Resampling, Time-based Dynamic Resampling, Rank-based and Rank-Time-based Dynamic Resampling.

In Figure 2 the results of the different resampling techniques together with R-NSGA-II are evaluated on the ZDT1-(0.05,0.5) problem with reference point (0.05, 0.5) and 2000 function evaluations. The results show that Static Resampling with 1 sample is both better than Time-based and Rank-based Dynamic Resampling. Static2-Resampling is worse than Static1-Resampling, which shows

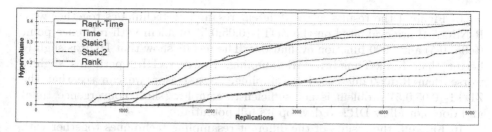

Fig. 3. Focused Hypervolume chart showing the R-NSGA-II progress development over time on ZDT4-(0.05,5) for resampling methods that do not rely on preference information. Reference point (0.05, 0.5).

that Dynamic Resampling is required to achieve a performance gain over the Static1 strategy. This is achieved by the hybrid strategy Rank-Time-based Resampling which outperforms all others.

In Figure 3 the results of the different resampling techniques together with R-NSGA-II are evaluated on the ZDT4-(0.05,5) problem with reference point (0.05, 0.5) and 5000 function evaluations. The results confirm the results shown in Figure 2, however, the differences between the different resampling algorithms become more clear, since the ZDT4-(0.05,5) problem is more difficult (many local Pareto-fronts) and needs more time to converge to the reference point.

5.6   Resampling with Reference Points

In this section the resampling algorithms from Section 3.2 are evaluated and compared: Distance-Progress-Time-based Dynamic Resampling DDR, and Distance-Rank(-Progress-Time)-based Dynamic Resampling DR2. The results for Rank-Time-based Resampling are included for comparison purposes.

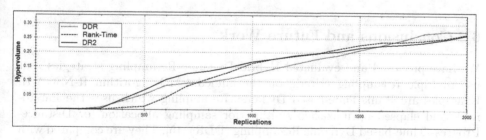

Fig. 4. Focused Hypervolume chart showing the R-NSGA-II progress development over time on ZDT1-(0.05,0.5) for resampling methods that use preference information. Reference point (0.05, 0.5). For comparison with the non-preference methods, the curve for Rank-Time-based resampling from Figure 2 is included.

In Figure 4 the results of the different resampling techniques together with R-NSGA-II are evaluated on the ZDT1-(0.05,0.5) problem with reference point (0.05, 0.5) and 2000 function evaluations. The results show that DDR is slightly better than Rank-Time-based Resampling. However, DR2 performs slightly worse than Rank-Time-based Resampling. As a reason we can see that the ZDT1-(0.05,0.5) problem is not sufficiently complex (short convergence time) and does not allow DR2 to develop its full potential.

In Figure 5 the results of the different resampling techniques together with R-NSGA-II are evaluated on the ZDT4-(0.05,5) problem with reference point (0.05, 0.5) and 5000 function evaluations. Since the the ZDT4-(0.05,5) problem is more difficult it allows for a more clear evaluation. Here, it can be seen very clearly that DR2 outperforms Rank-Time-based Resampling, and thereby all other resampling algorithms evaluated in Figure 3. DDR however, shows not to be very powerful on this problem. Yet, combined with the Pareto-rank criterion as DR2, it is superior to all others.

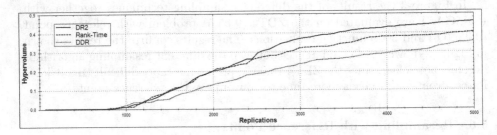

Fig. 5. Focused Hypervolume chart showing the R-NSGA-II progress development over time on ZDT4-(0.05,5) for resampling methods that use preference information. Reference point (0.05, 0.5). For comparison purposes, the curve for Rank-Time-based resampling from Figure 3 is included.

6   Conclusions and Future Work

We have proposed and evaluated Hybrid Dynamic Resampling strategies that use multiple resampling criteria on the guided EMO algorithm R-NSGA-II. Examples are Rank-Time-based Dynamic Resampling which uses the Pareto-rank and elapsed optimization runtime for sampling allocation, or Distance-Progress-Time-based Dynamic Resampling (DDR) [15]. They are compared with resampling techniques that base their sampling allocation on a single criterion, like Time-based Dynamic Resampling or Rank-based Dynamic Resampling. The results on benchmark functions and a reference point close to the Ideal point show that Hybrid Dynamic Resampling techniques are superior to single-criterion techniques and Static Resampling, given that the optimization problem is sufficiently complex. Furthermore, we proposed and evaluated a resampling algorithm that uses both the Pareto-rank and Reference point distance as a basis for sampling allocation. Both these criteria are used by the R-NSGA-II algorithm as fitness functions. Thus, we expected that the Distance-Rank Dynamic Resampling algorithm (DR2) is able to support the R-NSGA-II algorithm better than previous resampling algorithms that only consider one of the criteria, which we could prove in numerical benchmark experiments.

Future Work will cover the following studies:

- A future task will be to study the combination of a resampling algorithm that uses the objective variance and Distance-based Dynamic Resampling. Such a resampling strategy based on variance is Multi-objective Standard Error Dynamic Resampling [15].
- The resampling algorithms in this paper that are based on the Pareto-rank base their sampling allocation on a comparison of solutions. They have thereby an advantage over resampling algorithms that treat each solution individually. Slightly modified, the comparison approach could support the evolutionary optimization algorithm in comparing solutions for selection decisions, also called Selection Sampling. A study investigating the effect of Selection Sampling on guided EMO of stochastic systems will be performed.
- A parametric study will be performed that identifies guidelines for parameter configuration for different problems characteristics.
- A worthwhile future task will be to extend and evaluate the resampling and optimization algorithms for scenarios with feasible reference points.
- Extensions for existing EMO algorithms for guided search need to be proposed that allow for faster convergence to the preferred objective space area.

Acknowledgments. This study was partially funded by VINNOVA, Sweden, through the FFI-HSO project. The authors gratefully acknowledge their provision of research funding.

References

1. Bartz-Beielstein, T., Blum, D., and Branke, J.: Particle swarm optimization and sequential sampling in noisy environments. Metaheuristics - Progress in Complex Systems Optimization, 261–273 (2007)
2. Branke, J., Gamer, J.: Efficient sampling in interactive multi-criteria selection. In: Proceedings of the 2007 INFORMS Simulation Society Research Workshop, 42–46 (2007)
3. Branke, J., Schmidt, C.: Sequential Sampling in Noisy Environments. In: Yao, X., Burke, E.K., Lozano, J.A., Smith, J., Merelo-Guervós, J.J., Bullinaria, J.A., Rowe, J.E., Tiňo, P., Kabán, A., Schwefel, H.-P. (eds.) PPSN 2004. LNCS, vol. 3242, pp. 202–211. Springer, Heidelberg (2004)
4. Chen, C.H., He, D., Fu, M., Lee, L.H.: Efficient Simulation Budget Allocation for Selecting an Optimal Subset. Informs Journal on Computing **20**(4), 579–595 (2008)
5. Deb, K.: Multi-Objective Opimization using Evolutionary Algorithms. John Wiley & Sons (2001)

6. Deb, K., Agrawal, S., Pratap, A., Meyarivan, T.: A fast and elitist multi-objective genetic algorithm: NSGA-II. IEEE Transactions on Evolutionary Computation **6**(2), 182–197 (2002)

7. Deb, K., Sinha, A., Korhonen, P.J., Wallenius, J.: An interactive evolutionary multi-objective optimization method based on progressively approximated value functions. IEEE Transactions on Evolutionary Computation **14**(5), 723–739 (2010)

8. Deb, K., Siegmund, F., Ng, A.H.C.: R-HV : A metric for computing hyper-volume for reference point based EMOs. Accepted for publication at the International Conference on Swarm, Evolutionary, and Memetic Computing 2014, Bhubaneswar, Odisha, India (2014)

9. Deb, K., Sundar, J., Bhaskara Rao, N.U., Chaudhuri, S.: Reference point based multi-objective optimization using evolutionary algorithms. International Journal of Computational Intelligence Research **2**(3), 273–286 (2006)

10. Di Pietro, A.: Optimizing Evolutionary Strategies for Problems with Varying Noise Strength. University of Western Australia, Perth, PhD-thesis (2007)

11. Di Pietro, A., While, L., and Barone, L.: Applying evolutionary algorithms to problems with noisy, time-consuming fitness functions. Congress on Evolutionary Computation 2004, vol. 2, 1254–1261 (2004)

12. Jin, Y., and Branke, J.: Evolutionary optimization in uncertain environments - a survey. IEEE Transactions on Evolutionary Computation **9**(3), pp. 303–317 (2005). ISSN 1089–778X

13. Siegmund, F., Bernedixen, J., Pehrsson, L., Ng, A.H.C., Deb, K.: Reference point-based Evolutionary Multi-objective Optimization for Industrial Systems Simulation. In: Proceedings of the Winter Simulation Conference 2012, Berlin, Germany (2012). ISBN 978-1-4673-4781-5

14. Siegmund, F., Ng, A. H.C., Deb, K.: Finding a preferred diverse set of Pareto-optimal solutions for a limited number of function calls. In: Proceedings of the IEEE Congress on Evolutionary Computation 2012, Brisbane, Australia, pp. 2417–2424 (2012). ISBN 978-1-4673-1508-1

15. Siegmund, F., Ng, A. H.C., Deb, K.: A Comparative Study of Dynamic Resampling Strategies for Guided Evolutionary Multi-Objective Optimization. In: Proceedings of the IEEE Congress on Evolutionary Computation 2013, Cancún, Mexico, pp. 1826–1835 (2013). ISBN 978-1-4799-0454-9

16. Siegmund, F., Ng, A. H.C., Deb, K.: Dynamic Resampling Strategies for Guided EMO of Stochastic Systems - Part1. European Journal of Operational Research - EJOR, in preparation for submission (2015)

17. Stump, G., Simpson, T. W., Donndelinger, J.A., Lego, S., Yukish, M.: Visual Steering Commands for Trade Space Exploration: User-Guided Sampling With Example. Journal of Computing and Information Science in Engineering **9**(4), pp. 044501:1–10 (2009). ISSN 1530–9827

18. Zitzler, E., Deb, K., Thiele, L.: Comparison of multiobjective evolutionary algorithms: Empirical results. Evolutionary Computation **8**(2), 173–195 (2000)

19. Zitzler, E., Thiele, L.: Multiobjective Optimization Using Evolutionary Algorithms - A Comparative Case Study. Parallel Problem Solving from Nature V, 292–301 (1998)

A Comparison of Decoding Strategies for the 0/1 Multi-objective Unit Commitment Problem

Sophie Jacquin[2,3](✉), Lucien Mousin[3], Igor Machado[1,3],
El-Ghazali Talbi[2,3], and Laetitia Jourdan[2,3]

[1] University of Flumisende, Rio de Janeiro, Brazil
[2] DOLPHIN Project-Team, Inria Lille - Nord Europe, Lille, France
sophie.jacquin@inria.fr
[3] LIFL, UMR CNRS, Université Lille 1, 8022 Lille, France

Abstract. In the single objective Unit Commitment Problem (UCP) the problem is usually separated in two sub-problems : the commitment problem which aims to fix the on/off scheduling of each unit and the dispatching problem which goal is to schedule the production of each turned on unit. The dispatching problem is a continuous convex problem that can easily be solved exactly. For the first sub-problem genetic algorithms (GA) are often applied and usually handle binary vectors representing the solutions of the commitment problem.Then the solutions are decoded in solving the dispatching problem with an exact method to obtain the precise production of each unit. In this paper a multi-objective version of the UCP taking the emission of gas into account is presented. In this multi objective UCP the dispatching problem remains easy to solve whereas considering it separatly remains interesting. A multi-objective GA handling binary vectors is applied. However for a binary representation there is a set of solutions of the dispatching problem that are pareto equivalent. Three decoding strategies are proposed and compared. The main contribution of this paper is the third decoding strategy which attaches an approximation of the Pareto front from the associated dispatching problem to each genotypic solution. It is shown that this decoding strategy leads to better results in comparison to the other ones.

Keywords: UCP · Metaheuristics · Heuristic · Multi-objective optimization

1 Introduction

The UCP is used to find the scheduling of commissioning and production of generating units that minimizes the production cost. However, environmental protection has become a major issue. In response to the growth of the negative impacts on the environment and due to the growing importance of environmental interest in society, governments have developed and implemented laws or

A. Gaspar-Cunha et al. (Eds.): EMO 2015, Part I, LNCS 9018, pp. 381–395, 2015.
DOI: 10.1007/978-3-319-15934-8_26

technical standards in order to reduce the negative impacts of human activity on the environment. For this reason new modeling of the UCP taking into account some limitation constraints on gas emissions [2] has been proposed. A bi-objective model has also been proposed, but is usually solved by reducing the two objectives to one, using the weighted sum approach [13, 17]. In [15], a multiobjective version of the UCP is solved with a genetic algorithm, but it does not directly exploit the concept of Pareto dominance. The authors applied a classical genetic algorithm, with the exception that the selection process is based on a specific version of the tournament selection. Two individuals are randomly selected from the population and a stochastic competition of the objective that are chosen randomly, is performed to determine the winner that will survive into the next generation.

The UCP can be separated into two sub-problems. The commitment problem which is to give the on/off scheduling of each unit and then the dispatching problem which is to give the exact production for each turned on unit. In the single objective case, the dispatching problem is a quadratic continuous problem which is easy to solve exactly. For this reason evolutionary algorithms generally handle binary vectors giving the on/off scheduling of each unit to solve the single-objective UCP [3, 7, 8, 10, 16]. Then to obtain a complete description of the solution (the phenotypic solution), the production of each turned on unit is determined optimally using a λ-iteration method [14]. This binary representation takes advantage of a real one because the search space is considerably reduced. For the multi-objective the dispatching problem remains interesting to exploit separately because it has still good properties. All the solutions of the dispatching problem are supported and can be found by the scalarizing method. This method is to transform the problem with two objectives f_1 and f_2 into a single objective problem optimizing the function $\lambda f_1 + (1 - \lambda)f_2$, where $\lambda \in [0, 1]$. The quality of the results found in the single objective case and the fact that the multiobjective version of the dispatching problem has good properties lead us to a specific interest in considering a similar two level method of resolution for the multi-objective UCP.

In this paper a multi-objective GA based on NSGA-II [4] is proposed. This GA handle binary vectors representing the solutions of the commitment problem. Then the phenotypic solutions corresponding to the production of each unit has to be find by a decoding method. Since the dispatching problem is also a multi-objective problem there are many candidates of phenotypic solutions for one genotypic solution. Therefore three different decoding methods are proposed and compared. The first one is to construct the phenotypic solution by solving the dispatching problem by optimizing the function $f_1 + f_2$. The second is to add a real λ into the genotypic representation of a solution. The phenotypic solution is decoded by solving the dispatching problem minimizing $\lambda f_1 + (1 - \lambda)f_2$. The third is to associate a set of Pareto equivalent solutions of the corresponding multi-objective dispatching problem to each genotypic solution. For the last decoding system, the process of fitness and diversity assignment of NSGA-II has to be adapted. The fact that many phenotypic solutions are attached to a single genotypic solution must be taken into account. The main contribution of this paper is to show that the multi decoding embedded approach has an advantage over the two other less complex decoding systems that are proposed.

The paper is organized as follows. In the next section the multi-objective UCP is described in details. Then the three solving methods corresponding to the three binary genetic algorithms using the different decoding strategies are presented. Finally the experimental process and the obtained results are presented and discussed before the conclusion section.

2   Multi-objective Unit Commitment Problem - MO-UCP

In this section, the MO-UCP is presented in details. This problem is the same as the classical UCP but an objective is added to take into account the gas emission of SO_2 and CO_2.

2.1   Unit Commitment Problem

Unit Commitment Problem (UCP) [8] is to schedule generating units online or offline over a scheduling horizon. The goal is to minimize the power production cost while satisfying a set of operational constraints. The production cost includes the fuel and start-up costs. Constraints are capacity of production of each units, minimum up/down time and spinning reserve. UCP is usualy modeled as as a mixed integer non-linear problem. It consists of binary variables $u_{i,t}$ that takes value 1 if a unit i is turned on at time t and 0 otherwise, continuous variables $p_{i,t}$ that denotes their prodduction amounts. It is a very complex problem to solve because of its enormous dimension, a non-linear objective function, and time-dependent constraints. Indeed, the UCP is a NP-complete problem [6].

2.2   Multi-objective UCP

The first objective is the same as in single-objective UCP. It minimizes the cost of production. This production cost is divided into two components, the fuel cost and the start up cost. For a system of N units and a time horizon of T periods, the objective function can be described as follows:

$$f_1(u,p) = \sum_{t=1}^{T} \sum_{i=1}^{N} FC_i(p_{i,t}) \times u_{i,t} + CS_i(T_{i,t-1}^{off}) \times (1 - u_{i,t-1})u_{i,t},$$

where:

- FC_i is the fuel cost function of the unit i, which is modeled by a quadratic function:

$$FC(p_{i,t}) = a_{1,i} + a_{2,i} \times p_{i,t} + a_{3,i} \times p_{i,t}^2,$$

where $a_{1,i}$, $a_{2,i}$ and $a_{3,i}$ are real cost coefficients for the unit i.

– CS_i is the start-up cost for unit i, which depends on time $T_{i,t-1}^{off}$ the unit i has been turned off at time $t-1$:

$$CS_i(T_{i,t-1}^{off}) = \begin{cases} CS_{cold} & if \ T_{i,min}^{off} + T_{cs,i} \leq T_{i,t-1}^{off} \\ CS_{hot} & else \end{cases},$$

where $T_{i,min}^{off} + T_{cs,i}$ is the time it takes the unit i to become cold.

The second objective function measures the SO_2 and CO_2 emissions:

$$f_2(p,u) = \sum_{t=1}^{T} \sum_{i,u_{i,t}=1} b_{0,i} + b_{1,i}p_{i,t} + b_{2,i}p_{i,t}^2$$

The coefficients $b_{0,i}$, $b_{1,i}$, $b_{2,i}$ used in this paper are the ones proposed in [13].

The minimization of the objectives of the UCP is subject to the following system and unit constraints:

1. Power balance constraints:

$$\sum_{i=1}^{N} p_{i,t}u_{i,t} = D_t \quad \forall t,$$

where D_t is a real number giving the load demand at time t.

2. Spinning reserve constraints:

$$\sum_{i=1}^{N} p_{i,max}u_{i,t} \geq D_t + R_t \quad \forall t,$$

where R_t is a real giving the minimal reserve at time t.

3. Unit output constraints:

$$p_{i,min} \leq p_{i,t} \leq p_{i,max} \quad \forall t,$$

where $p_{i,min}$ and $p_{i,max}$ are the lower and upper bounds on the energy production of unit i respectively.

4. Minimum up time limit:

$$T_{i,t-1}^{on} \geq T_{i,min}^{on} \times (1 - u_{i,t})u_{i,t-1} \quad \forall t,$$

where $T_{i,t-1}^{on}$ is the time from which the unit i is turned on at time $t-1$ and $T_{i,min}^{on}$ is the minimal time during which unit i has to stay turned on.

5. Minimum down time limit:

$$T_{i,t-1}^{off} \geq T_{i,min}^{off} \times (1 - u_{i,t-1})u_{i,t} \quad \forall t,$$

where $T_{i,t-1}^{off}$ is the time from which the unit i is turned off at time $t-1$ and $T_{i,min}^{off}$ is the minimal time during which unit i has to stay off.

The feasible outcome vectors of the *objective space* are compared using the *Pareto dominance* $\succ$. In this minimization context, a solution $x \in \Omega$ is said to dominate a solution $y \in \Omega$, denoted by $x \succ y$, if they satisfy relation (1).

$$\forall i \in \{1,2\}, \ f_i(x) \leq f_i(y) \bigwedge \exists i \in \{1,2\}, \ f_i(x) < f_i(y) \tag{1}$$

3   Solving Methods

Many evolutionary algorithms involving a decoding system have been proposed to solve the single objective version of the UCP [3,7,8,10,16]. A binary vector u of size $T \times N$ is used to represent the solutions. In this representation each $u_{i,t}$ gives the state of a unit i (on or off) at a given time period t of the scheduling. Then the exact production of each unit is decoded by solving the dispatching sub-problem $(\mathscr{D}(u))$ using the lambda-iteration method [14]:

$$\min_{p} \sum_{t=1}^{T} \sum_{\substack{i \\ \text{s.t } u_i=1}} f_1((p_{i,t})_i)$$

such that:

$$\sum_{\substack{i \\ \text{s.t } u_i=1}} p_{i,t} = D_t \tag{2}$$

$$p_{i,min} \leq p_{i,t} \leq p_{i,max} \quad \forall i \text{ s.t } u_i = 1 \tag{3}$$

The advantage of the binary vector representation over a real vector representation where the productions are directly given is obvious: the search space is considerably reduced. In the multi-objective case, the dispatching problem is also a multi objective one as f_1 and f_2 have to be minimized. But this sub-problem has still very good properties that make it easy to solve. It is a convex and continuous bi-objective problem. As the objective functions are convex and the decision variables are defined in a convex set all the solutions are supported [5] . Then the Pareto front solution is convex and totally defined by the set:

$$\{f_{1,\lambda}, f_{2,\lambda} | \lambda \in [0,1]\},$$

where $f_{1,\lambda} = f_1(p*_\lambda)$ and $f_{1,\lambda} = f_1(p^*_\lambda)$, with p^*_λ solution of the dispatching problem $\mathscr{D}(u, \lambda)$ defined as follows:

$$p^*_\lambda = arg(\min_{p} \sum_{t=1}^{T} \sum_{\substack{i \\ \text{s.t } u_i=1}} \lambda f_1((p_{i,t})_i)) + (1-\lambda)f_2((p_{i,t})_i))$$

such that (2) and (3) are met.

It seems interesting to consider a two level based method in the multi-objective version which will be similar to those proposed for the single-objective UCP. The method proposed is a GA, NSGA-II, handling binary vectors representing the solutions of commitment problem. Then the production of the units are obtained by using a decoding method. However, since the dispatching problem is a multi-objective one, there are many pareto equivalent solution to fix the production values. It becomes difficult to choose a method to associate phenotypic solutions with genotypic ones. Then three approaches of decoding are proposed and compared. In the first approach the solution associated with a binary representation is the one obtained by solving the dispatching problem for

λ fixed to 0.5. This approach is a naive one because it might miss some possibly good solutions. The second approach consists of adding λ to the representation of an individual. Then a binary vector associated with an on/off scheduling can be present many times in the population with different value of λ and all the solutions are reachable. In the last approach, an approximation of the Pareto front of the dispatching sub-problem is associated with each individual. Therefore an adapted version of NSGA-II is proposed to manage the association of many phenotypic solutions to a single genotypic solution.

In the next subsection, the common components of the three methods, which are essentially evolutionary operators, are presented. Then each method are explained in details.

3.1 Common Components

Each multi-objective GA proposed hereafter are based on NSGA-II [4]. In each case the following operators are used:

Crossovers: Two crossovers are used. The first one is the classical one-point crossover. The second one is an intelligent two-points crossover. The principle is to randomly choose a window size and if the window size is smaller than the remaining portion of the solution, a new individual is created with the window portion of the worst parent and the remaining portion from the best parent. The reverse is done if the window size is larger than the number of genes in the remaining solution. In Figure 1 the size of the selected window S_W is larger than the remaining solution, then the offspring obtained C is composed with the window portion of the best parent P_2 and the remaining portions of P_1.

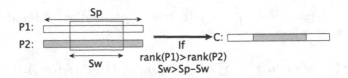

Fig. 1. Intelligent 2-points crossover between P_1 and P_2

Mutations: Two mutations are used. The first one is the standard 1-bit-flip-mutation. This operator randomly flips a bit of the vector with a low probability. The second one randomly chooses a window whose size is randomly chosen and flips all the bits of this window.

Repair Operator: The aim of this operator is to correct a solution if it does not meet the constraints of demand or of minimum on/off time. Firstly, it corrects the violations of time constraint hour by hour in modifying the states of unities if necessary. Then corrections are done on the constraints of power balance.

This is done hour-by-hour. If the maximal capacity of production of the turned on units is lower than the power balance then a unit is turned on. The unit to turn on is chosen randomly among the units that can be turned on depending on the past hour (i.e. if this unit was turned on the previous hour or if it was turned off for a time long enough). The reverse process is done if minimal capacity of the turned on units exceeds the load demand. Naturally this correction process does not guarantee obtaining a feasible solution but this operator speeds up the algorithm and hence increases the possibility of finding such solutions.

Objective Function: The objective values are computed by using the exact production $p_{i,t}$ at each time period t and for each unit i which is a phenotypic solution. The way to obtain phenotypic solutions from genotypic ones is different for each approach. The objective functions correspond to f_1, the cost production and f_2, the quantity of SO_2 and CO_2 emission. Nevertheless some penalties have to be added on the violation of time and load constraints. Therefore the objectives are:

$$\begin{cases} obj_1 = f_1 + c_p \sum_{t=1}^{T} (\sum_{i=1}^{N} p_{i,t} - D_t) + c_p(T_{i,off}^{t-1}(u_{i,t} - x_{i,t-1}) + T_{i,on}^{t-1}(u_{i,t-1} - u_{i,t})) \\ obj_2 = f_2 + c_p \sum_{t=1}^{T} (\sum_{i=1}^{N} p_{i,t} - D_t) + c_p(T_{i,off}^{t-1}(u_{i,t} - u_{i,t-1}) + T_{i,on}^{t-1}(u_{i,t-1} - u_{i,t})), \end{cases}$$

$$(4)$$

where c_p is a constant positive number.

3.2 Naive Approach

In this approach a genotypic solution is decoded in solving a dispatching problem that is reduced to a single objective one by scalarization. Then it is possible that some pareto optimal phenotypic solutions cannot be reached because they are not solution of the chosen scalarized sub-problem. The decoding process is explained in detail hereafter.

Decoding Process: The phenotypic solution $(p_{i,t})_{i,t}$ associated to a genotypic is computed by solving the dispatching problem $\mathcal{D}(u, 0.5)$ thank to the λ-iteration method.

3.3 Scalarized Decoding

In this approach an integer ($\lambda_u \in [\![0, 100]\!]$) is added to the genotypic representation of a solution. This value is chosen randomly during the initialization process and then can be modified by the evolutionary operators. It is used to define the coefficients of scalarization during the decoding process. Hence, the main difficulty of the previous approach becomes possible to overcome, which is the inaccessibility of some pareto optimal solutions. Actually as the pareto front solution of the dispatching problem will always be convex all the pareto optimal solutions are reatchable by the scalarization method.

Representation: A value λ_u is added at the end of the representation vector with a binary encoding. λ_u is an integer between 0 and 100. The representation is shown Fig. 2.

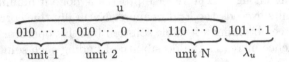

$$\overbrace{\underbrace{010 \cdots 1}_{\text{unit 1}} \quad \underbrace{010 \cdots 0}_{\text{unit 2}} \quad \cdots \quad \underbrace{110 \cdots 0}_{\text{unit N}}}^{u} \quad \underbrace{101 \cdots 1}_{\lambda_u}$$

Fig. 2. Representation genotypic of a solution: the scalar λ is included in the representation

Decoding Process: The phenotypic solution $(p_{i,t})_{i,t}$ associated to a genotypic is computed in solving the dispatching problem $\mathscr{D}(u, \frac{\lambda_u}{100})$ thank to the λ-iteration method.

Adaptation on the Crossover: The one point crossover is transformed in a two-points crossover with the first crossover point in a locus of the definition of u and the second point in a locus of definition of λ_u. It is done in order that the crossover can have a significant impact on λ_u.

Adaptation on the Mutation: The 1-bit-flip-mutation is applied only on the bits corresponding to u. λ_u is mutated by being replaced by a value chosen randomly between 0 and 100 with a normal distribution centered on its original value.

3.4   Multi Decoding Embedded Approach

In this approach a genotypic solution is associated with a set of phenotypic solutions. This set of solutions is from the optimal pareto front solution of the dispatching problem associated with the genotypic solution. Fig. 3 helps

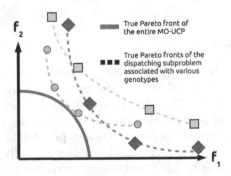

Fig. 3. Repressentation of the genotypic solutions in the objective space

to understand how genotypic solutions are represented in the objective space. The continuous line is the optimal pareto front of the entire MO-UCP. This front can be non-convex. Each one of the convex pareto fronts composed of the round, square or diamond points is derived from a single genotypic solution. They are composed of the phenotypic solutions found in solving the dispatching problem defined by the corresponding genotypic solution. The phenotypic solutions attached to a genotypic solution are pareto equivalent and form a convex front.

Decoding Process: In this case a pareto set P_u will be associated with a genotypic solution u. This set is:

$$P_u = \{p_u^{\frac{k}{n_\lambda}}, k = 0...n_\lambda\}, \tag{5}$$

where n_λ is a fixed integer and $p_u^{\frac{k}{n_\lambda}}$ the optimal solution of $\mathscr{D}(u, \frac{k}{n_\lambda})$.

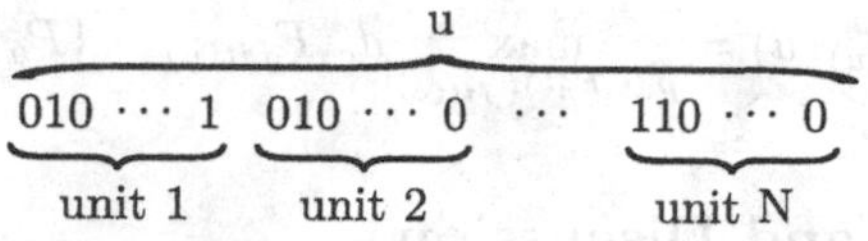

$$
\overbrace{\underbrace{010 \cdots 1}_{\text{unit 1}} \quad \underbrace{010 \cdots 0}_{\text{unit 2}} \quad \cdots \quad \underbrace{110 \cdots 0}_{\text{unit N}}}^{u}
$$

Fig. 4. representation genotypic of a solution u

$$p_u^0 : \underbrace{0p_{1,2}^0 0 \cdots p_{1,T}^0}_{\text{unit 1}} \underbrace{0p_{2,2}^0 0 \cdots 0}_{\text{unit 2}} \cdots \underbrace{p_{N,1}^0 p_{1,2}^0 0 \cdots 0}_{\text{unit N}}$$

$$p_u^{\frac{1}{3}} : \underbrace{0p_{1,2}^{\frac{1}{3}} 0 \cdots p_{1,T}^{\frac{1}{3}}}_{\text{unit 1}} \underbrace{0p_{2,2}^{\frac{1}{3}} 0 \cdots 0}_{\text{unit 2}} \cdots \underbrace{p_{N,1}^{\frac{1}{3}} p_{1,2}^{\frac{1}{3}} 0 \cdots 0}_{\text{unit N}}$$

$$p_u^{\frac{2}{3}} : \underbrace{0p_{1,2}^{\frac{2}{3}} 0 \cdots p_{1,T}^{\frac{2}{3}}}_{\text{unit 1}} \underbrace{0p_{2,2}^{\frac{2}{3}} 0 \cdots 0}_{\text{unit 2}} \cdots \underbrace{p_{N,1}^{\frac{2}{3}} p_{1,2}^{\frac{2}{3}} 0 \cdots 0}_{\text{unit N}}$$

$$p_u^1 : \underbrace{0p_{1,2}^1 0 \cdots p_{1,T}^1}_{\text{unit 1}} \underbrace{0p_{2,2}^1 0 \cdots 0}_{\text{unit 2}} \cdots \underbrace{p_{N,1}^1 p_{1,2}^1 0 \cdots 0}_{\text{unit N}}$$

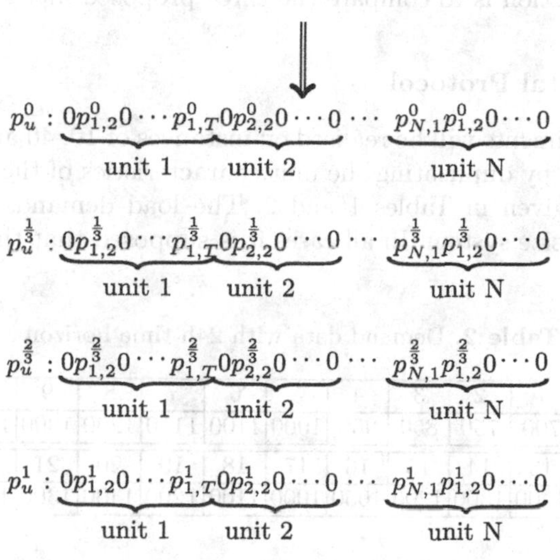

Fig. 5. representation phenotypic of a solution u for $n_\lambda = 3$

As many phenotypic solutions are attached to a single genotypic solution, the fitness assignment and diversity assignment methods of NSGA-II have to be adapted. This will be explained in detail in the following. The decoding process is represented in Fig. 4 and Fig. 5.

Adaptation of the Fitness Assignment Process: The fitness value assigned to a solution u is the best fitness value among the fitness values of the phenotypic solutions $p_u \in P_u$:

$$fit(u) = \underset{p_u \in P_u}{\text{opt}} \left(fit(p_u) \right) \tag{6}$$

In NSGA-II the fitness is the rank of the solution, then opt is the minimization operator. This process ensures that the genotypic solution from which a pareto optimal solution can be generated is not be discarded.

Adaptation of the Diversity Assignment Process: Let F_i be the set of the phenotypic solutions of rank i. In NSGA-II, the diversity measure used is the crowding distance between a solution x and the set of the other solutions having the same fitness, i.e. the same rank: $d_c(F_{fit(x)} - \{x\}, x)$. The adapted diversity assignment is the maximal diversity measurement among the ones of the individuals $p_u \in P_u \cap F_{fit(u)}$ computed without considering the elements of P_u:

$$d_c(F_{fit(u)}, u) = \underset{p_u \in P_u \cap F_{fit(u)}}{\max} d_c(F_{fit(u)} - \{P_u\}, p_u) \tag{7}$$

4   Experiments and Discussion

The aim of this section is to compare the three proposed methods.

4.1   Experimental Protocol

Instances: Experiments will be realised on instances of 10, 40 and 100-unit data that are generated by duplicating the unit characteristics of the ten-unit system and the demand given in Tables 1 and 2. The load demands are adjusted in proportion to the size system. In all cases it is supposed that the reserve is 10% of the demand.

Table 2. Demand data with 24h time horizon

hour	1	2	3	4	5	6	7	8	9	10	11	12
demand (MW)	700	750	850	950	1000	1100	1150	1200	1300	1400	1450	1500

hour	13	14	15	16	17	18	19	20	21	22	23	24
demand (MW)	1400	1300	1200	1050	1000	1100	1200	1400	1300	1100	900	800

Performance Assessment: The different methods of performance assessment that can be chosen to compare multi-objectives algorithms are explained in details in [9]. In our case the ε-indicator and the hypervolume difference indicator are selected as they are complementary. Let Z^{all} be the set of objective vectors from all the Pareto set approximations we obtained during all our experiments. Then, a reference set R contains the non-dominated points of Z^{all}.

Table 1. Generating unit data for the ten-unit base system

unit	unit1	unit2	unit3	unit4	unit5	unit6	unit7	unit8	unit9	unit10
P_{max}(MW)	455	455	130	130	162	80	85	55	55	55
P_{min}(MW)	150	150	20	20	25	20	25	10	10	10
a_1	1000	970	700	680	450	370	480	660	665	670
a_2	16.19	17.26	16.6	16.5	19.7	22.26	27.74	25.92	27.27	27.79
a_3 ($\times 10^{-5}$)	48	31	200	211	398	712	79	413	222	173
b_1	712	570	700	860	350	370	480	660	665	670
b_2	12.9	10.26	10.60	15.50	7.70	9.26	3.74	5.92	7.27	7.79
b_3($\times 10^{-4}$)	4	3	22	11	10	22	30	40	13	23
T_{min}^{up}	8	8	5	5	6	3	3	1	1	1
T_{min}^{down}	8	8	5	5	6	3	3	1	1	1
CS_{hot}	4500	50000	550	560	900	170	260	30	30	30
CS_{cold}	9000	10000	1100	1120	1800	340	520	60	60	60
T_{CS}	5	5	4	4	4	2	2	0	0	0
initial status (h)	8	8	-5	-5	-6	-3	-3	-1	-1	-1

ε-*indicator* $I_{\varepsilon+}^1$. The unary version of this indicator is computed using the binary version given by (8) and the reference set R, with $I_{\varepsilon+}^1(A) = I_{\varepsilon+}(A, R)$.

$$I_{\varepsilon+}(A, B) = \inf_{\varepsilon \in \mathbb{R}} \{\forall z^1 \in B, \exists z^2 \in A, \forall i \in 1 \ldots n, z_i^1 \leq \varepsilon + z_i^2\} \qquad (8)$$

Hypervolume difference indicator I_H^-. The hypervolume indicator I_H is computed by the measure of the hypervolume between a set of solutions and the point $z = (z_1, \ldots, z_n)$ where z_k is the upper bound of the kth objective regarding all the solutions of Z^{all}. The hypervolume difference indicator I_H^- is then computed with $I_H^-(A) = I_H(R) - I_H(A)$.

Experimental Design: All the implementations are realized under the ParadisEO 2.0 [11] software framework. A sensitivity analysis is carried out for each algorithm to determine the effect of the crossover rate and of the mutation rate. It is done thanks to the R statistical package *Irace* [12]. The population size is fixed to 100 individuals. A convergence criteria of 100 generations without improvement of the hypervolume is used as stopping criteria. For the multi decoding embedded approach the parameter n_λ is fixed to 10. For each case 20 runs are launched for each decoding system using the same seeds. Most of the performance assessment procedures are next achieved using PISA [1] platform and its performance assessment module. The existence of a significant difference between the result obtained by the different decoding systems is verified with the *Friedman* statistical test. Then a post-hoc test is carried out to compare the decoders by pairs. A p-value lower than 0.005 is used as a criterion for rejecting the null hypothesis.

Decoder	Naive	Scalarized	Multi
Naive	-	=	<
Scalarized	=	-	<
Multi	>	>	-

Fig. 6. Results optained from statistical comparisons with the ε-indicator and the hypervolume indicator

Indicator	$I_{\varepsilon+}^1$		I_H^-	
Decoder	best	mean	best	mean
Naive	0.736	0.738	0.451	0.455
Scalarized	0.719	0.738	0.433	0.454
Multi	**0.709**	**0.712**	**0.422**	**0.425**

Fig. 7. Best value obtained for each indicator and decoder over the 20 runs

4.2 Experimental Results

Results for the 10-units Case: In Table 6 the results of the statistical tests of comparisons are summarized. The results do not differ from one indicator to the other one. In this table and on all the following the column "Naive" indicates results of the first approach, the column "Scalarized" those of the second one and the column "multi" those of the multi decoding embedded approach. From this table we can see that for the 10-unit based case the difference between the naive approach and the scalarized approach is statistically significant. However the multi decoding embedded approach gives results significantly better than those obtained by the other approaches. Table 7 gives the best and average values obtained for each indicator and decoder over the 20 runs. The decoding embedded approach improves the ε-indicator value of 1.36% in comparison to the other method. It also improves the hypervolume indicator of 2.54%.

Results for the 40-units Case: In Table 8 the results of the statistical tests of comparisons are summarized. It can be observed that the multi decoding embedded approach gives significantly better results than the one obtained with the two other approaches. The scalarized decoding approach is better than the naive one.

In Tab. 9 the best and average values obtained for each indicator and decoder over the 20 runs are shown. It can be seen that that for the ε-indicator the scalarizing approach improves the results of more than 7% in comparison with the first naive approach. Then the multi decoding embedded approach improves the result of the scalarizing approach by more than 99%. For the hypervolume indicator there is an improvement of 31% when adding the λ value in the representation. Then the multi decoding embedded approach improves the result of the scalarizing approach by more than 99.5%.

Decoder	Naive	Scalarized	Multi
Naive	-	<	<
Scalirized	>	-	<
Multi	>	>	-

Fig. 8. Results obtained from statistical comparisons with the ε-indicator and the hypervolume indicator

Indicator	$I_{\varepsilon+}^1$		I_H^-	
Decoder	best	mean	best	mean
Naive	0.195	0.333	0.346	0.508
Scalarized	0.181	0.233	0.208	0.377
Multi	**0.00129**	**0.0840**	**0.000880**	**0.100**

Fig. 9. Best value obtained for each indicator and decoder over the 20 runs

Results for the 100-units Case:

Decoder	Naive	Scalarized	Multi
Naive	-	<	<
Scalarized	>	-	<
Multi	>	>	-

Fig. 10. Results obtained from statistical comparisons with the ε-indicator and the hypervolume indicator

In Table 10 the results of the statistical tests of comparisons are summarized. Again, the multi decoding embedded approach gives results significantly better than the ones obtained with the two other approaches. The scalarized decoding approach is better than the naive one.

Indicator	$I_{\varepsilon+}^1$		I_H^-	
Decoder	best	mean	best	mean
Naive	0.306	0.573	0.549	0.864
Scalarized	0.016	0.404	0.0169	0.636
Multi	**0.00389**	**0.150**	**0.000177**	**0.264**

Fig. 11. Best value obtained for each indicator and decoder over the 20 runs

In Table 11 the best and average values obtained for each indicator and decoder over the 20 runs are shown. In comparison with the naive approach, the scalarazing approach improves the results by 94.7% for the ε-indicator and by 97% for the hypervolume. Then the multi decoding embedded approach improves the result of the scalarizing approach by 75.69% for the ε-indicator and by 99% for the hypervolume-indicator .

This results and the statistical tests lead to conclude that the choice of the decoder system has a significant impact on the result. Results obtained with the last decoder are drastically better than the ones obtained with the other decoders.

5   Conclusion

In this article a binary genetic algorithm has been proposed to solve a multi -objectives UCP. The main difficulty is that for one genotypic solution many phenotypic solutions could be attached. These phenotypic solutions are those of the pareto front solution of the dispatching problem. Three original decoding systems associating phenotypic solutions with the genotypic ones have been presented and compared.

The multi decoding embedded system is the main contribution of this paper. The efficiency of this method has been shown on three data set of different size. In each case the results obtained are significantly better than those obtained by the two other strategies. The bigger the data, the better the improvement. This decoding system is then the one selected. In a future work, the objective will be to compare the proposed binary GA using this decoder to the GA proposed in [15] and to a more classical multi-objective GA using a real vector to encode the solutions. First of all, it will be interesting to make an analyse study of the impact of the choice of the n_λ parameter. An important advantage of this method is that it could be reused to any multi-objective problem that can be written as follow:

$$\operatorname*{opt}_{x,y}(f_1(x,y), f_2(x,y), ..., f_n(x,y)) \tag{9}$$

such that:

$$x \in X \tag{10}$$

$$y \in Y(x) \tag{11}$$

And such that for a fixed x the sub-problem $\mathscr{P}(x)$ finding the optimal pareto front of solutions with $y \in Y(x)$ is easy to solve. In this case the genetic algorithm will handle the x variables and the phenotypic solutions are found in solving $\mathscr{P}(x)$. The methodology chosen to solve $\mathscr{P}(x)$ does not matter. This is another advantage of this decoding system. Then, to test the multi decoding embedded approach on some other problem is one of our perspectives. We also believe that the multi decoding embedded approach can be generalized to any multi-objectives genetic algorithm. Then we plan to develop a generalized version of this approach that is suitable to any genetic algorithm.

References

1. Bleuler, S., Laumanns, M., Thiele, L., Zitzler, E.: PISA – A Platform and Programming Language Independent Interface for Search Algorithms. In: Fonseca, C.M., Fleming, P.J., Zitzler, E., Deb, K., Thiele, L. (eds.) EMO 2003. LNCS, vol. 2632, pp. 494–508. Springer, Heidelberg (2003)
2. Catalao, J., Mariano, S., Mendes, V., Ferreira, L.: A practical approach for profit-based unit commitment with emission limitations. International journal of electrical power & energy systems 32(3), 218–224 (2010)
3. Damousis, I.G., Bakirtzis, A.G., Dokopoulos, P.S.: A solution to the unit-commitment problem using integer-coded genetic algorithm. IEEE Transactions on Power Systems 19(2), 1165–1172 (2004)
4. Deb, K., Pratap, A., Agarwal, S., Meyarivan, T.: A fast and elitist multiobjective genetic algorithm: Nsga-II. IEEE Transactions on Evolutionary Computation 6(2), 182–197 (2002)
5. Ehrgott, M.: Multicriteria optimization, vol. 2. Springer (2005)
6. Guan, X., Zhai, Q., Papalexopoulos, A.: Optimization based methods for unit commitment: Lagrangian relaxation versus general mixed integer programming. In: Power Engineering Society General Meeting, vol. 2. IEEE (2003)
7. Jeong, Y.W., Park, J.B., Shin, J.R., Lee, K.Y.: A thermal unit commitment approach using an improved quantum evolutionary algorithm. Electric Power Components and Systems 37(7), 770–786 (2009)
8. Kazarlis, S.A., Bakirtzis, A., Petridis, V.: A genetic algorithm solution to the unit commitment problem. IEEE Transactions on Power Systems 11(1), 83–92 (1996)
9. Knowles, J., Thiele, L., Zitzler, E.: A Tutorial on the Performance Assessment of Stochastic Multiobjective Optimizers. TIK Report 214, Computer Engineering and Networks Laboratory (TIK), ETH Zurich, February 2006
10. Lau, T., Chung, C., Wong, K., Chung, T., Ho, S.: Quantum-inspired evolutionary algorithm approach for unit commitment. IEEE Transactions on Power Systems 24(3), 1503–1512 (2009)
11. Liefooghe, A., Jourdan, L., Talbi, E.G.: A software framework based on a conceptual unified model for evolutionary multiobjective optimization: Paradiseo-moeo. European Journal of Operational Research 209(2), 104–112 (2011)
12. López-Ibáñez, M., Dubois-Lacoste, J., Stützle, T., Birattari, M.: The irace package, iterated race for automatic algorithm configuration. Tech. rep, IRIDIA (2011)
13. de Moura Gomes Viana, A.M.M.: Metaheuristics for the Unit Commitment Problem The Constraint Oriented Neighbourhoods Search Strategy. Ph.D. thesis, Faculty of Engineering, University of Porto (2004)
14. Saramourtsis, A., Damousis, J., Bakirtzis, A., Dokopoulos, P.: Genetic algorithm solution to the economic dispatch problem–application to the electrical power grid of crete island. In: Proc. Workshop Machine Learning Applications to Power Systems (ACAI), pp. 308–317 (2001)
15. Srinivasan, D., Tettamanzi, A.G.: An evolutionary algorithm for evaluation of emission compliance options in view of the clean air act amendments. IEEE Transactions on Power Systems 12(1), 336–341 (1997)
16. Swarup, K, Yamashiro, S.: Unit commitment solution methodology using genetic algorithm. IEEE Transactions on Power Systems 17(1), 87–91 (2002)
17. Zhang, X.H., Zhao, J.Q., Chen, X.Y.: Multi-objective unit commitment fuzzy modeling and optimization for energy-saving and emission reduction. In: Proceedings of the CSEE 22, 71–76 (2010)

Comparing Decomposition-Based and Automatically Component-Wise Designed Multi-Objective Evolutionary Algorithms

Leonardo C.T. Bezerra[✉], Manuel López-Ibáñez, and Thomas Stützle

IRIDIA, Université, libre de Bruxelles (ULB) Brussels, Belgium
{lteonaci,manuel.lopez-ibanez,stuetzle}@ulb.ac.be

Abstract. A main focus of current research on evolutionary multi-objective optimization (EMO) is the study of the effectiveness of EMO algorithms for problems with many objectives. Among the several techniques that have led to the development of more effective algorithms, decomposition and component-wise design have presented particularly good results. But how do they compare? In this work, we conduct a systematic analysis that compares algorithms produced using the MOEA/D decomposition-based framework and the AutoMOEA component-wise design framework. In particular, we identify a version of MOEA/D that outperforms the best known MOEA/D algorithm for several scenarios and confirms the effectiveness of decomposition on problems with three objectives. However, when we consider problems with five objectives, we show that MOEA/D is unable to outperform SMS-EMOA, being often outperformed by it. Conversely, automatically designed AutoMOEAs display competitive performance on three-objective problems, and the best and most robust performance among all algorithms considered for problems with five objectives.

Keywords: Multi-objective optimization · Evolutionary algorithms · Decomposition · Component-wise design · Automatic configuration

1 Introduction

Over the past years, research on evolutionary multi-objective optimization (EMO) has focused on the development of effective algorithms for many-objective optimization, as evidenced by the number of recent publications on this topic [20]. Many are the reasons that stirred this interest. First, Pareto dominance becomes a weak relation as the number of objectives increases. As a result, the number of feasible solutions that are incomparable becomes too large to give algorithms that rely on Pareto dominance enough convergence pressure [1,13]. Second, the number of applications of many-objective optimization has demanded more effective algorithms for this scenario. In particular, many real-world engineering problems can be modeled as many-objective optimization problems, where constraints are

© Springer International Publishing Switzerland 2015
A. Gaspar-Cunha et al. (Eds.): EMO 2015, Part I, LNCS 9018, pp. 396–410, 2015.
DOI: 10.1007/978-3-319-15934-8_27

considered objectives [10]. Finally, the number of solutions needed to accurately approximate Pareto fronts grows exponentially with the number of objectives [13].

Among the many different search techniques proposed for improving the effectiveness of many-objective algorithms, indicator- and decomposition-based approaches have shown very good results [16,23,28]. In particular, decomposition is an old search paradigm originally applied by EMO already two decades ago [11], which has recently regained prominence with the proposal of the MOEA/D framework [26]. In this search paradigm, the original multi-objective problem is decomposed into simpler, single-objective subproblems by means of scalarizations. Originally, this approach was not pursued by the EMO community in general, particularly because decomposition-based algorithms may waste function evaluations searching in directions that do not present Pareto-optimal solutions. However, the best-known MOEA/D algorithm [27], which won the IEEE CEC 2009 competition on multi-objective optimization [28], uses a dynamic resource allocation strategy to overcome this drawback. Unfortunately, no performance assessment concerning this version of MOEA/D has been reported so far using large and representative benchmark sets on which other EMO algorithms have typically been tested.

More recently, another promising paradigm for devising effective EMO algorithms was proposed, namely the *component-wise design* [5]. This paradigm proposes reusing algorithmic components from well-known EMO algorithms in novel ways, thus leading to new designs. Concretely, given a flexible template, algorithms can be created by plugging in a set of desired components. In the original proposal, authors have automatically designed several algorithms using the component-wise design framework for continuous and combinatorial optimization [5]. We call the algorithms resulting from this automatic configuration process AutoMOEAs in what follows. The AutoMOEAs devised for many-objective optimization problems have shown competitive performance when compared to several Pareto- and indicator-based algorithms on a large set of three- and five-objective benchmark test problems, being able to match (and often surpass) the performance of the original algorithms from which the AutoMOEAs components were gathered.

In its current stage, the component-wise AutoMOEA framework only contains components from Pareto-based and indicator-based algorithms. However, given the interesting results the AutoMOEAs were already able to achieve, in this work we conduct a systematic performance assessment to understand how they compare to the effective decomposition-based approach. In particular, we consider several MOEA/D algorithms, including the version that won the IEEE CEC competition, and the AutoMOEAs designed in the original component-wise design paper [5]. To make this analysis more representative, we also include two effective indicator-based algorithms, SMS-EMOA [3] and IBEA [29], as well as the two best known Pareto-based algorithms, NSGA-II [8] and SPEA2 [30]. Furthermore, we consider a wide benchmark test set comprising the DLTZ [9] and WFG [12] benchmarks with three and five objectives, as well as several different problem sizes. In all scenarios considered, algorithms are properly tuned to perform at their best.

The investigation we conduct in this paper produces many interesting insights. First, we show that the MOEA/D algorithm that won the IEEE CEC competition is unable to outperform some of the other algorithms considered. Particularly for the five-objective WFG set, this version is clearly outperformed by SMS-EMOA. Second, we show that a straightforward alternative version of MOEA/D is able to consistently outperform the version used in the IEEE CEC competition, and also outperforms all other algorithms for the WFG set with three-objective problems. Nevertheless, SMS-EMOA still presents better results than this improved MOEA/D version on the five-objective WFG set. Finally, we show that the AutoMOEAs match the best-performing algorithms on all WFG scenarios, and outperform them on the 5-objective DTLZ set.

The remainder of this paper is organized as follows. We review the EMO search paradigms we consider in Section 2. Next, we describe the decomposition-based and the component-wise design paradigms in Sections 3 and 4, respectively. In particular, we detail the designs of the algorithms that are used in the experimental evaluation. The experimental setup is given in Section 5 followed by the presentation and discussion of the results in Section 6. We conclude and discuss future work in Section 7.

2   Search Paradigms in Multi-Objective Optimization

In this section, we briefly review the search paradigms found in the EMO literature that we use in this performance assessment. While this review is not exhaustive, the algorithms we highlight in each of the paradigms are the most representative and most effective in the literature for their corresponding paradigm [5]. In particular, this represents a major improvement over other recent experimental analysis conducted on the effectiveness of EMO algorithms for many-objective optimization [16], which have used representative but not the most effective algorithms for each paradigm.

Pareto-Based Approaches. Early EMO algorithms tried to find approximation fronts as diverse and close to the optimal front as possible mostly thanks to the convergence pressure provided by Pareto dominance. Among these, we highlight **NSGA-II** [8] and **SPEA2** [30]. Although these algorithms use different mechanisms, both are based on pushing the population towards convergence by favoring nondominated solutions, while simultaneously trying to maintain a population as diverse as possible. In EMO, diversity is a measure of the different trade-offs among the objectives considered, rather than an attempt to prevent stagnation as in the single-objective optimization literature. For most of the test problems considered then, these Pareto-based approaches were able to perform quite effectively [9,31]. However, the majority of these test cases considered two or three objectives only.

Indicator-Based Approaches. As the performance assessment of EMO algorithms reached a mature stage, researchers observed that quality indicators could be used within algorithms to direct their search in a Pareto-compliant way. More importantly, the convergence pressure provided by these quality indicators does not weaken as the number of objective increases. Within this paradigm, we highlight **IBEA** [29] and **SMS-EMOA** [3]. IBEA uses a binary quality indicator to compare solutions. In particular, the most effective version of IBEA uses the binary ϵ-indicator [5,29]. By contrast, SMS-EMOA uses the exclusive hypervolume contribution to direct its search. Although theoretical complexity analysis shows that this indicator can become exponentially costly as the number of objectives increases [2], empirical analysis has shown that recent efficient algorithms [2,24,25] give a runtime reasonable for practical purposes [19].

Decomposition-Based Approaches. Decomposition is one of the earliest search paradigms in EMO [11]. It is based on the principle that tackling single-objective subproblems is an easier task than facing the original multi-objective problem. However, in continuous optimization, decomposition was initially considered inefficient in comparison with other EMO algorithms, mostly due to the number of function evaluations that it may waste while searching along directions that do not present Pareto-optimal solutions. More recently, the **MOEA/D** [26] framework stirred the research on this paradigm, primarily when a variant of MOEA/D won the IEEE CEC 2009 competition on multi-objective optimization [27,28]. This variant improves over the original MOEA/D by using *dynamic resource allocation*, i.e., favoring search directions where the algorithm is progressing better. However, no performance assessment concerning this version of MOEA/D has been reported so far using a large and representative benchmark set where other EMO algorithms are typically tested.

Component-Wise Design. Proposed as a comprehensive design paradigm, the component-wise design aims at gathering the potential of the different existing EMO search paradigms. In its current version, the AutoMOEA framework [5] provides a flexible template and a collection of algorithmic components comprising both Pareto-based and indicator-based paradigms. Given an application, designers can then tailor algorithms to their target application. To demonstrate the potential of the component-wise design, the authors used an automatic configuration tool to automatically design various **AutoMOEAs** for the most-used continuous benchmarks [5], as well as for several combinatorial problems [6]. In particular, the AutoMOEAs designed for five-objective problems presented outstanding performance, matching the best-performing algorithms for the WFG benchmark, and outperforming all of them for the DTLZ benchmark [5].

In the following sections, we detail the specific variants of MOEA/D and AutoMOEA that are the focus of our performance assessment.

3   MOEA/D

Although it may be understood as an algorithmic framework, MOEA/D was originally proposed as a stand-alone algorithm [26]. Later, improved versions were also proposed as stand-alone algorithms [15,27]. For this reason, from now on we always refer to the different MOEA/D algorithms rather than to instantiations of a more general framework.

The common underlying structure shared by all MOEA/D algorithms considered in this work is the structure of the original MOEA/D algorithm, to which we will refer simply as **MOEA/D**. MOEA/D simultaneously explores the different search directions defined by the weight vectors of scalarization methods such as weighted linear sums or Tchebychev utility functions. Another particular feature presented by MOEA/D is the selection mechanism, namely, variation is applied to randomly selected parents from local neighborhoods, built for each search direction. Although the algorithm maintains a single global population, these local neighborhoods are meant to help the algorithm progress along the search directions employed.

A couple of years later, a new version of MOEA/D was proposed [27]. **MOEA/D$_{\text{DRA-DE}}$**, as we will call it, uses *dynamic resource allocation* (DRA) and the differential evolution (DE) variation operator. The DRA strategy works as follows. Initially, each of the N weight vectors is given the same utility value. At each iteration, MOEA/D$_{\text{DRA-DE}}$ selects a subset $\frac{N}{\nu}$ to explore via tournament selection based on the utility values of the weights. Once the weights have been selected, DE variation is applied to each search direction. In this version, however, a parameter δ regulates whether the target vector will be randomly chosen from the local neighborhood or from the whole population. Finally, a subset of the selection set (local neighborhood or population) is used to update the search reference point for the current weight. The size of this subset is regulated by an additional parameter ϕ. Every 50 iterations, the utility values of the weights are recomputed.

The differences between MOEA/D and MOEA/D$_{\text{DRA-DE}}$ are substantial, particularly given the number of parameters used to define the DRA strategy. In addition, since MOEA/D$_{\text{DRA-DE}}$ uses a different variation operator from all the other EMO algorithms we consider here, it is not really possible to assess whether improvements over the original MOEA/D (and other algorithms) could be explained solely by DRA, by DE, or by the combination of both components. For this reason, we consider an alternative version of MOEA/D$_{\text{DRA-DE}}$ that we call **MOEA/D$_{\text{DRA-SBX}}$**. The only difference between MOEA/D$_{\text{DRA-DE}}$ and MOEA/D$_{\text{DRA-SBX}}$ is how a trial vector (or offspring) is generated, that is, MOEA/D$_{\text{DRA-SBX}}$ uses the SBX crossover operator, instead of DE variation, to produce a single solution at a time.

Below we summarize all the MOEA/D algorithms we consider in this work. All versions use Tchebychev utility functions to search the objective space.

MOEA/D: original MOEA/D algorithm [26], with SBX crossover and no dynamic resource allocation.

MOEA/D$_{\text{DRA-DE}}$: MOEA/D algorithm used in the 2009 IEEE CEC competition [27]. This algorithm uses DE variation and dynamic resource allocation.

MOEA/D$_{DRA-SBX}$: alternative version of the MOEA/D algorithm used in the 2009 IEEE CEC competition [27]. This algorithm uses the SBX crossover operator and dynamic resource allocation.

4   AutoMOEA

The AutoMOEA component-wise design framework explores the concept that existing algorithmic components can lead to more effective designs than existing stand-alone algorithms if components are combined in more effective ways. This idea has been used in other multi-objective metaheuristics and led to the development of effective algorithms that significantly outperformed existing approaches from which algorithmic components were gathered [4,18]. Concerning EMO, the AutoMOEA framework is based on a template where components can be selected from existing Pareto- and indicator-based approaches.

The core structure of AutoMOEA algorithms are no different from traditional evolutionary algorithms. Starting from an initial population, select a mating pool of solutions from the population, apply variation operators to this pool, and replace solutions from the old population with these new offspring. AutoMOEA algorithms may also use an external bounded-size archive to store nondominated solutions, which is updated at the end of each iteration. The flexibility of the template relies heavily on the general preference relations used by the main components, namely mating, environmental, and external archive selection. For assembling a preference relation, AutoMOEA uses a tuple comprising a dominance-based set-partitioning, an indicator-based refinement, and a diversity metric. Concretely, solutions are partitioned in dominance-equivalent classes using a set-partitioning method, such as the ones originally proposed by Pareto-based approaches like NSGA-II or SPEA2. Since these partitions may contain incomparable solutions, indicator-based refinement relations are used, as in indicator-based approaches such as IBEA or SMS-EMOA. Finally, if solutions are still incomparable, diversity metrics are employed to ensure the population represents different trade-offs between the objectives.

Two other design concepts behind AutoMOEA provide additional flexibility to this framework. First, each of the main components may use a different preference relation, as proposed by more recent indicator-based algorithms like SMS-EMOA. Second, an AutoMOEA algorithm may use an internal bounded-size archive instead of a fixed-size population to increase the convergence pressure of the algorithms when required, as in algorithms such as PAES [14].

Below we summarize all the AutoMOEA algorithms we consider in this work. These algorithms are instantiations of the general AutoMOEA framework and have been automatically designed in [5] for the DTLZ and WFG benchmark sets with three and five objectives. The main components used by these algorithms are given in Table 1, where BuildMatingPool is the mating selection procedure, Replacement is the environmental selection procedure, and Replacement$_{Ext}$ is the external archive truncation method.

Table 1. Algorithm components of the AutoMOEAs used in this work. From top to bottom, AutoMOEA$_{D3}$, AutoMOEA$_{D5}$, AutoMOEA$_{W3}$, and AutoMOEA$_{W5}$.

BuildMatingPool				Replacement				Replacement$_{Ext}$	
Selection	SetPart	Quality	Diversity	SetPart	Quality	Diversity	Removal	Quality	Diversity
random	—	—	—	depth-rank	I_ϵ	—	—	I_H^1	sharing
tourn.	count	I_H^1	crowding	depth	I_ϵ	crowding	sequential	I_H^1	crowding
random	—	—	—	strength	I_H^h	kNN	—	I_H^1	kNN
tourn.	—	I_H^1	crowding	—	I_H^1	sharing	sequential	I_ϵ	kNN

AutoMOEA$_{D3}$ is an instantiation of AutoMOEA for 3-objective DTLZ problems. This algorithm uses a fixed-size population, random mating selection, and steady-state environmental selection based on dominance depth-rank and the binary ϵ-indicator (I_ϵ). In addition, AutoMOEA$_{D3}$ uses an external archive based on exclusive hypervolume contribution (I_H^1) and fitness sharing diversity.

AutoMOEA$_{D5}$ is an instantiation of AutoMOEA for 5-objective DTLZ problems. This algorithm uses a fixed-size population, mating selection based on deterministic tournament, and a mating preference relation that comprises dominance count set-partitioning, the exclusive hypervolume contribution as refinement, and crowding diversity. The environmental selection used by AutoMOEA$_{D5}$ is based on a preference relation that comprises dominance depth set-partitioning, the binary ϵ-indicator, crowding diversity, and sequential solution removal. In addition, AutoMOEA$_{D3}$ uses an external archive based on the exclusive hypervolume contribution and crowding diversity.

AutoMOEA$_{W3}$ is an instantiation of AutoMOEA for 3-objective WFG problems. This algorithm uses a fixed-size population, random mating selection, and steady-state environmental selection based on dominance strength, the shared hypervolume contribution (I_H^h), and nearest neighbor diversity. In addition, AutoMOEA$_{W3}$ uses an external archive based on the exclusive hypervolume contribution and nearest neighbor diversity.

AutoMOEA$_{W5}$ is an instantiation of AutoMOEA for 5-objective WFG problems. This algorithm uses a bounded internal archive, mating selection based on deterministic tournament, and a mating preference relation that comprises the exclusive hypervolume contribution and crowding diversity. The environmental selection used by AutoMOEA$_{W5}$ is based on a preference relation that comprises the exclusive hypervolume contribution, fitness sharing diversity, and sequential solution removal. In addition, AutoMOEA$_{D3}$ uses an external archive based on the binary ϵ-indicator and nearest neighbor diversity.

In the next section, we present the experimental setup we use in this work for the performance comparison of the different EMO paradigms.

5   Experimental Setup

The experimental setup we use in this work is the same used in the original component-wise design [5]. Since we use the same experimental setup, we use the same tuned settings for all algorithms except for the MOEA/D variants, which were not considered in the original paper and we tune them here. The benchmark sets we use are the DTLZ [9] and WFG [12] functions (DTLZ1–7 and WFG1–9), with three and five objectives. Concerning the number of variables n, we consider problems with $n \in \{20, 21, \ldots, 60\} \setminus n_{testing}$ for tuning, and $n_{testing} = \{30, 40, 50\}$ for testing. For both testing and tuning, algorithms are given 10 000 function evaluations per run, and all experiments are run on a single core of Intel Xeon E5410 CPUs, running at 2.33GHz with 6MB of cache size under Cluster Rocks Linux version 6.0/CentOS 6.3. For each problem instance, the approximation fronts produced by the algorithms are normalized to the range $[1, 2]$ to prevent issues due to dissimilar domains. Finally, we compute the hypervolume for each front using $r_i = 2.1$, $i = 1, 2, \ldots, M$ as reference point, where M is the number of objectives considered.

We tune the MOEA/D algorithms using irace [17] and the hypervolume as the quality measure, following the same procedure used for tuning all the other algorithms. In particular, for each tuning scenario, irace stops after 20 000 runs. For the original MOEA/D, the population size is given by the number of divisions in the objective space $N_{divisions}$ and the number of objectives. Since the population size can grow exponentially with the number of objectives, we use different ranges for each scenario: for the 3-objective problems, we use $N_{divisions} \in \{1, 2, \ldots, 30\}$, whereas for 5-objective problems we use $N_{divisions} \in \{1, 2, \ldots, 10\}$. For both MOEA/D$_{DRA}$ algorithms, the population size can be freely selected, and hence we use $\mu \in \{100, 200, \ldots, 500\}$. The remaining parameters tuned for the MOEA/D algorithms are given in Table 2. In particular, parameter ρ controls the size of the local neighborhoods ($\rho \cdot \mu$), parameter t_{size} is the size of the tournament used by the DRA strategy, and parameter η_m is the distribution index used by the polynomial mutation operator. For more details about any of the remaining parameters, we refer to Section 3 and to the original MOEA/D papers [26,27]. Finally, for the algorithms that use the SBX crossover, the tuning range of parameters p_c (crossover probability) and η_c (the distribution index) is the same used by all other algorithms. By contrast, when DE variation is used, there are two other parameters: the crossover probability $CR \in [0, 1]$ and the scale factor $F \in [0.1, 2]$. For brevity, the tuned settings selected for the MOEA/D algorithms are provided as supplementary material [7].

To compare algorithms, we run each algorithm 25 times and evaluate them based on the relative hypervolume of the approximation fronts they produce w.r.t. the actual Pareto optimal fronts. More precisely, we use the same Pareto fronts used by [5]. Given an approximation front A and the Pareto front for a problem instance P, the relative hypervolume of A equals $I_H(A)/I_H(P)$. A relative hypervolume of 1.0 means the algorithm was able to perfectly approximate the Pareto front for the problem considered. Algorithms are then compared based on boxplots of these relative hypervolumes. To draw overall conclusions,

Table 2. Parameter space for tuning all MOEA/D algorithms

	MOEAD$_{\text{DRA}}$					
Parameter	ρ	δ	ϕ	t_{size}	ν	η_m
Domain	$[0.1, 1]$	$[0, 1]$	$[0.01, 1]$	$\{1, 2, \ldots, 20\}$	$\{2, 3, \ldots, 10\}$	$\{1, \ldots, 50\}$

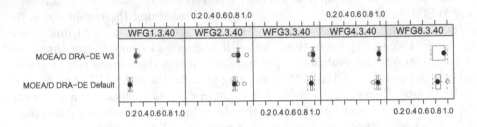

Fig. 1. Boxplots of the relative hypervolume achieved by MOEA/D$_{\text{DRA-DE}}$ using default or tuned parameter settings on selected 3-objective 40-variable WFG problems

we aggregate results through rank sums and test for significant differences using Friedman's test with 99% confidence level. Since we generate a large set of results, we only discuss the most representative ones here. The full set of results is provided as supplementary material [7].

6   Results and Discussion

Before proceeding to the actual comparison between the different search paradigms, we start this section with boxplots on selected 3-objective WFG problems (Fig. 1) to demonstrate the effect of the tuning on the performance of MOEA/D$_{\text{DRA-DE}}$, which can also be observed for other MOEA/D algorithms. In particular, the label "W3" indicates that the MOEA/D$_{\text{DRA-DE}}$ algorithm has been tuned for 3-objective WFG problems. This notation is also used in all remaining boxplots to make it explicit that all algorithms have been properly tuned for the scenarios in which they are compared. Concerning the parameter settings used by MOEA/D$_{\text{DRA-DE}}$, the most interesting remark is the very low δ values for both 3-objective benchmarks ($\delta \leq 0.1$), which indicate that local neighborhoods are rarely used by this algorithm in these scenarios. By contrast, for all DTLZ scenarios MOEA/D$_{\text{DRA-SBX}}$ uses extremely high δ values ($\delta \geq 0.93$), and MOEA/D uses extremely large niche sizes ($\rho \geq 0.97$). Altogether, these settings indicate that the effectiveness of the local neighborhood component is tightly related to the benchmark, number of objectives, and variation operator considered. We then proceed to an analysis per benchmark.

6.1   Analysis on the DTLZ Benchmark Set

The performance comparison of all algorithms on the DTLZ benchmark set is given in Fig. 2. Results for 3-objective problems are shown on the top row, while

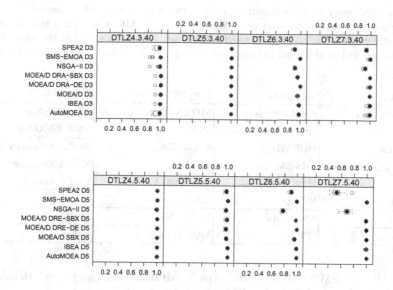

Fig. 2. Boxplots of the relative hypervolume achieved by all algorithms on selected DTLZ problems with 40 variables. Top: 3 objectives. Bottom: 5 objectives.

the bottom row depicts the performance assessment on the 5-objective problems. These results confirm insights previously identified in the literature [5]. First, the overall difficulty of this benchmark is low, as reflected by the very high relative hypervolumes achieved by most algorithms. In fact, we do not show the plots for DTLZ1–3 because they are identical to the plots shown for the 3-objective DTLZ5 problem, i.e., all algorithms are able to well approximate the Pareto optimal fronts used as reference. Second, the Pareto-based approaches are the ones that present worst-quality results among all algorithms. Although one can notice this already for 3-objective problems, it becomes far more evident when 5-objective problems are considered.

Regarding the performance of the remaining algorithms on the 3-objective problems, the only problem that actually poses difficulties for some algorithms is DTLZ6, where MOEA/D$_{DRA-DE}$ and SMS-EMOA presents results better than all other algorithms considered. The pattern observed in the boxplots is confirmed by the rank sum analysis given in Table 3. For the 3-objective DTLZ set, no difference can be observed between the four top-performing algorithms. Interestingly, MOEA/D$_{DRA-DE}$ ranks sixth, alongside IBEA. Concerning the 5-objective problems, AutoMOEA$_{D5}$, SMS-EMOA, and IBEA appear to always accurately approximate the Pareto fronts (Fig. 2, bottom), while the MOEA/D versions sometimes face difficulties, such as for problems DTLZ5 and DTLZ6. However, when we consider the rankings over the whole 5-objective DTLZ set (Table 3), AutoMOEA$_{D5}$ ranks first with much lower ranks than all other algorithms. No significant difference is observed between MOEA/D$_{DRA-DE}$,

Table 3. Rank sum analysis depicting overall performance on all scenarios. The best ranked algorithms are shown on top. Algorithms in boldface present rank sums not significantly worse than the best ranked algorithm. Algorithms within the same block are not significantly different, in terms of ranking, to the first algorithm of the same block.

3-obj DTLZ	5-obj DTLZ	3-obj WFG	5-obj WFG
SMS-EMOA$_{D3}$	**AutoMOEA$_{D5}$**	**MOEA/D$_{DRA-SBX W3}$**	**AutoMOEA$_{W5}$**
MOEA/D$_{D3}$	MOEA/D$_{DRA-DED5}$	MOEA/D$_{DRA-DEW3}$	**SMS-EMOA$_{W5}$**
AutoMOEA$_{D3}$	MOEA/D$_{D5}$	AutoMOEA$_{W3}$	MOEA/D$_{DRA-SBX W5}$
MOEA/D$_{DRA-SBX D3}$	SMS-EMOA$_{D5}$	SPEA2$_{W3}$	MOEA/D$_{DRA-DEW5}$
IBEA$_{D3}$	MOEA/D$_{DRA-SBX D5}$	SMS-EMOA$_{W3}$	MOEA/D$_{W5}$
MOEA/D$_{DRA-DED3}$	IBEA$_{D5}$	IBEA$_{W3}$	IBEA$_{W5}$
SPEA2$_{D3}$	NSGA-II$_{D5}$	NSGA-II$_{W3}$	SPEA2$_{W5}$
NSGA-II$_{D3}$	SPEA2$_{D5}$	MOEA/D$_{W3}$	NSGA-II$_{W5}$

MOEA/D, and SMS-EMOA, nor between MOEA/D$_{DRA-SBX}$, and IBEA. As expected, the Pareto-based algorithms rank last.

6.2   Analysis on the WFG Benchmark Set

Results for the WFG benchmark set are much more heterogeneous, confirming that this benchmark set is far more difficult for EMO algorithms than the DTLZ one. The performance comparison for 3-objective problems is given in Fig. 3. In fact, it is difficult to even find patterns on the performance of the algorithms. Given any pair of algorithms, one cannot visually identify the best approach when all problems are considered, which confirms that most algorithms perform very similarly in all problems. For this reason, we proceed to the rank sum analysis, also given in Table 3. Surprisingly, the algorithm that achieves lowest rank sums is MOEA/D$_{DRA-SBX}$, outperforming MOEA/D$_{DRA-DE}$ again for a 3-objective benchmark set. Ranking second, MOEA/D$_{DRA-DE}$, AutoMOEA$_{W3}$, and SPEA2 show equivalent rank sums. This indicates that some components from SPEA2 are indeed particularly effective for this benchmark, since AutoMOEA$_{W3}$ heavily relies on SPEA2 components. The indicator-based approaches come right after, and MOEA/D ranks last this time. This reinforces the contribution of our experimental analysis, since previous results led to the conclusion that MOEA/D was particularly effective for 3-objective benchmarks in general [16].

The performance assessment for the 5-objective WFG benchmark is given in Fig. 4. Again it is difficult to make an overall analysis since the results vary per problem instance, nonetheless both AutoMOEA$_{W5}$ and SMS-EMOA perform consistently well. Concerning the MOEA/D algorithms, all present very similar performances. The rank sum analysis (Table 3) confirms these observations: AutoMOEA$_{W5}$ and SMS-EMOA present nearly identical rank sums, outperforming all MOEA/D algorithms, which present equivalent rank sums.

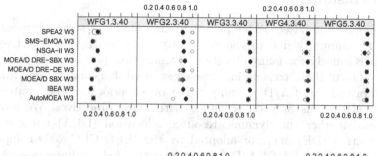

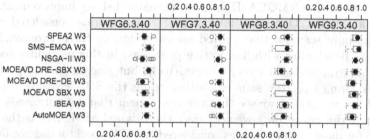

Fig. 3. Boxplots of the relative hypervolume achieved by all algorithms on WFG problems with 40 variables and 3 objectives

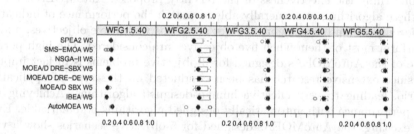

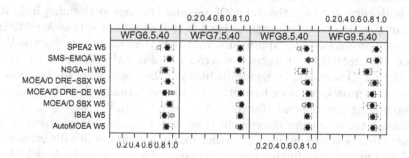

Fig. 4. Boxplots of the relative hypervolume achieved by all algorithms on WFG problems with 40 variables and 5 objectives

7   Conclusions

The decomposition-based EMO paradigm has drawn a strong interest from the EMO community due to the possibility of devising more effective algorithms, particularly for many-objective optimization problems. In this paper, we have shown that, considering the most used benchmark sets from the EMO literature, MOEA/D is competitive or superior to other state-of-the-art EMO algorithms only on scenarios with three objectives. We have also shown that, neither the dynamic resource allocation (DRA) nor the differential evolution (DE) operator adopted by the IEEE CEC 2009 competition MOEA/D algorithm (MOEA/D$_{\text{DRA-DE}}$) actually led to improvements over the original MOEA/D version for most of the scenarios considered in this paper. The only scenario that proved an exception to these two conclusions is the WFG benchmark with 3-objective problems. In this particular scenario, MOEA/D$_{\text{DRA-DE}}$ performed very competitively, but since it was outperformed by MOEA/D$_{\text{DRA-SBX}}$ (the same algorithm using the SBX crossover operator instead of DE variation), we see that the component that actually leads to this significant performance improvement over the original MOEA/D is the DRA. Moreover, for most of the scenarios considered, the DE operator did not improve the performance of the original MOEA/D. Since related work has shown that DE variation can often improve the performance of other algorithms [22], we hypothesize that the interaction between the decomposition approach and DE is responsible for this.

Concerning the effectiveness of the recently proposed AutoMOEAs, we see that these algorithms are generally able to match the performance of indicator- and decomposition-based algorithms for scenarios with three objectives, and to outperform most of them when five objectives are considered. The high performance of the AutoMOEAs designed for 3-objective problems is in fact impressive, since extensive research has been conducted on this type of application scenario, leading to very effective human-designed algorithms. Achieving the same performance with automatically designed algorithms is remarkable. Even more exciting, the AutoMOEAs designed for 5-objective scenarios show a very robust and competitive performance. For the DTLZ benchmark, the difference in the rank sums between the AutoMOEA$_{\text{D5}}$ and the best performing indicator- and decomposition-based algorithms is such that it indicates that AutoMOEA$_{\text{D5}}$ consistently produces better approximation fronts than the others. For the WFG benchmark, AutoMOEA$_{\text{W5}}$ matches the performance of SMS-EMOA, outperforming all MOEA/D versions. These results indicate the potential of the component-wise design approach, since we have attained this performance level by combining only two among the different effective EMO search paradigms.

Although the results for the current stage of the component-wise design approach are already quite convincing, further research efforts in this direction could potentially improve even further the performance of newly designed AutoMOEAs. Besides including algorithmic components proposed for algorithms from other paradigms, more effective automatic configuration tools (or longer tuning budgets) could lead to even more effective designs. However, it is also imperative to develop this research field towards practical application requirements. For instance, since many real-world problems are computationally demanding, the number of function evaluations desired might not allow offline tuning. One possible way to work around this problem is to devise several automatic designs for different benchmarks and learn problem features that could help understand better the effectiveness of individual algorithmic components.

Acknowledgments. The research presented in this paper has received funding from the COMEX project within the Interuniversity Attraction Poles Programme of the Belgian Science Policy Office. L.C.T. Bezerra, M. López-Ibáñez and T. Stützle acknowledge support from the Belgian F.R.S.-FNRS, of which they are a FRIA doctoral fellow, a postdoctoral researcher and a senior research associate, respectively.

References

1. Aguirre, H.: Advances on many-objective evolutionary optimization. In: Proceedings of the 15th Annual Conference Companion on Genetic and Evolutionary Computation, GECCO 2013 Companion, pp. 641–666. ACM (2013)
2. Beume, N., Fonseca, C.M., López-Ibáñez, M., Paquete, L., Vahrenhold, J.: On the complexity of computing the hypervolume indicator. IEEE Trans. Evol. Comput. **13**(5), 1075–1082 (2009)
3. Beume, N., Naujoks, B., Emmerich, M.: SMS-EMOA: Multiobjective selection based on dominated hypervolume. Eur. J. Oper. Res. **181**(3), 1653–1669 (2007)
4. Bezerra, L.C.T., López-Ibáñez, M., Stützle, T.: Automatic generation of multi-objective ACO algorithms for the bi-objective knapsack. In: Dorigo, M., Birattari, M., Blum, C., Christensen, A.L., Engelbrecht, A.P., Groß, R., Stützle, T. (eds.) ANTS 2012. LNCS, vol. 7461, pp. 37–48. Springer, Heidelberg (2012)
5. Bezerra, L.C.T., López-Ibáñez, M., Stützle, T.: Automatic component-wise design of multi-objective evolutionary algorithms. Tech. Rep. TR/IRIDIA/2014-012, IRIDIA, Université Libre de Bruxelles, Belgium, Brussels, Belgium (2014)
6. Bezerra, L.C.T., López-Ibáñez, M., Stützle, T.: Automatic design of evolutionary algorithms for multi-objective combinatorial optimization. In: Bartz-Beielstein, T., Branke, J., Filipič, B., Smith, J. (eds.) PPSN 2014. LNCS, vol. 8672, pp. 508–517. Springer, Heidelberg (2014)
7. Bezerra, L.C.T., López-Ibáñez, M., Stützle, T.: Comparing decomposition-based and automatically component-wise designed multi-objective evolutionary algorithms (2015). http://iridia.ulb.ac.be/supp/IridiaSupp2015-002/
8. Deb, K., Pratap, A., Agarwal, S., Meyarivan, T.: A fast and elitist multi-objective genetic algorithm: NSGA-II. IEEE Trans. Evol. Comput. **6**(2), 182–197 (2002)
9. Deb, K., Thiele, L., Laumanns, M., Zitzler, E.: Scalable test problems for evolutionary multiobjective optimization. In: Abraham, A., et al. (eds.) Evolutionary Multiobjective Optimization. Advanced Information and Knowledge Processing, pp. 105–145. Springer, London (2005)
10. Fleming, P.J., Purshouse, R.C., Lygoe, R.J.: Many objective optimization: an engineering design perspective. In: Coello Coello, C.A., Hernández Aguirre, A., Zitzler, E. (eds.) EMO 2005. LNCS, vol. 3410, pp. 14–32. Springer, Heidelberg (2005)
11. Hajela, P., Lin, C.Y.: Genetic search strategies in multicriterion optimal design. Structural Optimization **4**(2), 99–107 (1992)

12. Huband, S., Hingston, P., Barone, L., While, L.: A review of multiobjective test problems and a scalable test problem toolkit. IEEE Trans. Evol. Comput. **10**(5), 477–506 (2006)
13. Ishibuchi, H., Tsukamoto, N., Nojima, Y.: Evolutionary many-objective optimization: A short review. In: IEEE CEC, pp. 2419–2426. IEEE Press (2009)
14. Knowles, J.D., Corne, D.: Approximating the nondominated front using the Pareto archived evolution strategy. Evol. Comput. **8**(2), 149–172 (2000)
15. Li, H., Zhang, Q.: Multiobjective optimization problems with complicated Pareto sets, MOEA/D and NSGA-II. IEEE Trans. Evol. Comput. **13**(2), 284–302 (2009)
16. Li, M., Yang, S., Liu, X., Shen, R.: A comparative study on evolutionary algorithms for many-objective optimization. In: Purshouse et al. [20], pp. 261–275
17. López-Ibáñez, M., Dubois-Lacoste, J., Stützle, T., Birattari, M.: The irace package, iterated race for automatic algorithm configuration. Tech. Rep. TR/IRIDIA/2011-004, IRIDIA, Université Libre de Bruxelles, Belgium (2011)
18. López-Ibáñez, M., Stützle, T.: The automatic design of multi-objective ant colony optimization algorithms. IEEE Trans. Evol. Comput. **16**(6), 861–875 (2012)
19. Nowak, K., Märtens, M., Izzo, D.: Empirical performance of the approximation of the least hypervolume contributor. In: Bartz-Beielstein, T., Branke, J., Filipič, B., Smith, J. (eds.) PPSN 2014. LNCS, vol. 8672, pp. 662–671. Springer, Heidelberg (2014)
20. Purshouse, R.C., et al. (eds.): Evolutionary Multi-Criterion Optimization - 7th International Conference, EMO 2013, Proceedings. LNCS, vol. 7811, Sheffield, UK, March 19–22. Springer (2013)
21. Reed, P.M.: Many-objective visual analytics: Rethinking the design of complex engineered systems. In: Purshouse et al. [20], p. 1
22. Tušar, T., Filipič, B.: Differential evolution versus genetic algorithms in multiobjective optimization. In: Obayashi, S., Deb, K., Poloni, C., Hiroyasu, T., Murata, T. (eds.) EMO 2007. LNCS, vol. 4403, pp. 257–271. Springer, Heidelberg (2007)
23. Wagner, T., Beume, N., Naujoks, B.: Pareto-, aggregation-, and indicator-based methods in many-objective optimization. In: Obayashi, S., Deb, K., Poloni, C., Hiroyasu, T., Murata, T. (eds.) EMO 2007. LNCS, vol. 4403, pp. 742–756. Springer, Heidelberg (2007)
24. While, L., Bradstreet, L.: Applying the WFG algorithm to calculate incremental hypervolumes. In: IEEE CEC, pp. 1–8. IEEE Press (2012)
25. While, L., Bradstreet, L., Barone, L.: A fast way of calculating exact hypervolumes. IEEE Trans. Evol. Comput. **16**(1), 86–95 (2012)
26. Zhang, Q., Li, H.: MOEA/D: A multiobjective evolutionary algorithm based on decomposition. IEEE Trans. Evol. Comput. **11**(6), 712–731 (2007)
27. Zhang, Q., Liu, W., Li, H.: The performance of a new version of MOEA/D on CEC09 unconstrained MOP test instances. In: IEEE CEC, pp. 203–208. IEEE Press (2009)
28. Zhang, Q., Suganthan, P.N.: Special session on performance assessment of multiobjective optimization algorithms/CEC 2009 MOEA competition (2009). http://dces.essex.ac.uk/staff/qzhang/moeacompetition09.htm
29. Zitzler, E., Künzli, S.: Indicator-based selection in multiobjective search. In: Yao, X., Burke, E.K., Lozano, J.A., Smith, J., Merelo-Guervós, J.J., Bullinaria, J.A., Rowe, J.E., Tiňo, P., Kabán, A., Schwefel, H.-P. (eds.) PPSN 2004. LNCS, vol. 3242, pp. 832–842. Springer, Heidelberg (2004)
30. Zitzler, E., Laumanns, M., Thiele, L.: SPEA2: Improving the strength Pareto evolutionary algorithm for multiobjective optimization. In: Giannakoglou, K.C., et al. (eds.) EUROGEN, pp. 95–100. CIMNE, Barcelona (2002)
31. Zitzler, E., Thiele, L., Deb, K.: Comparison of multiobjective evolutionary algorithms: Empirical results. Evol. Comput. **8**(2), 173–195 (2000)

Upper Confidence Bound (UCB) Algorithms for Adaptive Operator Selection in MOEA/D

Richard A. Gonçalves[1]($\boxtimes$), Carolina P. Almeida[1], and Aurora Pozo[2]

[1] Department of Computer Science, UNICENTRO, Guarapuava, Brazil
{richard,carol}@unicentro.br
[2] Computer Science Department,
Federal University of Paraná (UFPR), Curitiba, Brazil
aurora@inf.ufpr.br

Abstract. Adaptive Operator Selection (AOS) is a method used to dynamically determine which operator should be applied in an optimization algorithm based on its performance history. Recently, Upper Confidence Bound (UCB) algorithms have been successfully applied for this task. UCB algorithms have special features to tackle the Exploration versus Exploitation (EvE) dilemma presented on the AOS problem. However, it is important to note that the use of UCB algorithms for AOS is still incipient on Multiobjective Evolutionary Algorithms (MOEAs) and many contributions can be made. The aim of this paper is to extend the study of UCB based AOS methods. Two methods are proposed: MOEA/D-UCB-Tuned and MOEA/D-UCB-V, both use the variance of the operators' rewards in order to obtain a better EvE tradeoff. In these proposals the UCB-Tuned and UCB-V algorithms from the multiarmed bandit (MAB) literature are combined with MOEA/D (MOEA based on decomposition), one of the most successful MOEAs. Experimental results demonstrate that MOEA/D-UCB-Tuned can be favorably compared with state-of-the-art adaptive operator selection MOEA/D variants based on probability (ENS-MOEA/D and ADEMO/D) and multi-armed bandits (MOEA/D-FRRMAB) methods.

Keywords: Adaptive Operator Selection (AOS) · MOEA/D · Upper Confidence Bound (UCB) Algorithms · UCB1 · UCB-Tuned · UCB-V

1 Introduction

Most real problems are multiobjective consisting of several conflicting objectives to optimize. Nowadays, Multiobjective Evolutionary Algorithms (MOEAs) are usually used to solve them. But obtaining the best performance of these algorithms in a specific domain needs setting a number of parameters and selects genetic operators that make hard their application for ordinary users.

Adaptive Operator Selection (AOS) is a paradigm to adaptively determine which operator should be applied in an optimization algorithm. An AOS is composed of two tasks: credit assignment and operator selection. The former decides

© Springer International Publishing Switzerland 2015
A. Gaspar-Cunha et al. (Eds.): EMO 2015, Part I, LNCS 9018, pp. 411–425, 2015.
DOI: 10.1007/978-3-319-15934-8_28

how much reward is attributed to an operator based on its performance and the latter selects which operator should be applied based on the accumulated reward.

A recent work ([11]) explored various aspects of using an UCB algorithm as an AOS method and proposed an algorithm called MOEA/D-FRRMAB that introduced an UCB based AOS method into the MOEA/D framework (MOEA based on Decomposition). UCB based algorithms are among the best algorithms for dealing with multi-armed bandit (MAB) problems. A fundamental issue of MAB and AOS is related to the exploration versus exploitation (EvE) dilemma: if it is apparently more advantageous to apply the best operator (exploitation), on the other hand it is important to keep applying other operators (exploration), since the performance of an operator may change at different search stages.

MOEA/D-FRRMAB obtained very good results focusing on the credit assignment task of the Adaptive Operator Selection [11]. It may be possible to improve the results obtained in [11] with more robust operator selection techniques which is the focus of this paper. Therefore, we propose and study two new AOS methods based on UCB algorithms to be used in combination with the MOEA/D framework and the FRRMAB credit assignment methodology. Namely, we propose UCB-Tuned and UCB-V AOS methods and their respectively multiobjective evolutionary algorithms, MOEA/D-UCB-Tuned and MOEA/D-UCB-V. Both AOS methods use the variance of the operators' rewards in order to obtain a tighter confidence interval of the estimated goodness of each operator, possibly reducing the application of suboptimal operators.

Although some recent works combining multiobjective evolutionary algorithms and adaptive operator selection have emerged in the specialized literature, such as ENS-MOEA/D [17], ADEMO/D [14] and MOEA/D-FRRMAB [11], the area still has various questions that need more investigation, such as the impact of different UCB based operator selection algorithms, which is the main focus of the present work. As a testbed, in this work, we consider the CEC 2009 multiobjective benchmark (10 instances) [15]. This set of instances permits a direct comparation between the proposed algorithm and state-of-the-art MOEAs, such as ENS-MOEA/D, ADEMO/D and MOEA/D-FRRMAB.

The remainder of this paper is organized as follows. Section 2 gives an overview of Multiobjective Optimization and the MOEA/D framework. The main concepts of Adaptive Operator Selection are described in Section 3. Our proposed approaches (MOEA/D-UCB-Tuned and MOEA/D-UCB-V) are presented in Section 4. Experimental results are given and analyzed in Section 5. Section 6 presents some conclusions.

2   Multiobjective Optimization and MOEA/D

A Multiobjective Optimization Problem (MOP) is defined as Min (or Max) $\mathbf{f}(\mathbf{x}) = (f_1(\mathbf{x}), ..., f_M(\mathbf{x}))$ subject to $g_i(\mathbf{x}) \leq 0$, i = {1, ..., G}, and $h_j(\mathbf{x}) = 0$, j = {1, ..., H} $\mathbf{x} \in \Omega$. A solution minimizes (or maximizes) the components of the objective vector $\mathbf{f}(\mathbf{x})$ where $\mathbf{x}$ is a n-dimensional decision variable vector $\mathbf{x} = (x_1, ..., x_n) \in \Omega$.

Usually, as long as the multiple objectives are conflicting, there is not a single solution that is optimal with respect to all objectives, so the solution of a MOP is a set of optimal solutions. In the absence of a priori preference information, the Pareto optimality concept is used to define the optimal solution set, which is called Pareto set. A solution is Pareto optimal if it is not dominated by any other feasible solution. A solution x dominates a solution y if $\forall i \in \{1, 2, 3, \ldots, M\}$: $f_i(x) \leq f_i(y) \wedge \exists j \in \{1, 2, 3, \ldots, M\} : f_j(x) < f_j(y)$.

MOEA/D is based on conventional aggregation approaches [4] as it decomposes a MOP into a number of single objective optimization subproblems. The objective of each subproblem is a linear (or nonlinear) weighted aggregation of all individual objectives in the MOP. Neighborhood relations among these subproblems depend on distances among their aggregation weight vectors. MOEA/D, generally, uses a set of N evenly spread weight vectors, where N is the number of subproblems. Each subproblem is simultaneously optimized using mainly information from its neighboring subproblems. There are various versions of MOEA/D and this paper extends the variant that won the CEC 2009 MOEA contest: MOEA/D-DRA (MOEA/D with Dynamical Resource Allocation) [16]. MOEA/D and its variants can use any decomposition approach for defining their subproblems. This work uses the Tchebycheff approach (Equation 1), where each subproblem can be formulated as:

$$\text{Min } g^{te}(\mathbf{x} \mid \boldsymbol{\lambda}, \mathbf{z}^*) = \max_{1 \leq j \leq M} \{\lambda_j \mid f_j(\mathbf{x}) - z_j^* \mid\} \tag{1}$$

$$\text{subject to } \mathbf{x} \in \Omega$$

where g^{te} is the Tchebycheff function, $\mathbf{f}(\mathbf{x}) = (f_1(\mathbf{x}), \ldots, f_M(\mathbf{x}))$ is the multiobjective function to be minimized, $\mathbf{z}^*$ is the empirical ideal point and $\boldsymbol{\lambda} = (\lambda_1, \ldots, \lambda_M)$ is the weight vector associated with subproblem i. One problem with using the empirical ideal point is the possible concentration of solutions in a specific region of the Pareto front [12]. In order to allocate the computational resources to the most appropriate subproblems, MOEA/D-DRA utilizes a tournament selection based on the utility value of each subproblem (π^i). The utility of each subproblem is calculated accordingly to Equation 2.

$$\pi^i = \begin{cases} 1, & \text{if } \Delta^i > 0.001 \\ (0.95 + 0.05 * \Delta^i/0.001) * \pi^i, & \text{otherwise} \end{cases} \tag{2}$$

where Δ^i is the relative decrease of the objective function value of subproblem i. The subproblems with greater Δ^i values have better chances of being selected.

3 Adaptive Operator Selection

Choosing suitable operators for generating new solutions in a Multiobjective Evolutionary Algorithms (MOEAs) may lead to high computational costs due to time-consuming trial-and-error processes and may be inefficient in cases where ideal operators usage vary during the evolutionary process [14]. The Adaptive

Operator Selection (AOS) paradigm was created in order to tackle this problem. AOS tries to determine the best operator to be selected at each stage of the search process, which usually is based on the performance of the operators [11]. The major procedures involved in AOS are: credit assignment and operator selection [11]. Credit assignment defines how to reward an operator based on its recent performance while operator selection decides which operator should be applied next based on the reward information collected during the search process [11].

3.1 Credit Assignment

The credit assignment procedure can be further divided in two parts: quality measurement and reward assignment. Quality measurement determines how good the application of a particular operator was while the reward assignment determines how to use the quality measured in order to update the information of the appropriateness of the operator [11]. The most commonly used metric for measuring the quality of an operator is based on the fitness improvement relative to a baseline solution [11]. The baseline solution can be the best solution of the population or a parent of the generated solution, for example.

In [6] and [8] four different reward techniques are investigated: Average Absolute Reward, Extreme Absolute Reward, Average Normalized Reward and Extreme Normalized Reward. The average techniques reward the operators accordingly to mean quality of the operator while the extreme techniques reward the operators accordingly to their best performance, i.e., it favours the rare occurance of large improvements. The difference between the abolute and normalized techniques lies on the fact that the former uses the raw measurement of the quality of an operator while the latter normalizes the quality value before using each in the reward calculation. These four reward techniques in conjunction with a quality measurement using a parent as the baseline solution were investigated in the MOEA/D context in [14], resulting in an algorithm called ADEMO/D.

In order to obtain more robust credit assignments, recent researches have focused on rank-based credit assignment procedures, such as AUC (Area Under the Curve), SR (Sum-of-Ranks) and FRR (Fitness-Rate-Rank) [5] [11]. The last two are investigated in [11] as credit assignments to AOS methods used in MOEA/D variants (MOEA/D-SRMAB and MOEA/D-FRRMAB, respectively). The proposed algorithms (MOEA/D-UCB-Tuned and MOEA/D-UCB-V) also uses the FRR credit assignment (described in Subsection 4.1).

3.2 Operator Selection

Operator selection procedures choose operators for generating new solutions based on the information collected by the credit assignment methods. The operator selection is usually based on probability or multiarmed bandit (MAB) methods. Probability methods use a roulette wheel-like process for selecting an operator while MAB methods use algorithms created to tackle the Exploration versus Exploitation (EvE) dilemma [11]. Two examples of probability based operator selection methods are Probability Matching (PM) and Adaptive

Pursuit (AP). The Probability Matching method calculates $p_{op}(g+1)$ (the probability of operator op being selected at generation $g + 1$) as follow [8], [13], [7]:

$$p_{op}(g + 1) = p_{min} + (1 - K * p_{min}) * \frac{q_{op}(g + 1)}{\sum_{s=1}^{K} q_s(g + 1)} \tag{3}$$

where K is the number of operators, p_{min} is the minimal probability of any operator and q_{op} is the quality associated with operator op.

Clearly, $\sum_{s=1}^{K} p_s(g+1) = 1$. From Eq. 3 we note that when only one strategy obtains a reward during a long period of time (with all other strategies receiving no rewards), its selection probability converges to $p_{max} = p_{min} + (1 - K * p_{min})$.

The Adaptive Pursuit method was introduced for AOS in the context of Genetic Algorithms [13]. AP calculates $p_{op}(g + 1)$ as follow:

$$p_{op^*}(g + 1) = p_{op^*}(g) + \beta * [p_{max} - p_{op^*}(g)] \tag{4}$$

and $\forall op \neq op^* : p_{op}(g+1) = p_{op}(g) + \beta * [p_{min} - p_{op}(g)]$, with $op^* = \text{argmax}_{op}$ $(q_{op}(g+1))$ and $p_{max} = p_{min}+1-S*p_{min}$, where op^* is the best operator during the current generation. This constraint makes sure that if $\sum_{s=1}^{K} p_s(g) = 1$, then $\sum_{s=1}^{S} p_s(g + 1) = 1$ [13]. The AP method has a learning rate $\beta \in (0,1]$, which controls how greedy the "winner-takes-all" strategy will behave. ADEMO/D [14] investigated the PM and AP methods as operator selection procedures in the MOEA/D context while ENS-MOEA/D [17] uses a procedure similar to PM with binary credit assignment in order to adaptively select its operators.

The most prominent MAB method is the UCB1, which provides asymptotic optimality guarantees [1][2]. In an UCB1-based operator selection procedure, the op operator has an empirical quality estimate q_{op} and a confidence interval that depends on the number of times (n_{op}) that it has been applied before. At each time point t, the operator maximizing Equation 5 is selected.

$$q_{op} + C * \sqrt{\frac{2 * \ln \sum_{s=1}^{K} n_s}{n_{op}}} \tag{5}$$

where C is a scaling factor used to regulate the tradeoff between exploitation (the first term, which favors the operators with best empirical rewards) and exploration (the square root term - confidence interval - which favors the infrequently tried operators) and K is the number of operators in the operator pool. MOEA/D-FRRMAB [11] combines UCB1 and MOEA/D to solve MOPs.

The proposed algorithms (MOEA/D-UCB-Tuned and MOEA/D-UCB-V) also uses UCB-based procedures, but the investigated procedure utilizes confidence interval that considers the variance of the operator qualities while the UCB1 confidence interval is just based on the number of times that each operator is applied. The proposed operator selection mechanisms are described in Subsection 4.2.

4   Proposed Approaches

This section describes our proposed approaches: MOEA/D-UCB-Tuned and MOEA/D-UCB-V. Both algorithms are based on the MOEA/D framework, use the FRR credit assignment scheme proposed in [11] and an UCB based operator selection procedure.

4.1   Credit Assignment

In order to better compare our proposed methods with MOEA/D-FRRMAB, in this work we use the same credit assignment used in [11]. The first step of credit assignment is to calculate the quality associated with the application of an operator. The main form of quality measurement is the absolute fitness improvement, but this method suffers from the drawback that fitness improvements tend to vary from problem to problem and, particularly, during different stages of the evolutionary search. So, to alleviate these drawbacks, our proposed approaches use the fitness improvement rate method (FIR) [11], defined in Equation 6 considering the Tchebycheff function[1].

$$FIR_{op,t} = \frac{pf_{op,t} - cf_{op,t}}{pf_{op,t}} = \frac{g^{te}(x^i|\lambda^i, z^*) - g^{te}(y|\lambda^i, z^*)}{g^{te}(x^i|\lambda^i, z^*)} \tag{6}$$

where $FIR_{op,t}$ is the quality associated with operator op in time t, $pf_{op,t}$ is the fitness associated with the parent in time t, $cf_{op,t}$ is the fitness associated with the generated child in time t, g^{te} is the Tchebycheff function, x^i is the parent and y is the generated child. Because in the MOEA/D framework more than one solution can be replaced during the population update (see Algorithm 3), more than one quality measurement can result from the same operator application (one for each replacement) [11].

The performance of an operator may vary accordingly to the current evolutionary stage of the algorithm and, in order to only consider recent FIR information during credit assignment, a sliding window is used [11].

As in [11], the Reward$_{op}$ associated with operator op is calculated as the sum of all FIR values for operator op in the sliding window. Afterwards, a decaying factor D is used to transform the initial reward accordingly to its relative rank with respect to the reward of the other operators (see steps 12 to 15 of Algorithm 1), resulting in the DecayedReward$_{op}$. Finally, the decayed rewards are normalized, resulting in the FRR$_{op}$ (Fitness-Rate-Rank) rewards that are used by the operator selection procedure. Algorithm 1 presents credit assignment procedure used in this work.

4.2   Operator Selection

The purpose of the operator selection procedure is to select operators to generate new individuals. This selection is based on the credit values calculated as described in Algorithm 1.

---

[1] Other aggregation functions can be used by substituting g^{te} by the desired function.

Algorithm 1. Pseudocode of the Credit Assignment Procedure

```
 1: for i = 0 to K do
 2:     Reward_i = 0.0
 3:     n_i = 0
 4: end for
 5: for i = 0 to Size of the Sliding Window do
 6:     op = SlidingWindow.getOperatorPos(i)
 7:     reward = SlidingWindow.getRewardPos(i)
 8:     Reward_op = Reward_op + reward
 9:     n_op++
10: end for
11: Rank the rewards in descending order
12: for op = 0 to K do
13:     DecayedReward_op = D^{Rank^op} * Reward_op
14:     n_i = 0
15: end for
16: TotalDecayedReward = Σ^K_{op=1} DecayedReward_op
17: for op = 0 to K do
18:     FRR_op = DecayedReward_op/TotalDecayedReward
19: end for
```

The main purpose of this paper is to compare the performance of different UCB-based algorithms as operator selection procedures. Three different UCB algorithms are investigated in this paper: UCB1, UCB-Tuned and UCB-V [2][1].

UCB1, or simple UCB, is the more classical MAB algorithm and is used in [11] as the operator selection of MOEA/D-FRRMAB. UCB1 and UCB-V provide asymptotic optimality guarantees while both UCB-Tuned and UCB-V use the rewards' variance (σ^2_{op}) in order to obtain a tigher confidence interval, i.e., to obtain a better exploration versus exploitation tradeoff. Accordingly to empirical experiments in the MAB literature [2], UCB-Tuned is supposed to have the better performance among the investigated algorithms. The investigated operator selection mechanism is described in Algorithm 2.

4.3   MOEA/D-UCB-Tuned and MOEA/D-UCB-V

The proposed approches (MOEA/D-UCB-Tuned and MOEA/D-UCB-V) are presented in Algorithm 3. The first steps of MOEA/D-UCB-Tuned and MOEA/-D-UCB-V initialize various data structures (steps 1 to 7), analogous to most MOEA/D variants. The weight vectors $\boldsymbol{\lambda}^i$, $i = 1, ..., N$, representing coefficients associated with each objective, are generated using a uniform distribution. The neighborhood ($B^i = \{i_1, \cdots, i_C\}$) of weight vector $\boldsymbol{\lambda}^i$ stores the indexes of the C weight vectors closest to $\boldsymbol{\lambda}^i$. The initial population is randomly generated and evaluated. Each individual ($\mathbf{x}^i$) is associated with the ith weight vector. The sliding window is initialized as an empty window. The empirical ideal point ($\mathbf{z}^*$) is initialized as the minimum value of each objective found in the initial population and the generation (g) is set to 1.

After initialization steps, the algorithm enters its main loop (steps 8 to 43). The first step of the main loop is to determine which individuals from the population will be processed. A 10-tournament selection based on the utility value of

Algorithm 2. Pseudocode of Multiarmed Bandit Based Operator Selection

1: **if** There are operators that have not been selected **then**
2:     Randomly choose an unselected operator
3: **else**
4:     **if** MAB_Method == UCB **then**
5:       //MOEA/D-FRRMAB
6:       $SelectedOperator = argmax_{op=1..K} \left(FRR_{op} + C * \sqrt{\frac{2*ln \sum_{i=1}^{K} n_i}{n_{op}}} \right)$

7:     **end if**
8:     **if** MAB_Method == UCB-Tuned **then**
9:       **for** (**do** $i = 0$ to K)
10:         $V_{op} = \sigma_{op}^2 + \sqrt{\frac{2*ln \sum_{i=1}^{K} n_i}{n_{op}}}$
11:       **end for**
12:       $SelectedOperator = argmax_{op=1..K} \left(FRR_{op} + C * \sqrt{\frac{ln \sum_{i=1}^{K} n_i}{n_{op}} * min(\frac{1}{4}, V_{op})} \right)$

13:     **end if**
14:     **if** MAB_Method == UCB-V **then**
15:       $SelectedOperator = argmax_{op=1..K} \left(FRR_{op} + C * \sqrt{\frac{2*ln \sum_{i=1}^{K} n_i * \sigma_{op}^2}{n_{op}}} + 3 * \frac{\sum_{i=1}^{K} n_i}{n_{op}} \right)$

16:     **end if**
17: **end if**

each subproblem (π^i, calculated accordingly to Equation 2) is used to determine the individuals (see Section 2).

The operator op used to generate a new individual is selected in step 11. The selection procedure is described in Algorithm 2 and uses a UCB based method (UCB-Tuned or UCB-V, depending on the algorithm chosen). This step differentiates between MOEA/D-UCB-Tuned and MOEA/D-UCB-V. It also distinguishes between the proposed approaches and MOEA/D-FRRMAB, which uses UCB1 in this step. The pool of operators used in this work contains four DE operators with distinct search characteristics:

- "DE/rand/1": $y^i = x^i + F * (x^{r_1} - x^{r_2})$;
- "DE/rand/2": $y^i = x^i + F * (x^{r_1} - x^{r_2}) + F * (x^{r_3} - x^{r_4})$;
- "DE/current-to-rand/1": $y^i = x^i + K * (x^i - x^{r_1}) + F * (x^{r_2} - x^{r_3})$; and
- "DE/current-to-rand/2": $y^i = x^i + K * (x^i - x^{r_1}) + F * (x^{r_2} - x^{r_3}) + F * (x^{r_4} - x^{r_5})$.

Next, the scope used during the generation of individuals and the population update is randomly chosen. DE mutation and crossover operators are applied considering individuals randomly selected from *scope*. In this work, *scope* can swap from the neighborhood to the entire population (and vice-versa) along the evolutionary process of MOEA/D-UCB-Tuned and MOEA/D-UCB-V. It is composed by the indexes of chromosomes from either the neighborhood B^i (with probability δ) or from the entire population (with probability $1 - \delta$).

A modified chromosome **y** is generated in step 18 using the chosen operator and modified by the polynomial mutation in step 21, generating $\mathbf{y'} = (y'_1, \cdots, y'_n)$ from **y**. If the generated solution is unfeseable, it is repaired. In step 25, if the new chromosome **y'** has an objective value better than the value stored in the empirical ideal point, $\mathbf{z}^*$ is updated with this value.

Algorithm 3. Pseudocode of MOEA/D-UCB-Tuned and MOEA/D-UCB-V

1: Generate N weight vectors $\boldsymbol{\lambda}^i = (\lambda_1^i, \lambda_2^i, ...\lambda_M^i), i = 1, ..., N$
2: Compute the Euclidean distances between any two weight vectors and select C closest weight
   vectors to each weight vector $\boldsymbol{\lambda}^i$. For each $i = 1, \cdots, N$, set B(i) $= i_1, \cdots, i_C$ where
   $\boldsymbol{\lambda}^{i_1}, \cdots, \boldsymbol{\lambda}^{i_C}$ are the C closest weight vectors to $\boldsymbol{\lambda}^i$
3: Generate an initial population $P^0 = \{\mathbf{x}^1, \cdots, \mathbf{x}^N\}$, $\mathbf{x}^i = (x_1^i, x_2^i, ...x_n^i)$
4: Evaluate each individual in the initial population
5: Initialize the Sliding Window
6: Initialize $\mathbf{z}^* = (z_1^*, \cdots, z_M^*)^T$ by setting $z_j^* = min_{1 \le i \le N} f_j(\mathbf{x}^i)$
7: $g = 1$
8: **repeat**
9:     Let all the indices of the subproblems whose objectives are MOP individual objectives f_i
       form the initial I. By using 10-tournament selection based on π^i, select other $N/5M$ indices
       and add them to I.
10:    **for** each individual x^i in I **do**
11:        Select an operator op accordingly to the Multiarmed Bandit Policy (Algorithm 2)
12:        Generate $rand$ in [0,1] //*Determining the scope*
13:        **if** $rand < \delta$ **then**
14:            $scope =$ B(i)
15:        **else**
16:            $scope = 1, \cdots, N$
17:        **end if**
18:        Generate a new solution $\mathbf{y}$ by applying operator op
19:        **if** $\mathbf{y}$ is unfeasible **then** Repair $\mathbf{y}$
20:        **end if**
21:        Apply polynomial mutation to produce $\mathbf{y}'$
22:        **if** $\mathbf{y}'$ is unfeasible **then** Repair $\mathbf{y}'$
23:        **end if**
24:        Evaluate $\mathbf{y}'$
25:        **if** $f_j(\mathbf{y}') < z_j^*$ **then** $z_j^* = f_j(\mathbf{y}')$ //*Updating* $\mathbf{z}^*$, *for each* $j = 1, \cdots, M$
26:        **end if**
27:        **for** each subproblem k in the neighborhood **do**
28:            With $k =$ randomly selected in scope
29:            **if** $g^{te}(\mathbf{y}' \mid \boldsymbol{\lambda}^k, \mathbf{z}^*) < g^{te}(\mathbf{x}^k \mid \boldsymbol{\lambda}^k, \mathbf{z}^*)$ **then**
30:                **if** less than n_r replacements **then**
31:                    Replace $\mathbf{x}^k$ by $\mathbf{y}'$
32:                    Calculate the fitness improvement rate using Equation 6
33:                **end if**
34:            **end if**
35:        **end for**
36:        Adjust the Sliding Window
37:        Calculate the Credit Assignment using a Decaying Factor (Algorithm 1)
38:    **end for**
39:    **if** g modulo 50 == 0 **then**
40:        Update the utility π^i of each subproblem i
41:    **end if**
42:    $g = g + 1$;
43: **until** $g >$MAX-EV

The next steps involve the population update process (steps 27 to 35) which
is based on the comparison of the fitness of individuals. In the MOEA/D framework, the fitness of an individual is measured accordingly to a decomposition
function. In this work the *Tchebycheff function* is used (Equation 1). Accordingly to what is selected for the *scope* (steps 14 or 16), the neighborhood or the
entire population is updated. To avoid the proliferation of $\mathbf{y}'$ to a great part of
the population, a maximum number of updates (NR) is used.

The population update is as follows: if a new replacement may occur, (i.e.
while $n_r < NR$ and there are unselected indexes in *scope*), a random index (k)
from *scope* is chosen. If $\mathbf{y}'$ has a better Tchebycheff value than $\mathbf{x}^k$ (both using

the kth weight vector - $\boldsymbol{\lambda}^k$) then $\mathbf{y}'$ replaces $\mathbf{x}^k$ and the number of updated chromosomes (n_r) is incremented. Also, the fitness improvement rate (FIR) is calculated accordingly to Equation 6 for each replacement.

Then, the sliding window is updated. All successful fitness improvements are inserted in the sliding window with their associated operator (op). If the sliding window is full, the least recent operator application is discarded from it.

Afterwards, the credit assignment procedure (Algorithm 1) is called to update the rewards and if the current generation is a multiple of 50, then the utility value of each subproblem is updated using Equation 2.

Finally, the evolutionary process stops when the maximum number of evaluations (MAX-EV) is reached. MOEA/D-UCB-Tuned and MOEA/D-UCB-V output the Pareto set and Pareto front approximations.

5   Experiments and Results

In this section we present the experiments conducted to evaluate our proposed approaches, considering all the unconstrained (bound constrained) instances from the CEC 2009 multiobjective benchmark (10 instances) [15]. The search space dimension n is defined as 30 for all the instances. Table 1 shows the characteristics of each instance.

Table 1. Characteristics of unconstrained functions considered

Function	Objectives	Search space range	Properties of Pareto Front
UF1	2	$[0, 1] \times [-1, 1]^{n-1}$	Concave
UF2	2	$[0, 1] \times [-1, 1]^{n-1}$	Concave
UF3	2	$[0, 1]^n$	Concave
UF4	2	$[0, 1] \times [-2, 2]^{n-1}$	Convex
UF5	2	$[0, 1] \times [-1, 1]^{n-1}$	21 points front
UF6	2	$[0, 1] \times [-1, 1]^{n-1}$	One isolated point and 2 disconnected parts
UF7	2	$[0, 1] \times [-1, 1]^{n-1}$	Continuous straight line
UF8	3	$[0, 1]^2 \times [-2, 2]^{n-2}$	Parabolic
UF9	3	$[0, 1]^2 \times [-2, 2]^{n-2}$	Planar
UF10	3	$[0, 1]^2 \times [-2, 2]^{n-2}$	Parabolic

Our experimental studies can be divided into two parts: (i) to investigate which UCB based operator selection mechanism among UCB, UCB-Tuned and UCB-V have the best performance and (ii) to compare the best UCB based algorithm with some recent MOEA/D variants and NSGA-II. A comparison against single operator versions was not conducted because it was already done in [11], where MOEA/D-FRRMAB achieved better results.

During the experiments, we consider the following performance metrics: IGD (Inverted Generational Distance) and HV (Hypervolume) [10]. Inverted Generational Distance (IGD) is as a way of estimating how far are the elements in the

Pareto front approximation (S) produced by one algorithm from those in the true Pareto front (P^*) of the problem. This measure is defined as:

$$IGD = \frac{\sum_{x \in P^*} d(x, S)}{|P^*|} \tag{7}$$

where d is the Euclidean distance (measured in objective space). The lower the IGD better. The hypervolume calculates the area covered by all the solutions in S using a reference point W:

$$hypervolume = volume \left(\bigcup_{i=1}^{|S|} v_i \right) \tag{8}$$

where for each solution $i \in S$, a hypercube v_i is constructed with reference to W. The HV is the union of all hypercubes. The higher the HV, the better. The parameter settings used by MOEA/D-FRRMAB, MOEA/D-UCB-Tuned and MOEA/D-UCB-V are the same. Table 2 presents the parameters' values.

Table 2. Parameters used in MOEA/D-FRRMAB, MOEA/D-UCB-Tuned and MOEA/D-UCB-V

	Values	Description
DE Parameters		
N	600	Population size (for instances with 2-objectives).
	1000	Population size (for instances with 3-objectives).
CR	1.0	Crossover rate.
F	0.5	Scaling factor.
p_m	$1/n$	polynomial mutation probability.
τ	20	distribution index of polynomial mutation.
MAX-EV	300,000	Maximum number of evaluations.
MOEA/D Parameters		
W	20	Number of weight vectors in the neighborhood.
n_r	2	Maximal number of solutions replaced by each offspring.
δ	0.9	Probability that parent solutions are selected from the neighborhood.
Multiarmed Bandit Parameters		
C	5.0	Coefficient to balance exploration and exploitation.
WS	0.5 * N	Sliding window size.
D	1.0	Decaying factor.

5.1   UCB Based Algorithms Comparison

We have first compared the three UCB based algorithms: MOEA/D-FRRMAB, MOEA/D-UCB-Tuned and MOEA/D-UCB-V. MOEA/D-FRRMAB results were obtained by executing the code provided by its authors [2]. Each algorithm was executed 30 times.

Table 3 and Table 4 show the results obtained by each algorithm on UF1 to UF10. Accordingly to the IGD metric, MOEA/D-UCB-Tuned was the best algorithm on UF2, UF3, UF5, UF6, UF7 and UF9 while MOEA/D-FRRMAB was the best on UF1, UF4 and UF10 and MOEA/D-UCB-V was the best only on UF8. Furthermore, the HV metric also indicated that MOEA/D-UCB-Tuned was the best algorithm on UF2, UF3, UF6, UF7 and UF9 while MOEA/D-FRRMAB was the best on UF1 and UF4 and MOEA/D-UCB-V was the best on UF5, UF8 and UF10. It is important to note that MOEA/D-UCB-Tuned was the most consistent algorithm when considering both metrics.

Table 3. Comparison of FRRMAB, UCB-Tuned and UCB-V on IGD. Mean and standard deviation. Dark gray cells indicate the best results. Light gray cells are equivalent accordingly to the Wilcoxon rank sum test [3] with 95% confiability.

	MOEA/D-FRRMAB	MOEA/D-UCB-Tuned	MOEA/D-UCB-V
UF1	$9.89e-04_{7.8e-05}$	$1.00e-03_{6.9e-05}$	$1.10e-03_{9.7e-05}$
UF2	$2.11e-03_{5.9e-04}$	$2.04e-03_{6.1e-04}$	$2.64e-03_{1.2e-03}$
UF3	$5.37e-03_{7.0e-03}$	$4.82e-03_{9.7e-03}$	$6.17e-03_{6.0e-03}$
UF4	$5.45e-02_{3.9e-03}$	$5.45e-02_{4.4e-03}$	$5.58e-02_{3.4e-03}$
UF5	$3.07e-01_{5.3e-02}$	$2.89e-01_{4.1e-02}$	$3.10e-01_{5.7e-02}$
UF6	$8.44e-02_{5.5e-02}$	$6.96e-02_{3.3e-02}$	$1.36e-01_{1.9e-01}$
UF7	$1.17e-03_{1.9e-04}$	$1.15e-03_{1.7e-04}$	$1.30e-03_{2.3e-04}$
UF8	$2.96e-02_{6.0e-03}$	$2.97e-02_{4.8e-03}$	$2.71e-02_{2.7e-03}$
UF9	$4.14e-02_{3.9e-02}$	$3.80e-02_{3.6e-02}$	$4.49e-02_{4.3e-02}$
UF10	$4.86e-01_{5.7e-02}$	$5.06e-01_{7.1e-02}$	$4.98e-01_{6.5e-02}$

Table 4. Comparison of FRRMAB, UCB-Tuned and UCB-V on HV. Mean and standard deviation. Dark gray cells indicate the best results. Light gray cells are equivalent accordingly to the Wilcoxon rank sum test [3] with 95% confiability.

	MOEA/D-FRRMAB	MOEA/D-UCB-Tuned	MOEA/D-UCB-V
UF1	$6.65e-01_{1.3e-04}$	$6.65e-01_{1.0e-04}$	$6.65e-01_{1.5e-04}$
UF2	$6.64e-01_{5.3e-04}$	$6.64e-01_{7.5e-04}$	$6.63e-01_{1.5e-03}$
UF3	$6.58e-01_{1.1e-02}$	$6.60e-01_{8.5e-03}$	$6.57e-01_{8.0e-03}$
UF4	$2.55e-01_{5.0e-03}$	$2.54e-01_{5.4e-03}$	$2.53e-01_{4.7e-03}$
UF5	$3.22e-02_{5.8e-02}$	$3.84e-02_{6.2e-02}$	$4.23e-02_{6.2e-02}$
UF6	$2.39e-01_{2.9e-02}$	$2.40e-01_{2.3e-02}$	$2.25e-01_{9.2e-02}$
UF7	$4.98e-01_{3.4e-04}$	$4.98e-01_{2.8e-04}$	$4.98e-01_{3.8e-04}$
UF8	$4.22e-01_{1.3e-02}$	$4.22e-01_{1.1e-02}$	$4.28e-01_{5.0e-03}$
UF9	$7.36e-01_{5.3e-02}$	$7.40e-01_{4.7e-02}$	$7.30e-01_{5.8e-02}$
UF10	$1.75e-02_{1.7e-02}$	$1.32e-02_{1.8e-02}$	$2.26e-02_{2.0e-02}$

The Friedman test [3] indicated with 95% confiability that, when considering all CEC 2009 instances, MOEA/D-UCB-Tuned was statistically better than the other UCB based algorithms in both metrics. So, it was possible to conclude that the proposed MOEA/D-UCB-Tuned was the best UCB based AOS method among the three (MOEA/D-FRRMAB, MOEA/D-UCB-Tuned and MOEA/D-UCB-V) compared in these experiments. The better performance obtained by

---

[2] Available in http://www.cs.cityu.edu.hk/~51888309/code/FRRMAB.rar

Table 5. The IGD statistics for MOEA/D-UCB-Tuned, ADEMO/D, MOEA/D-DRA, ENS-MOEA/D, MOEA/D-DRA-CMX+SPX and NSGA-II. Dark gray cells emphasize the best results while light gray cells emphasize the second best results. '-' indicates values that are not available in the original work.

CEC09	Algorithm	Median	Mean	Std	Min	Max
UF1	MOEA/D-UCB-Tuned	0.001000	0.001000	0.000071	0.000843	0.001139
	ADEMO/D	0.001053	0.001105	0.000134	0.000919	0.001505
	MOEA/D-DRA	0.001503	0.001526	0.000090	0.001417	0.001757
	MOEAD-CMX-SPX	0.004171	0.004292	0.000263	0.003985	0.005129
	ENS-MOEA/D	-	0.001642	0.000125	-	-
	NSGA-II	0.095186	0.094731	0.003249	0.088511	0.103222
UF2	MOEA/D-UCB-Tuned	0.001740	0.002043	0.000619	0.001560	0.003618
	ADEMO/D	0.003559	0.003692	0.001229	0.002156	0.007758
	MOEA/D-DRA	0.003375	0.003502	0.001039	0.002326	0.006638
	MOEAD-CMX-SPX	0.005472	0.005615	0.000412	0.005149	0.006778
	ENS-MOEA/D	-	0.004048	0.001005	-	-
	NSGA-II	0.035151	0.035071	0.001479	0.032968	0.039164
UF3	MOEA/D-UCB-Tuned	0.002960	0.004818	0.009852	0.001039	0.055887
	ADEMO/D	0.001640	0.002364	0.001348	0.001336	0.006504
	MOEA/D-DRA	0.001488	0.003948	0.004131	0.001086	0.014019
	MOEAD-CMX-SPX	0.005313	0.011165	0.013093	0.004155	0.068412
	ENS-MOEA/D	-	0.002591	0.000456	-	-
	NSGA-II	0.089894	0.090817	0.016815	0.062901	0.126556
UF4	MOEA/D-UCB-Tuned	0.053997	0.054511	0.004454	0.047378	0.067366
	ADEMO/D	0.040184	0.040846	0.001617	0.038520	0.045140
	MOEA/D-DRA	0.060765	0.060285	0.004757	0.051492	0.070912
	MOEAD-CMX-SPX	0.063524	0.064145	0.004241	0.055457	0.075361
	ENS-MOEA/D	-	0.042070	0.001325	-	-
	NSGA-II	0.080935	0.080737	0.002809	0.074034	0.084683
UF5	MOEA/D-UCB-Tuned	0.287825	0.289346	0.041842	0.215209	0.373328
	ADEMO/D	0.161385	0.162010	0.008849	0.141540	0.203285
	MOEA/D-DRA	0.220083	0.254930	0.089484	0.146933	0.511464
	MOEAD-CMX-SPX	0.379241	0.418508	0.135554	0.211058	0.707093
	ENS-MOEA/D	-	0.248110	0.042555	-	-
	NSGA-II	0.214958	0.220145	0.051622	0.154673	0.331853
UF6	MOEA/D-UCB-Tuned	0.064828	0.069553	0.033633	0.051310	0.244601
	ADEMO/D	0.063794	0.063468	0.002938	0.057753	0.068737
	MOEA/D-DRA	0.207831	0.326176	0.287571	0.053371	0.823381
	MOEAD-CMX-SPX	0.248898	0.327356	0.185717	0.056972	0.792910
	ENS-MOEA/D	-	0.060847	0.019840	-	-
	NSGA-II	0.080177	0.080728	0.006460	0.067996	0.090331
UF7	MOEA/D-UCB-Tuned	0.001113	0.001154	0.000174	0.000998	0.001704
	ADEMO/D	0.001663	0.001692	0.000263	0.001335	0.002574
	MOEA/D-DRA	0.001569	0.001945	0.001364	0.001336	0.008796
	MOEAD-CMX-SPX	0.004745	0.006262	0.003307	0.003971	0.014662
	ENS-MOEA/D	-	0.001728	0.000852	-	-
	NSGA-II	0.048873	0.048977	0.001959	0.044634	0.051968
UF8	MOEA/D-UCB-Tuned	0.029520	0.029741	0.004885	0.021318	0.042125
	ADEMO/D	0.081807	0.083363	0.010145	0.065271	0.118021
	MOEA/D-DRA	0.040352	0.040667	0.003788	0.033777	0.050412
	MOEAD-CMX-SPX	0.056872	0.057443	0.003366	0.051800	0.065620
	ENS-MOEA/D	-	0.031006	0.003005	-	-
	NSGA-II	0.112219	0.113226	0.002742	0.109836	0.121190
UF9	MOEA/D-UCB-Tuned	0.025449	0.038029	0.036903	0,021770	0,151891
	ADEMO/D	0.030595	0.038481	0.027877	0.025448	0.142735
	MOEA/D-DRA	0.137856	0.123078	0.039018	0.025008	0.139986
	MOEAD-CMX-SPX	0.144673	0.097693	0.054285	0.033314	0.151719
	ENS-MOEA/D	-	0.027874	0.009573	-	-
	NSGA-II	0.106841	0.106081	0.000681	0.105806	0.100729
UF10	MOEA/D-UCB-Tuned	0.493070	0.506181	0.072184	0.361202	0.661469
	ADEMO/D	0.549185	0.563805	0.088759	0.412789	0.784204
	MOEA/D-DRA	0.406094	0.408770	0.066770	0.210500	0.553572
	MOEAD-CMX-SPX	0.467715	0.462653	0.038698	0.391496	0.533234
	ENS-MOEA/D	-	0.21173	0.019866	-	-
	NSGA-II	0.257846	0.259851	0.012541	0.234047	0.288036

MOEA/D-UCB-Tuned was a consequence of it focusing more on the exploitation than the other algorithms while the poor performance of MOEA/D-UCB-V could be attributed to its emphasis on exploration[3]. This happened mainly because the variance influence on MOEA/D-UCB-Tuned was limited to 1/4 while it is unlimited in MOEA/D-UCB-V (see steps 12 and 15 of Algorithm 2).

5.2   Comparison with Literature

In the comparison with literature, we only use the IGD metric. This is due to a lack of data from some of the algorithms. We evaluate the IGD metric of the final approximation over 30 independent executions of MOEA/D-UCB-Tuned against ADEMO/D, MOEA/D-DRA, ENS-MOEA/D, MOEA/D-DRA-CMX+SPX and NSGA-II for each CEC09 test instance using the same reference set.

Table 5 presents median, mean, standard deviation (std), minimum (min) and maximum (max) of IGD metric values. The IGD-metric values of the ENS-MOEA/D and MOEA/D-DRA-CMX+SPX are those described in [17] and [9], respectively. With respect to the IGD values, MOEA/D-UCB-Tuned is the best algorithm on UF1, UF2, UF7, UF8 and UF9 and the second best algorithm on UF4 and UF6. Its performance is poor on UF3, UF5 and UF10. Similar results are obtained when considering the mean of the IGD values. These results show that MOEA/D-UCB-Tuned can be favorably compared with state-of-the-art algorithm from the literature.

6   Conclusions

In this paper, we proposed two new multiarmed bandit-based AOS methods, MOEA/D-UCB-Tuned and MOEA/D-UCB-V, to adaptively select appropriate operators in MOEA/D algorithms. As credit assignment, we used the FRR approach proposed in [11]. Our operator pool was constituted by four commonly used DE mutation operators. We conducted experimental studies on some test instance from the CEC 2009 MOEA Competition.

The main contribution of this work was the proposal and investigation of new combinations of UCB algorithms and the MOEA/D framework. The best proposed approach (MOEA/D-UCB-Tuned) used an operator selection mechanism that was based on the UCB-Tuned method from the multiarmed bandit literature. The better performance of MOEA/D-UCB-Tuned was expected based on the results presented on the multiarmed bandit literature. We also showed that the best proposed approach was favorably compared with state-of-the-art adaptive operator selection MOEA/D variants based on probability (ENS-MOEA/D and ADEMO/D) and multiarmed bandits (MOEA/D-FRRMAB) methods.

Acknowledgments. The authors acknowledge CNPq, CAPES and Fundação Araucária for the partial financial support. They also acknowledge the authors of the MOEA/D-FRRMAB for providing its source code.

---

[3] Data not shown due to a lack of space.

References

1. Auer, P.: Using confidence bounds for exploitation-exploration trade-offs. J. Mach. Learn. Res. **3**, 397–422 (2003)
2. Auer, P., Cesa-Bianchi, N., Fischer, P.: Finite-time analysis of the multiarmed bandit problem. Mach. Learn. **47**(2–3), 235–256 (2002)
3. Conover, W.J.: Practical Nonparametric Statistics, 3rd edn. Wiley (1999)
4. Ehrgott, M.: A discussion of scalarization techniques for multiple objective integer programming. Ann. Oper. Res. **147**, 343–360 (2006)
5. Fialho, A.: Adaptive operator selection for optimization. Ph.D. thesis, Comput. Sci. Dept. - Univ. Paris-Sud XI (2010)
6. Fialho, A., Schoenauer, M., Sebag, M.: Analysis of adaptive operator selection techniques on the royal road and long k-path problems. In: Conference on Genetic and Evolutionary Computation, pp. 779–786 (2009)
7. Goldberg, D.E.: Probability matching, the magnitude of reinforcement, and classifier system bidding. Mach. Learn. **5**, 407–425 (1990)
8. Gong, W., Fialho, l, Cai, Z., Li, H.: Adaptive strategy selection in differential evolution for numerical optimization: An empirical study. Inform. Sciences **181**(24), 5364–5386 (2011)
9. Mashwani, Khan: W., Salhi, A.: A decomposition-based hybrid multiobjective evolutionary algorithm with dynamic resource allocation. Appl. Soft Comput. **12**(9), 2765–2780 (2012)
10. Knowles, J., Thiele, L., Zitzler, E.: A Tutorial on the Performance Assessment of Stochastic Multiobjective Optimizers. TIK Report 214, Computer Engineering and Networks Laboratory (TIK), ETH Zurich (February 2006)
11. Li, K., Fialho, A., Kwong, S., Zhang, Q.: Adaptive operator selection with bandits for a multiobjective evolutionary algorithm based on decomposition. IEEE Transactions on Evolutionary Computation **18**(1), 114–130 (2014)
12. Sato, H.: Inverted PBI in MOEA/D and its impact on the search performance on multi and many-objective optimization. In: Proceedings of the 2014 Conference on Genetic and Evolutionary Computation, GECCO 2014, pp. 645–652. ACM, New York (2014). http://doi.acm.org/10.1145/2576768.2598297
13. Thierens, D.: An adaptive pursuit strategy for allocating operator probabilities. In: Conference on Genetic and Evolutionary Computation, pp. 1539–1546 (2005)
14. Venske, S.M., Gonalves, R.A., Delgado, M.R.: ADEMO/D: Multiobjective optimization by an adaptive differential evolution algorithm. Neurocomputing 127, 65–77 (2014), advances in Intelligent Systems Selected papers from the 2012 Brazilian Symposium on Neural Networks
15. Zhang, Q., Zhou, A., Zhao, S., Suganthan, P.N., Liu, W., Tiwari, S.: Multiobjective optimization test instances for the CEC 2009 special session and competition. Tech. rep., University of Essex and Nanyang Technological University, CES-487 (2008)
16. Zhang, Q., Liu, W., Li, H.: The performance of a new version of MOEA/D on CEC09 unconstrained mop test instances. In: IEEE Congress on Evolutionary Computation, CEC 2009, pp. 203–208 (May 2009)
17. Zhao, S.Z., Suganthan, P.N., Zhang, Q.: Decomposition-based multiobjective evolutionary algorithm with an ensemble of neighborhood sizes. IEEE Trans. Evol. Comput. **16**(3), 442–446 (2012)

Towards Understanding Bilevel Multi-objective Optimization with Deterministic Lower Level Decisions

Ankur Sinha[1]([✉]), Pekka Malo[1], and Kalyanmoy Deb[2]

[1] Aalto University School of Business, PO Box 21210, FIN-00076, Aalto, Finland
ankur.sinha@aalto.fi, pekka.malo@aalto.fi
[2] Michigan State University, East Lansing, MI 48824, USA
kdeb@egr.msu.edu

Abstract. Bilevel decision making and optimization problems are commonly framed as leader-follower problems, where the leader desires to optimize her own decision taking the decisions of the follower into account. These problems are known as Stackelberg problems in the domain of game theory, and as bilevel problems in the domain of mathematical programming. In a number of practical scenarios, both the leaders and the followers might be faced with multiple criteria bringing bilevel multi-criteria decision making aspects into the problem. In such cases, the Pareto-optimal frontier of the leader is influenced by the decision structure of the follower facing multiple objectives. In this paper, we analyze this effect by modeling the lower level decision maker using value functions. We study the problem using test cases and propose an algorithm that can be used to solve such problems.

Keywords: Stackelberg game · Bilevel optimization · Multi-objective optimization · Evolutionary algorithms · Quadratic approximations

1 Introduction

Bilevel optimization problems have been widely studied by both researchers as well as practitioners. The work has been driven by a number of applications that are bilevel in nature; for instance in transportation (network design, optimal pricing) [2,14], economics (Stackelberg games, principal-agent problem, policy decisions) [10,23,24,28], management (network facility location, coordination of multi-divisional firms) [1,27], engineering (optimal design, optimal chemical equilibria) [12,26]. The recent methodological and practical developments on bilevel optimization have been mostly directed towards problems with single objective at both levels. Apart from a few studies in classical optimization [8,9] and evolutionary optimization [7,11,18], little work has been done in the domain of multi-objective bilevel optimization. Most of these studies have not considered decision making intricacies that can arise from hierarchical decision interactions in the presence of multiple objectives.

© Springer International Publishing Switzerland 2015
A. Gaspar-Cunha et al. (Eds.): EMO 2015, Part I, LNCS 9018, pp. 426–443, 2015.
DOI: 10.1007/978-3-319-15934-8_29

While solving a bilevel optimization problem with multiple objectives at both levels, many a times the assumption is that the follower has little decision making power. This means the follower allows the leader to utilize any solution from her (follower's) frontier. This is an optimistic assumption and is often not realistic. A leader may anticipate the decisions of a follower and optimize her decisions accordingly, but it is unrealistic to assume that the leader can choose the solutions best suited to her from the follower's frontier. In this paper, we study cases where the lower level decision maker has sufficient power to make a deterministic decision from her own frontier. We analyze what kind of an impact deterministic lower level decisions have on the upper level frontier. We also highlight issues that need further attention.

To begin with, we provide a review of some of the recent work on multi-objective bilevel optimization. This is followed by the description of multi-objective bilevel optimization with and without decision making at the lower level. The lower level decision making aspects and its impact on the upper level Pareto-frontier are analyzed using two test problems. Thereafter, we propose an evolutionary algorithm to solve such multi-objective bilevel optimization problems where the lower level decisions are determined by a value function. The performance of the evolutionary algorithm is evaluated on test problems and comparisons have been drawn with an earlier approach [7].

2   Past Studies on Multi-objective Bilevel Optimization

There exists a significant amount of work on single objective bilevel problems, but little has been done on multi-objective bilevel problems primarily because of the computational and decision making complexities that such problems offer. In this section, we highlight the few studies available on multi-objective bilevel optimization. Studies by Eichfelder [8,9] utilize classical techniques to handle simple multi-objective bilevel problems. The lower level problems are handled using a numerical optimization technique, and the upper level problem is handled using an adaptive exhaustive search method. This makes the solution procedure computationally demanding and non-scalable to large-scale problems. The method is close to a nested strategy, where each of the lower level optimization problems are solved to Pareto-optimality. Shi and Xia [18] use ϵ-constraint method at both levels of multi-objective bilevel problem to convert the problem into an ϵ-constraint bilevel problem. The ϵ-parameter is elicited from the decision maker, and the problem is solved by replacing the lower level constrained optimization problem with its KKT conditions. The problem is solved for different ϵ-parameters, until a satisfactory solution is found.

One of the first studies, utilizing an evolutionary approach for bilevel multi-objective algorithms was by Yin [29]. The study involved multiple objectives at the upper lever, and a single objective at the lower level. The study suggested a nested genetic algorithm, and applied it on a transportation planning and management problem. Later Halter and Mostaghim [11] used a particle swarm optimization based nested strategy to solve a multi-component chemical system. The lower level problem in their application problem was linear for which they used a specialized linear multi-objective PSO approach. Recently, a hybrid bilevel evolutionary multi-objective optimization algorithm approach coupled with local search was proposed in [7] (For earlier versions, refer [4–6,20]). In the paper, the authors handled non-linear as well as discrete bilevel problems with relatively large r number of variables. The study also provided a suite of test problems for bilevel multi-objective optimization. Other recent work related to bilevel multi-objective optimization can be found in [13,15–17,30]. There has been some work done on decision making aspects primarily at the upper level. For example, in [19] an interaction with the upper level decision maker is performed during optimization to find the most preferred point instead of the entire Pareto-frontier. Since multi-objective bilevel optimization is computationally expensive, such an approach was justified as it led to enormous savings in computational expense. However, decision making at the lower level was ignored in this study.

3   Bilevel Multi-objective Optimization and Decision Making

In this section, we provide different formulations for a bilevel multi-objective optimization problem that contains two levels of optimization. The upper level optimization problem is the leader's problem (upper level decision maker) and the lower level optimization problem is the follower's problem (lower level decision maker). First, we consider a formulation, where there is no decision making involved at the lower level and all lower level Pareto-optimal solutions are considered at the upper level. Second, we consider a formulation, where the decision maker acts at the lower level and chooses a solution to her liking. This becomes the only possible feasible solution at the upper level.

3.1   Bilevel Multi-objective Optimization

Bilevel multi-objective optimization problems contain two levels of multi-objective optimization tasks. There are two types of variables in these problems; namely, the upper level variables $x_u \in X_U \subset \mathbb{R}^n$, and the lower level variables $x_l \in X_L \subset \mathbb{R}^m$. The lower level multi-objective problem is solved with respect to the lower level variables, while the upper level variables act as parameters to the optimization problem. The optimistic formulation of such problems requires that the Pareto-optimal solutions of the lower level optimization problem may be considered as possible feasible solutions for the upper level optimization problem. Below, we provide two equivalent definitions of a bilevel multi-objective optimization problem.

Definition 1. *def:bilevel1 For the upper-level objective function $F : \mathbb{R}^n \times \mathbb{R}^m \to \mathbb{R}^p$ and lower-level objective function $f : \mathbb{R}^n \times \mathbb{R}^m \to \mathbb{R}^q$*

$$\underset{x_u \in X_U, x_l \in X_L}{\text{minimize}} \quad F(x_u, x_l) = (F_1(x_u, x_l), \dots, F_p(x_u, x_l))$$

$$\text{subject to} \quad x_l \in \underset{x_l}{\text{argmin}} \{ f(x_u, x_l) = (f_1(x_u, x_l), \dots, f_q(x_u, x_l)) :$$

$$g_j(x_u, x_l) \le 0, j = 1, \dots, J \}$$

$$G_k(x_u, x_l) \le 0, k = 1, \dots, K$$

The above definition can be stated in terms of set-valued mappings as follows:

Definition 2. *def:bilevel2 Let* $\Psi : \mathbb{R}^n \rightrightarrows \mathbb{R}^m$ *be a set-valued mapping,*

$$\Psi(x_u) = \underset{x_l}{\text{argmin}} \{ f(x_u, x_l) = (f_1(x_u, x_l), \dots, f_q(x_u, x_l)) : g_j(x_u, x_l) \le 0, j = 1, \dots, J \},$$

which represents the constraint defined by the lower-level optimization problem, i.e. $\Psi(x_u) \subset X_L$ *for every* $x_u \in X_U$. *Then the bilevel multi-objective optimization problem can be expressed as a constrained multi-objective optimization problem as follows:*

$$\underset{x_u \in X_U, x_l \in X_L}{\text{minimize}} \quad F(x_u, x_l) = (F_1(x_u, x_l), \dots, F_p(x_u, x_l))$$

$$\text{subject to} \quad x_l \in \Psi(x_u)$$

$$G_k(x_u, x_l) \le 0, k = 1, \dots, K$$

where Ψ *can be interpreted as a parameterized range-constraint for the lower-level decision vector* x_l.

3.2 Decision Making at Lower Level

According to the formulation of a multi-objective bilevel problem in the previous sub-section, the follower provides all Pareto-optimal points to the leader, who chooses the most suitable point in accordance with the upper level objectives. However, this is rarely the case, as in reality it might often happen that the follower is interested in optimizing her own objectives and making her own decision for a given upper level vector. If the leader wants to solve such a problem where the follower has sufficient decision making power, then she needs to have a complete knowledge of the follower's decision structure. The decision structure of the follower may be represented in the form of a value function. If the value function of the lower level decision maker is known, then such an optimization problem can be formulated as follows:

Definition 3. *def:bilevel3 For the upper-level objective function* $F : \mathbb{R}^n \times \mathbb{R}^m \rightarrow \mathbb{R}^p$ *and lower-level objective function* $f : \mathbb{R}^n \times \mathbb{R}^m \rightarrow \mathbb{R}^q$

$$\underset{x_u \in X_U, x_l \in X_L}{\text{minimize}} \quad F(x_u, x_l) = (F_1(x_u, x_l), \dots, F_p(x_u, x_l))$$

$$\text{subject to} \quad x_l \in \underset{x_l}{\text{argmin}} \{ V(f_1(x_u, x_l), \dots, f_q(x_u, x_l); \omega) : g_j(x_u, x_l) \le 0, j = 1, \dots, J \}$$

$$G_k(x_u, x_l) \le 0, k = 1, \dots, K,$$

where V denotes the follower's value function, and ω is the parameter vector of the assumed value function form. For instance, if V is linear, such that $V(f_1(x_u, x_l), \ldots, f_q(x_u, x_l); \omega) = \sum_{i=1}^{q} \omega_i f_i(x_u, x_l)$, then $\omega_i \ \forall \ i \in \{1, \ldots, q\}$ represent the value function parameters.

If it is assumed that the lower level decision maker always returns a single point for a given x_u, then the definition gets modified as follows:

Definition 4. *def:bilevel4 For the upper-level objective function $F : \mathbb{R}^n \times \mathbb{R}^m \to \mathbb{R}^p$ and lower-level objective function $f : \mathbb{R}^n \times \mathbb{R}^m \to \mathbb{R}^q$*

$$\underset{x_u \in X_U, x_l \in X_L}{\text{minimize}} \quad F(x_u, x_l) = (F_1(x_u, x_l), \ldots, F_p(x_u, x_l))$$

$$\text{subject to} \quad x_l = \underset{x_l}{\text{argmin}}\{V(f_1(x_u, x_l), \ldots, f_q(x_u, x_l)) : \ g_j(x_u, x_l) \leq 0, j = 1, \ldots, J\}$$

$$G_k(x_u, x_l) \leq 0, k = 1, \ldots, K$$

In this paper, we aim to solve the problem formulated above. We assume that the leader has a complete knowledge of the follower's value function. Based on this information, we solve the bilevel problem to identify the upper level Pareto-frontier. Once the upper level Pareto-frontier is available to the leader, it becomes a multi-criteria decision making problem for the leader that we do not consider in this paper.

4   A Graphical Representation for Bilevel Multi-objective Optimization with Lower Level Decisions

Bilevel optimization problems are known to be computationally demanding. However, in case of multiple objectives at both levels of a bilevel optimization problem, an additional difficulty enters because the decision making aspects need to be considered. Even though the upper level decision maker is aware of the objectives of the lower level decision maker, she has little idea about the decisions the lower level decision maker might make from a multitude of lower level Pareto-optimal solutions. In order to handle the problem, the upper level decision maker needs to identify the preference structure of the lower level decision maker through studies or surveys.

Figure 1 shows the scenario, where the shaded region $(\Psi(x_u))$ represents the follower's Pareto-optimal solution for any given leader's decision (x_u). These are the rational actions, which the follower may make for a given leader's action. If the leader is aware of the follower's objectives, she will be able to identify the shaded region completely by solving the multi-objective optimization problem for the follower for all x_u. However, information about the follower's preferences on the lower level Pareto-optimal solutions is required by the leader to make an appropriate decision. If the preferences of the follower are perfectly known, then the lower level decision for any x_u is given by $\sigma(x_u)$, shown in the figure. In such a case, it is possible for the leader to solve the hierarchical optimization task completely, only when $\sigma(x_u)$ is available.

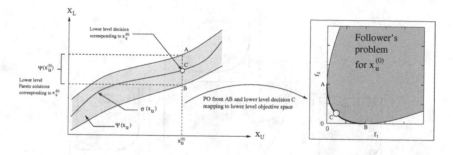

Fig. 1. Lower level Pareto-optimal solutions ($\psi(x_u)$) and corresponding decisions $\sigma(x_u)$

Next, we look at the multi-objective bilevel problem in the objective spaces of the leader and the follower. In the case when the leader solves the bilevel problem taking into account the actions of the follower, each point on the leader's Pareto-frontier corresponds to one of the points on the follower's Pareto-frontier. This has been shown in Figure 2, where points A_u, B_u and C_u are realized when the follower's choices are A_l, B_l and C_l. Points A_l, B_l and C_l lie on the lower level Pareto-optimal front corresponding to $x_u^{(1)}$, $x_u^{(2)}$ and $x_u^{(3)}$ respectively. If the follower decides to use a different preference structure, the Pareto-frontier for the leader may change. It may happen that the Pareto-frontier at the upper level improves, deteriorates or does not change. It may also happen that with change in the follower's preferences, upper level points $x_u^{(1)}$, $x_u^{(2)}$ and $x_u^{(3)}$ are no longer a part of the upper level frontier. Therefore, the Pareto-optimal solutions at the upper level are entirely dependent on the decision structure of the follower. In the next section, using examples we compare the leader's frontier corresponding to a follower with sufficient decision power and a follower with no decision power.

5   Examples

In this section, we consider two examples from the literature [7,8] and solve the problem analytically to show how the frontier at the upper level changes when the lower level decision maker exercises her decisions. A comparison has been made against the scenario when no lower level decision making is performed.

Example 1. Consider the following bilevel multi-objective optimization problem [8] with two objectives at each level. It contains three variables with y_1, y_2 belonging to x_l and x belonging to x_u.

$$\text{minimize } F(x, y_1, y_2) = \left\{ \begin{matrix} y_1 - x \\ y_2 \end{matrix} \right\},$$

$$\text{subject to } (y_1, y_2) \in \underset{(y_1, y_2)}{\text{argmin}} \left\{ f(x, y_1, y_2) = \begin{pmatrix} y_1 \\ y_2 \end{pmatrix} \middle| g_1(x, y_1, y_2) = x^2 - y_1^2 - y_2^2 \geq 0 \right\},$$

$$G_1(y_1, y_2) = 1 + y_1 + y_2 \geq 0,$$
$$-1 \leq y_1, y_2 \leq 1, \quad 0 \leq x \leq 1.$$

$$(1)$$

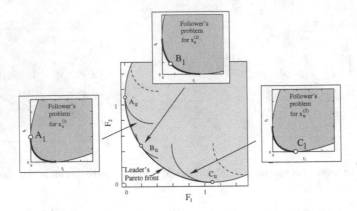

Fig. 2. The small figures show the follower's problem for different x_u. When the follower's preference structure is known, the leader optimizes the bilevel problem such that the follower's decisions corresponds to the leader's Pareto-frontier. A_l, B_l and C_l represent the follower's decisions for $x_u^{(1)}$, $x_u^{(2)}$ and $x_u^{(3)}$ respectively. A_u, B_u and C_u are the corresponding points for the leader in the leader's objective space.

For any fixed value of x, the feasible region of the lower-level problem is the area inside a circle with center at origin ($y_1 = y_2 = 0$) and radius equal to x. The Pareto-optimal set for the lower-level optimization task for a fixed x is the southwest quarter of the circle:

$$\{(y_1, y_2) \in \mathbf{R}^2 \quad | \quad y_1^2 + y_2^2 = x^2, y_1 \leq 0, y_2 \leq 0\}.$$

Let us first consider the upper level frontier with no lower level decision making. This represents the best possible frontier at the upper level as all the lower level Pareto-optimal members are available to the upper level decision maker. It is an ideal scenario for the upper level decision maker, where she freely chooses a suitable point from the lower level frontier. For the above example, such an upper level frontier is shown in Figure 3. The upper level Pareto-optimal set for this scenario can be generated as follows:

$$(x, y_1, y_2)^* = \left\{(y_1, y_2, x) \in \mathbf{R}^3 \quad | \quad x \in \left[\frac{1}{\sqrt{2}}, 1\right], y_1 = -1 - y_2, y_2 = -\frac{1}{2} \pm \frac{1}{4}\sqrt{8x^2 - 4}\right\}. (2)$$

From the Figure 3 it is clear that at most two members from the lower level frontiers corresponding to $x \in [\sqrt{0.5}, 1]$ participate in the upper level front. Note that for $x = 0.9$ points B and C are Pareto-optimal at the upper level and point A is infeasible because of the upper level constraint.

Next, let us consider the problem in the context of this paper, where the lower level decision maker has sufficient power to choose a point from her Pareto-optimal front. If one assumes a particular value function, say $V(f_1, f_2) = 5x^2 f_1 + f_2$, then the upper level frontier is given as shown in Figure 4. It is noteworthy that the assumed value function also contains x. This kind of dependency may not always exist, but has been considered here to show that the lower level value function may take any possible form. The theoretical upper level Pareto-optimal frontier corresponding to the deterministic lower level value function is theoretically given as:

$$V(f_1, f_2) = 5x^2 f_1 + f_2, \text{ then } y_1 = -\sqrt{\frac{25x^6}{(1 + 25x^4)}}, y_2 = -\sqrt{x^2 - y_1^2}, x \in [0.447, 0.797]$$

The leader's Pareto-optimal frontier corresponding to deterministic decisions of the follower is much worse as compared to the Pareto-optimal frontier corresponding to no decisions by the follower. From Figure 4 the leader can easily evaluate how worse she gets from the best possible frontier when she chooses a point on the Pareto-optimal frontier with deterministic lower level decisions.

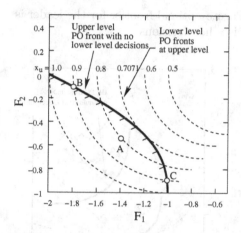

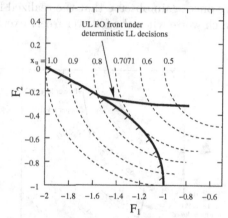

Fig. 3. Example 1: Upper level Pareto-optimal front (with no lower level decision making) and few representative lower level Pareto-optimal fronts in upper level objective space

Fig. 4. Example 1: Upper level (UL) Pareto-optimal front when lower level (LL) decisions are given by $V(f_1, f_2) = 5x^2 f_1 + f_2$

Example 2. Let us consider another simple multi-objective bilevel optimization problem that is discussed in [6,7]. The problem is scalable in terms of lower level variables, and contains a single upper level variable. For K variables at the lower level, $x_l = (y_1, \ldots, y_K)$ and $x_u = (x)$; the problem is defined as follows:

$$\text{Minimize } F(x_u, x_l) = \begin{pmatrix} (y_1 - 1)^2 + \sum_{i=2}^{K} y_i^2 + x^2 \\ (y_1 - 1)^2 + \sum_{i=2}^{K} y_i^2 + (x - 1)^2 \end{pmatrix},$$

$$\text{subject to}$$

$$(y_1, y_2, \ldots, y_K) \in \underset{(y_1, y_2, \ldots, y_K)}{\text{argmin}} \left\{ f(x_u, x_l) = \begin{pmatrix} y_1^2 + \sum_{i=2}^{K} y_i^2 \\ (y_1 - x)^2 + \sum_{i=2}^{K} y_i^2 \end{pmatrix} \right\}, \quad (3)$$

$$-1 \le (x, y_1, y_2, \ldots, y_K) \le 2.$$

For any x, the Pareto-optimal solutions of the lower level optimization problem are given as follows: $\{x_l \in \mathbf{R}^K \big| y_1 \in [0, x], y_i = 0, \text{ for } i = 2, \ldots, K\}$. In this paper, we choose $K = 14$, such that the problem contains 15 variables. The best possible frontier at the upper level may be obtained when there is no decision maker at the lower level, and all the lower level Pareto-optimal members are available at the upper level. In this example, such a frontier corresponds to the following conditions: $\{(x_u, x_l) \in \mathbf{R}^{K+1} \big| y_1 = x, y_i = 0, \text{ for } i = 2, \ldots, K, x \in [0.5, 1.0]\}$.

Now let us consider that there exists a decision maker at the lower level, whose value function is $V(f_1, f_2) = 2f_1 + f_2$. The upper level frontier for this case is shown in Figure 6. The theoretical upper level Pareto-optimal frontier for the deterministic lower level value function is given as:

$$V(f_1, f_2) = 2f_1 + f_2, \text{ then } y_1 = \frac{x}{3}, y_i = 0 \ \forall \ i = 2, \ldots, K, x \in [0.300, 1.201]$$

We once again observe that the realized Pareto-optimal frontier for the leader is much worse when the follower freely exercises her decisions.

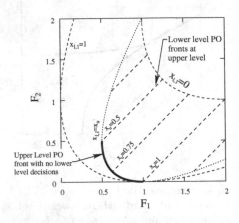

Fig. 5. Example 2: Upper level Pareto-optimal front (with no lower level decision making) and few representative lower level Pareto-optimal fronts in upper level objective space

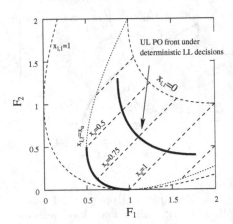

Fig. 6. Example 2: Upper level (UL) Pareto-optimal front when lower level (LL) decisions are given by $V(f_1, f_2) = 2f_1 + f_2$

6   Algorithm Description

In this section, we introduce an evolutionary algorithm for solving bilevel problems where the upper level has multiple objectives and the lower level decisions are modeled using a value function. This means that the algorithm solves a problem with multiple objectives at upper level and single objective at the lower level.

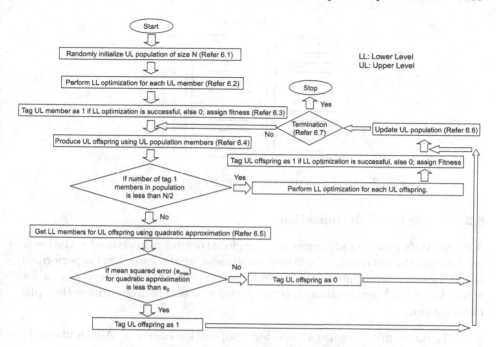

Fig. 7. Flowchart for m-BLEAQ

The approach is an extension of a recently proposed algorithm for single objective bilevel optimization [21, 22], and is referred as multi-objective bilevel evolutionary algorithm based on quadratic approximations (m-BLEAQ). The proposed approach is based on estimation of unknown lower level decisions using quadratic approximations, when lower level decisions corresponding to a few upper level vectors are known. The approximation helps in reducing the number of lower level optimization calls that leads to computational savings. The working of the algorithm has been shown through a flowchart in Figure 7.

6.1 Population Structure

The population structure at the upper level is shown in Figure 8. The first column represents the upper level population members and the second column represents the corresponding lower level population members that have been computed through lower level optimization or quadratic approximation. Based on the quality of the lower level members the upper level members are tagged as 0 or 1. For tag 1 upper level members the corresponding lower level members are expected to be close to lower level optimum.

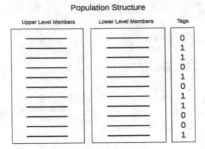

Fig. 8. Population structure at upper level of m-BLEAQ

6.2   Lower Level Optimization

A steady state evolutionary algorithm[1] for global optimization is used at the lower level to find the optimum. The fitness assignment at the lower level is performed based on lower level function value and constraints. The upper level vector for which lower level optimization is being performed is kept fixed during the optimization run.

Step 1. Randomly initialize a lower level population of size N. Assign fitness to the members based on lower level objective functions and constraints.

Step 2. Randomly choose 6 members from the population, and perform a tournament selection. This gives 3 parents for crossover.

Step 3. Create 2 offsprings from the parents using genetic operators on the lower level variables only.

Step 4. Randomly choose 2 members from the population, and pool them with 2 offsprings. The 2 best members from the pool replace the chosen members from the population.

Step 5. Perform a termination check. Proceed to next generation (Step 2), if the termination criteria is not satisfied, otherwise proceed to the next step.

Step 6. The best obtained lower level member is paired with the corresponding upper level member in the upper level population.

6.3   Fitness Evaluation

Fitness assignment for feasible upper level member is performed based on their non-domination rank and crowding distance [3]. For a given upper level member x, if the non-domination rank is given as $N_R(x)$ and crowding distance within its frontier is given as $C_D(x)$, then the fitness for the member is calculated as follows:

$$F_u(x) = \frac{1}{N_R(x) + e^{-C_D(x)}}, \tag{4}$$

---

[1] A classical optimization strategy can be used to replace the lower level evolutionary algorithm, if the lower level optimization problem adheres to the requirements of the classical approach.

Fitness for an infeasible upper level member is computed by subtracting the sum of upper level constraint violations from the fitness value of the worst feasible member.

The fitness during lower level optimization is given by lower level function values for the feasible members. For the infeasible lower level members, we subtract the sum of lower level constraint violations from the fitness value of the worst feasible member at that level.

6.4 Genetic Operations

A parent centric crossover and a polynomial mutation is performed to generate new parents. The crossover operator is similar to the PCX operator proposed in [25] with slight modifications. The operator requires 3 parents to create an offspring that are selected using tournament selection. The crossover operation is performed as show below:

$$c = x^{(p)} + \omega_\xi d + \omega_\eta \frac{p^{(2)} - p^{(1)}}{2} \tag{5}$$

The terms used in the above equation are defined as follows:

- $x^{(p)}$ is the *index* parent
- $d = x^{(p)} - g$, where g is the mean of μ parents
- $p^{(1)}$ and $p^{(2)}$ are the other two parents
- $\omega_\xi = 0.1$ and $\omega_\eta = \frac{dim(x^{(p)})}{||x^{(p)} - g||_1}$ are the two parameters.

Upper level crossovers and mutations are performed on upper level variables, while lower level crossovers and mutations are performed on lower level variables.

6.5 Quadratic Approximations

At any generation of the m-BLEAQ algorithm, we attempt to maintain at least $\frac{N}{2}$ tag 1 members. These are the upper level members for which the lower level optimal solutions are accurately known. We utilize these members to compute the lower level optimal solutions of the new upper level members. Based on the quality of the quadratic approximation, the estimated lower level optimum might be accurate or inaccurate. Figure 9 shows a scenario where there are three members for which lower level decisions are known. We utilize these members to construct a quadratic approximation that provides an estimate for the unknown lower level decision.

Figure 9 explains the approximation in the presence of a single lower and upper level variable. When multiple lower and upper level variables are present, we utilize all the upper level variables to construct the quadratic approximation for each lower level variable. Therefore, the number of quadratic approximations are as many as the number of lower level variables, and each lower level variable is a function of all the upper level variables. We choose the closest upper level members for quadratic approximation around the point for which we intend to estimate the

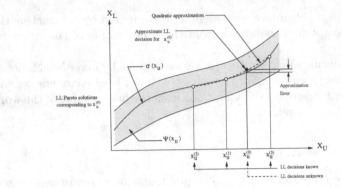

Fig. 9. Approximating decisions for an unknown upper level vector when decisions corresponding to few upper level vectors are known

lower level decision. Such an approximation is expected to provide a reliable local estimate. We propose to utilize at least $\frac{1}{2}[(dim(x_u)+1)(dim(x_u)+2)]+dim(x_u)$ upper level points for constructing the approximation.

6.6   Update at Upper Level

The upper level population is updated by choosing 2 worst members from the population. The members are pooled with 2 offsprings generated through genetic operations, and the best members from the pool are chosen to replace the selected population members.

6.7   Termination

At the upper level we terminate the algorithm based on maximum upper level function evaluations ($T_{\max}$). We use an improvement based termination at the lower level such that if the improvement in the lower level function value is less than $1e-5$ for 100 consecutive generations then we terminate the optimization.

6.8   Archiving

We store all the tag 1 upper level members produced by the algorithm in an archive. The final upper level Pareto-optimal solutions are presented to the user by providing the best frontier in the archive set.

6.9   Parameters

For all the computations in this paper we fix the algorithm parameters as $N = 50$. Crossover probability is fixed at 0.9 and the mutation probability is fixed at 0.1.

7 Results

We evaluate the m-BLEAQ algorithm on the examples that we discussed in an earlier section. Since we know the Pareto-optimal frontier for both problems, it is easy to test the performance using Inverted Generalization Distance (IGD) [31] metric. We compare our results against the H-BLEMO approach proposed in [7].

In order to compute the IGD value, we generate 500 evenly distributed points on the upper level Pareto-optimal front of the two problems. An average distance in some sense [31] is computed between these evenly distributed points and the points on the frontier achieved by the algorithm. The smaller the IGD value the better is the performance of the approach. The IGD metric is able to provide a measure for both convergence and diversity. While presenting the results for m-BLEAQ and H-BLEMO we fix the maximum number of upper level function evaluations (T_{max}) and then determine the IGD value achieved by both methods. H-BLEMO has been slightly modified at the lower level to incorporate a similar lower level termination criteria as in m-BLEAQ. The results are presented in Tables 1, 2, 3 and 4. Figures 10 and 11 show the Pareto-optimal fronts achieved by m-BLEAQ from one of the sample runs for the two test problems.

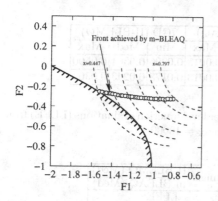

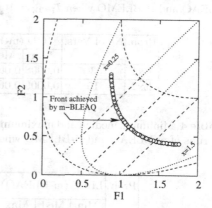

Fig. 10. Example 1: Pareto-optimal front obtained using m-BLEAQ from one of the runs when $T_{max} = 5000$

Fig. 11. Example 2: Pareto-optimal front obtained using m-BLEAQ from one of the runs when $T_{max} = 5000$

It is noteworthy that the H-BLEMO is capable of handling multiple objectives at both levels, but in the current formulation the lower level is represented by a value function, which means that H-BLEMO is handling a single objective problem at the lower level and a multi-objective problem at the upper level. On the other hand m-BLEAQ cannot directly handle multiple objectives at both levels. However, with multiple objectives at upper level and single objective at lower level m-BLEAQ is able to achieve much lower IGD values as compared to H-BLEMO for the same number of upper level function evaluations and much fewer lower level function evaluations.

Table 1. Minimum, median and maximum IGD values obtained from 21 runs of m-BLEAQ and H-BLEMO when $T_{max} = 5000$

Prob.	No of Vars.	IGD (m-BLEAQ)			IGD (H-BLEMO)		
		Min	Med	Max	Min	Med	Max
Ex1	3	0.0021	0.0026	0.0033	0.0425	0.0409	0.0980
Ex2	15	0.0017	0.0027	0.0036	0.0398	0.0683	0.0532

Table 2. Minimum, median and maximum lower level function evaluations (LLFE) from 21 runs of m-BLEAQ and H-BLEMO when $T_{max} = 5000$

Prob.	LLFE (m-BLEAQ)			Savings: $\dfrac{\text{H-BLEMO (Med)}}{\text{m-BLEAQ (Med)}}$
	Min	Med	Max	LLFE
Ex1	56043	73689	81201	5.12
Ex2	33054	47679	66533	5.38

Table 3. Minimum, median and maximum IGD values obtained from 21 runs of m-BLEAQ and H-BLEMO when $T_{max} = 10000$

Prob.	No of Vars.	IGD (m-BLEAQ)			IGD (H-BLEMO)		
		Min	Med	Max	Min	Med	Max
Ex1	3	0.0008	0.0011	0.0013	0.0206	0.0361	0.0500
Ex2	15	0.0009	0.0012	0.0013	0.0178	0.0243	0.0305

Table 4. Minimum, median and maximum lower level function evaluations (LLFE) from 21 runs of m-BLEAQ and H-BLEMO when $T_{max} = 10000$

Prob.	LLFE (m-BLEAQ)			Savings: $\dfrac{\text{H-BLEMO (Med)}}{\text{m-BLEAQ (Med)}}$
	Min	Med	Max	LLFE
Ex1	77735	94104	99187	8.57
Ex2	45377	65674	91898	7.63

8   Conclusions and Future Work

In this paper, we have analyzed bilevel optimization problems with multiple objectives. In order to account for deterministic decisions of the follower, multiple objectives at the lower level have been replaced by a value function. We have considered a realistic scenario in this paper where the follower has some decision power based on which she chooses a solution from her Pareto-optimal frontier. Through examples we have shown that decisions of the follower may have significant impact on the leader's frontier when compared against the case of not accounting the follower's decisions.

We have extended a recently proposed algorithm for single objective bilevel optimization (BLEAQ) to handle multi-objective bilevel problems with deterministic lower level decision. The extended algorithm (m-BLEAQ) is found to be computationally efficient when compared against an earlier proposed strategy (H-BLEMO). As a future research, we intend to study how the frontier at the upper level changes when the decision structure (value function) of the follower varies. It is not always possible to deterministically ascertain the value function of a decision maker. Therefore, future efforts will be directed towards handling problems with lower level decision uncertainty. It will also be interesting to consider cooperation between leader and follower, where the follower agrees to return a part of her frontier as possible lower level decisions to the leader. Such kind of cooperation by the follower may lead to an improved Pareto-optimal frontier at the upper level. Such a study will also allow to evaluate the extent of compromises and gains that can be made by the leader and the follower through mutual cooperation.

References

1. Bard, J.F.: Coordination of multi-divisional firm through two levels of management. Omega **11**(5), 457–465 (1983)
2. Brotcorne, L., Labbe, M., Marcotte, P., Savard, G.: A bilevel model for toll optimization on a multicommodity transportation network. Transportation Science **35**(4), 345–358 (2001)
3. Deb, K., Agrawal, S., Pratap, A., Meyarivan, T.: A fast and elitist multi-objective genetic algorithm: NSGA-II. IEEE Transactions on Evolutionary Computation **6**(2), 182–197 (2002)
4. Deb, K., Sinha, A.: Constructing test problems for bilevel evolutionary multi-objective optimization. In: 2009 IEEE Congress on Evolutionary Computation (CEC-2009), pp. 1153–1160. IEEE Press (2009)
5. Deb, K., Sinha, A.: An evolutionary approach for bilevel multi-objective problems. In: Shi, Y., Wang, S., Peng, Y., Li, J., Zeng, Y. (eds.) MCDM 2009. CCIS, vol. 35, pp. 17–24. Springer, Heidelberg (2009)
6. Deb, K., Sinha, A.: Solving bilevel multi-objective optimization problems using evolutionary algorithms. In: Ehrgott, M., Fonseca, C.M., Gandibleux, X., Hao, J.-K., Sevaux, M. (eds.) EMO 2009. LNCS, vol. 5467, pp. 110–124. Springer, Heidelberg (2009)
7. Deb, K., Sinha, A.: An efficient and accurate solution methodology for bilevel multi-objective programming problems using a hybrid evolutionary-local-search algorithm. Evolutionary Computation Journal **18**(3), 403–449 (2010)
8. Eichfelder, G.: Solving nonlinear multiobjective bilevel optimization problems with coupled upper level constraints. Technical Report Preprint No. 320, Preprint-Series of the Institute of Applied Mathematics, Univ. Erlangen-Nurnberg, Germany (2007)
9. Eichfelder, G.: Multiobjective bilevel optimization. Math. Program. **123**(2), 419–449 (2010)
10. Fudenberg, D., Tirole, J.: Game theory. MIT Press (1993)
11. Halter, W., Mostaghim, S.: Bilevel optimization of multi-component chemical systems using particle swarm optimization. In: Proceedings of World Congress on Computational Intelligence (WCCI-2006), pp. 1240–1247 (2006)

12. Kirjner-Neto, C., Polak, E., Der Kiureghian, A.: An outer approximations approach to reliability-based optimal design of structures. Journal of Optimization Theory and Applications **98**(1), 1–16 (1998)
13. Linnala, M., Madetoja, E., Ruotsalainen, H., Hämäläinen, J.: Bi-level optimization for a dynamic multiobjective problem. Engineering Optimization **44**(2), 195–207 (2012)
14. Migdalas, A.: Bilevel programming in traffic planning: Models, methods and challenge. Journal of Global Optimization **7**(4), 381–405 (1995)
15. Pieume, C.O., Fotso, L.P., Siarry, P.: Solving bilevel programming problems with multicriteria optimization techniques. OPSEARCH **46**(2), 169–183 (2009)
16. Pramanik, S., Dey, P.P.: Bi-level multi-objective programming problem with fuzzy parameters. International Journal of Computer Applications 30(10), 13–20 (2011), Published by Foundation of Computer Science, New York, USA
17. Ruuska, S., Miettinen, K.: Constructing evolutionary algorithms for bilevel multiobjective optimization. In: 2012 IEEE Congress on Evolutionary Computation (CEC), pp. 1–7 (June 2012)
18. Shi, X., Xia, H.S.: Model and interactive algorithm of bi-level multi-objective decision-making with multiple interconnected decision makers. Journal of Multi-Criteria Decision Analysis **10**(1), 27–34 (2001)
19. Sinha, A.: Bilevel multi-objective optimization problem solving using progressively interactive EMO. In: Takahashi, R.H.C., Deb, K., Wanner, E.F., Greco, S. (eds.) EMO 2011. LNCS, vol. 6576, pp. 269–284. Springer, Heidelberg (2011)
20. Sinha, A., Deb, K.: Towards understanding evolutionary bilevel multi-objective optimization algorithm. In: IFAC Workshop on Control Applications of Optimization (IFAC-2009), vol. 7. Elsevier (2009)
21. Sinha, A., Malo, P., Deb, K.: Efficient evolutionary algorithm for single-objective bilevel optimization. CoRR, abs/1303.3901 (2013)
22. Sinha, A., Malo, P., Deb, K.: An improved bilevel evolutionary algorithm based on quadratic approximations. In: 2014 IEEE Congress on Evolutionary Computation (CEC-2014), pp. 1870–1877. IEEE Press (2014)
23. Sinha, A., Malo, P., Frantsev, A., Deb, K.: Multi-objective stackelberg game between a regulating authority and a mining company: A case study in environmental economics. In: 2013 IEEE Congress on Evolutionary Computation (CEC-2013). IEEE Press (2013)
24. Sinha, A., Malo, P., Frantsev, A., Deb, K.: Finding optimal strategies in a multi-period multi-leader-follower stackelberg game using an evolutionary algorithm. Computers & Operations Research **41**, 374–385 (2014)
25. Sinha, A., Srinivasan, A., Deb, K.: A population-based, parent centric procedure for constrained real-parameter optimization. In: 2006 IEEE Congress on Evolutionary Computation (CEC-2006), pp. 239–245. IEEE Press (2006)
26. Smith, W.R., Missen, R.W.: Chemical Reaction Equilibrium Analysis: Theory and Algorithms. John Wiley & Sons, New York (1982)
27. Sun, H., Gao, Z., Jianjun, W.: A bi-level programming model and solution algorithm for the location of logistics distribution centers. Applied Mathematical Modelling **32**(4), 610–616 (2008)
28. Wang, F. J., Periaux, J.: Multi-point optimization using gas and Nash/Stackelberg games for high lift multi-airfoil design in aerodynamics. In: Proceedings of the 2001 Congress on Evolutionary Computation (CEC-2001), pp. 552–559 (2001)

29. Yin, Y.: Genetic algorithm based approach for bilevel programming models. Journal of Transportation Engineering **126**(2), 115–120 (2000)
30. Zhang, T., Hu, T., Zheng, Y., Guo, X.: An improved particle swarm optimization for solving bilevel multiobjective programming problem. Journal of Applied Mathematics (2012)
31. Zitzler, E., Thiele, L., Laumanns, M., Fonseca, C.M., da Fonseca, V.G.: Performance Assessment of Multiobjective Optimizers: An Analysis and Review. IEEE Transactions on Evolutionary Computation **7**(2), 117–132 (2003)

Author Index